Monografie Matematyczne
Instytut Matematyczny Polskiej Akademii Nauk (IMPAN)

Volume 63

(New Series)

Founded in 1932 by
S. Banach, B. Knaster, K. Kuratowski,
S. Mazurkiewicz, W. Sierpiński, H. Steinhaus

Volumes 31–62 of the series
Monografie Matematyczne were published by
PWN – Polish Scientific Publishers, Warsaw

Jörg Schürmann

Topology of Singular Spaces and Constructible Sheaves

Birkhäuser Verlag
Basel · Boston · Berlin

Author's address:

Jörg Schürmann
Westfälische Wilhelms-Universität
SFB 478
Hittorfstr. 27
48149 Münster
Germany
e-mail: jschuerm@math.uni-muenster.de

2000 Mathematics Subject Classification 14-02, 14F17, 14F25, 14F45, 14Pxx, 32-02, 32C18, 32F10, 32F27, 32B20, 32Sxx, 03C64, 55N25, 55N30, 55N33, 55N91

A CIP catalogue record for this book is available from the Library of Congress, Washington D.C., USA

Bibliographic information published by Die Deutsche Bibliothek
Die Deutsche Bibliothek lists this publication in the Deutsche Nationalbibliografie;
detailed bibliographic data is available in the Internet at <http://dnb.ddb.de>.

ISBN 3-7643-2189-X Birkhäuser Verlag, Basel – Boston – Berlin

Member of the BertelsmannSpringer Publishing Group
Cover design: Micha Lotrovsky, CH-4106 Therwil, Switzerland
Printed on acid-free paper produced of chlorine-free pulp. TCF ∞
Printed in Germany
ISBN 3-7643-2189-X
9 8 7 6 5 4 3 2 1

www.birkhauser.ch

For my Family: Friederike,
Philipp, David and Hendrik

Preface

Sheaf theory and homological algebra first arose in classical algebraic topology in terms of homology and cohomology of topological spaces, and was used as a language for describing properties of geometrical objects. This language was developed in different periods for applications to algebraic topology, complex analysis and algebraic geometry. For a beautiful account of the history of this subject we refer to the article "A Short History: Les débuts de la théorie des faisceaux" of C. Houzel (contained in the book "Sheaves on Manifolds" of M. Kashiwara and P. Schapira).

The last period deals with sheaf theory in the framework of derived and triangulated categories developed by A. Grothendieck and J.L. Verdier, especially for applications to algebraic geometry. But over the years, this very abstract language has found more and more applications in quite different fields of mathematics. Especially, the new "intersection (co)homology theory" and the theory of "perverse sheaves" have found spectacular applications to the topology of singular spaces.

There exists now a number of excellent books devoted to the introduction to the theory of derived and triangulated categories, especially also to sheaf theory in this framework. But applications of this theory to the topology of singular spaces are only worked out in a few places for some special examples. Moreover, a detailed translation of this abstract language into the concrete underlying geometrical ideas is often missing.

Assuming that the reader is familiar with some basic facts from sheaf theory, we give in this book a self-contained exposition of the theory of constructible sheaves related to many different kinds of singular spaces: cell-complexes, triangulated spaces, semialgebraic and subanalytic sets, complex algebraic or analytic sets, stratified spaces, quotient spaces, ...

We work out in detail its relation to the underlying geometrical ideas, and explain many applications of this abstract theory to the topology of such spaces. But as the reader will easily recognize by this list and the length of the book, this theory and its applications are related to many different fields of mathematics.

For this reason we have written up this book in different parts, whose statements and results (but not all of their proofs) themselves are (essentially) independent from one part to another. All chapters have their own detailed introduction so that the reader can switch to the part of the book, which is of interest for him or her.

But first we give a global introduction to the whole book, where we explain a little bit the history of this subject and the main ideas and results of the different chapters. Moreover, we try to explain our point of view about the underlying philosophy that holds all the different aspects together. We also give references to similar results in algebraic geometry, which are related to this viewpoint.

Acknowledgments

It is a pleasure to thank J.P. Brasselet, L. Bröcker, A. Dimca, H.A. Hamm, L. Kaup, Lê D.T., D. Massey, P. Schapira and S. Yokura for valuable conversations about this subjet and possible applications.

This book is an improved and expanded version of the author's Habilitationsschrift with the same title, which was submitted to the Fachbereich Mathematik der Universität Hamburg in the end of 2000. Here I express my gratitude to O. Riemenschneider for his encouragement during the completion of this habilitation.

This book was finished at the Sonderforschungsbereich 478 "Geometrische Strukturen in der Mathematik" in Münster, supported by the Deutsche Forschungsgemeinschaft.

Finally I want to thank P. Pragacz, the Institute of Mathematics of the Polish Academy of Sciences and the Birkhäuser Verlag for getting the book published in this form.

Münster, February 2003. Jörg Schürmann

Contents

Introduction

Sheaf theory has its origin in classical algebraic topology. One of the highlights of algebraic topology is the **Poincaré duality** isomorphism between the cohomology and homology of a compact oriented manifold. A powerful generalization appeared in the 1960s through the work of Verdier [SHS]. This **Poincaré-Verdier duality** can only be formulated in the abstract framework of derived and triangulated categories, and is related to the functorial formalism of sheaf theory, what Grothendieck calls the "six operations on sheaves", i.e. the six functors

$$Rf_*,\ f^*,\ Rf_!,\ f^!,\ \otimes^L,\ Rhom\,.$$

This is also related to the **Borel-Moore homology** theory [BM]. This Borel-Moore homology is the right homology theory for non-compact spaces, as shown by Borel-Haefliger [BH]. They showed that a complex (or real) analytic set has a **fundamental class** in the top-dimensional Borel-Moore homology group with coefficients in $\mathbb{Z}$ (or $\mathbb{Z}_2$), one of the first important results about the topology of singular spaces. This was later on extended by Schwartz [Schwa], Sullivan [Su] and MacPherson [MP] to a theory of **Stiefel-Whitney and Chern homology classes** of real and complex analytic spaces. The first approach of Schwartz uses "radial vector fields" in relation to **Poincaré-Hopf theorems** for singular complex spaces (cf. [Schwa2, Schwa3, BSchwa]). The other approach is based upon the observation of Sullivan [Su], that the link of a point in a complex (or real) analytic space has vanishing Euler characteristic (modulo 2). Note that these characteristic classes of a singular space lie in homology, not in cohomology, unlike the corresponding characteristic classes of a vector bundle.

This shows that questions about the topology of singular spaces are related to cohomology and homology. This viewpoint motivated the introduction of the new formalism of **bivariant theories** by Fulton-MacPherson [FM], which gives the categorical framework for the study of theorems of **Riemann-Roch type** in many different categories. In the context of constructible sheaves, this formalism is related to the "six operation on sheaves" (as above). Moreover, the relation to the characteristic homology classes is through the theory of **constructible functions**, which are closely related to constructible sheaves. Especially, the theory of Chern classes of complex analytic (or algebraic) spaces depends on the **Euler obstruction**, a constructible function introduced in [MP].

Another beautiful development was the introduction of **intersection (co)homology** by Goresky-MacPherson (compare [GM5, Klei]). This theory extends the classical Poincaré duality of manifolds to suitable singular spaces (e.g. stratified pseudomanifolds as in [IC], or locally conelike stratified spaces as in [HS]). This was first a geometric theory, but later on it was recreated in terms of sheaf theory (cf. [GM4, IC]). In this form it is deeply related to the theory of **perverse sheaves** of Beilinson, Bernstein, Deligne and Gabber [BBD]. The general theory of perverse sheaves makes sense only in the framework of derived and triangulated categories, and depends on the language of t-structures in triangulated categories.

Only with this abstract theory of perverse sheaves, not with the original geometric definition of the intersection (co)homology, was it possible to prove very deep results about complex algebraic varieties. These are **not of purely topological nature**, but use some "weight arguments", which where originally proved ([BBD]) by the corresponding theory for algebraic varieties over finite fields. Results of this type have found deep applications in representation theory.

Another bridge between intersection (co)homology, perverse sheaves and representation theory comes from the **Riemann-Hilbert correspondence** between regular holonomic D-modules and perverse sheaves (of complex vector spaces). Based on this correspondence, Saito [Sai] has developed his deep theory of "mixed Hodge modules", a kind of "Hodge sheaves". With this theory it is possible to use the above "weight arguments" directly for complex algebraic varieties.

A different aspect of the Riemann-Hilbert correspondence and the theory of D-modules is the relation to "micro-local analysis". This "micro-local" viewpoint was introduced in the general context of sheaf theory on manifolds by Kashiwara-Schapira [KS, KS2], with the help of their notion of the **micro-support** of a complex of sheaves. This is a powerful and very deep machinery. If one reduces the abstract theory of Kashiwara-Schapira to the context of **constructible sheaves**, then this theory is closely related to the **stratified Morse theory** of Goresky-MacPherson [GM]. This is a geometric Morse theory for subspaces of a manifold, which are endowed with a suitable stratification.

If one would like to apply the theory of constructible complexes of sheaves to topological problems of singular spaces, then one has to know the basic results of sheaf theory in the framework of derived categories (e.g. like [KS, chapter I-III] or [IC]). But one also needs the translation of this abstract language and functors into geometric terms.

This is especially the case for the construction and partially true also for the applications of the "mixed Hodge modules" of Saito [Sai], which are based on the theory of **complex analytically (or algebraically) constructible complexes** of sheaves. Important examples are given by the **nearby and vanishing cycle functor** of Deligne. In the context of intersection (co)homology one can find such a translation for example in [IC, GM4]. For more details on the history of this subject, we recommend the excellent survey article [Klei] by Kleiman, and the short note

[Bott] about the work of MacPherson, from which we borrow the following remark [Bott, p. 326]:

"... Intersection cohomology, as well as the personal interactions of MacPherson with so many principals, has therefore played a decisive part in these cumulative achievements of the sheaf theory of the 1980s. Of course, there is always a price to be paid for a deep understanding of so large and complicated an area of mathematics. It is paid in terms of the abstraction of the concepts and of the language. Thus, in this instance, the final clarity is only achieved in the realm of "perverse sheaves" or rather the derived category of complexes of such sheaves! ...".

Let us now explain the relation of our book to the above developments. We assume that the reader is familar with the usual language of sheaves in the framework of derived categories. For example knowledge of [KS, chapter I-III] would be more than enough. Other presentations can be found in [IC, Iv]. For an introduction to derived and triangulated categories, the reader may also consult [GelMa, Nee] and of course the original work of Verdier, now published in [Ver4].

Following new trends (compare [Nee]), we allow unbounded complexes! Most of the needed Grothendieck formalism about the "six operations" for unbounded complexes of sheaves are worked out in [Sp]. Nevertheless, as a basic reference we use [KS, chapter I-III]. If necessary, and only then, we work out in detail the modifications needed to work in this more general context of unbounded complexes.

A very important example is given by the **Künneth formula**, which is explained in section 1.4. Note that this Künneth formula is also one of the basic examples, mentioned in [Ver4, Introduction], for the advantages of the formalism of derived categories and functors. Let us only cite the following words of [Ver4, Introduction, p. 6]:

"... Même lorsque'on ne s'intéresse qu'aux valeurs des foncteurs dérivés traditionnels, i.e. aux seuls objects de cohomologie, les relations dans les catégories dérivés sont souvent des intermédiaires techniques indispensables pour arriver à étudier ces foncteurs et formuler pour eux certaines relations importantes ...".

This viewpoint underlies our whole book, and one of its aims is to show the power of this abstract formalism. Since we are not aware of all applications that follow, we develop the basic results and notions of the theory of **constructible sheaves** as general as possible. The only exception maybe is our treatment of Poincaré-Verdier duality, where we assume, for simplicity, that the base ring is commutative. This allows a uniform treatment of many different aspects, and shows in a very precise form the simple underlying principles. Of course, to get this clarity and simplicity, we have to "pay the price" mentioned by Bott, i.e. "... the abstraction of the concepts and of the language ...".

Nevertheless, many important applications of the abstract results can be formulated in "classical terms" of **homology or sheaf cohomology**! For the convenience of readers not familiar with the abstract language, we illustrate the general results in many **examples**, which are formulated in these "classical terms". Here we invite the reader to look first at these examples.

Similarly, the applications can often be reduced to a few main results, which are described in terms of "rules for the standard functors". So for the applications one does **not** need to know the deep and complicated proofs of these "rules". One only has to learn how to use these "rules as a black box". This can be compared with a similar approach to the applications of "mixed Hodge modules" [Sai2], or to the applications of the "micro-local sheaf theory" of Kashiwara-Schapira, which depend in many cases only on the general results about the functorial behavior of the **micro-support** (e.g. [ScSc, p. 64–66]).

For this reason, each chapter starts with a detailed (and therefore sometimes also long) **introduction**, containing the main results and definitions, illustrated in simple terms by a number of examples. The more difficult and technical details of the proof are then postponed to later sections.

As explained before, we prove all our results in the context of sheaf theory in the framework of derived categories. This abstract language gives us the **freedom that we need** in our approach to many results of this book. These results have a very clear geometric counterpart, but we only use cohomological techniques.

For example in almost all places of this book we use in our proofs suitable local or global direct image sheaves. This method cannot be used, if one studies more geometric properties like the homotopy type of our singular spaces (but the technique of "moving the wall" used in [GM] is close to our approach).

With this book we hope to convince the reader that the functorial theory of constructible sheaves is a very powerful and effective machinary, which can be applied to many classical questions about the geometry and topology of singular spaces. We understand these things just as two different aspects of the same story, and we would like to build up a "bridge" between these different viewpoints! In fact, it seems that this "bridge" for the beginner is missing at many places in the literature.

Let us give a list of the most important examples for this **correspondence** between the theory of constructible sheaves and the topology of singular spaces:

- For the Thom-Sebastini Theorem of chapter 1 we use a cohomological version of a **Milnor-fibration**. This codifies the geometry underlying the **nearby and vanishing cycle functors**.

  We also explain their counterparts in the context of "real geometry", which will be important later on in chapter 5 for our development of **Morse theory for constructible sheaves**. Especially, they give cohomological versions of the **local and normal Morse data** used in the **stratified Morse theory** of [GM].

- We introduce in chapter 2 the notion of a **"good cellular stratification"**, which is a cohomological generalization of the notion of a **regular CW-complex**.

- If a topological map is a **locally trivial fibration** (in a stratified sense), then this implies that suitable **direct images sheaves are locally constant**.

  In the context of "real geometry", such a locally trivial fibration comes from the so-called **cell decomposition theorem**. The translation of this into sheaf theory is given by our **cellular fibration lemma** of chapter 2.

  Another important source of such fibrations is the famous **first isotopy lemma of Thom** for proper stratified submersions on suitable stratified spaces. In chapter 4 we give a purely sheaf theoretic proof of the corresponding cohomological result, based on the important **"non-characteristic deformation lemma"** of Kashiwara. Here we only need very weak assumptions on the stratification, e.g. our cohomological result applies to **locally C-regular** stratifications in the sense of Bekka, for which the first isotopy lemma is unknown!

- Our definition of a **"cohomologically cone-like stratification"** given in chapter 4 uses suitable local direct image sheaves, which "simulate" the corresponding local geometry of **locally cone-like stratifications**. Similarly, we introduce in that chapter the notion of the **"local stratified acyclicity"** of a smooth stratification, which is a cohomological version of the geometric **"local stratified triviality"** studied recently by some authors.

- Geometric **transversality** assumptions for suitable stratifications correspond to **base change isomorphisms** in the language of constructible sheaves. This will be explained in chapter 4, and becomes very important in the applications in chapter 6.

Let us now go more into detail, starting with our definition of **constructible sheaves**.

We call a complex of sheaves $\mathcal{F}$ on a topological space X constructible with respect to a filtration by closed subsets

$$X_\bullet : \quad \emptyset = X_{-1} \subset X_0 \subset \cdots \subset X_m = X \ ,$$

or the filtration is called adapted to $\mathcal{F}$, if the restriction of the cohomology sheaves of $\mathcal{F}$ to the $X_i \backslash X_{i-1}$ are locally constant.

Note that this is **not** always the usual definition, since we allow unbounded complexes and make no additional assumption on the stalks of $\mathcal{F}$. The kind of filtration depends on which kind of singular spaces one uses, e.g. in the complex analytic context we assume for example that all X_i are complex analytic subsets of X. Sometimes it is important to work with a fixed filtration (e.g. on a triangulated or stratified space). But from the functorial viewpoint it is often simpler not to work with a fixed filtration (e.g. in the semialgebraic or complex analytic context).

The first important thing is to study the **stability** of this class of sheaf complexes under the usual Grothendieck functors (at least in special cases). We explain this for example in the following cases:

- In the **semialgebraic** context, or more generally in the context of **o-minimal structures on the real field** [vDr] and **analytic geometric categories** [vDrMi]. This is explained in chapter 2.

  At the same time we show similar things in the context of a fixed **"good cellular stratification"**.

- In the **equivariant context** of chapter 3. There we study the quotient map $\pi : X \to X/G$ for a proper action of a Lie group G. We prove in some important special cases that $R\pi_!\mathcal{F}, R\pi_*\mathcal{F}$ are constructible with respect to a cellular decomposition or "good cellular stratification".

- For a fixed **cohomologically cone-like stratification** in chapter 4. From this we deduce the usual stability results in the **complex algebraic or analytic context**. Especially the corresponding stability results for the famous **nearby and vanishing cycle functors** ψ_f and ϕ_f.

For other approaches to some of these stability results, compare also with:

1. [IC, HS] in the framework of stratified spaces and maps.
2. [Ver3] for the complex analytic or algebraic context, where resolution of singularities is used.
3. [KS, chapter VIII] for the embedded subanalytic or complex analytic context. That approach is based on "micro-local sheaf theory" with the help of the "micro-support".

One has especially a similar theory of Grothendieck functors

$$Rf_*,\ f^*,\ Rf_!,\ f^!,\ \otimes^L,\ Rhom$$

in other situations (which are not necessarily of topological nature):

- In the l-adique cohomology (e.g. compare [De3]).
- In the theory of algebraic holonomic D-modules [Bo].
- In the context of equivariant derived categories [BL].
- In the theory of algebraic mixed Hodge modules [Sai].
- In the context of semialgebraic geometry over a real closed base field. This theory exists at least in some parts [Schei].

One advantage of the abstract formalism is the fact that many arguments are purely formal (e.g. based on adjunction properties of these functors) and work in all the cases mentioned above. For a discussion of **Verdier duality** from this viewpoint, we recommend [Ill2]. Moreover, in many cases there exist suitable comparison or forgetful maps between these categories, compatible with the Grothendieck functors and the underlying triangulated categories.

This allows one for example, to **"lift up" important isomorphisms** from our topological context of constructible sheaves into some of these more sophisticated theories.

For the **equivariant derived category** this is worked out in chapter 3. Many results are there reduced to purely topological statements, for which we can use very simple arguments about "triangulations and cell-decompositions", based on the important **Vietoris-Begle theorem**!

Similarly, we prove in chapter 2 **Künneth formulae** for constructible sheaves in the framework of real geometry, using methods specific to this context. Nevertheless, this implies these formulae also for complex algebraically constructible complexes, which then can be "lifted up" into the theory of algebraic mixed Hodge modules.

The underlying principle is the following simple property of triangulated categories [Nee, cor. 1.2.6]:

- The map $f : X \to Y$ is an isomorphism, if and only if there exists a distinguished triangle
$$X \xrightarrow{\ f\ } Y \longrightarrow 0 \xrightarrow{[1]} \ .$$

Here we do not recall the definition of a triangulated category, but only point out that a reader not familiar with this notion should think of **distinguished triangles** as a "substitute" for **short exact sequences** of complexes, which behave better from the functorial viewpoint. Any short exact sequence of sheaf complexes induces a distinguished triangle in the derived category. Similarly, one gets long exact cohomology sequences from distinguished triangles just by the application of a (suitable) cohomology functor. Here is the most important example for us, that will be used almost everywhere in this book:

- Let $i : X \to Y$ be the inclusion of a **closed** subset of a topological space Y, with $j : U := Y\backslash X \to Y$ the inclusion of the **open** complement. Then there are two functorial distinguished triangles in the derived category $D(Y)$ of sheaves on Y:
$$Ri_*i^! \longrightarrow id \longrightarrow Rj_*j^* \xrightarrow{[1]} \ ,$$
$$Rj_!j^* \longrightarrow id \longrightarrow Ri_*i^* \xrightarrow{[1]} \ .$$

Application of the usual global (section) cohomology functor induces the usual long exact cohomology sequence of the pair (Y, U) or (Y, X). Other important distinguished triangles correspond to **Mayer-Vietoris sequences**.

Let us come back to our definition of a constructible complex of sheaves, where we do **not** make any (finiteness) assumption on the stalks of a constructible complex $\mathcal{F}$. From our point of view this is an **additional property**, that one should study in a second step. Of course, the aim of sheaf theory is always to obtain global information from local information. One of the leading ideas of this book is the following

program :

- If all **stalks** $\mathcal{F}_x$ of a constructible complex of sheaves have a property P, then the same is true (up to a shift) for $R\Gamma(X, \mathcal{F})$.
- If all **costalks** $i_x^! \mathcal{F}$ (with $i_x : \{x\} \to X$ the inclusion) of a constructible complex of sheaves have a property P, then the same is true (up to a shift) for $R\Gamma_c(X, \mathcal{F})$.
- If one has these results in the absolute case for $R\Gamma(X, \mathcal{F})$ (or $R\Gamma_c(X, \mathcal{F})$), then the same is true for other functors like $Rf_* \mathcal{F}$ (or $Rf_! \mathcal{F}$), i.e. the (co)stalk of these new (constructible) sheaf complexes also have the property P (up to a shift).

Of course we have to explain, what kind of property P we allow so that it makes sense to speak of this property "up to a shift". We introduce for this the following notion related to an abstract triangulated category:

Let D be a triangulated category with translation functor $T = [1]$. A subfamily B of $ob(D)$ satisfies the property "$\geq$" (or "$\leq$") if we have

I. $0 \in B$.

II. Let $X \to Y \to Z \overset{[1]}{\to}$ be a distinguished triangle in D. Then $X, Z \in B \Rightarrow Y \in B$ (i.e. B is stable by extensions [BBD, 1.2.6]).

III. $X \in B \Rightarrow X[-1] \in B$ (or $X \in B \Rightarrow X[1] \in B$).

We do not distinguish between a family B of objects of D and the corresponding full subcategory.

The most important examples are (compare with the beginning of chapter 6 for more details):

1. $(D^{\leq 0}, D^{\geq 0})$ is a **t-structure** on D.
2. $D^{\leq 0} = D^{\geq 0}$ is a **strict full triangulated subcategory** of D, also called a **null system** in [KS].

A subcategory with the property "$\leq$" is also a **suspended** subcategory of D in the sense of [KV] (and a subcategory with the property "$\geq$" is a suspended subcategory of the dual triangulated category D°), but we do not use this notion.

Since we are interested in **(co)stalk properties**, we look for corresponding subcategories of $D = D(R)$, with R the base-ring that we use. The two standard examples in this case are

1. $(D^{\leq 0}, D^{\geq 0})$ is the standard t-structure on $D(R)$, i.e. $D^{\geq 0}$ or $D^{\leq 0}$ consists of the complexes of R-modules with **vanishing cohomology** in negative or positive degrees.

2. The family $D_{perf}(R)$ of **perfect** complexes, i.e. which are isomorphic to a bounded complex of finitely generated projective R-modules, is a null system of $D(R)$, which is stable by direct summands. If R is Noetherian with finite homological dimension (like $\mathbb{Z}$ or a field), then a complex of R-module is perfect if and only if it is cohomologically bounded with **finitely generated** cohomology.

   Assume that R is commutative. Then the duality functor $D = Rhom(\cdot, R)$ induces an automorphism of $D_{perf}(R)$ with $D \circ D \simeq id$. It exchanges the subfamilies of $D_{perf}(R)$ with the property "$\geq$" or "$\leq$".

The perfect complexes will be important for **biduality** results of constructible sheaves. Moreover, our (co)stalk properties as above generalize and unify vanishing and finiteness conditions. Let us now explain the cases where we can **prove our program** above:

- We consider the case of a **null system** in $D(R)$ and work in the framework of **"real geometry"**, i.e. with semialgebraic sets, or more generally, sets definable in some o-minimal structure on the real field or analytic (Nash) geometric category.

  Then we prove in chapter 2 our program for the functors

$$Rf_*,\ Rf_!,\ f^*,\ f^!\ ,$$

  with f a continuous definable map, i.e. a map whose graph is definable. In the context of analytic (Nash) geometric categories, we have to assume that f is proper for the case of the functor $Rf_* \simeq Rf_!$. If we consider in addition as a stalk property the **perfect complexes** for a commutative base ring R, then we establish also the stability under the functors

$$\otimes^L \quad \text{and} \quad Rhom\,.$$

Moreover, we explain in section 2.3 a new approach to the question of how to deduce from this the corresponding transformations of **constructible functions**.

Consider a bounded constructible complex of sheaves $\mathcal{F}$ of vector spaces with **finite dimensional** stalks, and let χ be the usual **Euler characteristic**. Then the function

$$\alpha : X \to \mathbb{Z}\,;\ \alpha(x) := \chi(\mathcal{F}_x)$$

is of course constructible. This induces a surjective homomorphism

$$\chi_X : K_0(X) \to CF(X, \mathbb{Z})\,,$$

with $K_0(X)$ the corresponding **Grothendieck group** of constructible (complexes of) sheaves and $CF(X, \mathbb{Z})$ the group of **constructible functions**.

Then the kernel of χ_X is represented by the constructible complexes $\mathcal{F}$ with

$$\chi(\mathcal{F}_x) = 0 \quad \text{for all } x \in X.$$

But this condition defines a **null system** such that the corresponding stalk property is stable under the six Grothendieck functors as above! Therefore we get induced transformations on the level of **constructible functions**, with the same formalism, especially with a **duality transformation** such that $D \circ D = id$.

In the subanalytic or semialgebraic context, this is already due to Kashiwara-Schapira [KS, Scha], based on the deeper result that in this case the map χ_X is an isomorphism. Our approach is simpler and more general. Especially, it applies also in the framework of a **fixed stratification**, and in the **complex algebraic and analytic context** (cf. [Viro])! In the last case, the stability of such stalk properties (given by a null system) follows already from the "real context". The stability of the corresponding category of **complex algebraically and analytically constructible complexes** is worked out in chapter 4.

- We consider the case of a **null system** and work in the **equivariant context** for a Lie group G. Then we study in chapter 3 the quotient map $\pi : X \to X/G$ for a **suitable proper action** of G, and apply our program to

  $$R\pi_! \mathcal{F} \quad \text{and} \quad R\pi_* \mathcal{F}\,.$$

  As we will see, this implies generalizations of many different well-known **localization results**. Applications are also given to the complex algebraic context (where we do not assume that the action is proper).

So in this case the **points** of X/G correspond to the **orbits of** G, which is quite natural from the equivariant point of view. But the **(co)stalks** of

$$R\pi_! \mathcal{F} \quad \text{and} \quad R\pi_* \mathcal{F}$$

at these "points" are objects of the **equivariant derived category** $D_G(\{pt\})$. It is therefore important that we prove our results in the abstract framework of the six standard functors. Then one can easily "lift up" the results into the equivariant context.

- We consider **any** subfamiliy of $D(R)$ with the **property "** $\geq$**" or "** $\leq$**"**. Then we prove in chapter 6 our program in the **complex analytic or algebraic context**. As we explain there, the right way to formulate these (co)stalk properties is to **measure them with respect to the middle perversity function** m.

We prove our program in the complex analytic context for the important case of the (shifted) **nearby and vanishing cycle functors**

$$\psi_f[-1] \quad \text{and} \quad \phi_f[-1]\,.$$

This result has for example far reaching applications in the study of "hypersurface singularities" and "global fibers of compactifiable holomorphic functions".

In the complex algebraic context we prove in this generality a corresponding **Artin-Grothendieck type theorem**, i.e. the stability under

$$Rf_! \quad \text{and} \quad Rf_* \quad \text{for } f \text{ an } \textbf{affine} \text{ morphism.}$$

Typical applications of the absolute case for a constant map $f : X \to \{pt\}$, i.e. of these results for the functors

$$R\Gamma_c(X, \cdot) \quad \text{and} \quad R\Gamma(X, \cdot) \quad \text{with } X \text{ affine,}$$

are very general **Lefschetz type theorems**.

As far as possible, we show these results also in the complex analytic context, sometimes under the additional assumption of a suitable subanalytic compactification.

Also some important examples are worked out in more detail, especially for giving the reader some feeling about how to use the abstract machinery for concrete applications to the study of singular spaces. But the list of applications of these results is rapidly increasing, and seems to fill in their own books. Here we refer for example to the forthcoming books of Massey [Ma3] and Dimca [Dimca].

If we consider in this complex analytic or algebraic case the **standard vanishing conditions**, then our results reduce to well-known results about **perverse sheaves** (or the perverse t-structure with respect to the middle perversity). Compare for example with [BBD, sec. 4], [KS, chapter X] and [GM3]. Moreover, the notation that we use is motivated by this classical example.

Other results about constructible sheaves in algebraic geometry that are related to our point of view are the following ones:

- The proof of the **trace formula of Grothendieck** (for the Frobenius morphism) given in [SGA4$\frac{1}{2}$, p. 100–109] is close to our proof of the localization results (especially the corresponding trace formulae) given in chapter 3.

- The **Euler-Poincaré formula** of [SGA5, p. 396, thm. 7.1] for a constructible complex of sheaves on a proper connected smooth algebraic curve C (over an algebraically closed field). This seems to be one of the first places where it is proved that stalk properties of a constructible complex globalize to $R\Gamma(C, \cdot)$. In loc.sit. the stalk property is given by a null system, which is stable by direct summands.

- The most important example comes from the theory of **"mixed sheaves"**. Examples are given in [BBD, sec. 5], and by the theory of algebraic mixed Hodge modules of Saito [Sai]. These admit a theory of standard Grothendieck functors (corresponding to our first step). Moreover, one has in these "mixed theories" **a notion that an object is of weight** $\leq w$ **or** $\geq w$ $(w \in \mathbb{Z})$.

The condition of "weight $\leq w$" is a **stalk property**, and the other condition is defined by duality, i.e. corresponds to a **costalk property**. Denote the corresponding subcategories by

$$D^b_{\leq w} \quad \text{or} \quad D^b_{\geq w} .$$

Then $D^b_{\leq w}$ (or $D^b_{\geq w}$) has the property "$\geq$" or "$\leq$", and these categories are stable by the functors

$$Rf_!,\ f^* \quad \text{or} \quad Rf_*,\ f^! .$$

These are **very deep results, which are not of topological nature** (so these have nothing to do with our proofs). Compare for example with [DS] for some applications of this formalism to the topology of complex algebraic varieties.

Let us now give some remarks on the main ideas of the proofs of our results. There are basically two different approaches to globalize local (co)stalk properties of constructible complexes of sheaves:

1. The "classical" approach, using **triangulations and cell-decompositions**.
2. The "micro-local" approach, using **(stratified) Morse theory** on stratified spaces.

We use the first approach for our applications in "real geometry" given in chapter 2 and 3, and the second approach for the applications in "real and complex geometry" given in chapter 5 and 6. Here we have the following

Leitfaden

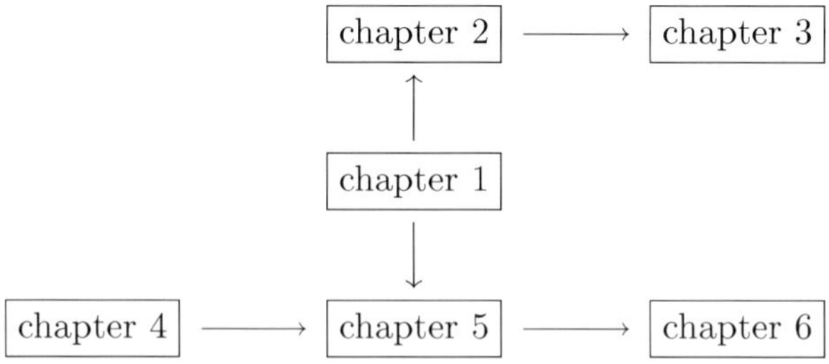

Roughly speaking, the first line develops the "classical" approach to our program in the context of "real geometry", and the last line developes the "micro-local" approach to our program in the context of "real and complex geometry". Finally, chapter 1 lays down the foundation for our program in the case of the nearby and vanishing cycle functors. In particular, there we explain important geometric descriptions of the stalks of these functors, which are completed in section 5.4 by the corresponding results for the costalks.

Here is a short summary of the different chapters (compare with the introduction of these chapters for more details):

Chapter 1 contains the relation between (cohomological versions) of Milnor fibrations and the nearby and vanishing cycle functors. At the same time, we give a detailed introduction to these important functors. As applications, different versions of Künneth type formulae for these functors are proved.

Chapter 2 develops the theory of constructible sheaves and functions in the context of real geometry, without the use of deep stratification techniques. We only use two basic results of the theory of o-minimal structures [vDr]: the "cell decomposition and the triangulation theorem". Otherwise this chapter is self-contained, except for the Künneth type theorems, which also depend on some results from chapter 1. As a byproduct, we also introduce the notion of a "Nash geometric category", which is a useful variant of the "analytic geometric categories" of [vDrMi].

Chapter 3 is an exposition about "localization and fixed point results" in the equivariant context. The basic results are proved for Lie group actions such that the quotient space has a suitable finite cell-decomposition. Important examples come from a "definable action" in the context of one of the "geometric categories" of chapter 2. Especially, in the semialgebraic context with an action of a compact group, no compactness assumptions about the underlying spaces are needed. Some applications to the complex algebraic context are also worked out.

Chapter 4 starts our second approach, where we lay down the foundation for the next chapter. We explain cohomological versions of some of the basic results of stratification theory. All results are proved in the language of sheaf theory. The whole theory is based on the "non-characteristic deformation lemma" of Kashiwara. From this we deduce a "micro-local" characterization of the sheaf complexes on a smooth manifold which have only locally constant cohomology sheaves. In this way we can avoid the use of the "first isotopy lemma of Thom". We also introduce a new regularity condition for a smooth stratification of a closed subset of a smooth manifold, which we call "d-regularity" or "distance-regularity". This notion is weaker than Whitney b-regularity, or the notion of a "locally C-regular" stratification in the sense of Bekka, but nevertheless it induces all the usual results about constructible sheaves on these stratified spaces.

Chapter 5 contains the deepest results of this book. Here we work out a new approach to "(stratified) Morse theory for constructible sheaves". We also explain in detail the relation of our Morse theory for constructible sheaves to the general "micro-local sheaf theory" of Kashiwara-Schapira [KS, KS2] and the "stratified Morse theory" of Goresky-MacPherson [GM]. But our approach to these results is independent of [GM, KS, KS2], and can also be used as an introduction to these references. As a first application, we give an introduction to the theory of characteristic cycles. We reduce the proof of one of the main results of this chapter to our general Thom-Sebastiani Theorem for constructible sheaves from chapter 1.

Chapter 6 finally develops the functorial theory of "generalized vanishing conditions of the (co)stalks" in the framework of complex analytic and algebraic geometry. Many important examples and applications are worked out, and all results are reduced to one "main result", which is proved with the help of our "Morse theory for constructible sheaves". But otherwise, this chapter is self-contained and no further use or knowledge of this Morse theory is needed! We also introduce suitable variants of (topologically) q-complete spaces and maps. Similarly, we define a generalization of the "middle perversity" relative to a holomorphic map between complex spaces, and explain especially the relation to the "(relative) rectified homological depth" introduced and studied by Grothendieck, Hamm and Lê.

Let us finally remark that this **Morse theory for constructible sheaves** is related to the new **micro-local** viewpoint on the geometry of singular subspaces of smooth manifolds. In particular, it gives a detailed translation between the general "micro-local sheaf theory" of Kashiwara-Schapira [KS, KS2] and the "stratified Morse theory" of Goresky-MacPherson [GM]. Note that these two theories are worked out in their own languages for different applications. But as we explain, in the framework of constructible sheaves both theories are more or less the same, if one uses the right translation!

The basic "micro-local notion" in the framework of Morse theory for constructible sheaves is the cohomologically counterpart of the **"normal Morse data"** of Goresky-MacPherson. As we will see, these correspond to the **"micro-local type"** in the language of Kashiwara-Schapira. As an application, we give in chapter 5 an explicit geometric description of the **micro-support** of a constructible complex of sheaves, which in this form is missing in [KS, KS2].

In this context one should also add a **third step to our program**:

- Local (co)stalk properties of constructible sheaves are **equivalent** to similar properties of the "normal Morse data" of these sheaves.

It will be worked out in chapter 6 for our "generalized vanishing conditions" of the (co)stalks. Here one has to measure them in the "complex context" with respect to the **middle perversity**, whereas in the "real context" one has to use the **minimum (or zero) perversity**!

If the corresponding (co)stalk property is given by a null system, then these results are related, by the **characteristic cycles** of constructible sheaves or functions, to the theory of (conic) **Lagrangian cycles**. The theory of (conic) Lagrangian cycles is the micro-local version of the theory of constructible functions. It is deeply related to Morse theory for constructible sheaves, but depends only on the "Euler characteristic version" of it, i.e. on a simpler "Morse theory for constructible functions"!

Note that this theory of **characteristic cycles** has its origin in the theory of **holonomic D-modules** and was only later on extended to the context of "real geometry". There are many approaches to this subject, often using quite different techniques and conventions. We therefore give in chapter 5 a detailed geometric introduction to these characteristic cycles from the viewpoint of "Morse theory for constructible functions", explaining the relation between all these conventions! Moreover, the more sophisticated language of constructible sheaves can be avoided (compare [GrM, Sch4]). This seems to be important if one looks at the geometric applications of this theory. Besides representation theory, this includes especially the theory of **characteristic classes and curvature measures** of singular spaces.

Due to lack of time and place, we only work out some "micro-local intersection formulae for Lagrangian cycles", but **not** the corresponding general functorial theory as in [KS, sec. 9.4]. The important cases of **exterior products and duality** are included, since these are just simple corollaries of related statements about our "normal Morse data"! Similarly, a special case of a **specialization result** is included to give a new geometric proof of the fact that these "characteristic cycles" are indeed geometric **cycles**. Of course, this is only non-trivial in the "real context".

Compared to other references, we work in greater generality, since we allow other **additive functions** of the (co)stalks, and not only the usual **Euler characteristic**. Even in the classical case of the "Euler characteristic", we get some refined information, since we consider a suitable fixed stratification. For example, we get in the complex context, "without any calculation", that the famous **Euler obstruction** is constructible with respect to **any** Whitney stratification. Moreover, our approach works also for weaker stratifications, and sometimes it applies to the "real context". This important relation between **obstruction theory** and **micro-local intersection formulae** is contained in section 5.2.

Let us finish this introduction with the following relation between the corresponding **characteristic cycles** of 1_X and $1_{X(R)}$, if X is a complex algebraic variety defined over $\mathbb{R}$:

$$CC_2(1_{X(R)}) = cl_R(CC(1_X)) .$$

Here CC (or CC_2) denotes the characteristic cycle map corresponding to the Euler characteristic (mod 2) with $cl_R(\cdot)$ being the **real cycle map** of [BH]. This relation implies the classical result of Sullivan that $X(\mathbb{R})$ is a *mod* 2 **Euler space**, i.e. that the Euler characteristic of a link is even for all $x \in X(\mathbb{R})$. So one should see it as a **micro-local refinement** of the result of Sullivan.

This relation has far reaching applications to the theory of characteristic classes. It implies, for example, that the **Stiefel-Whitney classes** of $X(\mathbb{R})$ **are represented by algebraic cycles**. This should be compared with the original "combinatorial definition" of Sullivan [Su] in terms of a triangulation. Our formula above will be proved in section 5.4, using "micro-local geometry", especially a description of the **"normal Morse data" in terms of vanishing cycles** !

Chapter 1

Thom-Sebastiani Theorem for constructible sheaves

Introduction

In [Mi] Milnor introduced his fibration for a holomorphic function germ

$$f : (\mathbb{C}^{n+1}, 0) \to (\mathbb{C}, 0)$$

(actually he required f to be a polynomial). For an **isolated** critical point 0 of f he showed that the **Milnor fiber**

$$M_f := \{z \in \mathbb{C}^{n+1} |\, f = w, |z| \leq \delta\} \quad (\, 0 < |w| << \delta << 1 \,)$$

has the homotopy type of a finite bouquet of n-spheres ($n > 0$). In particular, the reduced (co)homology of M_f is equal to $\mathbb{Z}^{\mu}$, concentrated in degree n (with $\mu :=$ the **Milnor number** of f). Of course this is also true for $n = 0$, where M_f consists of $\mu + 1$ points. For a beautiful introduction to many related ideas we recommend [Te3].

Let $f : (\mathbb{C}^{n+1}, 0) \to (\mathbb{C}, 0)$ and $g : (\mathbb{C}^{m+1}, 0) \to (\mathbb{C}, 0)$ be two holomorphic germs, with 0 as an **isolated** critical point. Then Sebastiani and Thom [ST] proved for

$$f \oplus g : (\mathbb{C}^{n+1} \times \mathbb{C}^{m+1}, 0) \to (\mathbb{C}, 0) \;,\; f \oplus g(x, y) := f(x) + g(y)$$

and $n, m > 0$:

$$H^{n+m+1}(M_{f \oplus g}, \mathbb{Z}) \simeq H^n(M_f, \mathbb{Z}) \otimes H^m(M_g, \mathbb{Z}) \tag{1.1}$$

(especially $\mu_{f \oplus g} = \mu_f \cdot \mu_g$), with the algebraic **monodromy** induced by the Milnor fibration given by

$$mon_{f \oplus g} \simeq mon_f \otimes mon_g \;.$$

Later on the first part of this result was re-proved in terms of the homotopy type of the Milnor fiber ([Oka, Sa, AVG]):

$$M_{f\oplus g} \simeq_{ht} M_f * M_g \,. \tag{1.2}$$

Here the join $X * Y$ of two topological spaces X, Y is the quotient of $X \times [0,1] \times Y$ by the equivalence relation : $(x,0,y) \sim (x,0,y')$ and $(x,0,y) \sim (x',0,y)$ for all $x, x' \in X; y, y' \in Y$. This implies (1.1) also in the case $n = 0$ or $m = 0$:

$$\tilde{H}^{n+m+1}(M_{f\oplus g}, \mathbb{Z}) \simeq \tilde{H}^n(M_f, \mathbb{Z}) \otimes \tilde{H}^m(M_g, \mathbb{Z}) \,. \tag{1.3}$$

Note that it is important to use **reduced** cohomology in this formulation ! Interestingly Sebastiani and Thom have to exclude this case in their proof (we will see the reason for this in the proof of our generalization of the Thom-Sebastiani Theorem).

Similar results for other invariants of $f \oplus g$ are obtained by Gabrielov [Ga] for the intersection matrix, by Sakamoto [Sa] for the Seifert form and by Deligne for the variation operator (see [Dem]).

In the end of his paper [Sa], Sakamoto remarks that his proof of the **homotopy type** result (1.2) works also in the case of **nonisolated** singularities. For a proof of the special case with $g(w) = w^j$ $(j \geq 2)$, see [Ma, prop. 8.1]. Another extension of the homotopy type result (1.2) is stated by Houston [Hou, lem. 2.8] for holomorphic function germs with isolated singular points with respect to Whitney stratifications. The general homotopy type result (1.2) is proved in [OY, appendix] for non-constant holomorphic functions germs on complex spaces.

In [De] Deligne introduced for a holomorphic function f his famous **nearby- and vanishing cycle functors** ψ_f, ϕ_f and explained for an **isolated** singularity $f : (\mathbb{C}^{n+1}, 0) \to (\mathbb{C}, 0)$ the isomorphism

$$H^*(\phi_f(\mathbb{Z}_{\mathbb{C}^{n+1}})_0) \simeq \tilde{H}^*(M_f, \mathbb{Z}) \,, \tag{1.4}$$

identifying the algebraic **monodromy** mon_f of the cohomology of the Milnor fiber with an automorphism m_f induced by an action on the functor ϕ_f.

It is now well known that the nearby- and vanishing cycle functors are deeply related to perverse sheaves (this will be explained in more detail in chapter 6 of this book). It is therefore important to look at them in the framework of complexes of constructible sheaves. Here we call a complex of sheaves $\mathcal{F}$ on a topological space X **constructible** with respect to a filtration

$$X_\bullet \,:\, \emptyset = X_{-1} \subset X_0 \subset \ldots \subset X_n = X$$

by closed subsets (or the filtration **adapted** to $\mathcal{F}$), if the restriction of the cohomology sheaves of $\mathcal{F}$ to the $X_i \backslash X_{i-1}$ are locally constant. If X is a complex (real) analytic space and the X_i can be chosen to be complex analytic (subanalytic) subsets, then we call $\mathcal{F}$ **complex analytically (subanalytically)** constructible.

The main result of this chapter (theorem 1.2.2) implies the

Theorem 1.0.1. *Let $f_i : X_i \to \mathbb{C}$ be holomorphic functions on the complex spaces X_i, $S_i \subset \{f_i = 0\}$ compact analytic subsets and $\mathcal{F}_i \in ob(D(X_i))$ be complex analytically constructible $(i = 1, 2)$. Then there exists for*

$$f := f_1 \oplus f_2 : X := X_1 \times X_2 \to \mathbb{C},\ f(x_1, x_2) := f_1(x_1) + f_2(x_2)\,,$$

$S := S_1 \times S_2 \subset \{f = 0\}$ and $\mathcal{F} := \mathcal{F}_1 \boxtimes^L \mathcal{F}_2 \in ob(D(X))$ an isomorphism

$$R\Gamma(S, \phi_f[-1](\mathcal{F})) \simeq R\Gamma(S_1, \phi_{f_1}[-1](\mathcal{F}_1)) \otimes^L R\Gamma(S_2, \phi_{f_2}[-1](\mathcal{F}_2))$$

such that the following diagram commutes :

$$\begin{array}{ccc} R\Gamma(S, \phi_f[-1](\mathcal{F})) & \xleftarrow{\sim} & R\Gamma(S_1, \phi_{f_1}[-1](\mathcal{F}_1)) \otimes^L R\Gamma(S_2, \phi_{f_2}[-1](\mathcal{F}_2)) \\ \Big\downarrow{\scriptstyle m_f[-1]} & & \Big\downarrow{\scriptstyle m_{f_1}[-1] \otimes^L m_{f_2}[-1]} \\ R\Gamma(S, \phi_f[-1](\mathcal{F})) & \xleftarrow{\sim} & R\Gamma(S_1, \phi_{f_1}[-1](\mathcal{F}_1)) \otimes^L R\Gamma(S_2, \phi_{f_2}[-1](\mathcal{F}_2))\,. \end{array}$$

If we take $S_1 = \{x_1\}$ and $S_2 = \{x_2\}$, we get an isomorphism

$$\big(\phi_f[-1](\mathcal{F})\big)_{(x_1,x_2)} \simeq \big(\phi_{f_1}[-1](\mathcal{F}_1)\big)_{x_1} \otimes^L \big(\phi_{f_2}[-1](\mathcal{F}_2)\big)_{x_2}, \tag{1.5}$$

commuting with the corresponding monodromy automorphisms. Especially, one gets

$$\big(\phi_f[-1](\mathcal{F})\big)_{(x_1,x_2)} \simeq 0$$

if one of the corresponding terms of the right-hand side is vanishing.

As an example let us assume that $\mathcal{F}_i$ is constructible with respect to a fixed complex analytic **Whitney stratification** $X_{\bullet,i}$ and $f_i : X_i \to \mathbb{C}$ has an **isolated** critical point x_i with respect to this stratification $(i = 1, 2)$. If the $\mathcal{F}_i$ are **perverse sheaves** with respect to the middle perversity, then we get for $k \neq 0$ (as shown in chapter 6) the **vanishing result**:

$$H^k\big((\phi_{f_i}[-1](\mathcal{F}_i))_{x_i}\big) = 0 \qquad (i = 1, 2)\,.$$

So if $H^0\big((\phi_{f_i}[-1](\mathcal{F}))_{x_i}\big)$ is torsion free for $i = 1$ or $i = 2$ (e.g. for $\mathcal{F}_i$ an intersection cohomology complex), we get

$$H^0\big((\phi_f[-1](\mathcal{F}))_{(x_1,x_2)}\big) \simeq H^0\big((\phi_{f_1}[-1](\mathcal{F}_1))_{x_1}\big) \otimes H^0\big((\phi_{f_2}[-1](\mathcal{F}_2))_{x_2}\big)\,.$$

Especially for $X_1 = \mathbb{C}^{n+1}, X_2 = \mathbb{C}^{m+1}$ and $\mathcal{F}_1 = \mathbb{Z}[n+1], \mathcal{F}_2 = \mathbb{Z}[m+1]$ (with $\mathbb{Z}$ the constant sheaf) and $f_i : X_i \to \mathbb{C}$ with an isolated critical point 0 (in the classical sense), we have an isomorphism

$$\begin{array}{ccc} \tilde{H}^{n+m+1}(M_{f_1 \oplus f_2}, \mathbb{Z}) & \xrightarrow{\sim} & H^0\big((\phi_{f_1 \oplus f_2}[-1]\ \mathbb{Z}[n+m+2])_{(0,0)}\big) \\ \Big\downarrow{\wr} & & \Big\downarrow{\wr} \\ \tilde{H}^n(M_{f_1}, \mathbb{Z}) \otimes \tilde{H}^m(M_{f_2}, \mathbb{Z}) & \xrightarrow{\sim} & H^0\big((\phi_{f_1}[-1]\ \mathcal{F}_1)_0\big) \otimes H^0\big((\phi_{f_2}[-1]\ \mathcal{F}_2)_0\big)\,, \end{array}$$

commuting with the monodromy action. So this case gives us the **classical** result (1.1) of Sebastiani and Thom (including the case $n = 0$ or $m = 0$) !

As we will see (and is also well known) the result (1.4) of Deligne is true for an arbitrary holomorphic function germ $f : (X, x) \to (\mathbb{C}, 0)$ on a complex space, with the **Milnor fiber** given by

$$M_f := \{z \in X \mid f = w, |z| \leq \delta\} \quad (0 < |w| << \delta << 1)$$

for a local model $(X, x) \hookrightarrow (\mathbb{C}^N, 0)$ so that theorem 1.0.1 implies the cohomological version (i.e. the commutativity of the following diagram)

$$\begin{array}{ccc} \widetilde{R\Gamma}(M_{f\oplus g}, \mathbb{Z})[-1] & \xrightarrow{\sim} & \widetilde{R\Gamma}(M_f, \mathbb{Z})[-1] \otimes^L \widetilde{R\Gamma}(M_g, \mathbb{Z})[-1] \\ \downarrow{\scriptstyle m_{f\oplus g}[-1]} & & \downarrow{\scriptstyle m_f[-1]\otimes^L m_g[-1]} \\ \widetilde{R\Gamma}(M_{f\oplus g}, \mathbb{Z})[-1] & \xrightarrow{\sim} & \widetilde{R\Gamma}(M_f, \mathbb{Z})[-1] \otimes^L \widetilde{R\Gamma}(M_g, \mathbb{Z})[-1] \end{array}$$

of the Thom-Sebastiani Theorem for arbitrary holomorphic function germs on complex spaces, with

$$\widetilde{R\Gamma}(M_f, \mathbb{Z}) := cone(\mathbb{Z} \to R\Gamma(M_f, \mathbb{Z}))$$

calculating the reduced cohomology of the Milnor fiber M_f !

Of course theorem 1.0.1 implies corresponding results for "additive invariants" of the automorphisms

$$(R\Gamma(S, \phi_f[-1](\mathcal{F})), m_f[-1]) \quad \text{and} \quad (\widetilde{R\Gamma}(M_f, \mathbb{Z})[-1], m_f[-1]) ,$$

if they are "multiplicative" with respect to $\otimes^L$-products. Typical examples are the zeta function (or **Lefschetz number**) of the monodromy or just the **Euler characteristic** of the Milnor fiber. As we will see later on, one can give in our topological setting for this type of results a short proof, **without** (!!) using the most difficult part of the proof of our main theorem. This corresponds to the effect that such "invariants" can often be treated in a more general situation. See for example

1. Némethi [Nem1, Nem2, Nem3] for the case of the zeta function of the monodromy.

2. Némethi-Steenbrink [NemSt] for spectral pairs in the framework of mixed Hodge modules (see also [SchSt] for the Hodge theoretic version of the classical Thom-Sebastiani Theorem).

3. Denef-Loeser [DenLoe] for a motivic version.

Very recently D. Massey proved in [Ma2] the following stronger form of theorem 1.0.1 by describing a natural isomorphism in the derived category:

Theorem 1.0.2. *Let $f_i : X_i \to \mathbb{C}$ be holomorphic functions on the complex spaces X_i and $\mathcal{F}_i \in ob(D(X_i))$ be complex analytically constructible $(i = 1, 2)$. Then there exists for*

$$f := f_1 \oplus f_2 : X := X_1 \times X_2 \to \mathbb{C},\ f(x_1, x_2) := f_1(x_1) + f_2(x_2)\,,$$

$\mathcal{F} := \mathcal{F}_1 \boxtimes^L \mathcal{F}_2 \in ob(D(X))$ *and* $k : \{f_1 = 0\} \times \{f_2 = 0\} \hookrightarrow \{f = 0\}$ *the inclusion a natural isomorphism*

$$k^* \phi_f[-1](\mathcal{F}) \simeq \phi_{f_1}[-1](\mathcal{F}_1) \boxtimes^L \phi_{f_2}[-1](\mathcal{F}_2)$$

commuting with the corresponding monodromies.

Note that this result implies theorem 1.0.1 by the **classical Künneth formula** (which we explain in the appendix). Interestingly, Massey uses a "real" version of the result in his proof. Note that we use in the proof of our main theorem only the weaker assumption about the **cohomological version of a Milnor fibration** (which is closer to the geometric picture in most applications, as we explain in section 1.1, and sometimes applies to more general sheaves). Nevertheless it implies generalizations of the result of Massey, as we explain in section 1.3 (see corollary 1.3.4).

If one forgets the assertion about the monodromy in the Thom-Sebastiani Theorem, a similar result makes sense in real geometry and our main result (theorem 1.2.2) implies for example:

Theorem 1.0.3. *Let $f_i : X_i \to \mathbb{R}$ be continuous subanalytic functions on the real analytic spaces X_i, $S_i \subset \{f_i = 0\}$ compact subanalytic subsets and $\mathcal{F}_i \in ob(D(X_i))$ be subanalytically constructible $(i = 1, 2)$. Then there exists for*

$$f := f_1 \oplus f_2 : X := X_1 \times X_2 \to \mathbb{R}, f(x_1, x_2) := f_1(x_1) + f_2(x_2)\,,$$

$S := S_1 \times S_2 \subset \{f = 0\}$ *and* $\mathcal{F} := \mathcal{F}_1 \boxtimes^L \mathcal{F}_2 \in ob(D(X))$ *an isomorphism :*

$$R\Gamma(S, R\Gamma_{\{f \geq 0\}}(\mathcal{F})) \simeq R\Gamma(S_1, R\Gamma_{\{f_1 \geq 0\}}(\mathcal{F}_1)) \otimes^L R\Gamma(S_2, R\Gamma_{\{f_2 \geq 0\}}(\mathcal{F}_2))\,.$$

The corresponding isomorphism in the derived category (as in theorem 1.0.2) follows from corollary 1.3.1.

Here the **local cohomology** functor $R\Gamma_{\{f \geq 0\}}(\cdot)$ is deeply related to (stratified) **Morse theory for constructible sheaves**, which we develop in chapter 5 of this book (see also [KS]). If $\mathcal{F}_i$ $(i = 1, 2)$ are bounded complexes of sheaves of vector spaces with finite dimensional stalk cohomology, then of course theorem 1.0.3 implies for the **global Euler characteristic** the formula :

$$\chi\big(R\Gamma(S, R\Gamma_{\{f \geq 0\}}(\mathcal{F}))\big) = \chi\big(R\Gamma(S_1, R\Gamma_{\{f_1 \geq 0\}}(\mathcal{F}_1))\big) \cdot \chi\big(R\Gamma(S_2, R\Gamma_{\{f_2 \geq 0\}}(\mathcal{F}_2))\big).$$

If one applies the corresponding stalk formula $(S_i = \{x_i\})$ to generic functions with respect to Whitney stratifications adapted to the $\mathcal{F}_i$ $(i = 1, 2)$, this gives a simple geometric proof for the **multiplicativity** of the characteristic cycles of Kashiwara-Schapira (compare [KS, formula (9.4.1)]) :

$$CC(\mathcal{F}_1 \boxtimes \mathcal{F}_2) = CC(\mathcal{F}_1) \times CC(\mathcal{F}_2)\,.$$

Theorem 1.0.3 itself implies the same result for the **micro-support** of the sheaves together with the **multiplicativity** of the "normal Morse datum" in the context of (stratified) Morse theory for constructible sheaves (as explained in chapter 5). This is not treated in [GM], but compare with [KS, prop. 7.5.10(i)].

This again is important for the multiplicativity of **Stiefel-Whitney and Chern classes** of singular spaces and also for **curvature measures** of singular spaces (see [BrBe, Fu, Sch]).

As we explain in the next chapter, the simplest example where theorem 1.0.3 applies is the case (of the cohomological version) of **classical Morse theory**. Consider the functions

$$f : (\mathbb{R}^n, 0) \to (\mathbb{R}, 0),\ f(x) := \sum_{i=1}^{n} x_i^2 \quad ; \quad g : (\mathbb{R}^m, 0) \to (\mathbb{R}, 0),\ g(y) := -\sum_{j=1}^{m} y_j^2\,.$$

Then one gets for $S_i := \{0\}$ $(i = 1, 2)$:

$$(R\Gamma_{\{f\geq 0\}}(\mathbb{Z}))_0 = \mathbb{Z} \quad \text{and} \quad (R\Gamma_{\{g\geq 0\}}(\mathbb{Z}))_0 = \mathbb{Z}[-m]\,,$$

where $\mathbb{Z}$ is the constant sheaf (or module) in degree 0. Then

$$(R\Gamma_{\{f\oplus g\geq 0\}}(\mathbb{Z}))_0 = \mathbb{Z}[-m]$$

describes the "local Morse datum", i.e. the reduced cohomology of the m-cell one has to attach in classical Morse theory (corresponding to the nondegenerate critical point 0 of $f \oplus g$ of index m).

The direct sum construction for **classical** Morse functions on smooth manifolds is used by Schwarz [Schw] in the context of **Morse homology** to explain corresponding "multiplicativity" properties of this Morse homology (compare with [Schw, p. 11, p. 167]).

The direct sum construction for a real version of a Milnor fibration is for example discussed by

1. Cornea [Cornea, cor. 2.7] for proper smooth functions on manifolds with "reasonable" critical submanifolds.

2. King [King, theorem 3] and [King2, thm. 8] for functions with isolated critical points in a very general topological context.

A very important application of our Thom-Sebastiani results (especially also of the "real versions") will be used and explained in chapter 5 of this book. It deals with a **generalization** to the framework of constructible sheaves of the cohomological version of the main result of Goresky and MacPherson about the "splitting of the local Morse datum" in their stratified Morse theory (compare with [GM, thm. 3.7] and [King, theorem 5]).

Let us finish this introduction with the following remark: It would be impossible to give our proofs or even to formulate the general results of this part of the book without the machinery of sheaf theory in the framework of derived categories. This language codifies in a beautiful way the geometry behind these results (see section 1.2) and moreover simplifies the discussion. By taking direct image complexes of sheaves we translate a possible difficult geometric situation into a very simple real two-dimensional picture. In this simple picture we then work with a constructible complex of sheaves. Even if we begin our discussion only with a constant sheaf, at this point all arguments also hold for a suitable constructible complex of sheaves.

Moreover, our proof of the Thom-Sebastiani result will show that this theorem is just a theorem of the type of a **Künneth formula** (for constant functions it reduces exactly to this). We make heavy use of the **Künneth formula** (in the context of derived categories), which we recall and explain in the appendix.

To illustrate this viewpoint of **Thom-Sebastiani results** as Künneth type formulae for the **vanishing cycle functor**, we explain also the corresponding results for the **nearby cycle functor**. Here one has to consider suitable **fiber products**, and these results follow directly from the **classical Künneth formula** applied to the corresponding **Milnor fibers**. In the complex analytic context one gets especially the

Theorem 1.0.4. *Let $f_i : X_i \to \mathbb{C}$ be holomorphic functions on the complex spaces X_i and $\mathcal{F}_i \in ob(D(X_i))$ be complex analytically constructible $(i = 1, 2)$. Consider the fiber product*

$$\begin{array}{ccc} X := X_1 \times_C X_2 & \xrightarrow{p_2} & X_2 \\ {\scriptstyle p_1}\downarrow & & \downarrow{\scriptstyle f_2} \\ X_1 & \xrightarrow{f_1} & \mathbb{C}\,, \end{array}$$

with $f := f_1 \circ p_1 = f_2 \circ p_2 : X \to \mathbb{C}$. Then there exists a natural isomorphism

$$\psi_f(\, \mathcal{F}_1 \boxtimes^L_C \mathcal{F}_2 \,) \simeq \psi_{f_1}(\mathcal{F}_1) \boxtimes^L \psi_{f_2}(\mathcal{F}_2)$$

commuting with the corresponding monodromies.

1.1 Milnor fibration

First we explain the relation between **cohomological versions of Milnor fibrations** and some functors in sheaf theory. Recall that all topological spaces are (assumed to be) locally compact with countable topology and of finite cohomological dimension (over $\mathbb{Z}$). For a ring R (associative with unit) let $D(X) := D(X, R)$ denote the derived category of complexes of sheaves of R-modules. If we have to specify left or right R-modules we use the notation $D(X, R-mod)$ for left R-modules and $D(X, mod-R)$ for right R-modules. Note that we allow unbounded complexes (see [Sp] and the appendix, especially for our convention about $\otimes^L$).

1.1.1 Cohomological version of a Milnor fibration

Let $f : X \to \mathbb{R}$ be a continuous function and consider the following cartesian diagram of maps induced by the inclusions $\{0\} \subset \mathbb{R} \supset \{x < 0\}$:

$$\begin{array}{ccccc} X_0 & \xrightarrow{i} & X & \xleftarrow{j} & \{f<0\} \\ \downarrow & & \downarrow f & & \downarrow \\ \{0\} & \longrightarrow & \mathbb{R} & \longleftarrow & \{x<0\}. \end{array} \tag{1.6}$$

The **specialization** ν_f and **micro-localization** μ_f of f are the functors

$$\nu_f := i^* Rj_* j^*,\ \mu_f := i^* R\Gamma_{\{f \geq 0\}}(\cdot) \ : D(X) \to D(X_0)\,.$$

So we have for $\mathcal{F} \in ob(D(X))$ the functorial distinguished triangle :

$$\mu_f(\mathcal{F}) \longrightarrow i^*(\mathcal{F}) \xrightarrow{sp} \nu_f(\mathcal{F}) \xrightarrow{[1]}\,, \tag{1.7}$$

with $sp = sp_f$ induced by the adjunction morphism $id \to Rj_* j^*$.

Remark 1.1.1. We use the notion specialization and micro-localization of f, because for a **smooth** function f on a smooth manifold X defining the submanifold X_0 we have the isomorphisms

$$\nu_f \simeq s^{-1}\nu_{X_0} \quad \text{and} \quad \mu_f \simeq s'^{-1}\mu_{X_0}$$

(with s' the section of $T^*_{X_0}X \to X_0$ given by df and $s = df^{-1}(1)$ for the function $df : T^*_{X_0}X \to \mathbb{R}$). Here ν_{X_0} and μ_{X_0} are the specialization and micro-localization functors as in [KS]. This follows for example from [KS, thm. 4.2.3.(v)] (compare with our discussion of the "micro local type" in chapter 5). Moreover, this shows the similarity to the nearby- and vanishing cycle functor for complex valued functions (compare with [KS, prop. 8.6.3]).

Let $f : X \to \mathbb{C}$ be a continuous function and consider the following cartesian diagram of maps induced by the inclusions $\{0\} \subset \mathbb{C} \supset \mathbb{C}^*$ and the universal covering $\widetilde{\mathbb{C}^*} \to \mathbb{C}^*$:

$$\begin{array}{ccccccc} X_0 & \xrightarrow{i} & X & \xleftarrow{j} & \{f \neq 0\} & \xleftarrow{p} & \widetilde{X^*} \\ \downarrow & & \downarrow f & & \downarrow & & \downarrow \tilde{f} \\ \{0\} & \longrightarrow & \mathbb{C} & \longleftarrow & \mathbb{C}^* & \longleftarrow & \widetilde{\mathbb{C}^*}. \end{array} \tag{1.8}$$

The **nearby cycle** ψ_f of f is the functor

$$\psi_f := i^* R\tilde{\pi}_* \tilde{\pi}^* : D(X) \to D(X_0) \,,$$

with $\tilde{\pi} := j \circ p$. The **vanishing cycle** ϕ_f of f is given as "the" cone of the natural morphism

$$sp : i^* \to \psi_f \,.$$

Let g be a generator of $\pi_1(\mathbb{C}^*) \simeq \mathbb{Z}$ (given by an orientation of $\mathbb{C}^*$) which we interpret as an automorphism of $\widetilde{\mathbb{C}^*}$. Then we get $\tilde{\pi} \circ a = \tilde{\pi}$ for the induced automorphism a of $\widetilde{X^*}$, so it acts on $R\tilde{\pi}_* \tilde{\pi}^*$ by

$$R\tilde{\pi}_* \tilde{\pi}^* \mathcal{F} \xrightarrow{ad_a} R\tilde{\pi}_*(Ra_* a^*)(\tilde{\pi}^* \mathcal{F}) \simeq R(\tilde{\pi} \circ a)_* (\tilde{\pi} \circ a)^* \mathcal{F} = R\tilde{\pi}_* \tilde{\pi}^* \mathcal{F}$$

and induces an automorphism m_f of ψ_f. Since

$$\begin{array}{ccc} R(\tilde{\pi} \circ a)_* (\tilde{\pi} \circ a)^* \mathcal{F} & \xrightarrow{\sim} & R\tilde{\pi}_*(Ra_* a^*)(\tilde{\pi}^* \mathcal{F}) \\ \uparrow {\scriptstyle ad_{\tilde{\pi} \circ a}} & & \uparrow {\scriptstyle ad_a} \\ \mathcal{F} & \xrightarrow{ad_{\tilde{\pi}}} & R\tilde{\pi}_* \tilde{\pi}^* \mathcal{F} \end{array}$$

commutes, we get the automorphism m_f of the distinguished triangle defining ϕ_f :

$$\begin{array}{ccccccc} i^*(\mathcal{F}) & \xrightarrow{sp} & \psi_f(\mathcal{F}) & \xrightarrow{can} & \phi_f(\mathcal{F}) & \xrightarrow{[1]} & \\ \downarrow {\scriptstyle id} & & \downarrow {\scriptstyle m_f} & & \downarrow {\scriptstyle m_f} & & \\ i^*(\mathcal{F}) & \xrightarrow{sp} & \psi_f(\mathcal{F}) & \xrightarrow{can} & \phi_f(\mathcal{F}) & \xrightarrow{[1]} & . \end{array} \tag{1.9}$$

But if we define $m_f : \phi_f(\mathcal{F}) \to \phi_f(\mathcal{F})$ in this way we have the usual problem of triangulated categories that this induced morphism of a cone is **non-canonical** so that it is in general not unique up to isomorphisms.

To deal with this problem (and to define ϕ_f really as a functor) we make the following **refinement** of the above construction. $\pi_1 := \pi_1(\mathbb{C}^*)$ acts already on $\tilde{\pi}^* \mathcal{F}$ which gets thereby the structure of a complex of equivariant sheaves. So $\tilde{\pi}_* \tilde{\pi}^* \mathcal{F}$ and $i^* \tilde{\pi}_* \tilde{\pi}^* \mathcal{F}$ are complexes of equivariant sheaves and this means complexes of sheaves of $R[\pi_1]$-modules (since π_1 acts trivially on the corresponding spaces, with $R[\pi_1] := R \otimes_Z \mathbb{Z}[\pi_1]$).

By using a flabby resolution one gets the derived functor

$$\tilde{\psi}_f := i^* R\tilde{\pi}_* \tilde{\pi}^* : D(X, R) \to D(X_0, R[\pi_1]) \,.$$

By construction we have $forget(\tilde{\psi}_f) = \psi_f$ for the exact forgetful functor

$$forget : D(X_0, R[\pi_1]) \to D(X_0, R) \,.$$

Moreover, this forgetful functor *forget* induces by the choice of a generator $1 \in \pi_1$ the automorphism m_f of ψ_f. The morphism

$$sp : i^* \to i^* R\tilde{\pi}_* \tilde{\pi}^*$$

is equivariant, if one uses the identity operation of π_1 on i^*. We define the functor

$$\tilde{\phi}_f : D(X, R) \to D(X_0, R[\pi_1])$$

as the cone of the morphism sp applied to (a suitable truncation of) the **canonical flabby resolution** of a complex $\mathcal{F}$. By construction we have

$$forget(\tilde{\phi}_f) = \phi_f$$

and the choice of a generator $1 \in \pi_1$ induces the automorphism m_f of ϕ_f (which is now unique up to isomorphisms).

Remark 1.1.2. We use the original definition of [De] for the vanishing cycle functor of f, in contrast to [KS, section 8.6], who use our $\phi_f[-1]$ as the vanishing cycle functor.

Now we give an important geometric interpretation of these functors. Let $f : X \to \mathbb{K}$ continuous (with $\mathbb{K} = \mathbb{R}, \mathbb{C}$), $S \subset X_0$ compact and $\mathcal{F} \in ob(D(X))$ be given. We make the following

Assumption 1.1.1 (cohomological version of a Milnor fibration). *There is a fundamental system $K_n \supset K_{n+1}$ of compact neighborhoods of S in X and a sequence of real numbers $\epsilon_n \searrow 0$ $(n \in \mathbb{N})$, with*

1. *The higher direct images sheaves $R^i f_*(\mathcal{F}|K_n^*)$ for the map*

$$f : K_n^* := K_n \cap \{|f| \leq \epsilon_n, f \neq 0\} \to \{w \in \mathbb{K}|\, 0 < |w| \leq \epsilon_n\}$$

are locally constant for all $i \in \mathbb{Z}$.

2. *$\forall n\ \exists \epsilon$ with $0 < \epsilon \leq \epsilon_{n+1}$ such that*

$$R\Gamma(K_n \cap \{f = w\}, \mathcal{F}) \to R\Gamma(K_{n+1} \cap \{f = w\}, \mathcal{F})$$

is an isomorphism for $|w| \leq \epsilon$.

We call $K_n \cap \{f = w\}$ a **Milnor fiber** of f $(0 < |w| \leq \epsilon_n)$. Of course for $\mathbb{K} = \mathbb{R}$ we have to distinguish between **left** and **right** Milnor fiber. Let also $K_{\epsilon_n} := K_n \cap \{|f| \leq \epsilon_n\}$.

For $\mathbb{K} = \mathbb{C}$ we introduce moreover the following notations. Let us denote the map $\pi : \widetilde{\mathbb{C}^*} \to \mathbb{C}$. For a subset $K \subset X$ (or $\mathbb{C}$) let $\tilde{K}$ be the inverse image under $\tilde{\pi}$ (or π). We identify $\widetilde{\mathbb{C}^*}$ with $\mathbb{C}$ by the exponential map. Then $\widetilde{D_r^*}$ is a halfspace for $D_r^* := \{z \in \mathbb{C}|\, 0 < |z| \leq r\}$ and $\widetilde{S_r}$ is a line for $S_r := \{z \in \mathbb{C}|\, |z| = r\}$. We call

$$\tilde{K}_n^{|w|} := \tilde{\pi}^{-1}(K_n^{|w|}) = \widetilde{K_n} \cap \{|\pi \circ \tilde{f}| = |w|\}$$

a **canonical Milnor fiber** (with $K_n^{|w|} := K_n \cap \{|f| = |w|\}$, and $0 < |w| \leq \epsilon_n$).

Lemma 1.1.1. *Under the assumption* 1.1.1, *there exists the following commutative diagrams and isomorphisms:*

$$\begin{array}{ccccc} R\Gamma(S,\mathcal{F}) & \xleftarrow{\sim} & R\Gamma(K_{\epsilon_n},\mathcal{F}) & \xrightarrow{\sim} & Rf_*(\mathcal{F}|K_n)_0 \\ \downarrow{\scriptstyle sp} & & \downarrow & & \downarrow{\scriptstyle sp} \\ R\Gamma(S,\nu_f(\mathcal{F})) & \xleftarrow{\sim} & R\Gamma(K_n\cap\{f=w\},\mathcal{F}) & \xrightarrow{\sim} & Rf_*(\mathcal{F}|K_n)_w\,, \end{array}$$

with $-\epsilon_n \le w < 0$ *in the case* $\mathbb{K}=\mathbb{R}$, *and similarly for* $\mathbb{K}=\mathbb{C}$:

$$\begin{array}{ccccc} R\Gamma(S,\mathcal{F}) & \xleftarrow{\sim} & R\Gamma(K_{\epsilon_n},\mathcal{F}) & \xrightarrow{\sim} & Rf_*(\mathcal{F}|K_n)_0 \\ \downarrow{\scriptstyle sp} & & \downarrow & & \downarrow{\scriptstyle sp} \\ R\Gamma(S,\tilde{\psi}_f(\mathcal{F})) & \xleftarrow{\sim} & R\Gamma\big(\tilde{K}_n^{|w|},\tilde{\pi}^*\mathcal{F}\big) & \xrightarrow{\sim} & R\Gamma\big(\widetilde{S_{|w|}}\,,\pi^*Rf_*(\mathcal{F}|K_n^{|w|})\big), \end{array}$$

with $0<|w|\le\epsilon_n$. *Especially, one has the commutative diagram*

$$\begin{array}{ccccc} R\Gamma(S,\psi_f(\mathcal{F})) & \xleftarrow{\sim} & R\Gamma(K_n\cap\{f=w\},\mathcal{F}) & \xrightarrow{\sim} & Rf_*(\mathcal{F}|K_n)_w \\ \downarrow{\scriptstyle m_f} & & \downarrow & & \downarrow{\scriptstyle mon} \\ R\Gamma(S,\psi_f(\mathcal{F})) & \xleftarrow{\sim} & R\Gamma(K_n\cap\{f=w\},\mathcal{F}) & \xrightarrow{\sim} & Rf_*(\mathcal{F}|K_n)_w\,, \end{array}$$

with $0<|w|\le\epsilon_n$ *and "mon" the induced* **monodromy** *action.*

Remark 1.1.3. 1. Note that lemma 1.1.1 implies:

$$\begin{array}{ccc} R\Gamma(S,\mu_f(\mathcal{F})) & \xleftarrow{\sim} & cone\big(R\Gamma(K_{\epsilon_n},\mathcal{F})\to R\Gamma(K_n\cap\{f=w\},\mathcal{F})\big)[-1] \\ \downarrow\wr & & \downarrow\wr \\ \mu_{id}\big(Rf_*(\mathcal{F}|K_n)\big)_0 & \xleftarrow{\sim} & cone\big(Rf_*(\mathcal{F}|K_n)_0 \overset{sp}{\to} Rf_*(\mathcal{F}|K_n)_w\big)[-1]\,, \end{array}$$

and

$$\begin{array}{ccc} R\Gamma(S,\tilde{\phi}_f(\mathcal{F})) & \xleftarrow{\sim} & cone\Big(R\Gamma(K_{\epsilon_n},\mathcal{F})\to R\Gamma\big(\tilde{K}_n^{|w|},\tilde{\pi}^*\mathcal{F}\big)\Big) \\ \downarrow\wr & & \downarrow\wr \\ \tilde{\phi}_{id}\big(Rf_*(\mathcal{F}|K_n)\big)_0 & \xleftarrow{\sim} & cone\Big(Rf_*(\mathcal{F}|K_n)_0 \overset{sp}{\to} R\Gamma\big(\widetilde{S_{|w|}},\pi^*Rf_*(\mathcal{F}|K_n^{|w|}\,)\big)\Big)\,. \end{array}$$

Especially, one gets the isomorphisms

$$\begin{array}{ccc} R\Gamma(S,\phi_f(\mathcal{F})) & \xleftarrow{\sim} & cone\big(R\Gamma(K_{\epsilon_n},\mathcal{F})\to R\Gamma(K_n\cap\{f=w\},\mathcal{F})\big) \\ \downarrow\wr & & \downarrow\wr \\ \phi_{id}\big(Rf_*(\mathcal{F}|K_n)\big)_0 & \xleftarrow{\sim} & cone\big(Rf_*(\mathcal{F}|K_n)_0 \overset{sp}{\to} Rf_*(\mathcal{F}|K_n)_w\big)\,. \end{array}$$

2. If K_n, ϵ_n ($n \in \mathbb{N}$) satisfy the assumption 1.1.1 for $\mathcal{F}, \mathcal{G} \in ob(D(X))$, then the above morphisms are **functorial** in $\mathcal{F}, \mathcal{G}$.

3. If $S' \subset S$ is a second compact subset, with a similar fundamental system $K'_n \subset K_n$ of compact neighborhoods such that assumption 1.1.1 is satisfied with the same ϵ_n ($n \in \mathbb{N}$), then the above morphisms **commute** with the induced **restriction morphisms**.

Before we begin the proof of the lemma, let us note the following fact (which we will often use without any comments):

(lc) Let $I \subset I' \subset \mathbb{R}$ be two compact intervals and let $\mathcal{F} \in ob(D(I'))$ have **locally constant** cohomology sheaves on $I' \backslash I$. Then the restriction morphism $R\Gamma(I', \mathcal{F}) \to R\Gamma(I, \mathcal{F})$ is an isomorphism.

Proof. We give the proof of the lemma in the following steps:

(1) The horizontal maps in the commutative diagram

$$\begin{array}{ccc} R\Gamma(K_{\epsilon_n}, \mathcal{F}) & \xrightarrow{\sim} & R\Gamma(K_n \cap \{f=0\}, \mathcal{F}) \\ \downarrow \wr & & \alpha \downarrow \wr \\ R\Gamma(\{x \in \mathbb{K} \,\big|\, |x| \leq \epsilon_n\}, Rf_*(\mathcal{F}|K_n)) & \xrightarrow{\sim} & Rf_*(\mathcal{F}|K_n)_0 \end{array}$$

are isomorphisms by the assumption 1.1.1(1) (α is an isomorphism since the restriction of f to the compact set K_n is proper). This implies the isomorphisms

$$\begin{array}{ccc} R\Gamma(K_{\epsilon_n}, \mathcal{F}) & \xrightarrow{\sim} & R\Gamma(K_{\epsilon_{n+1}}, \mathcal{F}) \\ \downarrow \wr & & \downarrow \wr \\ R\Gamma(K_n \cap \{f=0\}, \mathcal{F}) & \xrightarrow[\sim]{\beta} & R\Gamma(K_{n+1} \cap \{f=0\}, \mathcal{F}) \,, \end{array}$$

since β is an isomorphism by assumption 1.1.1(2) for $w = 0$ and

$$R\Gamma(K_{\epsilon_n}, \mathcal{F}) \xrightarrow{\sim} R\Gamma(S, \mathcal{F}) \,,$$

since the K_{ϵ_n} build a fundamental system of neighborhoods of S in X (see [KS, prop. 2.5.1, rem. 2.6.9]).

Moreover, the right morphism sp is defined by

$$Rf_*(\mathcal{F}|K_n)_0 \xleftarrow{\sim} R\Gamma(\{x \in \mathbb{K} \,\big|\, |x| \leq \epsilon_n\}, Rf_*(\mathcal{F}|K_n)) \longrightarrow Rf_*(\mathcal{F}|K_n)_w \,.$$

Let $U_n := \overset{\circ}{K}_n \cap \{|f| < \epsilon_n\}$ (with $\overset{\circ}{K}_n$ the interior of K_n) , $U_n^* := U_n \cap \{f < 0\}$ for $\mathbb{K} = \mathbb{R}$ and $U_n^* := U_n \cap \{f \neq 0\}$ for $\mathbb{K} = \mathbb{C}$. Assume now $\mathbb{K} = \mathbb{R}$.

(2) We get in the same way for $w \in [-\epsilon, 0[$ the isomorphisms

$$\begin{array}{ccc} R\Gamma(K_n \cap \{-\epsilon_n \le f < 0\}, \mathcal{F}) & \xrightarrow{\sim} & R\Gamma(K_n \cap \{f = w\}, \mathcal{F}) \\ \downarrow \wr & & \downarrow \wr \\ R\Gamma(K_{n+1} \cap \{-\epsilon_{n+1} \le f < 0\}, \mathcal{F}) & \xrightarrow{\sim} & R\Gamma(K_{n+1} \cap \{f = w\}, \mathcal{F}) \,, \end{array}$$

and $R\Gamma(K_n \cap \{f = w\}, \mathcal{F}) \xrightarrow{\sim} Rf_*(\mathcal{F}|K_n)_w$.

(3) Since U_n is open in X we have

$$R\Gamma(U_n, Rj_*j^*\mathcal{F}) \simeq R\Gamma(U_n^*, \mathcal{F}) \,.$$

So we get the commutative diagram

$$\begin{array}{ccccc} R\Gamma(S, \mathcal{F}) & \xleftarrow{\sim} & R\Gamma(K_{\epsilon_n}, \mathcal{F}) & \longrightarrow & R\Gamma(K_n \cap \{-\epsilon_n \le f < 0\}, \mathcal{F}) \\ \downarrow sp & & \downarrow & & \downarrow \\ R\Gamma(S, \nu_f(\mathcal{F})) & \longleftarrow & R\Gamma(U_n, Rj_*j^*\mathcal{F}) & \xleftarrow{\sim} & R\Gamma(U_n^*, \mathcal{F}) \,, \end{array}$$

and the induced map

$$R\Gamma(K_n \cap \{-\epsilon_n \le f < 0\}, \mathcal{F}) \to R\Gamma(S, \nu_f(\mathcal{F}))$$

is an isomorphism, since by (2) we get for $i \in \mathbb{Z}$:

$$\begin{array}{ccc} R^i\Gamma(K_n \cap \{-\epsilon_n \le f < 0\}, \mathcal{F}) & \xrightarrow{\sim} & \varinjlim R^i\Gamma(K_n \cap \{-\epsilon_n \le f < 0\}, \mathcal{F}) \\ \downarrow & & \downarrow \wr \\ R^i\Gamma(S, \nu_f(\mathcal{F})) & \xleftarrow{\sim} & \varinjlim R^i\Gamma(U_n, Rj_*j^*\mathcal{F}) \,. \end{array}$$

Note that U_n ($n \in \mathbb{N}$) is a fundamental system of open neighborhoods of S in X. Assume now $\mathbb{K} = \mathbb{C}$. Fix $\tilde{w} \in \widetilde{D^*_{\epsilon_n}}$, let $w := \pi(\tilde{w})$ and consider the cartesian diagram:

$$\begin{array}{ccc} (X, K_n^*) & \xleftarrow{\tilde{\pi}} & (\widetilde{X^*}, \widetilde{K_n^*}) \\ \downarrow f & & \downarrow \tilde{f} \\ (\mathbb{C}, D^*_{\epsilon_n}) & \xleftarrow{\pi} & (\widetilde{\mathbb{C}^*}, \widetilde{D^*_{\epsilon_n}}) \,. \end{array}$$

(2′) Since $f : K_n^* \to D^*_{\epsilon_n}$ is proper, the **proper base change theorem** ([KS, prop. 2.6.7]) implies $\pi^* Rf_*(\mathcal{F}|K_n^*) \simeq R\tilde{f}_*\tilde{\pi}^*(\mathcal{F}|K_n^*)$ and

$$R\Gamma(\widetilde{K_n^*} \cap \{\tilde{f} = \tilde{w}\}, \tilde{\pi}^*\mathcal{F}) \xrightarrow{\sim} \big(R\tilde{f}_*\tilde{\pi}^*(\mathcal{F}|K_n^*)\big)_{\tilde{w}} \xrightarrow{\sim} \cdots$$

$$\cdots \big(\pi^* Rf_*(\mathcal{F}|K_n^*)\big)_{\tilde{w}} \xleftarrow{\sim} Rf_*(\mathcal{F}|K_n^*)_w \xleftarrow{\sim} R\Gamma(K_n^* \cap \{f = w\}, \mathcal{F}) \,.$$

Therefore $R\tilde{f}_*\tilde{\pi}^*(\mathcal{F}|K_n^*)$ has (locally) constant cohomology sheaves on $\widetilde{D_{\epsilon_n}^*}$, with

$$R\Gamma\big(\widetilde{K_n^*}\cap\{\tilde{f}=\tilde{w}\},\tilde{\mathcal{F}}\big)\xrightarrow{\sim}R\Gamma\big(\widetilde{K_{n+1}^*}\cap\{\tilde{f}=\tilde{w}\},\tilde{\mathcal{F}}\big)$$

for $0<|\pi(\tilde{w})|\leq\epsilon$ and $\tilde{\mathcal{F}}:=\tilde{\pi}^*\mathcal{F}$. This implies as in (2):

$$\begin{array}{ccccc} R\Gamma\big(\widetilde{K_n^*},\tilde{\mathcal{F}}\big) & \xrightarrow{\sim} & R\Gamma\big(\tilde{K}_n^{|w|},\tilde{\mathcal{F}}\big) & \xrightarrow{\sim} & R\Gamma\big(\widetilde{K_n^*}\cap\{\tilde{f}=\tilde{w}\},\tilde{\mathcal{F}}\big) \\ \downarrow\wr & & \downarrow\wr & & \downarrow\wr \\ R\Gamma\big(\widetilde{K_{n+1}^*},\tilde{\mathcal{F}}\big) & \xrightarrow{\sim} & R\Gamma\big(\tilde{K}_{n+1}^{|w|},\tilde{\mathcal{F}}\big) & \xrightarrow{\sim} & R\Gamma\big(\widetilde{K_{n+1}^*}\cap\{\tilde{f}=\tilde{w}\},\tilde{\mathcal{F}}\big) \end{array}$$

(where we can consider the left square in $D(R[\pi_1]))$.

Moreover, we have in $D(R[\pi_1])$ the isomorphism

$$R\Gamma\big(\tilde{K}_n^{|w|},\tilde{\mathcal{F}}\big)\xrightarrow{\sim}R\Gamma\big(\widetilde{S_{|w|}},\pi^*Rf_*(\mathcal{F}|K_n^{|w|})\big)\,.$$

The automorphism m_f induced by the action of a generator of $\pi_1(S_{|w|})\simeq\pi_1(\mathbb{C}^*)$ corresponds by this isomorphism (after taking $forget$) to:

$$\begin{array}{ccccc} R\Gamma\big(\tilde{K}_n^{|w|},\tilde{\mathcal{F}}\big) & \xrightarrow{\sim} & R\Gamma\big(\widetilde{S_{|w|}},\pi^*Rf_*(\mathcal{F}|K_n^{|w|})\big) & \xrightarrow{\sim} & Rf_*(\mathcal{F}|K_n^*)_w \\ \downarrow m_f & & \downarrow m_f & & \downarrow \\ R\Gamma\big(\tilde{K}_n^{|w|},\tilde{\mathcal{F}}\big) & \xrightarrow{\sim} & R\Gamma\big(\widetilde{S_{|w|}},\pi^*Rf_*(\mathcal{F}|K_n^{|w|})\big) & \xrightarrow{\sim} & Rf_*(\mathcal{F}|K_n^*)_w\,. \end{array}$$

This automorphism is by definition the **monodromy** action on $Rf_*(\mathcal{F}|K_n^*)_w$.

(3′) Since U_n is open in X we have

$$R\Gamma(U_n,R\tilde{\pi}_*\tilde{\pi}^*\mathcal{F})\simeq R\Gamma(\widetilde{U_n^*},\tilde{\pi}^*\mathcal{F})\,.$$

So we get the commutative diagram

$$\begin{array}{ccccc} R\Gamma(S,\mathcal{F}) & \xleftarrow{\sim} & R\Gamma(K_{\epsilon_n},\mathcal{F}) & \longrightarrow & R\Gamma(\widetilde{K_n^*},\tilde{\pi}^*\mathcal{F}) \\ \downarrow sp & & \downarrow & & \downarrow \\ R\Gamma(S,\tilde{\psi}_f(\mathcal{F})) & \longleftarrow & R\Gamma(U_n,R\tilde{\pi}_*\tilde{\pi}^*\mathcal{F}) & \xleftarrow{\sim} & R\Gamma(\widetilde{U_n^*},\tilde{\pi}^*\mathcal{F})\,, \end{array}$$

and the induced map

$$R\Gamma(\widetilde{K_n^*},\tilde{\pi}^*\mathcal{F})\to R\Gamma(S,\tilde{\psi}_f(\mathcal{F}))$$

is an isomorphism, since by (2′) we get for $i\in\mathbb{Z}$:

$$\begin{array}{ccc} R^i\Gamma(\widetilde{K_n^*},\tilde{\pi}^*\mathcal{F}) & \xrightarrow{\sim} & \overset{\lim}{\to} R^i\Gamma(\widetilde{K_n^*},\tilde{\pi}^*\mathcal{F}) \\ \downarrow & & \downarrow\wr \\ R^i\Gamma(S,\tilde{\psi}_f(\mathcal{F})) & \xleftarrow{\sim} & \overset{\lim}{\to} R^i\Gamma(U_n,R\tilde{\pi}_*\tilde{\pi}^*\mathcal{F})\,. \end{array}$$

□

Remark 1.1.4. The reader should compare the above proof with [De] and [ST]. In the geometric version of a Milnor fibration, one usually assumes the condition:

1. $f : K_n^* \to \{w \in \mathbb{K} | \, 0 < |w| \leq \epsilon_n\}$ is a locally trivial topological fibration.

2. $\forall n \; \exists \epsilon$, with $0 < \epsilon \leq \epsilon_{n+1}$ such that the inclusion

$$K_{n+1} \cap \{f = w\} \to K_n \cap \{f = w\}$$

   is a homotopy equivalence for $0 < |w| \leq \epsilon$

(see for example [Ma, appendix]). Then the homotopy type of the Milnor fiber is well defined. If we similarly assume in the cohomological version instead of 2. only the condition:

(2′.) $\forall n \; \exists \epsilon$ with $0 < \epsilon \leq \epsilon_{n+1}$ such that

$$R\Gamma(K_n \cap \{f = w\}, \mathcal{F}) \to R\Gamma(K_{n+1} \cap \{f = w\}, \mathcal{F})$$

   is an isomorphism for $0 < |w| \leq \epsilon$,

then the proof of lemma 1.1.1 gives the same description of $R\Gamma(S, \nu_f(\mathcal{F}))$ or $R\Gamma(S, \psi_f(\mathcal{F}))$. Note, that for the isomorphism

$$\Gamma(S, \psi_f(\mathcal{F})) \xleftarrow{\sim} R\Gamma(K_n \cap \{f = w\}, \mathcal{F})$$

one has to choose a lift $\tilde{w}$ of w ! But the Thom-Sebastiani Theorem for sheaves is a result about

$$R\Gamma(S, \mu_f(\mathcal{F})) \quad \text{or} \quad R\Gamma(S, \phi_f[-1](\mathcal{F}))\,.$$

This corresponds in the classical case (1.3) to the use of reduced cohomology. In the proof of this theorem we use the isomorphisms of remark 1.1.3(1) so that it is important for us to make the assumptions as stated. Note that in the geometric case of a Milnor fibration with S equal to a point, one usually has a conic structure for K_{ϵ_n} or $K_n \cap \{f = 0\}$ (see for ex. [ST, p. 94] or [AVG, II p. 55]) !

The analogy between the real and complex valued case is not purely formal.

Corollary 1.1.1. *Let $f : X \to \mathbb{C}$, $S \subset X_0$ compact and $\mathcal{F} \in ob(D(X))$ be given such that the assumption 1.1.1 is true for K_n, ϵ_n $(n \in \mathbb{N})$. Then this is also true for $f' := re(f) : X \to \mathbb{R}$ (and $S, \mathcal{F}$), with $K_n' := K_{\epsilon_n}$ and $\epsilon_n' := 1/2 \cdot \epsilon_n$. All vertical maps in the commutative diagram*

$$\begin{array}{ccccc}
R\Gamma(K_n \cap \{f = 0\}, \mathcal{F}) & \longleftarrow & R\Gamma(K_{\epsilon_n}, \mathcal{F}) & \longrightarrow & R\Gamma(K_n \cap \{f = w\}, \mathcal{F}) \\
\uparrow\wr & & \downarrow\wr & & \uparrow\wr \\
R\Gamma(K_n' \cap \{f' = 0\}, \mathcal{F}) & \longleftarrow & R\Gamma(K_n'', \mathcal{F}) & \longrightarrow & R\Gamma(K_n' \cap \{f' = w'\}, \mathcal{F})
\end{array}$$

are isomorphisms (with $K''_n := K_{\epsilon_n} \cap \{|f'| \leq 1/2 \cdot \epsilon_n\}$, $0 < |w| \leq \epsilon_n, w' := re(w)$ *and* $0 < |w'| \leq 1/2 \cdot \epsilon_n$*). Especially*

$$R\Gamma(S, \nu_{f'}(\mathcal{F})) \simeq R\Gamma(S, \psi_f(\mathcal{F}))$$

and

$$R\Gamma(S, \mu_{f'}(\mathcal{F})) \simeq R\Gamma(S, \phi_f[-1](\mathcal{F}))\,.$$

Proof. Just look at $Rf_*(\mathcal{F}|K_{\epsilon_n})$ and the projection

$$\{|w| \leq \epsilon_n, |re(w)| \leq 1/2 \cdot \epsilon_n\} \to \{x \in \mathbb{R} \big| |x| \leq 1/2 \cdot \epsilon_n\}\,.$$

The details are left to the reader. □

1.1.2 Examples

Let us now explain some interesting examples, where the assumption 1.1.1 of the cohomological version of a Milnor fibration are satisfied.

Example 1.1.1 (global fibers of a proper map). *Suppose* $\mathcal{F} \in ob(D(X))$ *and* $f : X \to \{w \in \mathbb{K} \big| \, |w| < \delta\}$ *is a proper map such that* $R^i f_*(\mathcal{F}|X^*)$ *is locally constant for all* $i \in \mathbb{Z}$ *and*

$$f : X^* := X \cap \{f \neq 0\} \to \{w \in \mathbb{K} \big| \, 0 < |w| < \delta\}\,.$$

Take a sequence $\epsilon_n \searrow 0$*, and define* $K_n := \{|f| \leq \epsilon_n\}$*. Then assumption* 1.1.1 *is satisfied for* $S := \{f = 0\}$*, and*

$$R\Gamma(S, \nu_f(\mathcal{F})) \quad or \quad R\Gamma(S, \psi_f(\mathcal{F}))$$

describes the cohomology $R\Gamma(\{f = w\}, \mathcal{F})$ *of a* **global Milnor fiber** *of* f *with values in* $\mathcal{F}$*. Similarly,*

$$R\Gamma(S, \mu_f(\mathcal{F})) \quad or \quad R\Gamma(S, \phi_f[-1](\mathcal{F}))$$

describes the **relative** *cohomology* $R\Gamma(\{|f| \leq \epsilon_n\}, \{f = w\}; \mathcal{F})$ *of a* **"tube" modulo a generic global fiber**.

The assumption of example 1.1.1 is satisfied in the following cases:

1. $f : X^* \to \{w \in \mathbb{K} \big| \, 0 < |w| < \delta\}$ is a proper smooth submersion of smooth manifolds, and $\mathcal{F}|X^*$ has locally constant cohomology sheaves. This follows from the **Ehresmann fibration theorem** (as explained in chapter 4). For X smooth, this situation goes back to Picard and Lefschetz in the holomorphic context and to Morse in the case of classical Morse theory.

2. X is endowed with a stratification $X_\bullet$ (i.e. a filtration such that the $X_i \backslash X_{i-1}$ are smooth manifolds), $\mathcal{F}$ is constructible with respect to $X_\bullet$ and $f|X^*$ is a smooth submersion with respect to this stratification such that an **isotopy theorem of Thom** applies (e.g. a Whitney b-, Verdier w- or Bekka C-regular stratification). These notions will be discussed in chapter 4.

3. $X = V$ is the trivial vector bundle $B \times \mathbb{K}$ over a **compact** base space B, $\mathcal{F}$ is a **monodromic** complex of sheaves (with respect to the natural $\mathbb{K}^*$ action on V), and f is the projection onto $\mathbb{K}$ (compare Brylinski [Bry, VI]).

4. $X = V$ is a real vector bundle over a **compact** base B, $\mathcal{F}$ monodromic as before, and $-f = r : V \to \mathbb{R}^{\geq 0}$ is a **norm function** (on the fibers). Then one gets
$$R\Gamma(B, \mu_f(\mathcal{F})) = R\Gamma_c(V, \mathcal{F}) .$$

We included here the case of **monodromic** sheaves only for showing that sometimes the assumption 1.1.1 and therefore also our results apply to **more general sheaves** than just constructible ones.

Let us now be more precise about the **isotopy theorem** in (2). Usually X is a (locally) closed subset of a smooth manifold and $f : M \to N$ a smooth map of manifolds, whose restriction to X is proper. By integrating suitable vector fields, one shows that
$$f : (X, X_\bullet) \to N$$
is a **locally trivial fibration in a stratum preserving way**. See Mather [Mather] (or Shiota [Shiota, I.1]) for the case of Whitney stratifications, Bekka [Bekka1] for the case of C-regular stratifications and Verdier [Ver] for w-regular stratifications. Note that the last case also works in a non-embedded situation in the complex analytic or subanalytic context.

In chapter 4 of this book we give a purely sheaf theoretic proof (without using the technique of integrating vector fields) of the corresponding **cohomological result** that the higher direct image sheaves are locally constant. This is based on the **non-characteristic deformation lemma** of Kashiwara ([KS, prop. 2.7.2]).

The existence of such stratifications follows for example from

- [Hiro, Te, Ver] in the complex analytic (algebraic) situation.
- [Hiro, Ver, GM] in the subanalytic context.
- [Shiota, vDrMi, Ta1] in the context of o-minimal structures on the real field and analytic (Nash) geometric categories (these notions will be explained in chapter 2).

An interesting case of example 1.1.1 is given, if $f : X \to [0, \delta[$ is nonnegative so that f "defines" the compact subset $X_0 = \{f = 0\}$ and the discussion above applies. For the existence of such a smooth subanalytic function for a given compact subanalytic subset X_0 (in the more general context of analytic geometric categories) see [vDrMi]. By a **curve selection lemma** such a function has no critical values with respect to a subanalytic stratification in $]0, \delta[$ for small δ. Then
$$R\Gamma(X_0, \nu_{-f}(\mathcal{F}))$$

is the **global link cohomology** studied in [DS] and ν_{-f} is the **link cohomology functor** of [DS]. Compare also with [McPa] for applications of these functors (in relation to suitable **Milnor fibrations**) to the topology of real algebraic and analytic sets !

The case (2.) of the example applies also to the **non-proper** map

$$g := f : U \to \mathbb{K}\,, \quad \text{with } j : U \to X$$

the inclusion of an **open union of strata** of $X_\bullet$, and $\mathcal{F} \in ob(D(U))$ constructible with respect to the induced stratification. Then $Rj_*\mathcal{F}$ is also constructible (compare chapter 4), and

$$R\Gamma(\ \{f=0\}, \mu_f(Rj_*\mathcal{F})\ ) \quad \text{or} \quad R\Gamma(\ \{f=o\}, \phi_f[-1](Rj_*\mathcal{F})\ )$$

describes the **relative** cohomology

$$R\Gamma(\{|f| \leq \epsilon_n\}, \{f=w\}; Rj_*\mathcal{F}) \ \simeq \ R\Gamma(\{|g| \leq \epsilon_n\}, \{g=w\}; \mathcal{F})$$

of a "tube" modulo a global generic fiber **in the open subset** U ! For the last isomorphism one can use a suitable **base change isomorphism** coming from the "transversality condition" that 0 is the only critical value of f near the special fiber $\{f=0\}$ (compare chapter 4). An important special case is given by a **polynomial** function

$$g : U = \mathbb{K}^n \to \mathbb{K}\,,$$

with $\mathcal{F}$ a (locally) constant sheaf. Compare with [DiNe, thm. 4.2] for a similar Thom-Sebastiani result in this context.

Example 1.1.2 (isolated singularities). *Let $X_\bullet$ be a stratification as above (e.g. Whitney b, Verdier w, Bekka C), with $X \subset M$ (locally) closed and $f : M \to \mathbb{K}$ smooth. Suppose that f has an isolated singularity $\{x\}$ with respect to this stratification (with $f(x)=0$) such that $X^0 := \{f=0\}\backslash\{x\}$ with its induced stratification $X^0_\bullet$ is Whitney b regular over x. Define in local coordinates*

$$(M,x) \simeq (\mathbb{R}^n, 0)$$

the distance function r by $r(x) := \sum_{i=1}^n x_i^2$. Let $\mathcal{F}$ be constructible with respect to $X_\bullet$. If we take sequences $\epsilon_n, \delta_n \searrow 0$ with $0 < \epsilon_n << \delta_n << 1$ and define

$$K_n := \{|f| \leq \epsilon_n\ ,\ r \leq \delta_n\}\,,$$

then assumption 1.1.1 *is satisfied for $S := \{x\}$ and we get a cohomological version of a* **local** *Milnor fibration.*

The assumption of example 1.1.2 is satisfied in the following cases:

1. The classical case that X is a smooth manifold (with its trivial stratification $\emptyset \subset X$, so $\mathcal{F}$ has locally constant cohomology sheaves) and $f : X \to \mathbb{K}$ has an isolated critical point, which is "reasonable" in the sense that $\{f = 0\}\backslash\{x\}$ is Whitney b regular over x. For example f is subanalytic (in local coordinates) or x is **nondepraved** ($\mathbb{K} = \mathbb{R}$) in the sense of Goresky-MacPherson [GM, Part I, def. 2.3,lemma 2.5.1]. This includes our example of classical Morse theory with
$$f : (\mathbb{R}^n, 0) \to (\mathbb{R}, 0)\ ,\ f(x) := c \cdot \sum_{i=1}^{n} x_i^2 \quad (c \in \{-1, 1\})\,.$$
Then we get for $0 < \epsilon_n << \delta_n << 1$: $\{f = -\epsilon_n, r \leq \delta_n\}$ is empty ($c = 1$) or the n-sphere ($c = -1$).

2. X, the stratification and f are subanalytic (or more generally, are defined in an o-minimal structure on the real field or an analytic (Nash) geometric category). This case will be explained more directly in chapter 2.

3. f is **normally nondegenerate** ($\mathbb{K} = \mathbb{R}$) in the sense of [GM, Part I, def. 2.1(c)], the restriction of f to the stratum containing x has this as an isolated and "reasonable" critical point, and the stratification $X_\bullet$ is Whitney b regular over the point x (compare [GM, Part I, proof of lemma 6.2]).

4. $\{x\}$ is a stratum, $X_\bullet$ is Whitney b regular over the point x and
$$f = (f_1, f_2) : X \to \mathbb{C} = \mathbb{R}^2$$
has in x a **nondegenerate** critical point in the sense of [GM, Part I, def. 12.1] (see [GM, Part I, proof of lemma 12.2]).

Let us now explain why in example 1.1.2 the assumption 1.1.1 holds. Since $X^0_\bullet$ is Whitney b regular over x, there exists $\delta > 0$ such that r has with respect to $X^0_\bullet$ no critical values in $]0, \delta[$. Otherwise there would be a sequence b_i in some $X^0_k\backslash X^0_{k-1} =: B$ converging to x, with
$$T_{b_i}B \to \tau\ ,\quad \overline{b_i, x} \to l\ ,\quad \text{and } dr|T_{b_i}B \equiv 0\,.$$
Here $l_i := \overline{b_i, x}$ is the corresponding **secant line**, which is therefore perpendicular to $T_{b_i}B \subset ker(dr(b_i))$. So one gets $l \perp \tau$ in contradiction to the **Whitney b condition** $l \subset \tau$.

Take a sequence $\delta_n \searrow 0$ with $0 < \delta_n < \delta$. Since $X_\bullet$ is at least Whitney a regular, there is a sequence $\epsilon_n \searrow 0$ such that (r, f) has no critical points in
$$\{r = \delta_n,\ 0 < |f| \leq \epsilon_n\}\,.$$
Otherwise there would be a sequence b_i in some $X_k\backslash X_{k-1} =: B$ converging to $a \in A := X_j\backslash X_{j-1}$, with
$$T_{b_i}B \to \tau\ ,\quad r(b_i) = \delta_n\ ,\quad \text{and } df|T_{b_i}B \equiv 0\,.$$

But $T_aA \subset \tau$ by the **Whitney a condition**. Therefore one gets $df|T_aA \equiv 0$, contrary to the assumption that f has an **isolated** singularity.

Therefore $K'_n := X \cap \{r \leq \delta_n\}$ gets an induced stratification such that

$$f : K'_n \cap \{0 < |f| \leq \epsilon_n\} \to \{w \in \mathbb{K}|\ 0 < |w| \leq \epsilon_n\}$$

is a proper stratified submersion. Then property 1.1.1(1) follows from (the cohomological version of) the fibration theorem. Similarly $X \cap \{f = 0\}$ gets an induced stratification such that

$$r : X \cap \{f = 0\} \to \mathbb{R}$$

is a proper stratified submersion over $]0, \delta_n]$ and this implies the property 1.1.1(2) for $w = 0$. By the same argument as before, there exists $1 >> \epsilon > 0$ such that (r, f) has no critical values in

$$[\delta_n, \delta_{n-1}] \times \{0 < |f| \leq \epsilon\}\,.$$

So $X \cap \{f = w\}$ gets an induced stratification such that $r : X \cap \{f = w\} \to \mathbb{R}$ is a proper stratified submersion over $[\delta_n, \delta_{n-1}]$, and this implies as before the property 1.1.1(2) for $0 < |w| \leq \epsilon$ (the above argument shows the relation between the assumption 1.1.1 and the notion of a "fringed subset of type $0 < y << x$" used by Goresky-MacPherson [GM, Part I, def. 5.1, lem. 5.2, 6.2, 12.2]).

Example 1.1.3 (nonisolated singularities with a_f-condition). *Consider a smooth function $f : (X, X_0) \to (\mathbb{K}, 0)$. Let $X_\bullet$ be a stratification as above (Whitney b, Verdier w, Bekka C), adapted to X_0 (i.e. X_0 is a union of the strata := connected components of the $X_i \backslash X_{i-1}$), that satisfies the following a_f-***condition of Thom***:*

1. *f restricted to $X \backslash X_0$ is a stratified submersion.*
2. *Let b_i be a sequence of points in a stratum B of $X \backslash X_0$ converging to some point a in a stratum A of X_0 such that $ker(df(b_i)|T_{b_i}B)$ converges to some limiting plane τ, then $\tau \supset T_aA$.*

Moreover let $S \subset X_0$ be compact and $r : X \to \mathbb{R}^{\geq 0}$ continuous with $S = \{r = 0\}$ such that r restricted to $X \backslash S$ is smooth and a stratified submersion on

$$X_0 \cap \{0 < r < \delta\}$$

for some $\delta > 0$. If we take sequences $\epsilon_n, \delta_n \searrow 0$ with $0 < \epsilon_n << \delta_n < \delta$ and define

$$K_n := \{|f| \leq \epsilon_n\ ,\ r \leq \delta_n\} \cap U$$

for a suitable open neighborhood U of S, then assumption 1.1.1 is satisfied for $\mathcal{F}$ constructible with respect to $X_\bullet$.

The proof is the same as before, since by the a_f-**condition** there exists $1 >> \epsilon > 0$ such that $(r, f)|U$ has no critical values in

$$[\delta_n, \delta_{n-1}] \times \{0 < |f| \leq \epsilon\}\,.$$

Otherwise there would be a sequence $b_i \to a$ as in (2.), with

$$r(a) \in [\delta_n, \delta_{n-1}] \ , \ ker(df(b_i)|T_{b_i}B) \to \tau$$

and

$$dr\big|ker(df(b_i)|T_{b_i}B) \equiv 0 \ .$$

By the a_f-condition and the assumptions on r, one gets the contradiction $dr|T_aA = 0$ (compare also with [GM, proof of lemma 2.A.2, p. 171] and [Le2]).

Note that

$$r : U \cap \{r < \delta\} \to [0, \delta[$$

is proper for small δ and a suitable U, since $S = \{r = 0\}$ is compact.

The assumption of example 1.1.3 is satisfied in the following cases:

1. X is a complex space, the stratification is complex analytic, S is a complex subset and $f : X \to \mathbb{C}$ is holomorphic. For the existence of an a_f-stratification (refining a given complex analytic stratification) see [Hiro, Cor. 1 to thm. 2, p. 248], [LeTe, cor. 1.3.5.1] and compare also with [KS, prop. 8.5.9]. For the existence of a smooth subanalytic function

$$r : X \to [0, \infty[$$

defining S (i.e. with $S = \{r = 0\}$) see [vDrMi, 1.20, D.19]. Since S is compact, we can realize a neighborhood of S in X as a real analytic subspace of a bounded open set $U \subset \mathbb{R}^N (N >> 0)$ so that $S \subset U \subset \mathbb{R}^N$ is **globally subanalytic** and we also can apply [vDrMi, thm. C.11]).

2. X is a real analytic space, the stratification is subanalytic, S is a subanalytic subset and $f : X \to \mathbb{R}$ is continuous subanalytic. After taking the graph embedding we can assume that f is also smooth. For the existence of an a_f-stratification (refining a given subanalytic stratification) see [Bekka2, Ta1, Ta2]. The existence of r follows as before.

3. $X = \mathbb{R}^{2n}$, the stratification is given by a central real subspace arrangement with even dimensional flats (see [GM, part III]), the $\mathbb{R}$-linear function

$$f = (f_1, f_2) : X \to \mathbb{C} = \mathbb{R}^2$$

defines a subspace $\{f = 0\}$ of this arrangement, we take $S = \{0\}$ and r is the distance function given by $r(x) := \sum_{i=1}^n x_i^2$.

Remark 1.1.5. By the examples and references above, the constructible complexes of sheaves in theorem 1.0.1 and theorem 1.0.3 satisfy our assumption 1.1.1 so that we can apply our **main theorem**. Moreover theorem 1.0.3 holds also in the o-minimal or analytic (Nash) geometric context and the assumption 1.1.1 can in this context already be deduced from the **cell decomposition theorem** for o-minimal structures (as explained in chapter 2 of this book). The "real case" of our results can therefore be proved without stratification techniques !

1.2 Thom-Sebastiani Theorem

For applications to **stratified Morse theory** in chapter 5 of this book, we prove first some preliminary results in a slightly more general form than needed for the Thom-Sebastiani theorem.

1.2.1 Preliminaries and Thom-Sebastiani for additive functions

As remarked at the end of the introduction, we start with the following simple two-dimensional picture (see Figure 1.1 on page 80).

Lemma 1.2.1. *Let $a : \mathbb{K}^2 \to \mathbb{K}$ be the addition map $a(x, y) := x + y$,*

$$B := B(\epsilon') := \{(x, y) \in \mathbb{K}^2 |\, |x| \leq 2\epsilon', |y| \leq 2\epsilon'\} \quad (\epsilon' > 0)$$

and $\mathcal{G} \in ob(D(B))$ be constructible with respect to the filtration

$$B_\bullet := B(\epsilon')_\bullet : \{0\} \subset B \cap \{xy = 0\} \subset B\,.$$

Then we have for $0 < \epsilon < \epsilon'$:

1. *$R^i a_*(\mathcal{G}|B \cap \{0 < |a| \leq \epsilon\})$ is locally constant on $\{0 < |w| \leq \epsilon\}$ for all $i \in \mathbb{Z}$.*

2. *All maps in the commutative diagram*

$$\begin{array}{ccc} R\Gamma(B \cap \{a = 0\}, \mathcal{G}) & \xleftarrow{\sim} & R\Gamma(B \cap \{|a| \leq \epsilon\}, \mathcal{G}) \\ \downarrow \wr & & \uparrow \wr \\ R\Gamma(\{(0,0)\}, \mathcal{G}) & \xleftarrow{\sim} & R\Gamma(B, \mathcal{G}) \end{array}$$

   are isomorphisms.

3. *The natural restriction map*

$$R\Gamma(B \cap \{a = w\}, \mathcal{G}) \xrightarrow{\sim} R\Gamma([(w, 0), (0, w)], \mathcal{G})$$

   is an isomorphism for $|w| \leq \epsilon$. Here $[(w, 0), (0, w)]$ is by definition the line segment $\{(tw, (1-t)w)|\ t \in [0, 1]\}$.

4. *For $0 < |w| \leq \epsilon$, there exists a distinguished triangle*

$$(Ra_*\mathcal{G})_w \longrightarrow \mathcal{G}_{(w,0)} \oplus \mathcal{G}_{(0,w)} \longrightarrow \mathcal{G}_{(w/2,w/2)} \xrightarrow{[1]}\ , \qquad (1.10)$$

   which for $\mathbb{K} = \mathbb{C}$ is functorial in the canonical monodromy action.

Proof. The reader should first look at Figure 1.1 on page 80.
Note that by multiplication of the coordinates we have a natural action of

$$D := \{\lambda \in \mathbb{K} |\ 0 < |\lambda| \leq 1\}$$

on B (respecting the filtation $B_\bullet$). Fix $w \in \mathbb{K}$ with $0 < |w| = \epsilon < \epsilon'$. Then one has the homeomorphism

$$h : B \cap \{a = w\} \times D \to B \cap \{0 < |a| \leq \epsilon\},\ h(x, y, \lambda) := (\lambda x, \lambda y)\ .$$

Let

$$p : B \cap \{a = w\} \times D \to B \cap \{a = w\}$$

be the first projection and let

$$q : B \cap \{a = w\} \times D \to \{0 < |z| \leq \epsilon\}$$

be the second projection multiplied by w. Then

$$q \circ h^{-1} = a : B \cap \{0 < |a| \leq \epsilon\} \to \{0 < |z| \leq \epsilon\}$$

so that a is a trivial fibration (in a filtration preserving way).

1. Fix a **compact contractible** subset $K \subset \{0 < |z| \leq \epsilon\}$, let $K' := q^{-1}(K)$ and consider the cartesian diagram

$$\begin{array}{ccc} B \cap \{a = w\} & \xleftarrow{p} & K' \\ \downarrow{\scriptstyle q'} & & \downarrow{\scriptstyle q} \\ \{pt\} & \xleftarrow{p'} & K\ . \end{array}$$

Since the cohomology sheaves of $h^*\mathcal{G}$ are locally constant on the compact contractible fibers of this p, one has by the **Vietoris-Begle theorem**

$$h^*\mathcal{G}|K' =: \mathcal{G}' \xrightarrow{\sim} p^* Rp_*\mathcal{G}'$$

(see [KS, cor. 2.7.7(iv)]). By the **proper base change theorem** ([KS, prop. 2.6.7]) we finally get

$$Ra_*(\mathcal{G}|B \cap a^{-1}(K)) \simeq Rq_*\mathcal{G}' \simeq Rq_*p^*\mathcal{G}'' \simeq p'^* Rq'_*\mathcal{G}''$$

(with $\mathcal{G}'' := Rp_*\mathcal{G}'$). So $Ra_*\mathcal{G}$ has constant cohomology sheaves on K and this implies (1.).

2. Moreover, (1.) gives

$$R\Gamma(B \cap \{a = 0\}, \mathcal{G}) \xleftarrow{\sim} R\Gamma(B \cap \{|a| \leq \epsilon\}, \mathcal{G})$$

and the isomorphism

$$R\Gamma(B \cap \{a=0\}, \mathcal{G}) \xrightarrow{\sim} R\Gamma(\{(0,0)\}, \mathcal{G})$$

is obvious. So (3.) is true for $w = 0$. Similarly one gets

$$R\Gamma(B, \mathcal{G}) \xrightarrow{\sim} R\Gamma(\{(0,0)\}, \mathcal{G})$$

by looking at the direct image sheaves under the projection

$$B \to \{|x| \leq 2\epsilon'\}$$

onto the first coordinate.

3. Similarly, (3.) for $w \neq 0$ follows for $\mathbb{K} = \mathbb{R}$ from our condition (lc) on page 28. For the case $\mathbb{K} = \mathbb{C}$, we look again at the projection pr onto the first coordinate

$$(B \cap \{a = w\}, [(w,0),(0,w)]) \xrightarrow{\sim} (A, [w,0])$$

with $A := \{|x| \leq 2\epsilon'\} \cap \{|x-w| \leq 2\epsilon'\}$. Then we have

$$R\Gamma(A \cap \{x = tw|\ t \in \mathbb{R}\}, Rpr_*\mathcal{G}) \xrightarrow{\sim} R\Gamma([w,0], Rpr_*\mathcal{G})$$

again by condition (lc) and

$$R\Gamma(A, Rpr_*\mathcal{G}) \xrightarrow{\sim} R\Gamma(A \cap \{x = tw|\ t \in \mathbb{R}\}, Rpr_*\mathcal{G}) ,$$

by looking at the direct image sheaves under the orthogonal projection onto $\{x = tw|\ t \in \mathbb{R}\}$ and applying (lc) to the fibers (see Figure 1.2 on page 80).

4. By (3.) we have

$$(Rq_*\mathcal{G})_w \simeq R\Gamma(B \cap \{a = w\}, \mathcal{G}) \simeq R\Gamma([(w,0),(0,w)], \mathcal{G}) .$$

Consider the inclusions

$$\{(w,0),(0,w)\} \overset{i}{\hookrightarrow} [(w,0),(0,w)] \overset{j}{\hookleftarrow}](w,0),(0,w)[,$$

with $](w,0),(0,w)[:= \{(tw,(1-t)w)|\ t \in]0,1[\}$, and

$$\{(w/2,w/2)\} \overset{i'}{\hookrightarrow} [(w,0),(0,w)] .$$

Applying $R\Gamma([(w,0),(0,w)], \cdot)$ to the distinguished triangle

$$Rj_!j^*\mathcal{G}' \to \mathcal{G}' \to Ri_*i^*\mathcal{G}' \xrightarrow{[1]}$$

(with $\mathcal{G}' := \mathcal{G}|[(w,0),(0,w)]$) gives us the triangle (1.10), since

$$R\Gamma([(w,0),(0,w)], Rj_!j^*\mathcal{G}') \simeq R\Gamma_c(](w,0),(0,w)[, j^*\mathcal{G}') \simeq \mathcal{G}'_{(w/2,w/2)}[-1]\,.$$

The last isomorphism is well known (see [IC, V, prop. 3.7]). We deduce it here from the **projection formula** (see appendix): Let

$$p' :](w,0),(0,w)[\to \{pt\}$$

be the constant map. Since $\mathcal{G}'|](w,0),(0,w)[=: \mathcal{G}''$ has (locally) constant cohomology sheaves, we get by a **second version of the Vietoris-Begle theorem** ([KS, prop. 2.7.8])

$$\mathcal{G}'' \xrightarrow{\sim} p'^*Rp'_*\mathcal{G}''\,.$$

Applying i'^* gives $i'^*\mathcal{G}'' \simeq i'^*p'^*Rp'_*\mathcal{G}'' \simeq Rp'_*\mathcal{G}''$. Then the **projection formula** implies

$$R\Gamma_c(](w,0),(0,w)[,\mathcal{G}'') \simeq R\Gamma_c(](w,0),(0,w)[, p'^*i'^*\mathcal{G}'') \simeq$$

$$R\Gamma_c(](w,0),(0,w)[,\mathbb{Z}) \otimes^L i'^*\mathcal{G}'' \simeq \mathbb{Z}[-1] \otimes^L i'^*\mathcal{G}''\,.$$

5. For $\mathbb{K} = \mathbb{C}$ we deduce the **monodromy** statement by a relative version of the argument above. First remark that we have a canonical monodromy action on $(Ra_*\mathcal{G})_w$, since by (1.)

$$(Ra_*\mathcal{G})|\{|z| = |w|\}$$

has locally constant cohomology sheaves. Moreover we have by (3.):

$$Ra_*(\mathcal{G}|B \cap \{|z| = |w|\}) \simeq Rq_*(h^*\mathcal{G}|B \cap \{a = w\} \times \{|\lambda| = 1\}) \xrightarrow{\sim}$$

$$Rq_*(h^*\mathcal{G}|[(w,0),(0,w)] \times \{|\lambda| = 1\}).$$

Since $h^*\mathcal{G}$ has locally constant cohomology sheaves on

$$\{(x,y)\} \times \{|\lambda| = 1\} \quad (x + y = w)\,,$$

we also have a canonical monodromy action on the other terms in the triangle (1.10). Consider the inclusions $\{(w,0),(0,w)\} \times \{|\lambda| = 1\} \overset{i}{\hookrightarrow} \cdots$

$$\cdots \to [(w,0),(0,w)] \times \{|\lambda| = 1\} \overset{j}{\hookleftarrow}](w,0),(0,w)[\, \times \{|\lambda| = 1\}$$

and

$$\{(w/2,w/2)\} \times \{|\lambda| = 1\} \overset{i'}{\hookrightarrow} [(w,0),(0,w)] \times \{|\lambda| = 1\}\,.$$

Applying Rq_* to the distinguished triangle

$$Rj_!j^*\mathcal{G}' \to \mathcal{G}' \to Ri_*i^*\mathcal{G}' \xrightarrow{[1]}\,,$$

with

$$\mathcal{G}' := h^*\mathcal{G}|(\ [(w,0),(0,w)] \times \{|\lambda| = 1\}\ )\,,$$

gives a triangle of complexes of sheaves on $\{|z| = |w|\}$, whose cohomology sheaves are locally constant, and whose pullback under

$$i_w : \{w\} \hookrightarrow \{|z| = |w|\}$$

gives the triangle (1.10). It remains only to show that the above application of the projection formula commutes with the monodromy action: Let

$$q' := q \circ j,\ p' :](w,0),(0,w)[\ \times\{|\lambda| = 1\} \to (w/2, w/2) \times \{|\lambda| = 1\}$$

be induced by the second projection and let

$$\kappa : (w/2, w/2) \times \{|\lambda| = 1\} \to \{|z| = |w|\}$$

be the isomorphism $\kappa(w/2, w/2, \lambda) := w\lambda$, with $q' = \kappa \circ p'$. Since

$$\mathcal{G}'|(\ ](w,0),(0,w)[\times\{|\lambda| = 1\}\ ) =: \mathcal{G}''$$

has locally constant cohomology sheaves, we get as before by the **Vietoris-Begle theorem** ([KS, prop. 2.7.8]):

$$\mathcal{G}'' \xrightarrow{\sim} p'^* Rp'_* \mathcal{G}'' \simeq p'^* i'^* \mathcal{G}'' \simeq q'^* \tilde{\mathcal{G}}$$

(with $\tilde{\mathcal{G}} := \kappa_* i'^* \mathcal{G}''$). Then the **projection formula** implies

$$Rq_* Rj_! j^* \mathcal{G}' \simeq Rq'_! \mathcal{G}'' \simeq Rq'_! q'^* \tilde{\mathcal{G}} \simeq (Rq'_! \mathbb{Z}) \otimes^L \tilde{\mathcal{G}} \simeq \tilde{\mathcal{G}}[-1]\,.$$

The last isomorphism $(Rq'_!\mathbb{Z}) \otimes^L \tilde{\mathcal{G}} \simeq \tilde{\mathcal{G}}[-1]$ follows from the **base change isomorphism** for the proper direct image functor ([KS, prop. 2.6.7]), applied to the cartesian diagram:

$$\begin{array}{ccc}](w,0),(0,w)[\ \times\{|\lambda| = 1\} & \xrightarrow{r'} &](w,0),(0,w)[\\ \downarrow{\scriptstyle q'} & & \downarrow{\scriptstyle q''} \\ \{|z| = |w|\} & \xrightarrow{r} & \{pt\}, \end{array}$$

with

$$Rq'_!\mathbb{Z} \simeq Rq'_! r'^* \mathbb{Z} \simeq r^* Rq''_! \mathbb{Z} \simeq r^* \mathbb{Z}[-1]\,.$$

□

Remark 1.2.1. 1. We proved in (1.) the following result: Let

$$p : K' := Y \times K \to Y\ ,\ q : K' \to K$$

be the projections, with K **compact contractible**, and let $\mathcal{F} \in ob(D(K'))$ have (locally) constant cohomology sheaves on the fibers of p. Then $Rq_!\mathcal{F}$ (with $Rq_! = Rq_*$ for Y compact) has constant cohomology sheaves (compare with [IC, V, lemma 3.8]). Especially, this explains why in the examples of section 1.1 the **Thom isotopy theorem** implies the assumption 1.1.1(1).

2. The proof of (3.-4.) is just an abstraction of the arguments of Sebastiani-Thom [ST]. The last part of the proof of (3.) is our substitute for a retraction argument in the proof of the **classical** Thom-Sebastiani Theorem (see [ST, p. 94] and [AVG, II p. 56]).

3. In the same way one can show the following generalization of condition (lc): Let $K \subset K' \subset \mathbb{R}^{n+1}$ be **compact** and **convex**. Let $l : \mathbb{R}^{n+1} \to \mathbb{R}^n$ be a linear surjection with $l(K) = l(K')$. Let $\mathcal{F} \in ob(D(K'))$ have locally constant cohomology sheaves on $(K' \backslash K) \cap \{l = x\}$ for all $x \in l(K)$. Then the restriction morphism $R\Gamma(K', \mathcal{F}) \to R\Gamma(K, \mathcal{F})$ is an isomorphism.

Lemma 1.2.2. *Let $g : X \to \mathbb{K}^2$ continuous, $S \subset \{g = 0\}$ compact and $\mathcal{F} \in ob(D(X))$ be given. We make the following assumption: There is a fundamental system $K_n \supset K_{n+1}$ of compact neighborhoods of S in X and a sequence of real numbers $\epsilon_n \searrow 0$ $(n \in \mathbb{N})$ with*

1. *$Rg_*\big(\mathcal{F}|K_n \cap g^{-1}(B(\epsilon_n))\big)$ is constructible with respect to $B(\epsilon_n)_\bullet$.*

2. *$\forall n\ \exists \epsilon$ with $0 < \epsilon < \epsilon_{n+1}$ such that*

$$R\Gamma(K_n \cap \{g = (x_1, x_2)\}, \mathcal{F}) \to R\Gamma(K_{n+1} \cap \{g = (x_1, x_2)\}, \mathcal{F})$$

is an isomorphism for $(x_1, x_2) \in B(\epsilon)$.

Then $a \circ g : X \to \mathbb{K}$, $S \subset \{a \circ g = 0\}$ and $\mathcal{F}$ satisfy the assumption 1.1.1 *with*

$$K'_n := K_n \cap g^{-1}(B(\epsilon_n))$$

and $\epsilon'_n \searrow 0$ any sequence with $\epsilon'_n < \epsilon_n$ $(n \in \mathbb{N})$.

Proof. Property (1) of assumption 1.1.1 follows by $R(a \circ g)_* = Ra_* \circ Rg_*$ from lemma 1.2.1(1). Property (2) of assumption 1.1.1 follows from

$$R\Gamma(K'_n \cap \{a \circ g = w\}, \mathcal{F}) \xrightarrow{\sim} R\Gamma(B(\epsilon_n) \cap \{a = w\}, \mathcal{G}_n) ,$$

with $\mathcal{G}_n := Rg_*\big(\mathcal{F}|K_n \cap g^{-1}(B(\epsilon_n))\big)$ and lemma 1.2.1(3) applied to the commutative diagram:

$$\begin{array}{ccc} R\Gamma(B(\epsilon_n) \cap \{a = w\}, \mathcal{G}_n) & \xrightarrow{\sim} & R\Gamma([(w,0),(0,w)], \mathcal{G}_n) \\ \downarrow & & \downarrow \alpha \\ R\Gamma(B(\epsilon_{n+1}) \cap \{a = w\}, \mathcal{G}_{n+1}) & \xrightarrow{\sim} & R\Gamma([(w,0),(0,w)], \mathcal{G}_{n+1}) . \end{array}$$

Here α is an isomorphism for $|w| \le \epsilon$, because

$$\mathcal{G}_n|[(w,0),(0,w)] \to \mathcal{G}_{n+1}|[(w,0),(0,w)]$$

is an isomorphism by assumption (2). Note that

$$g : K_n \cap g^{-1}(B(\epsilon_n)) \to B(\epsilon_n)$$

is proper. □

Corollary 1.2.1. *Let $f_i : X_i \to \mathbb{K}$, $S_i \subset \{f_i = 0\}$ compact and $\mathcal{F}_i \in ob(D(X_i))$ satisfy the assumption* 1.1.1 *($i = 1, 2$). Then*

$$f := f_1 \oplus f_2 := a \circ (f_1, f_2) : X := X_1 \times X_2 \to \mathbb{K}\,,$$

$S := S_1 \times S_2 \subset \{f = 0\}$ and $\mathcal{F} := \mathcal{F}_1 \boxtimes^L \mathcal{F}_2 \in ob(D(X))$ satisfy also the assumption 1.1.1. *Moreover*

$$R\Gamma(S, \mathcal{F}) \simeq R\Gamma(S_1, \mathcal{F}_1) \otimes^L R\Gamma(S_2, \mathcal{F}_2)\,,$$

and there is for $\mathbb{K} = \mathbb{R}$ a distinguished triangle of the type:

$$R\Gamma(S, \nu_f \mathcal{F}) \longrightarrow R\Gamma(S_1, \nu_{f_1}\mathcal{F}_1) \otimes^L R\Gamma(S_2, \mathcal{F}_2) \oplus R\Gamma(S_1, \mathcal{F}_1) \otimes^L \cdots$$

$$\cdots \otimes^L R\Gamma(S_2, \nu_{f_2}\mathcal{F}_2) \longrightarrow R\Gamma(S_1, \nu_{f_1}\mathcal{F}_1) \otimes^L R\Gamma(S_2, \nu_{f_2}\mathcal{F}_2) \xrightarrow{[1]} .$$

In the case $\mathbb{K} = \mathbb{C}$ one gets the same as before, with ψ instead of ν such that the following diagram commutes:

$$\begin{array}{ccc} R\Gamma(S, \psi_f \mathcal{F}) & \longrightarrow & R\Gamma(S_1, \psi_{f_1}\mathcal{F}_1) \otimes^L R\Gamma(S_2, \mathcal{F}_2) \oplus R\Gamma(S_1, \mathcal{F}_1) \otimes^L \cdots \\ \downarrow{\scriptstyle m_f} & & {\scriptstyle m_{f_1} \otimes^L id}\downarrow{\scriptstyle \oplus id \otimes^L \cdots} \\ R\Gamma(S, \psi_f \mathcal{F}) & \longrightarrow & R\Gamma(S_1, \psi_{f_1}\mathcal{F}_1) \otimes^L R\Gamma(S_2, \mathcal{F}_2) \oplus R\Gamma(S_1, \mathcal{F}_1) \otimes^L \cdots \end{array}$$

$$\begin{array}{ccc} \cdots \otimes^L R\Gamma(S_2, \psi_{f_2}\mathcal{F}_2) & \longrightarrow & R\Gamma(S_1, \psi_{f_1}\mathcal{F}_1) \otimes^L R\Gamma(S_2, \psi_{f_2}\mathcal{F}_2) \xrightarrow{[1]} \\ \cdots \downarrow{\scriptstyle \otimes^L m_{f_2}} & & \downarrow{\scriptstyle m_{f_1} \otimes^L m_{f_2}} \\ \cdots \otimes^L R\Gamma(S_2, \psi_{f_2}\mathcal{F}_2) & \longrightarrow & R\Gamma(S_1, \psi_{f_1}\mathcal{F}_1) \otimes^L R\Gamma(S_2, \psi_{f_2}\mathcal{F}_2) \xrightarrow{[1]} . \end{array}$$

Proof. (1) Let $K_{i,n}, \epsilon_{i,n}$ ($n \in \mathbb{N}$) be given by assumption 1.1.1 for f_i ($i = 1, 2$). If we define $\epsilon_n := min\{\epsilon_{1,n}, \epsilon_{2,n}\}$, then $K_{i,n}, \epsilon_n$ also satisfies the assumption 1.1.1 for f_i ($i = 1, 2$). Let $K_n := K_{1,n} \times K_{2,n}$, $g := (f_1, f_2)$ and ϵ_n as before. Then the **Künneth isomorphism** (see appendix) implies:

$$Rg_*\big(\mathcal{F}|K_n \cap g^{-1}(B(\epsilon_n))\big) \simeq Rf_{1*}(\mathcal{F}_1|K_{1,\epsilon_n}) \boxtimes^L Rf_{2*}(\mathcal{F}_2|K_{2,\epsilon_n})$$

so that assumption (1) of lemma 1.2.2 is satisfied. Fix $n \in \mathbb{N}$ and let ϵ_i (with $0 < \epsilon_i \leq \epsilon_{i,n+1}$) be given by assumption 1.1.1(2) for f_i ($i = 1, 2$). Fix ϵ with $0 < \epsilon < min\{\epsilon_1, \epsilon_2\}$. By the **Künneth formula** (see lemma 1.4.2) we have a commutative diagram

$$\begin{array}{ccc} R\Gamma(K_n \cap \{g = x\}, \mathcal{F}) & \xleftarrow{\sim} & R\Gamma(K(\epsilon_n, x_1), \mathcal{F}_1) \otimes^L R\Gamma(K(\epsilon_n, x_2), \mathcal{F}_2) \\ \downarrow & & \downarrow{\scriptstyle \alpha_1 \otimes^L \alpha_2} \\ R\Gamma(K_{n+1} \cap \{g = x\}, \mathcal{F}) & \xleftarrow{\sim} & R\Gamma(K(\epsilon_{n+1}, x_1), \mathcal{F}_1) \otimes^L R\Gamma(K(\epsilon_{n+1}, x_2), \mathcal{F}_2)\,, \end{array}$$

with

$$K(\epsilon_n, x_i) := K_{i,\epsilon_n} \cap \{f_i = x_i\}\,,\ x := (x_1, x_2)$$

and

$$\alpha_i : R\Gamma(K(\epsilon_n, x_i), \mathcal{F}_i) \xrightarrow{\sim} R\Gamma(K(\epsilon_{n+1}, x_i), \mathcal{F}_i)$$

the restriction morphism for $|x_i| \leq \epsilon$ $(i = 1, 2)$. Therefore also assumption (2) of lemma 1.2.2 is satisfied and the first statement of the corollary follows from lemma 1.2.2.

(2) The formula

$$R\Gamma(S, \mathcal{F}) \simeq R\Gamma(S_1, \mathcal{F}_1) \otimes^L R\Gamma(S_2, \mathcal{F}_2)$$

is just the **classical Künneth formula** (S_1, S_2 are compact).

(3) Moreover

$$R\Gamma(K_n \cap \{f = w\}, \mathcal{F}) \simeq R\Gamma(B(\epsilon_n) \cap \{a = w\}, Rg_*(\mathcal{F}|K_n))\,,$$

and again by the **classical Künneth formula** :

$$\begin{aligned}(Rg_*(\mathcal{F}|K_n))_{(x_1,x_2)} &\simeq R\Gamma(K_n \cap \{g = (x_1, x_2)\}, \mathcal{F}) \\ &\simeq R\Gamma(K_1(\epsilon_n, x_1), \mathcal{F}_1) \otimes^L R\Gamma(K_2(\epsilon_n, x_2), \mathcal{F}_2) \\ &\simeq \left(Rf_{1*}(\mathcal{F}_1|K_{1,\epsilon_n})\right)_{x_1} \otimes^L \left(Rf_{2*}(\mathcal{F}_2|K_{2,\epsilon_n})\right)_{x_2}.\end{aligned}$$

Therefore the distinguished triangles of the corollary follow from lemma 1.1.1 and lemma 1.2.1(4).

(4) For the monodromy statement ($\mathbb{K} = \mathbb{C}$) look at the inclusions

$$\kappa_i : \{\lambda_i x_i |\ \lambda_i \in S^1\} \hookrightarrow \{|z_i| \leq \epsilon_n\}\ \ (i = 1, 2)$$

(with $S^1 := \{z \in \mathbb{C} |\ |z| = 1\}$), $\kappa : \{(\lambda x_1, \lambda x_2) |\ \lambda \in S^1\} \hookrightarrow D(\epsilon_n)$ and

$$d : \{(\lambda x_1, \lambda x_2) |\ \lambda \in S^1\} \to \{\lambda_1 x_1 |\ \lambda_1 \in S^1\} \times \{\lambda_2 x_2 |\ \lambda_2 \in S^1\}\ ,$$

with $\kappa = d \circ (\kappa_1 \times \kappa_2)$. Then the **Künneth isomorphism** implies:

$$\kappa^* Rg_*(\mathcal{F}|K_n) \simeq d^*\left((\kappa_1^* Rf_{1*}(\mathcal{F}_1|K_{1,\epsilon_n})) \boxtimes^L (\kappa_2^* Rf_{2*}(\mathcal{F}_2|K_{2,\epsilon_n}))\right),$$

from which the **monodromy** statement easily follows. For example the case $x_1 \neq 0$, $x_2 \neq 0$ corresponds to a commutative diagram of covering spaces of type

$$\begin{array}{ccc} \mathbb{R} & \xrightarrow{\tilde{d}} & \mathbb{R} \times \mathbb{R} \\ \downarrow{\scriptstyle \pi} & & \downarrow{\scriptstyle \pi \times \pi} \\ S^1 & \xrightarrow{d} & S^1 \times S^1\,, \end{array}$$

with $\pi := exp(2\pi i \cdot\,)$ and $d, \tilde{d}$ the diagonal maps. Then $\pi^* d^* = \tilde{d}^*(\pi \times \pi)^*$ and $\tilde{d}$ is equivariant with respect to the diagonal inclusion

$$\pi_1(S^1) \to \pi_1(S^1 \times S^1) = \pi_1(S^1) \times \pi_1(S^1)\,. \qquad \square$$

Corollary 1.2.1 gives a description of

$$R\Gamma(S, \nu_f \mathcal{F}) \quad \text{and} \quad R\Gamma(S, \psi_f \mathcal{F})$$

in terms of $\mathcal{F}_1$ and $\mathcal{F}_2$. But the **Thom-Sebastiani Theorem** will be a result about

$$R\Gamma(S, \mu_f \mathcal{F}) \quad \text{or} \quad R\Gamma(S, \phi_f[-1]\mathcal{F}) .$$

In the classical case of an **isolated** critical point

$$f : (\mathbb{C}^{n+1}, 0) \to (\mathbb{C}, 0) ,$$

the difference between $(\psi_f \mathbb{Z})_0$ and $(\phi_f \mathbb{Z})_0$ is not so important, since

$$H^i((\psi_f \mathbb{Z})_0) = H^i((\phi_f \mathbb{Z})_0)$$

for $i > 0$ and $n > 0$! So in this case one can conclude the **classical** Thom-Sebastiani theorem from corollary 1.2.1 (compare [ST]).

But in general $R\Gamma(S, \psi_f \mathcal{F})$ and $R\Gamma(S, \phi_f \mathcal{F})$ are quite different:

$$R\Gamma(S, \phi_f \mathcal{F}) \simeq cone\big(R\Gamma(S, \mathcal{F}) \stackrel{sp_f}{\to} R\Gamma(S, \psi_f \mathcal{F})\big) .$$

For the case of the **specialization** and **nearby cycle functor** one has to consider **fiber products** (as already remarked in the introduction). Here one gets for example the

Theorem 1.2.1. *Let $f_i : X_i \to \mathbb{K}$, $S_i \subset \{f_i = 0\}$ compact and $\mathcal{F}_i \in ob(D(X_i))$ satisfy the assumption 1.1.1 ($i = 1, 2$). Consider the fiber product*

$$\begin{array}{ccc} X := X_1 \times_K X_2 & \xrightarrow{p_2} & X_2 \\ {\scriptstyle p_1}\downarrow & & \downarrow{\scriptstyle f_2} \\ X_1 & \xrightarrow{f_1} & \mathbb{K} , \end{array}$$

with $f := f_1 \circ p_1 = f_2 \circ p_2 : X \to \mathbb{K}$. Then f, $S := S_1 \times S_2 \subset \{f = 0\}$ and $\mathcal{F} := \mathcal{F}_1 \boxtimes^L_K \mathcal{F}_2 \in ob(D(X))$ satisfy also the assumption 1.1.1. For $\mathbb{K} = \mathbb{R}$ one has an isomorphism

$$R\Gamma(S, \nu_f \mathcal{F}) \simeq R\Gamma(S_1, \nu_{f_1} \mathcal{F}_1) \otimes^L R\Gamma(S_2, \nu_{f_2} \mathcal{F}_2) .$$

For $\mathbb{K} = \mathbb{C}$ one has an isomorphism

$$R\Gamma(S, \psi_f \mathcal{F}) \simeq R\Gamma(S_1, \psi_{f_1} \mathcal{F}_1) \otimes^L R\Gamma(S_2, \psi_{f_2} \mathcal{F}_2)$$

commuting with the corresponding monodromies.

Proof. Let $K_{i,n}, \epsilon_{i,n}$ ($n \in \mathbb{N}$) be given by assumption 1.1.1 for f_i ($i = 1, 2$). Let $K_n := K_{1,n} \times K_{2,n}$, $g := (f_1, f_2)$ and $\epsilon_n := min\{\epsilon_{1,n}, \epsilon_{2,n}\}$ be as before. Then the claim follows as in the proof of corollary 1.2.1 from the **Künneth isomorphism** and the **proper base change theorem** [KS, prop. 2.6.7] applied to the cartesian diagram

$$\begin{array}{ccc} K'_n := K_{1,n} \times_K K_{2,n} & \longrightarrow & K_n := K_{1,n} \times K_{2,n} \\ {\scriptstyle f}\downarrow & & \downarrow{\scriptstyle g=(f_1,f_2)} \\ \mathbb{K} & \xrightarrow{d} & \mathbb{K} \times \mathbb{K}\,, \end{array}$$

with d the diagonal map. We leave the details to the reader. □

To see that one has to use μ_f or $\phi_f[-1]$ in the **Thom-Sebasiani Theorem**, let us look at the case of "nice additive" invariants, which can already be deduced from corollary 1.2.1.

Let R be a commutative ring, T be a full **triangulated** subcategory of $D(R)$, **stable under isomorphisms** (i.e. $F \simeq G$ with $G \in ob(T)$ implies $F \in ob(T)$) and $\otimes^L$ (i.e. $F, G \in ob(T)$ implies $F \otimes^L G \in ob(T)$). We assume that the isomorphism classes $\{[F]\}$ of $F \in ob(T)$ form a set.

1. Let $\chi : \{[F]\} \to R'$ be an **additive** function with values in the commutative ring R', i.e.
$$\chi(F) = \chi(F') + \chi(F'')$$
for any distinguished triangle
$$F' \to F \to F'' \xrightarrow{[1]}$$
with objects in T, which is also **multiplicative** with respect to $\otimes^L$:
$$\chi(F \otimes^L G) = \chi(F) \cdot \chi(G)$$
for all $F, G \in ob(T)$.

2. Similarly, let $\{[F, \alpha]\}$ be the set of isomorphism classes of endomorphisms $\alpha : F \to F$ with $F \in ob(T)$. Consider an **additive** function $\zeta : \{[F, \alpha]\} \to R'$ with values in the commutative ring R', i.e.
$$\zeta(F, \alpha) = \zeta(F', \alpha') + \zeta(F'', \alpha'')$$
for any endomorphism
$$\begin{array}{ccccccc} F' & \longrightarrow & F & \longrightarrow & F'' & \xrightarrow{[1]} & \\ \downarrow{\scriptstyle \alpha'} & & \downarrow{\scriptstyle \alpha} & & \downarrow{\scriptstyle \alpha''} & & \\ F' & \longrightarrow & F & \longrightarrow & F'' & \xrightarrow{[1]} & \end{array}$$

of a distinguished triangle with objects in T, which is also **multiplicative** with respect to $\otimes^L$:

$$\zeta(F \otimes^L F', \alpha \otimes^L \alpha') = \zeta(F, \alpha) \cdot \zeta(F', \alpha')$$

for any two such endomorphisms $\alpha : F \to F$, $\alpha' : F' \to F'$.

The standard example is $R = k$ a field, T denotes the (cohomological) bounded complexes with **finite dimensional cohomology**,

$$\chi = \sum (-1)^i \cdot dim_k H^i(\cdot)$$

is the usual **Euler characteristic** and

$$\zeta = \sum (-1)^i \cdot tr(\alpha^i : H^i(\cdot) \to H^i(\cdot))$$

the **Lefschetz number**.

Corollary 1.2.2. *Let $f_i : X_i \to \mathbb{K}$, $S_i \subset \{f_i = 0\}$ compact and $\mathcal{F}_i \in ob(D(X_i))$ satisfy the assumption 1.1.1 ($i = 1, 2$). Define*

$$f := f_1 \oplus f_2 := a \circ (f_1, f_2) : X := X_1 \times X_2 \to \mathbb{K}\,,$$

$S := S_1 \times S_2 \subset \{f = 0\}$ and $\mathcal{F} := \mathcal{F}_1 \boxtimes^L \mathcal{F}_2 \in ob(D(X))$.

1. *Let $\mathbb{K} = \mathbb{R}$ and assume*

$$R\Gamma(S_i, \mathcal{F}_i),\ R\Gamma(S_i, \nu_{f_i}\mathcal{F}_i) \in ob(T) \quad (i = 1, 2)\,.$$

Then also

$$R\Gamma(S_i, \mu_{f_i}\mathcal{F}_i),\ R\Gamma(S, \mathcal{F}),\ R\Gamma(S, \nu_f\mathcal{F}),\ R\Gamma(S, \mu_f\mathcal{F}) \in ob(T)\,,$$

with

$$\chi\big(R\Gamma(S, \mu_f\mathcal{F})\big) = \chi\big(R\Gamma(S_1, \mu_{f_1}\mathcal{F}_1)\big) \cdot \chi\big(R\Gamma(S_2, \mu_{f_2}\mathcal{F}_2)\big)\,.$$

2. *Let $\mathbb{K} = \mathbb{C}$ and make the same assumption as in (1.), with ψ instead of ν. Then the same conclusion as in (1.) holds, with $\phi[-1]$ instead of μ. Moreover for an additive function ζ of the second type, one has the relation:*

$$\zeta\big(R\Gamma(S, \phi_f[-1]\,\mathcal{F}), m_f[-1]\big) =$$

$$\zeta\big(R\Gamma(S_1, \phi_{f_1}[-1]\,\mathcal{F}_1), m_{f_1}[-1]\big) \cdot \zeta\big(R\Gamma(S_2, \phi_{f_2}[-1]\,\mathcal{F}_2), m_{f_2}[-1]\big)\,.$$

Proof. We only consider the case $\mathbb{K} = \mathbb{R}$ (the other case follows in exactly the same way). The first statement follows from corollary 1.2.1 and the defining triangle of the micro-localization functor. In the same way, one gets from the distinguished triangle of corollary 1.2.1 by additivity and multiplicativity (with the short-cut $\chi(S,\mathcal{F}) := \chi\big(R\Gamma(S,\mathcal{F})\big)$ and so on ...):

$$\begin{aligned}
\chi(S,\mu_f\mathcal{F}) &= \chi(S,\mathcal{F}) - \chi(S,\nu_f\mathcal{F}) \\
&= \chi(S_1,\mathcal{F}_1)\cdot\chi(S_2,\mathcal{F}_2) - \chi(S_1,\nu_{f_1}\mathcal{F}_1)\cdot\chi(S_2,\mathcal{F}_2) \\
&\quad -\chi(S_1,\mathcal{F}_1)\cdot\chi(S_2,\nu_{f_2}\mathcal{F}_2) + \chi(S_1,\nu_{f_1}\mathcal{F}_1)\cdot\chi(S_2,\nu_{f_2}\mathcal{F}_2) \\
&= \big(\chi(S_1,\mathcal{F}_1) - \chi(S_1,\nu_{f_1}\mathcal{F}_1)\big)\cdot\big(\chi(S_2,\mathcal{F}_2) - \chi(S_2,\nu_{f_2}\mathcal{F}_2)\big) \\
&= \chi(S_1,\mu_{f_1}\mathcal{F}_1)\cdot\chi(S_2,\nu_{f_2}\mathcal{F}_2).
\end{aligned}$$
□

Remark 1.2.2. The distinguished triangle of corollary 1.2.1 implies in the case $\mathbb{K} = \mathbb{C}$ the same formula as in corollary 1.2.2 also for the **powers** $(m_f[-1])^n$ of $m_f[-1]$ ($n \in \mathbb{N}$). This implies for example in our standard example the multiplicativity of the Lefschetz numbers of the powers of the (shifted) monodromy transformation. From this one can deduce the relation of the corresponding zeta-functions!

Example 1.2.1. *As a very simple example, consider the function*

$$f = z^a : \mathbb{C} \to \mathbb{C} \quad (a \in \mathbb{N})\,.$$

Then the **Milnor fiber** M_f *consists of* a **points**. *Let* $\mathcal{F}$ *be a sheaf of vector spaces on* $X = \mathbb{C}$ *with finite dimensional stalks, constructible with respect to the filtration* $\{0\} \subset \mathbb{C}$ *(which we consider as a complex placed in degree zero). Let*

$$k := dim(\mathcal{F}_0) \quad , \quad d := dim(\mathcal{F}_w) \quad (w \neq 0)$$

so that we get for the **Euler characteristic***:*

$$\chi\big((\phi_{z^a}[-1]\mathcal{F}[1])_0\big) = \chi\big((\phi_{z^a}\mathcal{F})_0\big) = a \cdot d - k\,.$$

If $\mathcal{F}[1]$ *is a* **perverse** *sheaf (with respect to the middle perversity), then this is the* **dimension** *of* $H^0((\phi_{z^a}\mathcal{F})_0)$, *since all other cohomology groups are vanishing (as explained in chapter* 6*). For example one can take* $j_*j^*\mathcal{F}$, *with* $j : \mathbb{C}^* \to \mathbb{C}$ *the inclusion. Then*

$$k = dim(\ ker(id - mon : \mathcal{F}_w \to \mathcal{F}_w)\ )\,,$$

with "mon" the corresponding **monodromy** *action, and* $j_*j^*\mathcal{F}[1]$ *is the* **intersection cohomology complex** *with coefficients in the local system* $j^*\mathcal{F}$. *If we take* n *such sheaves* $\mathcal{F}_i$, *then iterated application of corollary* 1.2.2 *implies for*

$$f := z_1^{a_1} + \cdots + z_n^{a_n} : \mathbb{C}^n \to \mathbb{C}\,,\ \mathcal{F} := \mathcal{F}_1 \boxtimes \cdots \boxtimes \mathcal{F}_n$$

the formula (compare [Ma2, sec. 2.4]*):*

$$\chi\big((\phi_f[-1]\mathcal{F}[n])_0\big) = \prod (a_i \cdot d_i - k_i)\,.$$

Especially, the **Milnor number** *of the Brieskorn-Pham singularity of* f *in* 0 *is equal to* $\prod(a_i - 1)$.

1.2.2 Thom-Sebastiani Theorem for sheaves

For the Thom-Sebastiani Theorem, we have to take care of the **specialization morphismus** sp_f in the **cone** defining μ_f or $\tilde{\phi}_f[-1]$. This is not so easy to control in the approach taken up to now, because $B \cap \{a = w\}$ is **not** adapted to the product situation. Therefore we look at a "deformation", which fits nicely with the product situation and the Künneth formula. First we treat the case $\mathbb{K} = \mathbb{R}$.

Lemma 1.2.3. *Under the assumption of lemma* 1.2.1, *there exists for* $-\epsilon \leq w < 0$ *(with* $0 < \epsilon < \epsilon'$*) a distinguished triangle:*

$$(Ra_*\mathcal{G})_w \xrightarrow{\alpha} \mathcal{G}_{(w,0)} \oplus \mathcal{G}_{(0,w)} \xrightarrow{(\alpha_1, -\alpha_2)} \mathcal{G}_{(w,w)} \xrightarrow{[1]} ,$$

with

$$\alpha_1 : \mathcal{G}_{(w,0)} \xleftarrow{\sim} R\Gamma([(w,0),(w,w)],\mathcal{G}) \to \mathcal{G}_{(w,w)}$$

and

$$\alpha_2 : \mathcal{G}_{(0,w)} \xleftarrow{\sim} R\Gamma([(0,w),(w,w)],\mathcal{G}) \to \mathcal{G}_{(w,w)} .$$

Moreover, there exists a commutative diagram:

$$\begin{array}{ccc} (Ra_*\mathcal{G})_0 & \xrightarrow{sp_a} & (Ra_*\mathcal{G})_w \\ \downarrow \wr & & \downarrow \alpha \\ \mathcal{G}_{(0,0)} & \xrightarrow{(\beta_1,\beta_2)} & \mathcal{G}_{(w,0)} \oplus \mathcal{G}_{(0,w)} , \end{array}$$

with

$$\beta_1 : \mathcal{G}_{(0,0)} \xleftarrow{\sim} R\Gamma([(w,0),(0,0)],\mathcal{G}) \to \mathcal{G}_{(w,0)}$$

and

$$\beta_2 : \mathcal{G}_{(0,0)} \xleftarrow{\sim} R\Gamma([(0,w),(0,0)],\mathcal{G}) \to \mathcal{G}_{(0,w)} .$$

Proof. (1) Let $-\epsilon \leq w < 0$. We already have

$$(Ra_*\mathcal{G})_w \simeq R\Gamma(B \cap \{a = w\},\mathcal{G}) \xrightarrow{\sim} R\Gamma([(w,0),(0,w)],\mathcal{G}) .$$

By remark 1.2.1(3), we get (see Figure 1.1):

$$R\Gamma([(w,0),(0,w)],\mathcal{G}) \xleftarrow{\sim} R\Gamma(\triangle(w),\mathcal{G}) \xrightarrow{\sim} R\Gamma(I_1(w) \cup I_2(w),\mathcal{G})$$

with $I_1(w) := [(w,0),(w,w)]$, $I_2(w) := [(w,w),(0,w)]$ and $\triangle(w)$ the affine triangle spanned by $(w,0),(w,w),(0,w)$. From the covering

$$\{I_1(w),\ I_2(w)\} \quad \text{of} \quad I(w) := I_1(w) \cup I_2(w) ,$$

we get an isomorphism of distinguished triangles of **Mayer-Vietoris type** (see [KS, prop. 2.3.6(vi), formula (2.6.31)]):

$$
\begin{array}{ccccccc}
R\Gamma(I(w),\mathcal{G}) & \longrightarrow & R\Gamma(I_1(w),\mathcal{G}) \oplus R\Gamma(I_2(w),\mathcal{G}) & \longrightarrow & \mathcal{G}_{(w,w)} & \xrightarrow{[1]} & \\
{\scriptstyle\gamma}\uparrow{\scriptstyle\wr} & & \downarrow{\scriptstyle\wr} & & \| & & \\
(Ra_*\mathcal{G})_w & \xrightarrow{\alpha} & \mathcal{G}_{(w,0)} \oplus \mathcal{G}_{(0,w)} & \xrightarrow{(\alpha_1,-\alpha_2)} & \mathcal{G}_{(w,w)} & \xrightarrow{[1]} & ,
\end{array}
$$

with α defined by commutativity of the diagram and γ the composition of the above isomorphisms. Remark that there is a **choice of sign** in this triangle. Our convention follows [KS] and fits with other sign conventions later on.

(2) Consider the following commutative diagram of maps (induced by inclusions of subsets, with $B_0 := B \cap \{a = 0\}$):

$$
\begin{array}{ccccc}
(Ra_*\mathcal{G})_0 & \xleftarrow{\sim} & R\Gamma(\{|a| \le |w|\}, Ra_*\mathcal{G}) & \longrightarrow & (Ra_*\mathcal{G})_w \\
\downarrow{\scriptstyle\wr} & & \downarrow{\scriptstyle\wr} & & \downarrow{\scriptstyle\wr} \\
R\Gamma(B_0,\mathcal{G}) & \xleftarrow{\sim} & R\Gamma(B \cap \{|a| \le |w|\},\mathcal{G}) & \longrightarrow & R\Gamma(B \cap \{a = w\},\mathcal{G}) \\
\downarrow{\scriptstyle\wr} & & \uparrow{\scriptstyle\wr} & & \downarrow{\scriptstyle\wr} \\
\mathcal{G}_{(0,0)} & \xleftarrow{\sim} & R\Gamma(B,\mathcal{G}) & \longrightarrow & R\Gamma([(w,0),(0,w)],\mathcal{G}) \\
\| & & \downarrow{\scriptstyle\gamma_2} & & \uparrow{\scriptstyle\wr} \\
\mathcal{G}_{(0,0)} & \xleftarrow{\gamma_1} & R\Gamma([w,0]\times[w,0],\mathcal{G}) & \longrightarrow & R\Gamma(\triangle(w),\mathcal{G}) \\
\| & & \| & & \downarrow{\scriptstyle\wr} \\
\mathcal{G}_{(0,0)} & \xleftarrow{\sim} & R\Gamma([w,0]\times[w,0],\mathcal{G}) & \longrightarrow & R\Gamma(I(w),\mathcal{G}) \\
\| & & \| & & \downarrow \\
\mathcal{G}_{(0,0)} & \xleftarrow{\sim} & R\Gamma([w,0]\times[w,0],\mathcal{G}) & \longrightarrow & \mathcal{G}_{(w,0)} \oplus \mathcal{G}_{(0,w)} ,
\end{array}
$$

whose top row describes the specialization morphism and whose right column is the morphism α. (The first two rows are isomorphic, since the restriction of a to B is proper. The isomorphism of the second and third row follows from lemma 1.2.1(2). Finally, γ_1 is an isomorphism by the next argument, and then also γ_2 by commutativity.) Then the claim follows, since the following diagram commutes:

$$
\begin{array}{ccccc}
\mathcal{G}_{(0,0)} & \xleftarrow{\sim} & R\Gamma([w,0]\times[w,0],\mathcal{G}) & \longrightarrow & R\Gamma(I_1(w),\mathcal{G}) \\
\| & & \downarrow{\scriptstyle\wr} & & \downarrow{\scriptstyle\wr} \\
\mathcal{G}_{(0,0)} & \xleftarrow{\sim} & R\Gamma([(w,0),(0,0)],\mathcal{G}) & \longrightarrow & \mathcal{G}_{(w,0)} ,
\end{array}
$$

and

$$\begin{array}{ccccc} \mathcal{G}_{(0,0)} & \xleftarrow{\sim} & R\Gamma([w,0]\times[w,0],\mathcal{G}) & \longrightarrow & R\Gamma(I_2(w),\mathcal{G}) \\ \| & & \downarrow \wr & & \downarrow \wr \\ \mathcal{G}_{(0,0)} & \xleftarrow{\sim} & R\Gamma([(0,0),(0,w)],\mathcal{G}) & \longrightarrow & \mathcal{G}_{(0,w)} \, . \end{array}$$

The indicated isomorphisms come from remark 1.2.1(3). □

We consider now the case $\mathbb{K} = \mathbb{C}$. Before we can formulate the similar statements, we have to introduce some notation. Let $B_R := B \cap \mathbb{R}^2$ (with $\mathbb{R}^2 \subset \mathbb{C}^2$ the standard inclusion) and fix $w \in B_R$ with $-\epsilon \leq w < 0$. Consider the cartesian diagram

$$\begin{array}{ccc} B_R \times \mathbb{R} & \xrightarrow{pr} & \mathbb{R} \\ \downarrow p & & \downarrow \pi \\ B_R \times S^1 & \xrightarrow{q} & S_{|w|} \, , \end{array}$$

with $S^1 := \{\lambda \in \mathbb{C} | |\lambda| = 1\}$, pr the projection, $p = id \times exp(2\pi i \, \cdot \,)$, $\pi = w \cdot exp(2\pi i \, \cdot \,)$ and q the projection multiplied by w. We also use the map

$$h : B_R \times S^1 \to B, \; (x, y, \lambda) \mapsto (\lambda x, \lambda y) \, .$$

If $A \subset B_R$ is a subset with $x \in A$ implies $-x \notin A$ (especially $0 \notin A$), then

$$h : A \times S^1 \to B$$

is injective:

$$(\lambda x, \lambda y) = (\lambda' x', \lambda' y') \Rightarrow \lambda/\lambda' \in \mathbb{R} \Rightarrow \lambda/\lambda' = -1, +1 \Rightarrow \lambda = \lambda' \, .$$

This applies for example to $\triangle(w)$ and $([w, 0] \times [w, 0]) \backslash \{(0, 0)\}$. Finally, let us introduce the maps

$$s_1 : S_{|w|} \simeq S_{|w|} \times \{0\} \hookrightarrow B \text{ and } s_2 : S_{|w|} \simeq \{0\} \times S_{|w|} \hookrightarrow B \, .$$

Lemma 1.2.4. *Under the assumption of lemma 1.2.1, there exists for $-\epsilon \leq w < 0$ (with $0 < \epsilon < \epsilon'$) a distinguished triangle:*

$$(Ra_* \mathcal{G})|S_{|w|} \xrightarrow{\alpha} Rq_*(h^* \mathcal{G}|\{(w,0)\} \times S^1) \oplus Rq_*(h^* \mathcal{G}|\{(0,w)\} \times S^1) \cdots$$

$$\cdots \xrightarrow{(\alpha_1, -\alpha_2)} Rq_*(h^* \mathcal{G}|\{(w,w)\} \times S^1) \xrightarrow{[1]} \, ,$$

with α_1:

$$Rq_*(h^* \mathcal{G}|\{(w,0)\} \times S^1) \xleftarrow{\sim} Rq_*(h^* \mathcal{G}|I_1(w) \times S^1) \to Rq_*(h^* \mathcal{G}|\{(w,w)\} \times S^1)$$

and α_2:

$$Rq_*(h^* \mathcal{G}|\{(0,w)\} \times S^1) \xleftarrow{\sim} Rq_*(h^* \mathcal{G}|I_2(w) \times S^1) \to Rq_*(h^* \mathcal{G}|\{(w,w)\} \times S^1) \, .$$

All these sheaf complexes have locally constant cohomology sheaves. Moreover, there exists a commutative diagram:

$$\begin{array}{ccc}
(Ra_*\mathcal{G})_0 & \xrightarrow{sp_a} & R\Gamma\big(\mathbb{R}, \pi^*((Ra_*\mathcal{G})|S_{|w|})\big) \\
\downarrow \wr & & \downarrow \pi^*\alpha \\
\mathcal{G}_{(0,0)} & \xrightarrow{(\beta_1,\beta_2)} & R\Gamma\big(\mathbb{R}, \pi^* Rq_*\mathcal{G}'\big) \oplus R\Gamma\big(\mathbb{R}, \pi^* Rq_*\mathcal{G}''\big) ,
\end{array}$$

with $\mathcal{G}' := h^*\mathcal{G}|\{(w,0)\} \times S^1$, $\mathcal{G}'' := h^*\mathcal{G}|\{(0,w)\} \times S^1$, β_1*:*

$$\begin{array}{ccccc}
\mathcal{G}_{(0,0)} & \xleftarrow{\sim} & R\Gamma([(w,0),(0,0)] \times \mathbb{R}, p^*h^*\mathcal{G}) & \longrightarrow & R\Gamma\big(\mathbb{R}, \pi^* Rq_*\mathcal{G}'\big) \\
\| & & \uparrow \wr & & \uparrow \wr \\
\mathcal{G}_{(0,0)} & \xleftarrow{\sim} & R\Gamma(\{|x| \le |w|\} \times \{0\}, \mathcal{G}) & \longrightarrow & R\Gamma(\mathbb{R}, \pi^* s_1^*\mathcal{G})
\end{array}$$

and β_2*:*

$$\begin{array}{ccccc}
\mathcal{G}_{(0,0)} & \xleftarrow{\sim} & R\Gamma([(0,w),(0,0)] \times \mathbb{R}, p^*h^*\mathcal{G}) & \longrightarrow & R\Gamma\big(\mathbb{R}, \pi^* Rq_*\mathcal{G}''\big) \\
\| & & \uparrow \wr & & \uparrow \wr \\
\mathcal{G}_{(0,0)} & \xleftarrow{\sim} & R\Gamma(\{0\} \times \{|y| \le |w|\}, \mathcal{G}) & \longrightarrow & R\Gamma(\mathbb{R}, \pi^* s_2^*\mathcal{G}) .
\end{array}$$

Proof. (1) $h^*\mathcal{G}$ has locally constant cohomology sheaves on the fibers of the projection $B_R \times S^1 \to B_R$. Moreover q is a trivial fibration so that

$$Rq_*(h^*\mathcal{G}|A \times S^1)$$

has locally constant cohomology sheaves by remark 1.2.1(1) (with $A \subset B_R$ compact). Note that the restriction of h, q to $[(w,0),(0,w)] \times S^1$ is the same map h, q as in the proof of lemma 1.2.1 ! So we already have

$$(Ra_*\mathcal{G})|\{|z| = |w|\} \xrightarrow{\sim} Rq_*(h^*\mathcal{G}|\{|z| = |w|\} \times S^1)$$

by (5) of the proof of lemma 1.2.1. Since q is proper, the isomorphisms of part (1) of the proof of lemma 1.2.3 imply:

$$\begin{array}{ccc}
Rq_*(h^*\mathcal{G}| \triangle (w) \times S^1) & \xrightarrow{\sim} & Rq_*(h^*\mathcal{G}|I(w) \times S^1) \\
\downarrow \wr & & \\
Rq_*(h^*\mathcal{G}|[(w,0),(0,w)] \times S^1) . & &
\end{array}$$

Similarly, the **Mayer-Vietoris sequence** translates into:

$$\begin{array}{ccc}
Rq_*(h^*\mathcal{G}|I(w) \times S^1) & \longrightarrow & Rq_*(h^*\mathcal{G}|I_1(w) \times S^1) \oplus Rq_*(h^*\mathcal{G}|I_2(w) \times S^1) \cdots \\
\uparrow \wr & & \downarrow \wr \\
(Ra_*\mathcal{G})|\{|z| = |w|\} & \xrightarrow{\alpha} & Rq_*(h^*\mathcal{G}|\{(w,0)\} \times S^1) \oplus Rq_*(h^*\mathcal{G}|\{(0,w)\} \times S^1)
\end{array}$$

$$\begin{array}{ccc} \cdots \longrightarrow & Rq_*(h^*\mathcal{G}|(w,w)\times S^1) & \xrightarrow{[1]} \\ & \Vert & \\ \cdots \longrightarrow & Rq_*(h^*\mathcal{G}|(w,w)\times S^1) & \xrightarrow{[1]} \end{array} .$$

(2) Since $p^*h^*\mathcal{G}$ has (locally) constant cohomology sheaves on the fibers of the projection $B_R \times \mathbb{R} \to B_R$, the **Vietoris-Begle theorem** ([KS, prop. 2.7.8]) implies

$$R\Gamma(A,\mathcal{G}) \xrightarrow{\sim} R\Gamma(A\times\{0\}, p^*h^*\mathcal{G}) \xleftarrow{\sim} R\Gamma(A\times\mathbb{R}, p^*h^*\mathcal{G})$$

functorial in the subset $A \subset B_R$. So part (2) of the proof of lemma 1.2.3 implies the following commutative diagram:

$$\begin{array}{ccccc}
(Ra_*\mathcal{G})_0 & \xleftarrow{\sim} & R\Gamma(\{|a|\le|w|\}, Ra_*\mathcal{G}) & \longrightarrow & \mathcal{H}_1 \\
\downarrow\wr & & \downarrow\wr & & \downarrow\wr \\
\mathcal{G}_{(0,0)} & \xleftarrow{\sim} & R\Gamma(B,\mathcal{G}) & \longrightarrow & \mathcal{H}_2 \\
\downarrow\wr & & \downarrow & & \Vert \\
R\Gamma(\{(0,0)\}\times\mathbb{R}, p^*h^*\mathcal{G}) & \xleftarrow{\sim} & R\Gamma(B_R\times\mathbb{R}, p^*h^*\mathcal{G}) & \longrightarrow & \mathcal{H}_2 \\
\Vert & & \downarrow\wr & & \downarrow\wr \\
R\Gamma(\{(0,0)\}\times\mathbb{R}, p^*h^*\mathcal{G}) & \xleftarrow{\sim} & R\Gamma(sq(w)\times\mathbb{R}, p^*h^*\mathcal{G}) & \longrightarrow & \mathcal{H}_3 ,
\end{array}$$

with $sq(w) := [w,0]\times[w,0]$,

$$R\Gamma\big(\mathbb{R}, \pi^*((Ra_*\mathcal{G})|S_{|w|})\big) =: \mathcal{H}_1 ,$$

$$R\Gamma\big(\mathbb{R}, \pi^* Rq_*(h^*\mathcal{G}|[(w,0),(0,w)]\times S^1)\big) =: \mathcal{H}_2 ,$$

$$R\Gamma\big(\mathbb{R}, \pi^* Rq_*(h^*\mathcal{G}|\triangle(w)\times S^1)\big) =: \mathcal{H}_3 ,$$

and

$$\begin{array}{c}
\mathcal{H}_3 \xrightarrow{\sim} R\Gamma\big(\mathbb{R}, \pi^* Rq_*(h^*\mathcal{G}|(I_1(w)\cup I_2(w))\times S^1)\big) \longrightarrow \\
R\Gamma\big(\mathbb{R}, \pi^* Rq_*(h^*\mathcal{G}|I_1(w)\times S^1)\big) \oplus R\Gamma\big(\mathbb{R}, \pi^* Rq_*(h^*\mathcal{G}|I_2(w)\times S^1)\big) \xrightarrow{\sim} \\
R\Gamma\big(\mathbb{R}, \pi^* Rq_*(h^*\mathcal{G}|\{(w,0)\}\times S^1)\big) \oplus R\Gamma\big(\mathbb{R}, \pi^* Rq_*(h^*\mathcal{G}|\{(0,w)\}\times S^1)\big) .
\end{array}$$

In the same way, one gets the commutative diagrams

$$\begin{array}{ccc}
R\Gamma(\{(0,0)\}\times\mathbb{R}, p^*h^*\mathcal{G}) & \xleftarrow{\sim} & R\Gamma(sq(w)\times\mathbb{R}, p^*h^*\mathcal{G}) \\
\Vert & & \downarrow\wr \\
R\Gamma(\{(0,0)\}\times\mathbb{R}, p^*h^*\mathcal{G}) & \xleftarrow{\sim} & R\Gamma([(w,0),(0,0)]\times\mathbb{R}, p^*h^*\mathcal{G}) \\
\uparrow\wr & & \uparrow \\
\mathcal{G}_{(0,0)} & \xleftarrow{\sim} & R\Gamma(\{|x|\le|w|\}\times\{0\}, \mathcal{G})
\end{array}$$

and

$$\begin{array}{ccc} R\Gamma(\{(0,0)\} \times \mathbb{R}, p^*h^*\mathcal{G}) & \xleftarrow{\sim} & R\Gamma(sq(w) \times \mathbb{R}, p^*h^*\mathcal{G}) \\ \| & & \downarrow \wr \\ R\Gamma(\{(0,0)\} \times \mathbb{R}, p^*h^*\mathcal{G}) & \xleftarrow{\sim} & R\Gamma([(0,0),(0,w)] \times \mathbb{R}, p^*h^*\mathcal{G}) \\ \uparrow \wr & & \uparrow \\ \mathcal{G}_{(0,0)} & \xleftarrow{\sim} & R\Gamma(\{0\} \times \{|y| \leq |w|\}, \mathcal{G})\,. \end{array}$$

Note that

$$h : [(w,0),(0,0)] \times S^1 \to \{|x| \leq |w|\} \times \{0\}$$

or

$$h : [(0,0),(0,w)] \times S^1 \to \{0\} \times \{|y| \leq |w|\}$$

is just the "real blowing up" of this disc. Finally

$$S_{|w|} \times \{0\} \underset{\sim}{\xleftarrow{h}} \{(w,0)\} \times S^1 \underset{\sim}{\xrightarrow{q}} S_{|w|}$$

implies $Rq_*(h^*\mathcal{G}|\{(w,0)\} \times S^1) \simeq s_1^*\mathcal{G}$, and

$$\{0\} \times S_{|w|} \underset{\sim}{\xleftarrow{h}} \{(0,w)\} \times S^1 \underset{\sim}{\xrightarrow{q}} S_{|w|}$$

implies $Rq_*(h^*\mathcal{G}|\{(0,w)\} \times S^1) \simeq s_2^*\mathcal{G}$. □

Now we can prove the **main result** of this section.

Theorem 1.2.2 (Thom-Sebastiani Theorem for sheaves). *Let $f_i : X_i \to \mathbb{K}$, $S_i \subset \{f_i = 0\}$ compact and $\mathcal{F}_i \in ob(D(X_i))$ satisfy the assumption* 1.1.1 *($i = 1, 2$). Define*

$$f := f_1 \oplus f_2 := a \circ (f_1, f_2) : X := X_1 \times X_2 \to \mathbb{K}\,, \quad S := S_1 \times S_2 \subset \{f = 0\}$$

and $\mathcal{F} := \mathcal{F}_1 \boxtimes^L \mathcal{F}_2 \in ob(D(X))$.

1. *Let $\mathbb{K} = \mathbb{R}$. Then there is an isomorphism*

$$R\Gamma(S, \mu_f\mathcal{F}) \simeq R\Gamma(S_1, \mu_{f_1}\mathcal{F}_1) \otimes^L R\Gamma(S_2, \mu_{f_2}\mathcal{F}_2)\,.$$

2. *Let $\mathbb{K} = \mathbb{C}$. Then there is an isomorphism*

$$R\Gamma(S, \tilde{\phi}_f[-1]\mathcal{F}) \simeq R\Gamma(S_1, \tilde{\phi}_{f_1}[-1]\mathcal{F}_1) \otimes^L R\Gamma(S_2, \tilde{\phi}_{f_2}[-1]\mathcal{F}_2)\,.$$

Especially, we have an isomorphism as in (1.), with $\phi[-1]$ instead of μ such that the diagram

$$\begin{array}{ccc} R\Gamma(S, \phi_f[-1]\mathcal{F}) & \xrightarrow{\sim} & R\Gamma(S_1, \phi_{f_1}[-1]\mathcal{F}_1) \otimes^L R\Gamma(S_2, \phi_{f_2}[-1]\mathcal{F}_2) \\ \downarrow m_f[-1] & & \downarrow m_{f_1}[-1]\otimes^L m_{f_2}[-1] \\ R\Gamma(S, \phi_f[-1]\mathcal{F}) & \xrightarrow{\sim} & R\Gamma(S_1, \phi_{f_1}[-1]\mathcal{F}_1) \otimes^L R\Gamma(S_2, \phi_{f_2}[-1]\mathcal{F}_2) \end{array}$$

commutes.

Proof. One should first look to the appendix for our conventions about $\otimes^L$.

Let $K_{i,n}, \epsilon_{i,n}$ ($n \in \mathbb{N}$) be given by assumption 1.1.1 for f_i ($i = 1, 2$). Then $K_n := K_{1,n} \times K_{2,n}$ and $\epsilon_n := min\{\epsilon_{1,n}, \epsilon_{2,n}\}$ satisfy also the assumption 1.1.1 for f, by the beginning of the proof of corollary 1.2.1. Let $g := (f_1, f_2)$, with $f = a \circ g$.

(1) Let $\mathbb{K} = \mathbb{R}$. Then we have by remark 1.1.3(1) the isomorphisms (for $-\epsilon_n \leq w < 0$):

$$R\Gamma(S, \mu_f \mathcal{F}) \xleftarrow{\sim} cone\big(Rf_*(\mathcal{F}|K_n)_0 \xrightarrow{sp_f} Rf_*(\mathcal{F}|K_n)_w\big)[-1]$$
$$\xrightarrow{\sim} cone\big(Ra_*(Rg_*(\mathcal{F}|K_n))_0 \xrightarrow{sp_a} Ra_*(Rg_*(\mathcal{F}|K_n))_w\big)[-1]\,.$$

Now we apply lemma 1.2.3, with $\epsilon := \epsilon_n$, $\mathcal{G} := Rg_*(\mathcal{F}|K_n)$, and identify the morphisms $(\alpha_1, \alpha_2), (\beta_1, \beta_2)$ with the help of the **Künneth formula** (see lemma 1.4.2) in the following commutative diagrams (with $\mathcal{F}'_i := \mathcal{F}_i|K_{i,n}$ for $i = 1, 2$).

(i) $sp_{f_1} \otimes^L id \simeq \beta_1$:

$$\begin{array}{ccc} (Rf_{1*}\mathcal{F}'_1)_0 \otimes^L (Rf_{2*}\mathcal{F}'_2)_0 & \xrightarrow{\sim} & \mathcal{G}_{(0,0)} \\ \uparrow\wr & & \uparrow\wr \\ R\Gamma([w,0], Rf_{1*}\mathcal{F}'_1) \otimes^L (Rf_{2*}\mathcal{F}'_2)_0 & \xrightarrow{\sim} & R\Gamma([(w,0),(0,0)], \mathcal{G}) \\ \downarrow & & \downarrow \\ (Rf_{1*}\mathcal{F}'_1)_w \otimes^L (Rf_{2*}\mathcal{F}'_2)_0 & \xrightarrow{\sim} & \mathcal{G}_{(w,0)}\,. \end{array}$$

(ii) $id \otimes^L sp_{f_2} \simeq \beta_2$:

$$\begin{array}{ccc} (Rf_{1*}\mathcal{F}'_1)_0 \otimes^L (Rf_{2*}\mathcal{F}'_2)_0 & \xrightarrow{\sim} & \mathcal{G}_{(0,0)} \\ \uparrow\wr & & \uparrow\wr \\ (Rf_{1*}\mathcal{F}'_1)_0 \otimes^L R\Gamma([0,w], Rf_{2*}\mathcal{F}'_2) & \xrightarrow{\sim} & R\Gamma([(0,w),(0,0)], \mathcal{G}) \\ \downarrow & & \downarrow \\ (Rf_{1*}\mathcal{F}'_1)_0 \otimes^L (Rf_{2*}\mathcal{F}'_2)_w & \xrightarrow{\sim} & \mathcal{G}_{(0,w)}\,. \end{array}$$

(iii) $id \otimes^L sp_{f_2} \simeq \alpha_1$:

$$\begin{array}{ccc}
(Rf_{1*}\mathcal{F}'_1)_w \otimes^L (Rf_{2*}\mathcal{F}'_2)_0 & \xrightarrow{\sim} & \mathcal{G}_{(w,0)} \\
\uparrow\wr & & \uparrow\wr \\
(Rf_{1*}\mathcal{F}'_1)_w \otimes^L R\Gamma([0,w], Rf_{2*}\mathcal{F}'_2) & \xrightarrow{\sim} & R\Gamma([(w,0),(w,w)], \mathcal{G}) \\
\downarrow & & \downarrow \\
(Rf_{1*}\mathcal{F}'_1)_w \otimes^L (Rf_{2*}\mathcal{F}'_2)_w & \xrightarrow{\sim} & \mathcal{G}_{(w,w)}\ .
\end{array}$$

(iv) $sp_{f_1} \otimes^L id \simeq \alpha_2$:

$$\begin{array}{ccc}
(Rf_{1*}\mathcal{F}'_1)_0 \otimes^L (Rf_{2*}\mathcal{F}'_2)_w & \xrightarrow{\sim} & \mathcal{G}_{(0,w)} \\
\uparrow\wr & & \uparrow\wr \\
R\Gamma([w,0], Rf_{1*}\mathcal{F}'_1) \otimes^L (Rf_{2*}\mathcal{F}'_2)_w & \xrightarrow{\sim} & R\Gamma([(0,w),(w,w)], \mathcal{G}) \\
\downarrow & & \downarrow \\
(Rf_{1*}\mathcal{F}'_1)_w \otimes^L (Rf_{2*}\mathcal{F}'_2)_w & \xrightarrow{\sim} & \mathcal{G}_{(w,w)}\ .
\end{array}$$

(2) Let $\mathbb{K} = \mathbb{C}$. Then we have by remark 1.1.3(1) the isomorphisms (for $-\epsilon_n \leq w < 0$):

$$R\Gamma(S, \tilde{\phi}_f(\mathcal{F})) \xleftarrow{\ \sim\ } cone\Big(Rf_*(\mathcal{F}|K_n)_0 \overset{sp}{\to} R\Gamma\big(\widetilde{S_{|w|}}, \pi^* Rf_*(\mathcal{F}|K_n^{|w|})\big)\Big)$$

$$\xrightarrow{\ \sim\ } cone\Big(Ra_*(Rg_*(\mathcal{F}|K_n))_0 \overset{sp}{\to} R\Gamma\big(\widetilde{S_{|w|}}, \pi^* Ra_*(Rg_*(\mathcal{F}|K_n^{|w|}))\big)\Big)\,.$$

Now we apply lemma 1.2.4, with $\epsilon := \epsilon_n$, $\mathcal{G} := Rg_*(\mathcal{F}|K_n)$, and identify the morphisms $(\pi^*\alpha_1, \pi^*\alpha_2)$ and (β_1, β_2) with the help of the **Künneth morphisms** (see appendix) in the following commutative diagrams (with $\mathcal{F}'_i := \mathcal{F}_i|K_{i,n}$ and $\mathcal{F}''_i := \mathcal{F}_i|K_{i,n}^{|w|}$ for $i = 1, 2$).

(i′) $sp_{f_1} \otimes^L id \simeq \beta_1$:

$$\begin{array}{ccc}
(Rf_{1*}\mathcal{F}'_1)_0 \otimes^L (Rf_{2*}\mathcal{F}'_2)_0 & \xrightarrow{\sim} & \mathcal{G}_{(0,0)} \\
\uparrow\wr & & \uparrow\wr \\
R\Gamma(\{|x| \leq |w|\}, Rf_{1*}\mathcal{F}'_1) \otimes^L (Rf_{2*}\mathcal{F}'_2)_0 & \xrightarrow{\sim} & R\Gamma(\{|x| \leq |w|\} \times \{0\}, \mathcal{G}) \\
\downarrow & & \downarrow \\
R\Gamma(\mathbb{R}, \pi^* Rf_{1*}\mathcal{F}''_1) \otimes^L (Rf_{2*}\mathcal{F}'_2)_0 & \xrightarrow{\sim} & R\Gamma(\mathbb{R}, \pi^* s_1^*\mathcal{G})\ .
\end{array}$$

The last horizontal morphism is the **projection morphism** (see appendix) applied to

$$\pi^* s_1^* \mathcal{G} \simeq (\pi^* Rf_{1*}\mathcal{F}''_1) \otimes^L k^*((Rf_{2*}\mathcal{F}'_2)_0)\,,$$

with k the constant map $\mathbb{R} \to \{0\}$. Note that this morphism commutes in our case with the natural action of $\pi_1 = \pi_1(S_{|w|}) \simeq \mathbb{Z}$ so that we consider the above diagram in $D(R_3[\pi_1])$. Moreover, it commutes with the forgetful functor and is therefore an isomorphism by the commutative diagram

$$\begin{array}{ccc} R\Gamma(\mathbb{R}, \pi^* Rf_{1*}\mathcal{F}_1'') \otimes^L (Rf_{2*}\mathcal{F}_2')_0 & \longrightarrow & R\Gamma(\mathbb{R}, \pi^* s_1^* \mathcal{G}) \\ \downarrow \wr & & \downarrow \wr \\ (\pi^* Rf_{1*}\mathcal{F}_1'')_0 \otimes^L (Rf_{2*}\mathcal{F}_2')_0 & \xrightarrow{\sim} & (\pi^* s_1^* \mathcal{G})_0 \,. \end{array}$$

In the same way one gets

(ii′) $id \otimes^L sp_{f_2} \simeq \beta_2$:

$$\begin{array}{ccc} (Rf_{1*}\mathcal{F}_1')_0 \otimes^L (Rf_{2*}\mathcal{F}_2')_0 & \xrightarrow{\sim} & \mathcal{G}_{(0,0)} \\ \uparrow \wr & & \uparrow \wr \\ (Rf_{1*}\mathcal{F}_1')_0 \otimes^L R\Gamma(\{|y| \leq |w|\}, Rf_{2*}\mathcal{F}_2') & \xrightarrow{\sim} & R\Gamma(\{0\} \times \{|y| \leq |w|\}, \mathcal{G}) \\ \downarrow & & \downarrow \\ (Rf_{1*}\mathcal{F}_1')_0 \otimes^L R\Gamma(\mathbb{R}, \pi^* Rf_{2*}\mathcal{F}_2'') & \xrightarrow{\sim} & R\Gamma(\mathbb{R}, \pi^* s_2^* \mathcal{G}) \,. \end{array}$$

For the description of α_1 we look at the inclusion

$$i : h(I_1(w) \times S^1) \to S_{|w|} \times \{|y| \leq |w|\} \,.$$

Note that

$$h : I_1(w) \times S^1 \to h(I_1(w) \times S^1)$$

is an isomorphism with $q = pr \circ i \circ h$. Here pr is by definition the projection

$$S_{|w|} \times \{|y| \leq |w|\} \to S_{|w|} \,,$$

and

$$i \circ h : \{(w, 0)\} \times S^1 \xrightarrow{\sim} S_{|w|} \times \{0\} \,.$$

Then we have

$$(Rf_{1*}\mathcal{F}_1'') \otimes^L (Rf_{2*}\mathcal{F}_2'') \simeq Rq_*(h^* \mathcal{G} | \{(w, w)\} \times S^1)$$

and

$$\begin{array}{ccc} Rpr_*\big(Rf_{1*}\mathcal{F}_1''\big) \boxtimes^L (Rf_{2*}\mathcal{F}_2')\big) & \xrightarrow{\sim} & Rpr_*(\mathcal{G} | S_{|w|} \times \{|y| \leq |w|\}) \\ \downarrow \wr & & \\ (Rf_{1*}\mathcal{F}_1'') \otimes^L k^*(R\Gamma(\{|y| \leq |w|\}, Rf_{2*}\mathcal{F}_2')) \,, & & \end{array}$$

together with the commutative diagram

$$\begin{array}{ccc} (Rf_{1*}\mathcal{F}_1'') \otimes^L k^*((Rf_{2*}\mathcal{F}_2')_0) & \xrightarrow{\sim} & Rq_*(h^*\mathcal{G}|\{(w,0)\} \times S^1) \\ \wr\uparrow id\otimes^L k^* ad & & \uparrow\wr \\ (Rf_{1*}\mathcal{F}_1'') \otimes^L k^*(R\Gamma(\{|y| \le |w|\}, Rf_{2*}\mathcal{F}_2')) & \xrightarrow{ad_i} & Rq_*(h^*\mathcal{G}|I_1(w) \times S^1) \end{array}$$

(with k the constant map on $S_{|w|}$). Applying $R\Gamma(\mathbb{R}, \pi^* \cdot)$ and using the **projection (iso)morphism** as before one gets the commutative diagram

(iii′) $id \otimes^L sp_{f_2} \simeq R\Gamma(\mathbb{R}, \pi^*\alpha_1)$:

$$\begin{array}{ccc} R\Gamma(\mathbb{R}, \pi^*(Rf_{1*}\mathcal{F}_1'')) \otimes^L (Rf_{2*}\mathcal{F}_2')_0 & \xrightarrow{\sim} & R\Gamma(\mathbb{R}, \pi^* Rq_*\mathcal{G}_1) \\ \uparrow\wr & & \uparrow\wr \\ R\Gamma(\mathbb{R}, \pi^*(Rf_{1*}\mathcal{F}_1'')) \otimes^L R\Gamma(\{|y| \le |w|\}, Rf_{2*}\mathcal{F}_2') & \xrightarrow{\sim} & R\Gamma(\mathbb{R}, \pi^* Rq_*\mathcal{G}_2) \\ \downarrow & & \downarrow \\ R\Gamma(\mathbb{R}, \pi^*(Rf_{1*}\mathcal{F}_1'')) \otimes^L R\Gamma(\mathbb{R}, \pi^*(Rf_{2*}\mathcal{F}_2'')) & \xrightarrow{\sim} & R\Gamma(\mathbb{R}, \pi^* Rq_*\mathcal{G}_3)\ , \end{array}$$

with

$$\mathcal{G}_1 := h^*\mathcal{G}|\{(w,0)\} \times S^1\ , \quad \mathcal{G}_2 := h^*\mathcal{G}|I_1(w) \times S^1$$

and $\mathcal{G}_3 := h^*\mathcal{G}|\{(w,w)\} \times S^1$. Here we also use that the natural morphism

$$R\Gamma(\mathbb{R}, \pi^* Rf_{1*}\mathcal{F}_1'') \otimes^L R\Gamma(\mathbb{R}, \pi^* Rf_{2*}\mathcal{F}_2'') \to R\Gamma\big(\mathbb{R}, \pi^*(Rf_{1*}\mathcal{F}_1'' \otimes^L Rf_{2*}\mathcal{F}_2'')\big)$$

is an isomorphism (by the same argument as for the projection morphism).

In the same way one finally gets the commutative diagram

(iv′) $sp_{f_1} \otimes^L id \simeq R\Gamma(\mathbb{R}, \pi^*\alpha_2)$:

$$\begin{array}{ccc} (Rf_{1*}\mathcal{F}_1')_0 \otimes^L R\Gamma(\mathbb{R}, \pi^*(Rf_{2*}\mathcal{F}_2'')) & \xrightarrow{\sim} & R\Gamma(\mathbb{R}, \pi^* Rq_*\mathcal{G}_4) \\ \uparrow\wr & & \uparrow\wr \\ R\Gamma(\{|x| \le |w|\}, Rf_{1*}\mathcal{F}_1') \otimes^L R\Gamma(\mathbb{R}, \pi^*(Rf_{2*}\mathcal{F}_2'')) & \xrightarrow{\sim} & R\Gamma(\mathbb{R}, \pi^* Rq_*\mathcal{G}_5) \\ \downarrow & & \downarrow \\ R\Gamma(\mathbb{R}, \pi^*(Rf_{1*}\mathcal{F}_1'')) \otimes^L R\Gamma(\mathbb{R}, \pi^*(Rf_{2*}\mathcal{F}_2'')) & \xrightarrow{\sim} & R\Gamma(\mathbb{R}, \pi^* Rq_*\mathcal{G}_3)\ , \end{array}$$

with $\mathcal{G}_4 := h^*\mathcal{G}|\{(0,w)\} \times S^1, \mathcal{G}_5 := h^*\mathcal{G}|I_2(w) \times S^1$ and $\mathcal{G}_3 := h^*\mathcal{G}|\{(w,w)\} \times S^1$.

(3) Now all the important data are fixed and we leave (the "paradise" of) the derived category of sheaves (which describes the geometry in such a nice way) by the following trick: by a result of Spaltenstein ([Sp, thm.C]), we can choose for each complex $C \in D(R)$ a K-**projective resolution** P (if C is bounded above, this is just a classical projective resolution). These have the following important properties:

1. If two vertices of a distinguished triangle are K-projective, then so is the third and P is K-projective $\Leftrightarrow$ $P[1]$ is K-projective ([Sp, prop. 1.3]).

2. Every morphism $P \to A$ in $D(R)$ comes from a **true morphism** $P \to A$ of complexes, and any two lifts of the same morphism are homotopic ([Sp, prop.1.4]).

3. A K-projective complex is K-flat so that it can be used for calculating $\otimes^L$ (see [Sp, prop. 5.8] and the appendix).

Let $A_0 \xrightarrow{a} A_1$ be a true morphism of K-projective complexes, isomorphic in $D(R_1)$ to

$$(Rf_{1*}\mathcal{F}'_1)_0 \xrightarrow{sp_{f_1}} (Rf_{1*}\mathcal{F}'_1)_w$$

for $\mathbb{K} = \mathbb{R}$ and isomorphic in $D(R_1[\pi_1])$ to

$$(Rf_{1*}\mathcal{F}'_1)_0 \xrightarrow{sp_{f_1}} R\Gamma(\mathbb{R}, \pi^*(Rf_{1*}\mathcal{F}''_1))$$

for $\mathbb{K} = \mathbb{C}$. Let similarly $B_0 \xrightarrow{b} B_1$ be a true morphism of K-projective complexes, isomorphic in $D(R_2)$ to

$$(Rf_{2*}\mathcal{F}'_2)_0 \xrightarrow{sp_{f_2}} (Rf_{2*}\mathcal{F}'_2)_w$$

for $\mathbb{K} = \mathbb{R}$ and isomorphic in $D(R_2[\pi_1])$ to

$$(Rf_{2*}\mathcal{F}'_2)_0 \xrightarrow{sp_{f_2}} R\Gamma(\mathbb{R}, \pi^*(Rf_{2*}\mathcal{F}''_2))$$

for $\mathbb{K} = \mathbb{C}$.

By remark 1.1.3(1) we therefore have for $\mathbb{K} = \mathbb{R}$ isomorphisms:

$$R\Gamma(S_1, \mu_{f_1}\mathcal{F}_1) \simeq cone(A_0 \xrightarrow{a} A_1)[-1]\,,$$

$$R\Gamma(S_2, \mu_{f_2}\mathcal{F}_2) \simeq cone(B_0 \xrightarrow{b} B_1)[-1]\,.$$

Similarly, we have for $\mathbb{K} = \mathbb{C}$ isomorphisms:

$$R\Gamma(S_1, \tilde{\phi}_{f_1}[-1]\mathcal{F}_1) \simeq cone(A_0 \xrightarrow{a} A_1)[-1]\,,$$

$$R\Gamma(S_2, \tilde{\phi}_{f_2}[-1]\mathcal{F}_2) \simeq cone(B_0 \xrightarrow{b} B_1)[-1]\,.$$

(4) By definition one has

$$cone(A_0 \xrightarrow{a} A_1)[-1] = s(A_0 \xrightarrow{-a} A_1)\,,$$

with A_α in degree α. Here we use the following convention for the **simple complex** $s(\cdot)$ of a **naive double complex** $(A_{\alpha,i}, d_\alpha, d_i)$ (naive means $d_\alpha \circ d_i = d_i \circ d_\alpha$):

$$s(A_{\alpha,i})_n := \oplus_{\alpha+i=n} A_{\alpha,i}\,,$$

with differential $d_\alpha + (-1)^\alpha d_i$.

Therefore we get for $\mathbb{K} = \mathbb{R}$ an isomorphism in $D(R_3)$:

$$R\Gamma(S_1, \mu_{f_1}\mathcal{F}_1) \otimes^L R\Gamma(S_2, \mu_{f_2}\mathcal{F}_2) \simeq s(A_0 \xrightarrow{-a} A_1) \otimes s(B_0 \xrightarrow{-b} B_1) =: S_1 .$$

Similarly, we get for $\mathbb{K} = \mathbb{C}$ an isomorphism in $D(R_3[\pi_1])$:

$$R\Gamma(S_1, \tilde{\phi}_{f_1}[-1]\mathcal{F}_1) \otimes^L R\Gamma(S_2, \tilde{\phi}_{f_2}[-1]\mathcal{F}_2) \simeq s(A_0 \xrightarrow{-a} A_1) \otimes s(B_0 \xrightarrow{-b} B_1) =: S_1 .$$

But by lemma 1.2.3 and the isomorphisms (i)-(iv), we also get for $\mathbb{K} = \mathbb{R}$ the next isomorphisms in $D(R_3)$:

$$R\Gamma(S, \nu_f\mathcal{F}) \xleftarrow{\sim} cone\big(A_1 \otimes B_0 \oplus A_0 \otimes B_1 \xrightarrow{id\otimes b - a\otimes id} A_1 \otimes B_1\big)[-1] =: C ,$$

$$R\Gamma(S, \mu_f\mathcal{F}) \xleftarrow{\sim} cone\big(A_0 \otimes B_0 \xrightarrow{(a\otimes id + id\otimes b, 0)} C\big)[-1] .$$

Similarly, by lemma 1.2.4 and the isomorphisms (i')-(iv'), we get for $\mathbb{K} = \mathbb{C}$ the next isomorphisms in $D(R_3[\pi_1])$:

$$R\Gamma(S, \tilde{\psi}_f\mathcal{F}) \xleftarrow{\sim} cone\big(A_1 \otimes B_0 \oplus A_0 \otimes B_1 \xrightarrow{id\otimes b - a\otimes id} A_1 \otimes B_1\big)[-1] =: C ,$$

$$R\Gamma(S, \tilde{\phi}_f[-1]\mathcal{F}) \xleftarrow{\sim} cone\big(A_0 \otimes B_0 \xrightarrow{(a\otimes id + id\otimes b, 0)} C\big)[-1] .$$

Using the identity

$$s(K_0 \xrightarrow{k} K_1 \xrightarrow{-k'} K_2) = s\big(K_0 \xrightarrow{(k,0)} s(K_1 \xrightarrow{k'} K_2)\big) ,$$

with K_α in degree α, we finally get that $R\Gamma(S, \mu_f\mathcal{F})$ or $R\Gamma(S, \tilde{\phi}_f[-1]\mathcal{F})$ is isomorphic to

$$s\big(A_0 \otimes B_0 \xrightarrow{-a\otimes id - id\otimes b} A_1 \otimes B_0 \oplus A_0 \otimes B_1 \xrightarrow{id\otimes b - a\otimes id} A_1 \otimes B_1\big) .$$

Denote this complex by S_2. Then there is a **canonical sign-isomorphism**

$$\tau : S_1 \to S_2 ,$$

because both complexes are **simple complexes** (with respect to different **orderings**) of the same **naive fourfold complex** (compare [De2, 1.1.4]):

$$\begin{array}{ccc} A_1 \otimes B_0 & \xrightarrow{-id\otimes b} & A_1 \otimes B_1 \\ \uparrow{\scriptstyle -a\otimes id} & & \uparrow{\scriptstyle -a\otimes id} \\ A_0 \otimes B_0 & \xrightarrow{-id\otimes b} & A_0 \otimes B_1 \end{array}$$

(with $A^i_\alpha \otimes B^j_\beta$ in degree (α, β, i, j)). S_1 is the simple complex with respect to the ordering $\alpha < i < \beta < j$ and S_2 is the simple complex with respect to the ordering $\alpha < \beta < i < j$. So τ is explicitly given by

$$\tau | A^i_\alpha \otimes B^j_\beta = (-1)^{i\beta} \cdot id . \qquad \square$$

1.3 The Thom-Sebastiani Isomorphism in the derived category

In the last section we explained "absolute versions" of the Thom-Sebastiani Theorem based on **cohomological versions of a Milnor fibration**. Following ideas of Massey [Ma2], we show now under slightly **different assumptions**, that the isomorphism of the Thom-Sebastiani Theorem comes from a **natural morphism** in the derived category. This observation has important applications (for a specific example compare with [Ma2, sec. 2.3]).

First we treat the real valued case of the **micro-localization functor** (this is only implicitly used in [Ma2]). We explain why in the **local case** ($S_i = \{x_i\}$ a point for $i = 1, 2$), the isomorphism of theorem 1.2.2(1) is indeed induced by a **natural morphism** in the derived category.

Let $i_A : A \to X$ and $i_B : B \to Y$ be two inclusions of closed subsets. Denote the inclusions

$$j_A : U_A := X\backslash A \to X \; , \; j_B : U_B := Y\backslash B \to Y \; ,$$

$$i := i_A \times i_B : A \times B \to X \times Y \quad \text{and} \quad j : X \times Y\backslash A \times B \to X \times Y \; .$$

Fix $\mathcal{F} \in ob(D(X))$, $\mathcal{G} \in ob(D(Y))$ with $\mathcal{H} := \mathcal{F} \boxtimes^L \mathcal{G} \in ob(D(X \times Y))$. Then the composed morphism

$$R\Gamma_A(\mathcal{F}) \boxtimes^L R\Gamma_B(\mathcal{G}) \longrightarrow \mathcal{F} \boxtimes^L \mathcal{G} \xrightarrow{ad_j} Rj_*j^*\mathcal{H}$$

is the zero morphism, since

$$(R\Gamma_A(\mathcal{F}) \boxtimes^L R\Gamma_B(\mathcal{G}))_{(x,y)} \simeq (R\Gamma_A(\mathcal{F}))_x \otimes^L (R\Gamma_B(\mathcal{G}))_y \simeq 0$$

for $x \notin A$ or $y \notin B$. Therefore one gets the induced morphism

$$R\Gamma_A(\mathcal{F}) \boxtimes^L R\Gamma_B(\mathcal{G}) \xrightarrow{\cup} R\Gamma_{A\times B}(\mathcal{F} \boxtimes^L \mathcal{G}) \; ,$$

which is uniquely defined (see [BBD, prop. 1.1.9]) by the commutativity of

$$\begin{array}{ccc} R\Gamma_A(\mathcal{F}) \boxtimes^L R\Gamma_B(\mathcal{G}) & \xrightarrow{\cup} & R\Gamma_{A\times B}(\mathcal{F} \boxtimes^L \mathcal{G}) \\ \| & & \downarrow \\ R\Gamma_A(\mathcal{F}) \boxtimes^L R\Gamma_B(\mathcal{G}) & \longrightarrow & \mathcal{F} \boxtimes^L \mathcal{G} \; . \end{array}$$

We use for the corresponding stalk morphism a description in terms of **relative cohomology**. Since

$$R\Gamma_A(\mathcal{F}) \simeq cone(\mathcal{F} \to Rj_{A*}j_A^*\mathcal{F})[-1]$$

and

$$R\Gamma_B(\mathcal{G}) \simeq cone(\mathcal{G} \to Rj_{B*}j_B^*\mathcal{G})[-1] \; ,$$

we get as in (4) of the proof of theorem 1.2.2:

$$(R\Gamma_A(\mathcal{F}) \boxtimes^L R\Gamma_B(\mathcal{G}))_{(x,y)} \simeq S_1 ,$$

with S_1 the **simple complex** with respect to the ordering $\alpha < i < \beta < j$ of the **naive fourfold complex** (with $\alpha, \beta \in \{0,1\}$):

$$\begin{array}{ccc}
(Rj_{A*}j_A^*\mathcal{F})_x \otimes^L \mathcal{G}_y & \xrightarrow{-id\otimes^L ad_{j_B}} & (Rj_{A*}j_A^*\mathcal{F})_x \otimes^L (Rj_{B*}j_B^*\mathcal{G})_y \\
\uparrow{\scriptstyle -ad_{j_A}\otimes^L id} & & \uparrow{\scriptstyle -ad_{j_A}\otimes^L id} \\
\mathcal{F}_x \otimes^L \mathcal{G}_y & \xrightarrow{-id\otimes^L ad_{j_B}} & \mathcal{F}_x \otimes^L (Rj_{B*}j_B^*\mathcal{G})_y .
\end{array}$$

Of course one has to represent everything by **true morphisms** of K-**projective** complexes. Similarly

$$R\Gamma_{A\times B}(\mathcal{H}) \simeq cone(\mathcal{H} \to Rj_*j^*\mathcal{H})[-1] .$$

Let us denote the inclusions of open subsets

$$k_A := j_A \times id : U_A \times Y \to X \times Y , \; k_B := id \times j_B : X \times U_B \to X \times Y$$

and $k := j_A \times j_B : U_A \times U_B \to X \times Y$. Then we have a distinguished triangle of **Mayer-Vietoris type** (see [KS, prop. 2.3.9(v), formula (2.6.28)]):

$$Rj_*j^*\mathcal{H} \longrightarrow Rk_{A*}k_A^*\mathcal{H} \oplus Rk_{B*}k_B^*\mathcal{H} \xrightarrow{(+,-)} Rk_*k^*\mathcal{H} \xrightarrow{[1]} \tag{1.11}$$

(all maps are induced by restriction morphisms).

Let $j_U : U \to X$ and $j_V : V \to Y$ be two inclusions with

$$j_{U\times V} : U \times V \to X \times Y .$$

Then the isomorphism

$$j_{U\times V}^*((Rj_{U*}j_U^*\mathcal{F}) \boxtimes^L (Rj_{V*}j_V^*\mathcal{G})) \simeq j_U^*\mathcal{F} \boxtimes^L j_V^*\mathcal{G} \simeq j_{U\times V}^*\mathcal{H}$$

induces by **adjunction** a natural morphism

$$(Rj_{U*}j_U^*\mathcal{F}) \boxtimes^L (Rj_{V*}j_V^*\mathcal{G}) \to Rj_{U\times V*}j_{U\times V}^*\mathcal{H} .$$

If we apply this to $(j_U, j_V) = (j_A, id), (id, j_B)$ and (j_A, j_B) we get a commutative diagram:

$$\begin{array}{ccccc}
(R\Gamma_A(\mathcal{F}) \boxtimes^L R\Gamma_B(\mathcal{G}))_{(x,y)} & \longrightarrow & \mathcal{F}_x \otimes^L \mathcal{G}_y & \longrightarrow & \cdots \\
\downarrow{\scriptstyle \cup} & & \downarrow{\scriptstyle \wr} & & \\
(R\Gamma_{A\times B}(\mathcal{H}))_{(x,y)} & \longrightarrow & \mathcal{H}_{(x,y)} & \longrightarrow & \cdots
\end{array} \tag{1.12}$$

$$\begin{array}{ccc}
(Rj_{A*}j_A^*\mathcal{F})_x \otimes^L \mathcal{G}_y \oplus \mathcal{F}_x \otimes^L (Rj_{B*}j_B^*\mathcal{G})_y & \xrightarrow{(+,-)} & (Rj_{A*}j_A^*\mathcal{F})_x \otimes^L (Rj_{B*}j_B^*\mathcal{G})_y \\
\downarrow{\scriptstyle ad\oplus ad} & & \downarrow{\scriptstyle ad} \\
\cdots (Rk_{A*}k_A^*\mathcal{H})_{(x,y)} \oplus (Rk_{B*}k_B^*\mathcal{H})_{(x,y)} & \xrightarrow{(+,-)} & (Rk_*k^*\mathcal{H})_{(x,y)}\,.
\end{array}$$

We also need the following **supplement** to lemma 1.2.3 (the proof of this lemma is left to the reader. Use remark 1.2.1(3) and a **Mittag-Leffler argument** [KS, prop. 2.7.1(iii)]). With the notation of lemma 1.2.1 (with $\mathbb{K} = \mathbb{R}$), let

$$U_1 := B \cap \{x < 0\}\,,\ U_2 := B \cap \{y < 0\} \quad \text{and} \quad U := U_1 \cup U_2\,.$$

Lemma 1.3.1. *Under the assumption and with the notation of lemma 1.2.3, we have the following properties:*

1. *$R\Gamma(U,\mathcal{G}) \to R\Gamma(B \cap \{a < 0\},\mathcal{G})$ is an isomorphism.*
2. *There exists a natural isomorphism of distinguished triangles*

$$\begin{array}{ccccccc}
(Ra_*\mathcal{G})_w & \xrightarrow{\alpha} & \mathcal{G}_{(w,0)} \oplus \mathcal{G}_{(0,w)} & \xrightarrow{(\alpha_1,-\alpha_2)} & \mathcal{G}_{(w,w)} & \xrightarrow{[1]} & \\
\uparrow\wr & & \uparrow\wr & & \uparrow\wr & & \\
R\Gamma(U,\mathcal{G}) & \longrightarrow & R\Gamma(U_1,\mathcal{G}) \oplus R\Gamma(U_2,\mathcal{G}) & \xrightarrow{(+,-)} & R\Gamma(U_1 \cap U_2,\mathcal{G}) & \xrightarrow{[1]} &
\end{array}$$

(the lower triangle is the **Mayer-Vietoris triangle** *with respect to the open cover $\{U_1, U_2\}$ of U).*

3. *There exists a commutative diagram*

$$\begin{array}{ccc}
R\Gamma(B,\mathcal{G}) & \longrightarrow & R\Gamma(U_1,\mathcal{G}) \oplus R\Gamma(U_2,\mathcal{G}) \\
\downarrow\wr & & \downarrow\wr \\
\mathcal{G}_{(0,0)} & \xrightarrow{(\beta_1,\beta_2)} & \mathcal{G}_{(w,0)} \oplus \mathcal{G}_{(0,w)}
\end{array}$$

(the morphisms different from (β_1, β_2) are restriction morphisms).

Now we can prove the

Proposition 1.3.1. *Let $f_i : X_i \to \mathbb{R}$, $S_i := \{x_i\} \subset \{f_i = 0\}$ and $\mathcal{F}_i \in ob(D(X_i))$ satisfy the assumption 1.1.1 ($i = 1, 2$). Define*

$$f := f_1 \oplus f_2 := a \circ (f_1, f_2) : X := X_1 \times X_2 \to \mathbb{R}$$

and $\mathcal{F} := \mathcal{F}_1 \boxtimes^L \mathcal{F}_2 \in ob(D(X))$.

1. *The natural morphism*

$$(R\Gamma_{\{f_1 \geq 0\} \times \{f_2 \geq 0\}}(\mathcal{F}))_{(x_1,x_2)} \to (R\Gamma_{\{f \geq 0\}}(\mathcal{F}))_{(x_1,x_2)}$$

is an isomorphism.

2. *The morphism*

$$(R\Gamma_{\{f_1\geq 0\}}(\mathcal{F}_1))_{x_1} \otimes^L (R\Gamma_{\{f_2\geq 0\}}(\mathcal{F}_2))_{x_2} \xrightarrow{\cup} (R\Gamma_{\{f_1\geq 0\}\times\{f_2\geq 0\}}(\mathcal{F}))_{(x_1,x_2)}$$

is isomorphic by (1.) *to the isomorphism of theorem* 1.2.2(1).

Proof. Let $K_{i,n}, \epsilon_{i,n}$ ($n \in \mathbb{N}$) be given by assumption 1.1.1 for f_i ($i = 1, 2$). Define $K_n := K_{1,n} \times K_{2,n}$ and $\epsilon_n := min\{\epsilon_{1,n}, \epsilon_{2,n}\}$. Now we apply lemma 1.3.1 with

$$B_n := B(\epsilon_n),\ U_{1,n} := B_n \cap \{x < 0\},\ U_{2,n} := B_n \cap \{y < 0\},\ U_n := U_{1,n} \cup U_{2,n}$$

and $\mathcal{G}_n := Rg_*(\mathcal{F}|K_n)$ (with $g := (f_1, f_2)$). Then the **vertical maps** in

$$\begin{array}{ccccc}
R\Gamma(B_n, \mathcal{G}_n) & \longrightarrow & R\Gamma(U_n, \mathcal{G}_n) & \xrightarrow{\sim} & R\Gamma(B_n \cap \{a < 0\}, \mathcal{G}_n) \\
\downarrow & & \downarrow & & \downarrow \\
R\Gamma(B_{n+1}, \mathcal{G}_{n+1}) & \longrightarrow & R\Gamma(U_{n+1}, \mathcal{G}_{n+1}) & \longrightarrow & R\Gamma(B_{n+1} \cap \{a < 0\}, \mathcal{G}_{n+1}) ,
\end{array}$$

and the **horizontal maps** in

$$\begin{array}{ccc}
R\Gamma(U_n, \mathcal{G}_n) & \longrightarrow & R\Gamma(U_{n+1}, \mathcal{G}_{n+1}) \\
\downarrow & & \downarrow \\
R\Gamma(U_{1,n}, \mathcal{G}_n) \oplus R\Gamma(U_{2,n}, \mathcal{G}_n) & \longrightarrow & R\Gamma(U_{1,n+1}, \mathcal{G}_{n+1}) \oplus R\Gamma(U_{2,n+1}, \mathcal{G}_{n+1}) \\
\downarrow{\scriptstyle (+,-)} & & \downarrow{\scriptstyle (+,-)} \\
R\Gamma(U_{1,n} \cap U_{2,n}, \mathcal{G}_n) & \longrightarrow & R\Gamma(U_{1,n+1} \cap U_{2,n+1}, \mathcal{G}_{n+1})
\end{array}$$

are isomorphisms. Therefore, one gets the commutative diagrams (with

$$i_A : A := \{f_1 \geq 0\} \to X_1,\ i_B : B := \{f_2 \geq 0\} \to X_2,\ l := \{f < 0\} \to X$$

and j, k_A, k_B, k as before):

$$\begin{array}{ccccc}
R\Gamma(B_n, \mathcal{G}_n) & \longrightarrow & R\Gamma(U_{1,n} \cup U_{2,n}, \mathcal{G}_n) & \xrightarrow{\sim} & R\Gamma(B_n \cap \{a < 0\}, \mathcal{G}_n) \\
\downarrow\wr & & \downarrow\wr & & \downarrow\wr \\
\mathcal{F}_{(x_1,x_2)} & \longrightarrow & (Rj_*j^*\mathcal{F})_{(x_1,x_2)} & \xrightarrow{\sim} & (Rl_*l^*\mathcal{F})_{(x_1,x_2)} ,
\end{array}$$

and

$$\begin{array}{ccc}
R\Gamma(U_{1,n} \cup U_{2,n}, \mathcal{G}_n) & \xrightarrow{\sim} & (Rj_*j^*\mathcal{F})_{(x_1,x_2)} \\
\downarrow & & \downarrow \\
R\Gamma(U_{1,n}, \mathcal{G})_n \oplus R\Gamma(U_{2,n}, \mathcal{G}_n) & \xrightarrow{\sim} & (Rk_{A*}k_A^*\mathcal{F})_{(x_1,x_2)} \oplus (Rk_{B*}k_B^*\mathcal{F})_{(x_1,x_2)} \\
\downarrow{\scriptstyle (+,-)} & & \downarrow{\scriptstyle (+,-)} \\
R\Gamma(U_{1,n} \cap U_{2,n}, \mathcal{G}_n) & \xrightarrow{\sim} & (Rk_*k^*\mathcal{F})_{(x_1,x_2)} .
\end{array}$$

The first diagram implies already the first statement of the proposition. Moreover, we have the following commutative diagrams:

$$\begin{array}{ccc} (Rf_{1*}(\mathcal{F}_1|K_{1,n}))_w \otimes^L (Rf_{2*}(\mathcal{F}_2|K_{2,n}))_0 & \xrightarrow[\sim]{K\ddot{u}} & (\mathcal{G}_n)_{(w,0)} \\ \downarrow\wr & & \downarrow\wr \\ (Rj_{A*}j_A^*\mathcal{F}_1)_{x_1} \otimes^L (\mathcal{F}_2)_{x_2} & \xrightarrow{ad} & (Rk_{A*}k_A^*\mathcal{F})_{(x_1,x_2)} \ , \end{array}$$

with $j_A : \{f_1 < 0\} \to X_1$.

$$\begin{array}{ccc} (Rf_{1*}(\mathcal{F}_1|K_{1,n}))_0 \otimes^L (Rf_{2*}(\mathcal{F}_2|K_{2,n}))_w & \xrightarrow[\sim]{K\ddot{u}} & (\mathcal{G}_n)_{(0,w)} \\ \downarrow\wr & & \downarrow\wr \\ (\mathcal{F}_1)_{x_1} \otimes^L (Rj_{B*}j_B^*\mathcal{F}_2)_{x_2} & \xrightarrow{ad} & (Rk_{B*}k_B^*\mathcal{F})_{(x_1,x_2)} \ , \end{array}$$

with $j_B : \{f_2 < 0\} \to X_2$.

$$\begin{array}{ccc} (Rf_{1*}(\mathcal{F}_1|K_{1,n}))_w \otimes^L (Rf_{2*}(\mathcal{F}_2|K_{2,n}))_w & \xrightarrow[\sim]{K\ddot{u}} & (\mathcal{G}_n)_{(w,w)} \\ \downarrow\wr & & \downarrow\wr \\ (Rj_{A*}j_A^*\mathcal{F}_1)_{x_1} \otimes^L (Rj_{B*}j_B^*\mathcal{F}_2)_{x_2} & \xrightarrow{ad} & (Rk_*k^*\mathcal{F})_{(x_1,x_2)} \ . \end{array} \tag{1.13}$$

This implies with the notation of part (4) of the proof of theorem 1.2.2, that the corresponding complex C is isomorphic to

$$C' := cone\big((Rk_{A*}k_A^*\mathcal{H})_{(x,y)} \oplus (Rk_{B*}k_B^*\mathcal{H})_{(x,y)} \xrightarrow{(+,-)} (Rk_*k^*\mathcal{H})_{(x,y)}\big)[-1] \ .$$

And similarly

$$S_2 \xrightarrow[\sim]{ad} cone\big(\mathcal{H}_{(x,y)} \longrightarrow C'\big)[-1] \ .$$

Then the second statement of the proposition follows from the distinguished triangle (1.11) and the commutative diagram (1.12). □

Remark 1.3.1. Proposition 1.3.1(1) corresponds to [Ma2, lem. 1.2] (compare also with [KS, (3.7.12-13), prop. 3.7.15]) and proposition 1.3.1(2) corresponds to [Ma2, prop. 1.3, 1.4], but is more precise. We showed under the assumption 1.1.1 (in the local case), that the "cup-product" induces an isomorphism on the stalks (by reducing this to the Thom-Sebastiani Theorem in the **local** case). Massey established first an abstract isomorphism of these stalk-complexes (using biduality in the proof of [Ma2, prop. 1.3], which doesn't apply in our context) and deduced from this the Thom-Sebastiani Theorem.

We explain the power of proposition 1.3.1 by the **Thom-Sebastiani Isomorphism** ("real case"):

Corollary 1.3.1. *Let $f_i : X_i \to \mathbb{R}$, $K_i \subset \{f_i = 0\}$ and $\mathcal{F}_i \in ob(D(X_i))$ be given. Define*

$$f := f_1 \oplus f_2 := a \circ (f_1, f_2) : X := X_1 \times X_2 \to \mathbb{R} \,,$$

and

$$K := K_1 \times K_2 \ ; \ \mathcal{F} := \mathcal{F}_1 \boxtimes^L \mathcal{F}_2 \in ob(D(X)) \,.$$

Assume that $f_i, \mathcal{F}_i$ satisfy for all $x_i \in K_i$ the assumption 1.1.1, *with $S_i := \{x_i\}$ $(i = 1, 2)$. Denote the inclusions*

$$i_1 : K_1 \to X_1 \ , \ i_2 : K_2 \to X_2$$

and $i := i_1 \times i_2 : K \to X$. Then there is an isomorphism

$$i_1^*(R\Gamma_{\{f_1 \geq 0\}}(\mathcal{F}_1)) \boxtimes^L i_2^*(R\Gamma_{\{f_2 \geq 0\}}(\mathcal{F}_2)) \xrightarrow{\sim} i^*(R\Gamma_{\{f \geq 0\}}(\mathcal{F})) \,.$$

Proof. If we apply i^* to the composed morphism

$$R\Gamma_{\{f_1 \geq 0\}}(\mathcal{F}_1) \boxtimes^L R\Gamma_{\{f_2 \geq 0\}}(\mathcal{F}_2) \xrightarrow{\cup} R\Gamma_{\{f_1 \geq 0\} \times \{f_2 \geq 0\}}(\mathcal{F}) \to R\Gamma_{\{f \geq 0\}}(\mathcal{F}) \,,$$

we get by proposition 1.3.1 the isomorphism

$$\begin{aligned} i_1^*(R\Gamma_{\{f_1 \geq 0\}}(\mathcal{F}_1)) \boxtimes^L i_2^*(R\Gamma_{\{f_2 \geq 0\}}(\mathcal{F}_2)) &\xrightarrow{\sim} \\ i^*\big(R\Gamma_{\{f_1 \geq 0\}}(\mathcal{F}_1) \boxtimes^L R\Gamma_{\{f_2 \geq 0\}}(\mathcal{F}_2)\big) &\xrightarrow{\sim} i^*(R\Gamma_{\{f \geq 0\}}(\mathcal{F})) \,. \end{aligned}$$

□

By the **classical Künneth formula** (see appendix), this implies the

Corollary 1.3.2. *Under the assumption of corollary* 1.3.1, *there is for locally closed K_i $(i = 1, 2)$ an isomorphism*

$$R\Gamma_c(K, R\Gamma_{\{f \geq 0\}}(\mathcal{F})) \simeq R\Gamma_c(K_1, R\Gamma_{\{f_1 \geq 0\}}(\mathcal{F}_1)) \otimes^L R\Gamma_c(K_2, R\Gamma_{\{f_2 \geq 0\}}(\mathcal{F}_2)) \,.$$

As we have already explained, the assumptions of corollary 1.3.1 are satisfied in the subanalytic (o-minimal or analytic (Nash) geometric) context (these notions will be explained in chapter 2 of the book). So corollary 1.3.2 is a **generalization** of theorem 1.0.3.

One also has similar results for the **specialization** and **nearby cycle functor**.

Theorem 1.3.1. *Let $f_i : X_i \to \mathbb{K}$ (with $\mathbb{K} = \mathbb{R}, \mathbb{C}$), $C_i \subset \{f_i = 0\}$ and $\mathcal{F}_i \in ob(D(X_i))$ be given. Consider the fiber product*

$$\begin{array}{ccc} X := X_1 \times_K X_2 & \xrightarrow{p_2} & X_2 \\ {\scriptstyle p_1}\downarrow & & \downarrow{\scriptstyle f_2} \\ X_1 & \xrightarrow{f_2} & \mathbb{K} \,, \end{array}$$

with $f := f_1 \circ p_1 = f_2 \circ p_2 : X \to \mathbb{K}$ *and*

$$C := C_1 \times C_2 \ ; \ \mathcal{F} := \mathcal{F}_1 \boxtimes^L_K \mathcal{F}_2 \in ob(D(X)) \, .$$

Assume that $f_i, \mathcal{F}_i$ *satisfy for all* $x_i \in C_i$ *the assumption* 1.1.1, *with* $S_i := \{x_i\}$ $(i = 1, 2)$. *Denote the inclusions*

$$i_1 : C_1 \to X_1 \ , \ i_2 : C_2 \to X_2 \quad and \quad i := i_1 \times_K i_2 : C \to X \, .$$

1. *Let* $\mathbb{K} = \mathbb{R}$. *Then there is a natural isomorphism*

$$i_1^* \nu_{f_1}(\mathcal{F}_1) \boxtimes^L i_2^* \nu_{f_2}(\mathcal{F}_2)) \xrightarrow{\sim} i^* \nu_f(\mathcal{F}) \, .$$

2. *Let* $\mathbb{K} = \mathbb{C}$. *Then there is a natural isomorphism*

$$i_1^* \psi_{f_1}(\mathcal{F}_1) \boxtimes^L i_2^* \psi_{f_2}(\mathcal{F}_2)) \xrightarrow{\sim} i^* \psi_f(\mathcal{F})$$

   commuting with the corresponding monodromies.

Proof. Let us first remark that there are natural morphisms

$$\nu_{f_1}(\mathcal{F}_1) \boxtimes^L \nu_{f_2}(\mathcal{F}_2) \to \nu_f(\mathcal{F})$$

and

$$\psi_{f_1}(\mathcal{F}_1) \boxtimes^L \psi_{f_2}(\mathcal{F}_2) \to \psi_f(\mathcal{F}) \, .$$

These are induced from the **Künneth morphism** (1.18) (see appendix). The corresponding morphism of stalks is for $(x_1, x_2) \in C$ an isomorphism by theorem 1.2.1. This implies the theorem. □

Remark 1.3.2. Theorem 1.3.1 implies especially theorem 1.0.4, and also the corresponding "real version". For similar results in the algebraic context of **étale cohomology**, the reader should compare with [Schr], [BB, lem. 5.1.1] and [Ill, cor. 4.8].

Now we treat the case of the **vanishing cycle functor**. We follow an idea of Massey [Ma2] and **reduce** the results to the real valued case of the **microlocalization functor**.

Consider the cartesian diagram from the definition of the nearby and vanishing cycle functor:

$$\begin{array}{ccccc} X_0 & \xrightarrow{i} & X & \xleftarrow{\tilde{\pi}} & \widetilde{X^*} \\ \downarrow & & \downarrow f & & \downarrow \tilde{f} \\ \{0\} & \longrightarrow & \mathbb{C} & \xleftarrow{\pi} & \widetilde{\mathbb{C}^*} . \end{array}$$

Since $\{re(z) < 0\} \subset \mathbb{C}$ is simply connected, we can choose a section

$$s : \{re(z) < 0\} \to \widetilde{\mathbb{C}^*}$$

of π (i.e. $\pi \circ s = id$). By base change we get an induced section

$$\tilde{s} : \{re(f) < 0\} \to \widetilde{X^*}$$

of $\tilde{\pi}$, with $\tilde{j} : \{re(f) < 0\} \hookrightarrow X$. If we denote the inclusions

$$k : X_0 \to \{re(f) = 0\} \text{ and } \tilde{i} : \{re(f) = 0\} \to X \,,$$

then we get the morphism of functors

$$r_\psi : \psi_f = i^* R\tilde{\pi}_* \tilde{\pi}^* \to i^* R\tilde{\pi}_* R\tilde{s}_* \tilde{s}^* \tilde{\pi}^* \simeq i^* R\tilde{j}_* \tilde{j}^* = k^* \tilde{i}^* R\tilde{j}_* \tilde{j}^* = k^* \nu_{re(f)} \,. \quad (1.14)$$

Since the diagram

$$\begin{array}{ccc} \psi_f & \xrightarrow{r_\psi} & k^* \nu_{re(f)} \\ \uparrow{\scriptstyle sp_f} & & \uparrow{\scriptstyle k^* sp_{re(f)}} \\ i^* & \xrightarrow{\sim} & k^* \tilde{i}^* \end{array}$$

commutes, we get the morphism of functors

$$r_\phi : \phi_f[-1] \to k^* \mu_{re(f)} \,.$$

Note that this morphism can only be defined after the choice of a section!

Lemma 1.3.2. *Let $l : K \hookrightarrow \{f = 0\}$ and $\mathcal{F} \in ob(D(X))$ be given. Assume that $f, \mathcal{F}$ satisfy for all $x \in K$ the assumption* 1.1.1, *with $S := \{x\}$. Then*

$$l^* r_\phi : l^* \phi_f[-1](\mathcal{F}) \to l^* k^* \mu_{re(f)}(\mathcal{F})$$

is an isomorphism.

Proof. This can be checked on the stalks and follows then from corollary 1.1.1. □

To describe also the monodromy action under this isomorphism, we proced as follows (compare [Ma2, p. 2]. He claimed without a proof that the monodromy isomorphism can be recovered by "moving the halfspace $\{z \in \mathbb{C} |\; re(z) \geq 0\}$ once around"). Consider the cartesian diagram

$$\begin{array}{ccccc} K \times \mathbb{R} & \xrightarrow{i_2} & X \times \mathbb{R} & \xleftarrow{j_2} & \{(x,\theta) \in X \times \mathbb{R} |\; re(f(x)exp(-2\pi i\theta)) < 0\} \\ \downarrow{\scriptstyle q_0} & & \downarrow{\scriptstyle q} & & \downarrow{\scriptstyle q_1} \\ K \times S^1 & \xrightarrow{i_1} & X \times S^1 & \xleftarrow{j_1} & \{(x,\lambda) \in X \times S^1 |\; re(f(x)/\lambda) < 0\} \\ \downarrow{\scriptstyle p_0} & & \downarrow{\scriptstyle p} & & \\ K & \xrightarrow{iol} & X \,, & & \end{array}$$

with p the projection, $q = id \times exp(2\pi i \cdot\ )$ and $S^1 := \{\lambda \in \mathbb{C} |\ |\lambda| = 1\}$). Let $r := p \circ q$, $r_0 := p_0 \circ q_0$ and identify

$$\pi = exp(2\pi i \cdot\ ) : \widetilde{\mathbb{C}^*} = \mathbb{C} \to \mathbb{C}^*$$

so that the action of $1 \in \mathbb{Z} \simeq \pi_1(\mathbb{C}^*)$ corresponds to $w \mapsto w + 1$.

Let t be the composition of the inverse of $exp(2\pi i \cdot\ ) \times id$:

$$\{(w, \theta) \in \mathbb{C} \times \mathbb{R} | 1/2 < re(w) - \theta < 3/2\} \to \{(z, \theta) \in \mathbb{C} \times \mathbb{R} | re(z \cdot exp(-2\pi i\theta)) < 0\}$$

and the projection to $\mathbb{C}$. By base change we get an induced map $\tilde{t}$ such that the following diagram commutes:

$$\begin{array}{ccc} X \times \mathbb{R} & \xleftarrow{j_2} & \{(x, \theta) \in X \times \mathbb{R} |\ re(f(x)exp(-2\pi i\theta)) < 0\} \\ \downarrow{\scriptstyle r} & & \downarrow{\scriptstyle \tilde{t}} \\ X & \xleftarrow{\tilde{\pi}} & \widetilde{X^*} \,. \end{array}$$

Note that j_2 and $\tilde{t}$ are equivariant with respect to the action of

$$1 \in \mathbb{Z} =: \pi_1 \simeq \pi_1(\mathbb{C}^*) \quad \text{on} \quad \widetilde{X^*} \quad \text{and} \quad \theta \mapsto \theta + 1 \quad \text{on} \quad \mathbb{R}\,.$$

Now we can construct a morphism of functors (from $D(X, R)$ to $D(K, R[\pi_1])$)

$$\tilde{r}_\psi : l^*\tilde{\psi} \to Rp_{0*}Rq_{0*}q_0^*i_1^*Rj_{1*}j_1^*p^*$$

such that the diagram

$$\begin{array}{ccc} l^*\psi & \xrightarrow{forget(\tilde{r}_\psi)} & Rp_{0*}Rq_{0*}q_0^*i_1^*Rj_{1*}j_1^*p^* \\ \downarrow{\scriptstyle m_f} & & \downarrow{\scriptstyle m_{q_0}} \\ l^*\psi & \xrightarrow{forget(\tilde{r}_\psi)} & Rp_{0*}Rq_{0*}q_0^*i_1^*Rj_{1*}j_1^*p^* \end{array}$$

commutes in $D(K, R)$, with m_{q_0} the action on $Rq_{0*}q_0^*$ induced by $\theta \mapsto \theta + 1$ on $\mathbb{R}$. The natural morphism

$$q_0^*i_1^*Rj_{1*}j_1^*p^* \simeq i_2^*q^*Rj_{1*}j_1^*p^* \to i_2^*Rj_{2*}q_1^*j_1^*p^* \simeq i_2^*Rj_{2*}j_2^*r^*$$

is an isomorphism, since q is a **covering** (i.e. smooth of relative dimension 0). Then $\tilde{r}_\psi$ is induced by the morphism

$$l^*i^*R\tilde{\pi}_*\tilde{\pi}^* \to l^*i^*R\tilde{\pi}_*R\tilde{t}_*\tilde{t}^*\tilde{\pi}^* \simeq l^*i^*Rr_*Rj_{2*}j_2^*r^* \to Rp_{0*}Rq_{0*}i_2^*Rj_{2*}j_2^*r^*\,.$$

Lemma 1.3.3. *Let $l : K \hookrightarrow \{f = 0\}$ and $\mathcal{F} \in ob(D(X))$ be given. Assume that $f, \mathcal{F}$ satisfy for all $x \in K$ the assumption* 1.1.1, *with $S := \{x\}$. Then we have:*

1. $\tilde{r}_\psi : l^*\tilde{\psi}(\mathcal{F}) \to Rp_{0*}Rq_{0*}q_0^*i_1^*Rj_{1*}j_1^*p^*(\mathcal{F})$ *is an isomorphism.*

2. *The cohomology sheaves of* $i_1^*Rj_{1*}j_1^*p^*(\mathcal{F})$ *are locally constant on the fibers of* p_0.

3. *Let* s_1 *be the section* $id \times \{1\}$ *of* p_0 *and* s_0 *be the section* $id \times \{0\}$ *of* r_0 *(with* $q_0 \circ s_0 = s_1$*). The natural morphisms*

$$Rr_{0*}q_0^*i_1^*Rj_{1*}j_1^*p^*(\mathcal{F}) \xrightarrow{ad_{s_0}} Rr_{0*}Rs_{0*}s_0^*q_0^*i_1^*Rj_{1*}j_1^*p^*(\mathcal{F}) \dashrightarrow{\sim} \cdots$$

$$\cdots s_0^*q_0^*i_1^*Rj_{1*}j_1^*p^*(\mathcal{F}) \xrightarrow{\sim} s_1^*i_1^*Rj_{1*}j_1^*p^*(\mathcal{F}) \longrightarrow l^*k^*\tilde{i}^*R\tilde{j}_*\tilde{j}^*(\mathcal{F})$$

are isomorphisms so that the following diagram commutes:

$$\begin{array}{ccc} l^*\psi(\mathcal{F}) & \xrightarrow[\sim]{l^*r_\psi} & l^*k^*\nu_{re(f)}(\mathcal{F}) \\ \| & & \uparrow\wr \\ l^*\psi(\mathcal{F}) & \xrightarrow[\sim]{s_0^*\tilde{r}_\psi} & s_1^*i_1^*Rj_{1*}j_1^*p^*(\mathcal{F}) \\ \downarrow m_f & & \downarrow mon \\ l^*\psi(\mathcal{F}) & \xrightarrow[\sim]{s_0^*\tilde{r}_\psi} & s_1^*i_1^*Rj_{1*}j_1^*p^*(\mathcal{F}) \end{array}$$

(with mon the monodromy action induced by 2.).

Proof. The same idea as in the proof of corollary 1.1.1, together with the **Vietoris-Begle theorem** ([KS, cor. 2.7.7(iv)]) imply the isomorphism

$$r_0^*(forget(\tilde{r}_\psi)) : r_0^*l^*\psi(\mathcal{F}) \to q_0^*i_1^*Rj_{1*}j_1^*p^*(\mathcal{F}) .$$

This implies 2. (since q_0 is a covering) and 1. by **another version of the Vietoris-Begle theorem** ([KS, prop. 2.7.8]), since it is enough to show that $forget(\tilde{r}_\psi)$ is an isomorphism. It implies also the isomorphism

$$Rr_{0*}Rs_{0*}s_0^*q_0^*i_1^*Rj_{1*}j_1^*p^*(\mathcal{F}) \xleftarrow[\sim]{ad_{s_0}} Rr_{0*}q_0^*i_1^*Rj_{1*}j_1^*p^*(\mathcal{F}) .$$

Moreover, the action of m_{q_0} gets identified with the monodromy action mon. The remaining statements follow from the commutative diagram

$$\begin{array}{ccc} l^*\psi(\mathcal{F}) & \xrightarrow[\sim]{l^*r_\psi} & l^*k^*\nu_{re(f)}(\mathcal{F}) \\ \| & & \uparrow \\ l^*\psi(\mathcal{F}) & \xrightarrow[\sim]{s_0^*\tilde{r}_\psi} & s_1^*i_1^*Rj_{1*}j_1^*p^*(\mathcal{F}) . \end{array}$$

□

Since the diagram

$$\begin{array}{ccccc}
l^*\tilde{\psi} & \xrightarrow{\tilde{r}_\psi} & Rp_{0*}Rq_{0*}q_0^*i_1^*Rj_{1*}j_1^*p^* & \xrightarrow{s_0^*} & s_1^*i_1^*Rj_{1*}j_1^*p^* \\
\uparrow{\scriptstyle sp_f} & & \uparrow{\scriptstyle ad_{j_1}} & & \uparrow{\scriptstyle ad_{j_1}} \\
l^*i^* & \xrightarrow{ad_{r_0}} & Rp_{0*}Rq_{0*}q_0^*i_1^*p^* & \xrightarrow{s_0^*} & s_1^*i_1^*p^*
\end{array}$$

commutes (where we can consider the left square in $D(K, R[\pi_1])$), one gets the

Corollary 1.3.3. *Under the assumption of lemma* 1.3.3, *the cohomology sheaves of*

$$i_1^*R\Gamma_{\{re(f/\lambda)\geq 0\}}(p^*\mathcal{F})$$

are locally constant on the fibers of p_0, *and there are isomorphisms such that the following diagram commutes*

$$\begin{array}{ccc}
l^*\phi[-1](\mathcal{F}) & \xrightarrow[\sim]{l^*r_\phi} & l^*k^*\mu_{re(f)}(\mathcal{F}) \\
\| & & \uparrow \wr \\
l^*\phi[-1](\mathcal{F}) & \xrightarrow{\sim} & s_1^*i_1^*R\Gamma_{\{re(f/\lambda)\geq 0\}}(p^*\mathcal{F}) \\
\downarrow{\scriptstyle m_f} & & \downarrow{\scriptstyle mon} \\
l^*\phi[-1](\mathcal{F}) & \xrightarrow{\sim} & s_1^*i_1^*R\Gamma_{\{re(f/\lambda)\geq 0\}}(p^*\mathcal{F})\,.
\end{array}$$

Finally, we have the **Thom-Sebastiani Isomorphism** (compare [Ma2]):

Corollary 1.3.4. *Let* $f_i : X_i \to \mathbb{C}$, $K_i \subset \{f_i = 0\}$ *and* $\mathcal{F}_i \in ob(D(X_i))$ *be given. Define*

$$f := f_1 \oplus f_2 := a \circ (f_1, f_2) : X := X_1 \times X_2 \to \mathbb{C}\,,\ K := K_2 \times K_2$$

and

$$\mathcal{F} := \mathcal{F}_1 \boxtimes^L \mathcal{F}_2 \in ob(D(X))\,.$$

Assume that $f_i, \mathcal{F}_i$ *satisfy for all* $x_i \in K_i$ *the assumption* 1.1.1, *with* $S_i := \{x_i\}$ $(i = 1, 2)$. *Let* $l_i : K_i \to X_i$ $(i = 1, 2)$ *and* $l : K \to X$ *be the inclusions. Then there is an isomorphism*

$$(l_1^*\phi_{f_1}[-1](\mathcal{F}_1)) \boxtimes^L (l_2^*\phi_{f_2}[-1](\mathcal{F}_2)) \xrightarrow{\sim} l^*\phi_f[-1](\mathcal{F})$$

such that the diagram

$$\begin{array}{ccc}
(l_1^*\phi_{f_1}[-1](\mathcal{F}_1)) \boxtimes^L (l_2^*\phi_{f_2}[-1](\mathcal{F}_2)) & \xrightarrow{\sim} & l^*\phi_f[-1](\mathcal{F}) \\
\downarrow{\scriptstyle m_{f_1}[-1]\boxtimes^L m_{f_2}[-1]} & & \downarrow{\scriptstyle m_f[-1]} \\
(l_1^*\phi_{f_1}[-1](\mathcal{F}_1)) \boxtimes^L (l_2^*\phi_{f_2}[-1](\mathcal{F}_2)) & \xrightarrow{\sim} & l^*\phi_f[-1](\mathcal{F})
\end{array}$$

commutes.

Proof. By corollary 1.2.1, $f, \mathcal{F}$ satisfy for all $x \in K$ the assumption 1.1.1, with $S := \{x\}$. Therefore we can apply lemma 1.3.2 to $f_i, \mathcal{F}_i, K_i$ $(i = 1, 2)$ and $f, \mathcal{F}, K$. The isomorphism follows then from corollary 1.1.1 and corollary 1.3.1. The monodromy statement follows from corollary 1.3.3 and the next argument: Consider the commutative diagram

$$\begin{array}{ccccc} K & \xrightarrow{s_1} & K \times S^1 & \xrightarrow{i_1} & X \times S^1 \\ \| & & \downarrow d' & & \downarrow d \\ K_1 \times K_2 & \xrightarrow{s_{1,1} \times s_{2,1}} & K_1 \times S^1 \times K_2 \times S^1 & \xrightarrow{i_{1,1} \times i_{2,1}} & X_1 \times S^1 \times X_2 \times S^1 \end{array}$$

with d, d' the obvious "diagonal" maps and $s_1, s_{1,1}, s_{2,1}$ the corresponding section $id \times \{1\}$. Let

$$p : X \times S^1 \to X \ , \ p_i : X_i \times S^1 \to X_i \quad (i = 1, 2)$$

be the projections. Then there is a natural morphism

$$d^* \left(R\Gamma_{\{re(f_1/\lambda_1) \geq 0\}}(p_1^* \mathcal{F}_1) \boxtimes^L R\Gamma_{\{re(f_2/\lambda_2) \geq 0\}}(p_2^* \mathcal{F}_2) \right) \to R\Gamma_{\{re(f/\lambda) \geq 0\}}(p^* \mathcal{F})$$

(this is constructed as the "cup-product" at the beginning of this section) such that the following diagram commutes:

$$\begin{array}{ccc} s_1^* i_1^* d^* \left(R\Gamma_{\{re(f_1/\lambda_1) \geq 0\}}(p_1^* \mathcal{F}_1) \boxtimes^L R\Gamma_{\{re(f_2/\lambda_2) \geq 0\}}(p_2^* \mathcal{F}_2) \right) & \longrightarrow & \mathcal{H}_1 \\ \downarrow \wr & & \| \\ (s_{1,1}^* i_{1,1}^* R\Gamma_{\{re(f_1/\lambda_1) \geq 0\}}(p_1^* \mathcal{F}_1)) \boxtimes^L (s_{2,1}^* i_{2,1}^* R\Gamma_{\{re(f_2/\lambda_2) \geq 0\}}(p_2^* \mathcal{F}_2)) & \longrightarrow & \mathcal{H}_1 \\ \downarrow \wr & & \downarrow \wr \\ (l_1^* \phi_{f_1}[-1](\mathcal{F}_1)) \boxtimes^L (l_2^* \phi_{f_2}[-1](\mathcal{F}_2)) & \xrightarrow{\sim} & \mathcal{H}_2 \ , \end{array}$$

with $\mathcal{H}_1 := s_1^* i_1^* R\Gamma_{\{re(f/\lambda) \geq 0\}}(p^* \mathcal{F})$ and $\mathcal{H}_2 := l^* \phi_f[-1](\mathcal{F})$. □

By the **classical Künneth formula** (see appendix) this implies the following **generalization** of theorem 1.0.1:

Corollary 1.3.5. *Under the assumption of corollary* 1.3.4, *there is for locally closed* K_i $(i = 1, 2)$ *an isomorphism*

$$R\Gamma_c(K, \phi_f[-1](\mathcal{F})) \simeq R\Gamma_c(K_1, \phi_{f_1}[-1](\mathcal{F}_1)) \otimes^L R\Gamma_c(K_2, \phi_{f_2}[-1](\mathcal{F}_2))$$

such that the diagram

$$\begin{array}{ccc} R\Gamma_c(K, \phi_f[-1](\mathcal{F})) & \xrightarrow{\sim} & R\Gamma_c(K_1, \phi_{f_1}[-1](\mathcal{F}_1)) \otimes^L R\Gamma_c(K_2, \phi_{f_2}[-1](\mathcal{F}_2)) \\ \downarrow m_f[-1] & & \downarrow m_{f_1}[-1] \otimes^L m_{f_2}[-1] \\ R\Gamma_c(K, \phi_f[-1](\mathcal{F})) & \xrightarrow{\sim} & R\Gamma_c(K_1, \phi_{f_1}[-1](\mathcal{F}_1)) \otimes^L R\Gamma_c(K_2, \phi_{f_2}[-1](\mathcal{F}_2)) \end{array}$$

commutes.

As we already explained, the assumptions of corollary 1.3.4 are satisfied in the complex analytic or algebraic context, if the sheaf complexes $\mathcal{F}_i$ $(i = 1, 2)$ are complex analytically or algebraically constructible. As we explain in section 4.2, the **nearby and vanishing cycle** complexes

$$\psi_{f_i}(\mathcal{F}_i) \quad \text{and} \quad \phi_{f_i}(\mathcal{F}_i) \qquad (i = 1, 2)$$

are then also **complex analytically or algebraically constructible**.

The description of the **Thom-Sebastiani Isomorphism** as an isomorphism in the derived category contains then much more information than that of the stalk information given by the **local Thom-Sebastini Theorem**. For example, let R be a field and assume that the $\mathcal{F}_i$ are perverse sheaves (with respect to the middle perversity). Then $\mathcal{F} := \mathcal{F}_1 \boxtimes \mathcal{F}_2$ and (as explained in chapter 6) also

$$\phi_{f_1}[-1](\mathcal{F}_1) \ , \ \phi_{f_2}[-1](\mathcal{F}_2) \quad \text{and} \quad \phi_f[-1](\mathcal{F})$$

are perverse sheaves. So the **Thom-Sebastiani Isomorphism** of corollary 1.3.4 describes an isomorphism in the **abelian category of perverse sheaves** (compare [Ma2, sec. 2.3]). Moreover, if the **support** of the $\phi_{f_i}[-1](\mathcal{F}_i)$ $(i = 1, 2)$ is at most **one**-dimensional, these perverse sheaves are quite well understood. So (iterated application of) the **Thom-Sebastiani Isomorphism** gives a description of

$$\phi_f[-1](\mathcal{F}) \ ,$$

whose support is (in general) **higher**-dimensional !

1.4 Appendix: Künneth formula

Recall, that all topological spaces are locally compact with countable topology and of finite cohomological dimension (over $\mathbb{Z}$).

Since we do not assume that the base ring R is commutative, we have to explain our convention about the (derived) tensor product:

$$\otimes^L : D(X, R_1) \times D(X, R_2) \to D(X, R_3) \ .$$

For our applications, it is sufficient to assume that we are in one of the following cases:

1. R is not necessarily commutative, with C a subring of the center of R and we use $R_i = R\ (i = 1, 2)$, $R_3 = C$ with

$$\otimes^L = \otimes^L_R : D(X, mod - R) \times D(X, R - mod) \to D(X, C) \ .$$

Recall that $R - mod$ denotes left and $mod - R$ denotes right R-modules. If R is commutative we can take $R_i = R\ (i = 1, 2, 3)$.

2. R is not necessarily commutative and we use $R_1 = \mathbb{Z},\ R_i = R\,(i = 2, 3)$ with

$$\otimes^L = \otimes^L_Z : D(X, \mathbb{Z}) \times D(X, R) \to D(X, R)\,.$$

1*. We take a group-ring extension of the case (1): $R_3 = C[G],\ R_i = R[G]$ $(i = 1, 2)$ with

$$\otimes^L = \otimes^L_R : D(X, mod - R[G]) \times D(X, R[G] - mod) \to D(X, C[G])\,.$$

Here we consider on the tensor product the diagonal action of G.

2*. We take a group-ring extension of the case (2): $R_1 = \mathbb{Z}[G],\ R_i = R[G]$ $(i = 2, 3)$ with

$$\otimes^L = \otimes^L_Z : D(X, \mathbb{Z}[G]) \times D(X, R[G]) \to D(X, R[G])\,.$$

Note that in all cases $\otimes^L$ is defined for the whole derived category of **unbounded** complexes of sheaves:

Spaltenstein [Sp, def. 5.1] calls a complex $\mathcal{F} \in D(X, mod - R)$ K-**flat** , if for every **acyclic** complex $\mathcal{G} \in D(X, R - mod)$ also

$$\mathcal{F} \otimes_R \mathcal{G} \in D(X, C)$$

is **acyclic**. He actually assumed R commutative, but this definition applies in general. Then one can use K-**flat resolutions** to define $\otimes^L$, and by [Sp, cor. 5.5, prop. 5.6], there are **enough** K-flat complexes (i.e. every complex has a K-flat resolution). These are constructed as **special limits** of bounded above complexes, whose objects are direct sums of sheaves of the form $R_{Z\subset X}$,with $Z \subset X$ open or locally closed. The point is, that $R_{Z\subset X}$ is then K-**flat** (so that the arguments of [Sp] work for general R).

This explains the definition of $\otimes^L$ in our case (1). For the case (2), we apply these results to $R = \mathbb{Z}$ and use a K-**flat Resolution** of $\mathcal{F} \in D(X, \mathbb{Z})$ to define $\otimes^L$ (this is of course well known, since $\mathbb{Z}$ has **finite global homological dimension**, compare [KS, IC]). In the cases (1*) and (2*) one can use a resolution by a complex $\mathcal{F}$ such that $forget(\mathcal{F})$ is K-flat. Since $R[G]$ is a **free** and therefore flat R-module, one has the implication: $\mathcal{F}$ K-flat as $R[G]$ -module implies $forget(\mathcal{F})$ is K-flat as R-module.

We already used in the proof of the **Thom-Sebastiani Theorem for sheaves** the fact that a K-**projective** complex is K-flat (so we consider modules , corresponding to the case X a point) so that these complexes could be used for calculating $\otimes^L$. This works perfectly in our cases (1), (1*). But in the cases (2), (2*), the situation is **not symmetric**. Nevertheless our proof works (!), because it is enough that the corresponding complexes

$$A_0, A_1 \in ob(D(\mathbb{Z})) \quad \text{or} \quad ob(D(\mathbb{Z}[G]))$$

are K-**projective** and therefore K-flat with respect to $\otimes_Z^L$, and we only considered products $A_\alpha \otimes_Z B_\beta$ containing at least one of the complexes A_0, A_1.

Now we come to the

Proposition 1.4.1 (projection formula). *Let $f : X \to Y$ be a continuous map. Then there are functorial isomorphisms:*

$$(Rf_!\mathcal{F}) \otimes^L \mathcal{G} \xrightarrow{\sim} Rf_!(\mathcal{F} \otimes^L (f^*\mathcal{G})) \quad \textit{for} \quad \mathcal{F} \in D(X, R_1)\ ,\ \mathcal{G} \in D(Y, R_2)\ .$$

$$\mathcal{F} \otimes^L (Rf_!\mathcal{G}) \xrightarrow{\sim} Rf_!((f^*\mathcal{F}) \otimes^L \mathcal{G}) \quad \textit{for} \quad \mathcal{F} \in D(Y, R_1)\ ,\ \mathcal{G} \in D(X, R_2)\ .$$

Note that only in cases (2), (2*) of our conventions, these are really **two** formulae.

For a proof in the cases (1) and (1*), see [Sp, prop. 6.18]. Note that his proof also works for non-commutative rings R. One reduces for example the first formula to the case $\mathcal{F}$ c-**soft** and $\mathcal{G}$ a **special limit** of bounded above complexes of direct sums of sheaves of the form $R'_{Z\subset X}$, with $R' = R$, $R[G]$ and $Z \subset X$ open or locally closed. Then it is enough, to prove

$$(f_!\mathcal{F}) \otimes \mathcal{G} \xrightarrow{\sim} f_!(\mathcal{F} \otimes (f^*\mathcal{G}))$$

for $\mathcal{F}$ c-**soft** and $\mathcal{G} = R'_{Z\subset X}$.

The same argument proves also the second formula in cases (2) and (2*), because we can use such a **special limit** for $\mathcal{F}$. For a proof of the first formula in case (2), see [Gerg, prop. 4.4(ρ)]. The point is, that $\mathbb{Z}$ is a **regular noetherian ring** and therefore every $\mathcal{F} \in D(X, \mathbb{Z})$ has a resolution which is c-**soft and flat** ! A similar argument works in the case (2*).

Finally let us **remark:**

(a) The isomorphism pr_f of the **projection formula** is compatible with **composition** of maps, i.e. for continuous maps $g : X \to Y$ and $f : Y \to Z$, the following diagram commutes:

$$\begin{array}{ccc} (R(f\circ g)_!\mathcal{F}) \otimes^L \mathcal{G} & \xrightarrow{pr_{f\circ g}} & R(f\circ g)_!(\mathcal{F} \otimes^L (f\circ g)^*\mathcal{G}) \\ \downarrow \wr & & \uparrow pr_g \\ (Rf_!(Rg_!\mathcal{F})) \otimes^L \mathcal{G} & \xrightarrow{pr_f} & Rf_!(Rg_!\mathcal{F} \otimes^L f^*\mathcal{G})\ . \end{array}$$

(b) The isomorphism pr_f of the **projection formula** is compatible with **pullback**. For a cartesian diagram of continuous maps

$$\begin{array}{ccc} X' & \xrightarrow{g'} & X \\ \downarrow f' & & \downarrow f \\ Y' & \xrightarrow{g} & Y\ , \end{array}$$

the following diagram commutes:

$$\begin{array}{ccc} g^*((Rf_!\mathcal{F}) \otimes^L \mathcal{G}) & \xrightarrow{pr_f} & g^*(Rf_!(\mathcal{F} \otimes^L f^*\mathcal{G})) \\ \downarrow \wr & & \downarrow \wr \\ (Rf'_!(g'^*\mathcal{F})) \otimes^L g^*\mathcal{G} & \xrightarrow{pr_{f'}} & Rf'_!(g'^*\mathcal{F} \otimes^L (f'^*g^*\mathcal{G})) \,, \end{array}$$

where we use the following rules for the vertical isomorphisms:

(i) $g^*Rf_! \simeq Rf'_!g'^*$ is the **base change isomorphism** for the proper direct image functor.

(ii) $g^*(\cdot \otimes^L \cdot) = g^*(\cdot) \otimes^L g^*(\cdot)$.

(iii) $f'^*g^* = g'^*f^*$.

For the **usual direct image functor** one has the

Lemma 1.4.1 (projection morphism). *Let $f : X \to Y$ be a continuous map. Then there are functorial morphisms:*

$$(Rf_*\mathcal{F}) \otimes^L \mathcal{G} \to Rf_*(\mathcal{F} \otimes^L (f^*\mathcal{G})) \quad \textit{for} \quad \mathcal{F} \in D(X, R_1)\,,\ \mathcal{G} \in D(Y, R_2)\,.$$

$$\mathcal{F} \otimes^L (Rf_*\mathcal{G}) \to Rf_*((f^*\mathcal{F}) \otimes^L \mathcal{G}) \quad \textit{for} \quad \mathcal{F} \in D(Y, R_1)\,,\ \mathcal{G} \in D(X, R_2)\,.$$

Its construction uses only that f^* and Rf_* are **adjoint**, and that f^* **commutes** with $\otimes^L$. For example, the first morphism follows by **adjunction** from

$$f^*((Rf_*\mathcal{F}) \otimes^L \mathcal{G}) \xrightarrow{\sim} (f^*Rf_*\mathcal{F}) \otimes^L (f^*\mathcal{G}) \xrightarrow{ad_f \otimes^L id} \mathcal{F} \otimes^L (f^*\mathcal{G})\,.$$

In the same way one gets the natural morphism

$$(Rf_*\mathcal{F}) \otimes^L (Rf_*\mathcal{G}) \to Rf_*(\mathcal{F} \otimes^L \mathcal{G})\,. \tag{1.15}$$

Moreover, the following diagram commutes:

$$\begin{array}{ccc} (Rf_!\mathcal{F}) \otimes^L \mathcal{G} & \xrightarrow{can} & (Rf_*\mathcal{F}) \otimes^L \mathcal{G} \\ \downarrow \wr & & \downarrow \\ Rf_!(\mathcal{F} \otimes^L (f^*\mathcal{G})) & \xrightarrow{can} & Rf_*(\mathcal{F} \otimes^L (f^*\mathcal{G})) \end{array}$$

(and similarly in the other case). Especially, the projection morphism for a **proper** f is nothing else but the isomorphism of the projection formula.

With the help of the projection formula, we can now define the **Künneth morphism**. For a cartesian diagram of continuous maps over the **base space** S :

$$\begin{array}{ccccc} X' \times_S Y & \xrightarrow{h} & X \times_S Y & \xrightarrow{q} & Y \\ {\scriptstyle p'}\downarrow & & \downarrow{\scriptstyle p} & & \downarrow \\ X' & \xrightarrow{f} & X & \longrightarrow & S\,, \end{array}$$

with $h =: f \times_S id$ and $q' =: q \circ h$, there exists the following isomorphism:

$$(Rf_!\mathcal{F}) \boxtimes_S^L \mathcal{G} := (p^*(Rf_!\mathcal{F})) \otimes^L q^*\mathcal{G} \xrightarrow{\sim} (Rh_!(p'^*\mathcal{F})) \otimes^L q^*\mathcal{G}$$
$$\xrightarrow{\sim} Rh_!\big((p'^*\mathcal{F}) \otimes^L (h^*q^*\mathcal{G})\big) = Rh_!\big((p'^*\mathcal{F}) \otimes^L (q'^*\mathcal{G})\big) = Rh_!(\mathcal{F} \boxtimes_S^L \mathcal{G}) .$$

The first isomorphism is the **base change isomorphism** as in (i), and the second is the **projection formula**.

As usual, the cartesian product

$$f \times_S g : X' \times_S Y' \to X \times_S Y$$

of two maps $f : X' \to X$ and $g : Y' \to Y$ over the **base space** S (i.e. with $X \to S$, $Y \to S$ fixed) has a factorization as

$$(f \times_S id_Y) \circ (id_{X'} \times_S g) .$$

We finally get the **Künneth isomorphism**

$$(Rf_!\mathcal{F}) \boxtimes_S^L (Rg_!\mathcal{G}) \xrightarrow{\sim} R(f \times_S g)_!(\mathcal{F} \boxtimes_S^L \mathcal{G}) \tag{1.16}$$

by composing the above isomorphisms:

$$(Rf_!\mathcal{F}) \boxtimes_S^L (Rg_!\mathcal{G}) \xrightarrow{\sim} R(f \times_S id_Y)_!(\mathcal{F} \boxtimes_S^L (Rg_!\mathcal{G}))$$
$$\xrightarrow{\sim} R(f \times_S id_Y)_! R(id_{X'} \times_S g)_!(\mathcal{F} \boxtimes_S^L \mathcal{G}) = R(f \times_S g)_!(\mathcal{F} \boxtimes_S^L \mathcal{G}) .$$

One can show, that this **Künneth isomorphism** does not depend on the chosen factorization

$$(f \times_S id_Y) \circ (id_{X'} \times_S g)$$

or

$$(id_X \times_S g) \circ (f \times_S id_{Y'})$$

(compare [De2, lem. 5.4.3.5]), but this is not needed. Since the Künneth isomorphism was constructed with the help of the projection formula and the base change isomorphism for the proper direct image functor, it has the same properties with respect to **composition** of maps or **pullbacks** !

In the special case that the base space S is a point, we omit the corresponding index in our notation as above. For example, if one takes for f, g the constant map, then one gets the **classical Künneth formula**:

$$R\Gamma_c(X', \mathcal{F}) \otimes^L R\Gamma_c(Y', \mathcal{G}) \simeq R\Gamma_c(X' \times Y', \mathcal{F} \boxtimes^L \mathcal{G}) . \tag{1.17}$$

In the same way, one gets the **Künneth morphism** for the **usual direct image functor**:

$$(Rf_*\mathcal{F}) \boxtimes_S^L (Rg_*\mathcal{G}) \to R(f \times_S g)_*(\mathcal{F} \boxtimes_S^L \mathcal{G}) . \tag{1.18}$$

For another construction of this morphism, see [De2, 5.4.1].

Of course this is in general not an isomorphism. But the commutative diagram (1.13) of the proof of proposition 1.3.1 implies the

Corollary 1.4.1. *Let* $f_i : X_i \to \mathbb{R}$, $S_i := \{x_i\} \subset \{f_i = 0\}$ *and* $\mathcal{F}_i \in ob(D(X_i))$ *satisfy the assumption* 1.1.1 *(*$i = 1, 2$*). Denote the inclusions*

$$j_i : \{f_i < 0\} \hookrightarrow X_i \qquad (i = 1, 2)$$

and $k := j_1 \times j_2$. *Then this Künneth morphism induces an isomorphism*

$$((Rj_{1*}j_1^*\mathcal{F}_1) \boxtimes^L (Rj_{2*}j_2^*\mathcal{F}_2))_{(x_1,x_2)} \xrightarrow[\sim]{K\ddot{u}} (Rk_*(j_1^*\mathcal{F}_1 \boxtimes^L j_2^*\mathcal{F}_2))_{(x_1,x_2)}\,.$$

Corollary 1.4.1 applies for example in the subanalytic and o-minimal context (see the next chapter 2).

Let us finish this appendix with the

Lemma 1.4.2. *Let* $f : X' \to X$, $g : Y' \to Y$ *be continuous and*

$$i : X'' \hookrightarrow X',\ j : Y'' \hookrightarrow Y'$$

be the inclusion of **closed** *subsets. Let us denote the inclusion*

$$k := i \times j : X'' \times Y'' \hookrightarrow X' \times Y'\,.$$

Then there exists for $h := f \times g$ *and*

$$\mathcal{F} \in ob(D(X', R_1))\ ,\ \mathcal{G} \in ob(D(Y', R_2))$$

a commutative diagram:

$$\begin{array}{ccc} (Rf_!\mathcal{F}) \boxtimes^L (Rg_!\mathcal{G}) & \xrightarrow{f_!ad_i \boxtimes^L g_!ad_j} & (Rf_!(Ri_!i^*\mathcal{F})) \boxtimes^L (Rg_!(Rj_!j^*\mathcal{G})) \\ \wr\downarrow K\ddot{u} & & \downarrow\wr \\ Rh_!(\mathcal{F} \boxtimes^L \mathcal{G}) & \xrightarrow{h_!ad_k} & Rh_!(Rk_!k^*(\mathcal{F} \boxtimes^L \mathcal{G}))\,. \end{array}$$

Especially, for X, Y *a point and* X', Y' **compact**, *there exists a commutative diagram:*

$$\begin{array}{ccc} R\Gamma(X', \mathcal{F}) \otimes^L R\Gamma(Y', \mathcal{G}) & \xrightarrow{ad_i \otimes^L ad_j} & R\Gamma(X'', \mathcal{F}) \otimes^L R\Gamma(Y'', \mathcal{G}) \\ \wr\downarrow K\ddot{u} & & \wr\downarrow K\ddot{u} \\ R\Gamma(X' \times Y', \mathcal{F} \boxtimes^L \mathcal{G}) & \xrightarrow{ad_k} & R\Gamma(X'' \times Y'', \mathcal{F} \boxtimes^L \mathcal{G})\,. \end{array}$$

Proof. The lemma follows from

$$k^*(\mathcal{F} \boxtimes^L \mathcal{G}) \simeq k^*((Ri_!i^*\mathcal{F}) \boxtimes^L (Rj_!j^*\mathcal{G}))$$

and the commutative diagram:

$$\begin{array}{ccc}
(Rf_!\mathcal{F}) \boxtimes^L (Rg_!\mathcal{G}) & \xrightarrow{f_! ad_i \boxtimes^L g_! ad_j} & (Rf_!(Ri_!i^*\mathcal{F})) \boxtimes^L (Rg_!(Rj_!j^*\mathcal{G})) \\
\wr \downarrow K\ddot{u} & & \wr \downarrow K\ddot{u} \\
Rh_!(\mathcal{F} \boxtimes^L \mathcal{G}) & \xrightarrow{h_!(ad_i \boxtimes^L ad_j)} & Rh_!((Ri_!i^*\mathcal{F}) \boxtimes^L (Rj_!j^*\mathcal{G})) \\
\downarrow h_! ad_k & & \uparrow \wr \\
Rh_!(Rk_!k^*(\mathcal{F} \boxtimes^L \mathcal{G})) & \xleftarrow{\sim} & Rh_!\big(Rk_!k^*((Ri_!i^*\mathcal{F}) \boxtimes^L (Rj_!j^*\mathcal{G}))\big) .
\end{array} \quad \square$$

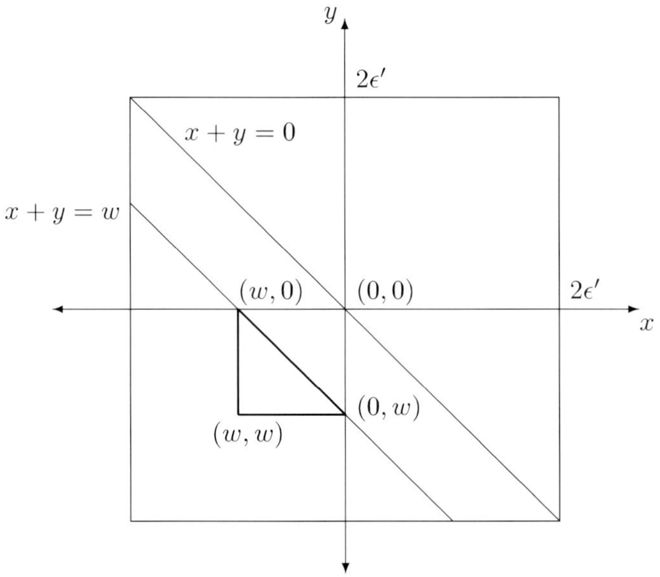

Figure 1.1: Milnor fiber of $x + y$

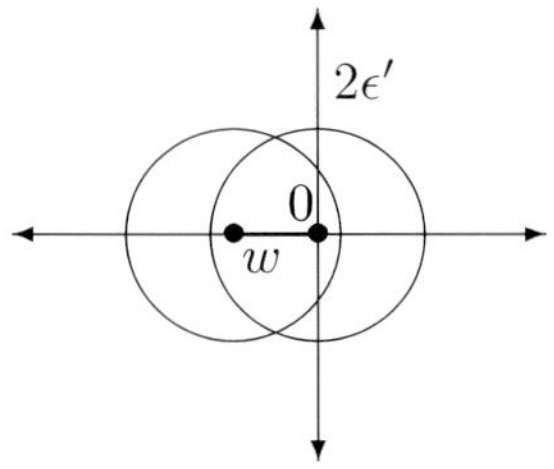

Figure 1.2: Projection of the Milnor fiber of $x + y$

Chapter 2

Constructible sheaves in geometric categories

Introduction

The aim of this chapter is to develop in an **elementary way** some basic results in the theory of constructible sheaves in the framework of nice "geometric categories". So we **do not** use deep results from **stratification theory** or **micro-local geometry** in this chapter. For results of this type, one should compare with chapter 4 and 5 of this book and with [KS, SchVi]. Many of the results of this chapter can be interpreted as the counterparts in real geometry of the "finiteness theorems" in l-adic cohomology as described for example by Deligne [De3] in [SGA4$\frac{1}{2}$].

2.0.1 The basic results

The theory of o-**minimal structures** and its relation to "tame topology" (as in Grothendieck's "Esquisse d'un Programme" [Groth]) is developed from scratch by van den Dries in [vDr] (compare also with the survey articles [Te2, vDr2]). For the basic results we recommend especially his introduction and overview [vDr, p. 1-9].

An o-**minimal structure** $\mathfrak{S}$ on $\mathbb{R}$ is a sequence $\mathfrak{S}_n$ $(n \in \mathbb{N})$ such that $\forall\, n$:

1. $\mathfrak{S}_n$ is a boolean algebra of subsets of $\mathbb{R}^n$.
2. $A \in \mathfrak{S}_n \Rightarrow A \times \mathbb{R},\ \mathbb{R} \times A \in \mathfrak{S}_{n+1}$.
3. $\{(x_1, \dots, x_n) \in \mathbb{R}^n |\ x_i = x_j\} \in \mathfrak{S}_n$ for $1 \le i < j \le n$.
4. $A \in \mathfrak{S}_{n+1} \Rightarrow \pi(A) \in \mathfrak{S}_n$, where $\pi : \mathbb{R}^{n+1} \to \mathbb{R}^n$ is the projection on the first coordinates.
5. $\{r\} \in \mathfrak{S}_1$ for each $r \in \mathbb{R}$, and $\{(x, y) \in \mathbb{R}^2 |\ x < y\} \in \mathfrak{S}_2$.
6. The only sets in $\mathfrak{S}_1$ are the finite unions of open intervals and points.

This notion is an elegant generalization of **semialgebraic geometry** (a standard reference for this theory is [BCR]) and the semilinear or semialgebraic sets give the simplest examples of an o-minimal structure. A subset $A \in \mathbb{R}^n$ is **definable** with respect to $\mathfrak{S}$, if it belongs to $\mathfrak{S}_n$.

For the construction of other o-minimal structures, the reader should look at [KarMac, LiRo, PeSpSt, Speiss, RoSpWi, vDrSp, vDrSp2, Wilkie] and the references in [vDrMi]. We use the notation $\mathbb{R}_{an}$ for the o-minimal structure of **globally subanalytic** sets (i.e. subsets of $\mathbb{R}^n$, that are subanalytic as subsets of the larger projective space $\mathbb{P}^n(\mathbb{R})$).

The basic result is the **cell decomposition theorem** ([vDr, p. 4; thm. 3.2.11]). Before we state its sheaf theoretic counterpart, we make the following definitions. A complex of sheaves $\mathcal{F}$ on a topological space X is called **constructible** with respect to a filtration

$$X_\bullet \ : \emptyset = X_{-1} \subset X_0 \subset \ldots \subset X_m = X$$

by closed subsets (or the filtration is called **adapted** to $\mathcal{F}$), if the restriction of the cohomology sheaves of $\mathcal{F}$ to the $X_i \backslash X_{i-1}$ are locally constant. If $X \subset \mathbb{R}^n$ belongs to $\mathfrak{S}_n$ and the X_i can also be chosen in this way, then we call $\mathcal{F}$ **constructible with respect to** $\mathfrak{S}$ (or $\mathfrak{S}$**-constructible** for short).

Remark 2.0.1. This condition is **stronger** than the assumption, that all cohomology sheaves are constructible in this sense (since we allow **unbounded** complexes). All of the following stability results of this category with respect to functors of finite cohomological dimension are then also true for this weaker notion of constructibility (just apply the standard truncation arguments, e.g. the distinguished triangles [KS, (1.7.2-4)]).

We call a continuous map $f : X \to Y$ of locally compact spaces a **cellular fibration**, if X is equipped with a filtration

$$X_\bullet \ : \emptyset = X_{-1} \subset X_0 \subset \ldots \subset X_m = X$$

by closed subsets such that the following holds (for $i = 1, \ldots, m$):

(cf) All $X_i \backslash X_{i-1}$ are the disjoint union of **finitely** many open subsets $U_{i,j}$ such that $f : U_{i,j} \to Y$ is a locally trivial topological **fibration**, with fiber homeomorphic to some **affine space** $\mathbb{R}^{n_{i,j}}$.

Especially for Y a point, we call $X_\bullet$ a **finite cellular decomposition** of X. For example, a compact CW-complex has a canonical finite cellular decompositon. Then one has the

Lemma 2.0.1 (cellular fibration lemma). *Let $f : X \to Y$ be a cellular fibration with respect to the filtration $X_\bullet$. Assume that $X_\bullet$ is adapted to the complex of sheaves $\mathcal{F}$. Then all $R^i f_! \mathcal{F}$ $(i \in \mathbb{Z})$ are locally constant.*

Therefore, the **cell decomposition theorem** implies (by using the graph embedding):

Theorem 2.0.1. *Let $A \subset \mathbb{R}^n, B \subset \mathbb{R}^m$ be locally closed and $f : A \to B$ be continuous and definable with respect to the o-minimal structure $\mathfrak{S}$ (i.e. A, B and the graph of f are definable). If $\mathcal{F}$ is a $\mathfrak{S}$-constructible complex of sheaves on A, then $Rf_!\mathcal{F}$ is an $\mathfrak{S}$-constructible complex of sheaves on B (in particular, all $R^i f_! \mathcal{F}$ are $\mathfrak{S}$-constructible).*

By **property (6)** of o-minimal structures this implies (comp. [KS, lem. 8.4.7]):

Corollary 2.0.1. *Under the assumption of theorem 2.0.1, let in addition*

$$f : A \to B \subset \mathbb{R} \quad \text{be proper}\,.$$

1. *Suppose $B = \,]-c, c[$ (with $c > 0$). Then one has for $0 < \epsilon << c$ the following isomorphisms:*
   (a) $R\Gamma(\{|f| \leq \epsilon\}, \mathcal{F}) \xrightarrow{\sim} R\Gamma(\{|f| < \epsilon\}, \mathcal{F}) \xrightarrow{\sim} R\Gamma(\{f = 0\}, \mathcal{F})\,.$
   (b) $R\Gamma_{\{f=0\}}(A, \mathcal{F}) \xrightarrow{\sim} R\Gamma_c(\{|f| < \epsilon\}, \mathcal{F})\,.$
2. *Suppose $B = \mathbb{R}$. Then one has for $c >> 0$ the following isomorphisms:*
   (a) $R\Gamma(A, \mathcal{F}) \xrightarrow{\sim} R\Gamma(\{|f| \leq c\}, \mathcal{F}) \xrightarrow{\sim} R\Gamma(\{|f| < c\}, \mathcal{F})\,.$
   (b) $R\Gamma_c(\{|f| < c\}, \mathcal{F}) \xrightarrow{\sim} R\Gamma_c(A, \mathcal{F})\,.$

As an example, one can take the **distance function** $f = dist(\cdot, K)$ to a compact $\mathfrak{S}$-set K, with

$$dist(x, K) := inf(\{\| x - y \|_\infty \mid y \in K\}$$

and $\| \cdot \|_\infty$ the supremum norm. Especially for K a point, corollary 2.0.1(1) is the cohomological counterpart corresponding to the **local conic structure** of $\mathfrak{S}$-sets ([vDr, thm. 8.2.3]).

Moreover, theorem 2.0.1 together with the **cell decomposition theorem** implies already the following cohomological version about the **bifurcation set** of continuous definable functions (compare [TaZah]):

Corollary 2.0.2. *Under the assumption of theorem 2.0.1, let in addition B be a subset of $\mathbb{R}$. Then $Rf_*\mathcal{F}$ is an $\mathfrak{S}$-constructible complex of sheaves on B (in particular, all $R^i f_* \mathcal{F}$ are $\mathfrak{S}$-constructible). Moreover, there exists a* **finite** *subset $\Sigma \subset B$ such that the canonical map*

$$(Rf_*\mathcal{F})_x \to R\Gamma(\{f = x\}, \mathcal{F})$$

is an isomorphism for all $x \in B \backslash \Sigma$.

In a similar way, one shows that a continuous definable function together with a $\mathfrak{S}$-constructible complex admits a **cohomological version of a Milnor fibration** (see section 2.2 for details). Therefore, one can apply the results of chapter 1 of this book and gets the following **Künneth theorems**:

Proposition 2.0.1. *Let $\mathcal{F}_i$ be $\mathfrak{S}$-constructible complexes on the locally closed $\mathfrak{S}$-sets B_i $(i = 1,2)$. Let A_i be closed $\mathfrak{S}$-subsets of B_i, and denote the open inclusions $j_i : B_i \backslash A_i \hookrightarrow B_i$ $(i = 1,2)$ and $k := j_1 \times j_2$. Then the natural morphisms*

$$R\Gamma_{A_1}(\mathcal{F}_1) \boxtimes^L R\Gamma_{A_2}(\mathcal{F}_2) \to R\Gamma_{A_1 \times A_2}(\mathcal{F}_1 \boxtimes^L \mathcal{F}_2)$$

and

$$(Rj_{1*}j_1^*\mathcal{F}_1) \boxtimes^L (Rj_{2*}j_2^*\mathcal{F}_2) \to Rk_*(j_1^*\mathcal{F}_1 \boxtimes^L j_2^*\mathcal{F}_2)$$

are isomorphisms.

A basic **definable invariant** is the **Euler characteristic** ([vDr, chap. 4]). Before we can explain its relation to $\mathfrak{S}$-constructible sheaves, we have to introduce some notions.

Let us fix a base ring R so that we use sheaves of R-modules. Consider the derived category $D(R)$ of complexes of R-modules together with a **strict full triangulated subcategory** T, i.e. T is a full subcategory, with

$$(1) \quad 0 \in ob(T) \ , \ (2) \quad M \in ob(T) \Leftrightarrow M[1] \in ob(T) \ ,$$

and (3) T is stable under **extensions**: for a distinguished triangle

$$M' \to M \to M'' \overset{[1]}{\to}$$

in $D(R)$, with $M', M'' \in ob(T)$, we have $M \in ob(T)$.

Let μ be an **additive function** on T, i.e. μ is a function from the isomorphism-classes of objects of T to an abelian group G, with

$$\mu(M) = \mu(M') + \mu(M'')$$

for each distinguished triangle $M' \to M \to M'' \overset{[1]}{\to}$ in T.

The **standard example** is $R = k$ a field, T denotes the (cohomological) bounded complexes with **finite dimensional** cohomology, and

$$\mu = \sum (-1)^i \cdot dim_k H^i(\cdot)$$

is the usual **Euler characteristic**.

Let $\mathcal{F}$ be a complex of sheaves of R-modules, constructible with respect to the **finite cellular decompositon** $X_\bullet$ of X, with $\mathcal{F}_x \in ob(T)$ for all $x \in X$. Denote by $U_{i,j} \simeq \mathbb{R}^{n_{i,j}}$ the open "cells" of $X_i \backslash X_{i-1}$ and let $\mu_{i,j} := \mu(\mathcal{F}_x)$, with $x \in U_{i,j}$.

Lemma 2.0.2. *Under the assumptions above, we have $R\Gamma_c(X, \mathcal{F}) \in ob(T)$, with*

$$\mu(R\Gamma_c(X, \mathcal{F})) = \sum_{i,j} (-1)^{n_{i,j}} \cdot \mu_{i,j} \ .$$

Let $\mathcal{F}$ be an $\mathfrak{S}$-constructible complex of sheaves of R-modules on the $\mathfrak{S}$-set $A \subset \mathbb{R}^n$, with $\mathcal{F}_x \in ob(T)$ for all $x \in A$. Then the **function**

$$\mu(\mathcal{F}) : A \to G, \ x \mapsto \mu(\mathcal{F}_x)$$

is also $\mathfrak{S}$-**constructible**: it takes only finitely many values g_i ($i = 1, \ldots, m$), and the sets

$$A_i := \{x \in A | \ \mu(\mathcal{F})(x) = g_i\}$$

are $\mathfrak{S}$-sets (for $i = 1, \ldots, m$).

Moreover, lemma 2.0.2 together with [vDr, prop. 4.1.13] implies the

Proposition 2.0.2. *Under the assumptions above, we have for a locally closed A: $R\Gamma_c(A, \mathcal{F}) \in ob(T)$, with*

$$\mu(R\Gamma_c(A, \mathcal{F})) = \sum_{i=1}^{m} g_i \cdot \chi(A_i) \,.$$

Here $\chi(\cdot)$ is the **Euler characteristic** *of a $\mathfrak{S}$-set.*

Especially, for $\mu(\mathcal{F})$ a **constant** function with value g one gets

$$\mu(R\Gamma_c(A, \mathcal{F})) = g \cdot \chi(A) \,.$$

If we take the standard example for T, μ and $\mathcal{F}$ the **constant sheaf** k_A on the locally closed $\mathfrak{S}$-set A, we get that the **cohomology** with compact support $H^*_c(A, k)$ is **finite dimensional**, with

$$\chi(A) = \sum (-1)^i \cdot dim_k \ H^i_c(A, k) \,.$$

So we get a topological **finiteness result** about (locally closed) $\mathfrak{S}$-sets, together with the **topological interpretation** of the Euler characteristic.

Together with theorem 2.0.1, one can extend proposition 2.0.2 to a **relative** context. Let us denote by

$$D(A, \mathfrak{S}) \qquad (\text{ or } \quad D_T(A) := D_T(A, \mathfrak{S}) \,)$$

the **strict triangulated subcategory** of the derived category $D(A) := D(A, R)$ of sheaves of R-modules on A, whose objects are the $\mathfrak{S}$-constructible complexes (with $\mathcal{F}_x \in ob(T)$ for all $x \in A$). Moreover, denote by

$$CF(A) := CF(A, G)$$

the abelian group of G-**valued** $\mathfrak{S}$-**constructible functions** so that μ induces the additive map

$$\mu_A : D_T(A) \to CF(A), \ \mathcal{F} \mapsto \mu(\mathcal{F}) := (x \mapsto \mu(\mathcal{F}_x)) \,.$$

Note that $CF(A, G)$ is nothing else but $G \otimes_Z CF(A, \mathbb{Z})$. It is therefore generated by

$$\{g \otimes 1_X \mid g \in G, X \text{ a } \mathfrak{S} - \text{subset of } A\} \,,$$

with 1_X the indicator function of the subset X.

Let $f : A \to B$ be a definable function between definable sets. Then f induces by [vDr, 4.2.2, 4.2.10] (and the graph embedding) a **pushdown** map

$$f_! : CF(A) \to CF(B) ,$$

uniquely defined by

$$f_!(g \otimes 1_X)(b) := g \cdot \chi(X \cap \{f = b\}) \quad (b \in B) .$$

Corollary 2.0.3. *Let $f : A \to B$ be a continuous definable function between locally closed definable sets. Then $Rf_!$ maps $D_T(A)$ into $D_T(B)$ and the diagram*

$$\begin{array}{ccc} D_T(A) & \xrightarrow{Rf_!} & D_T(B) \\ \downarrow{\scriptstyle \mu_A} & & \downarrow{\scriptstyle \mu_B} \\ CF(A) & \xrightarrow{f_!} & CF(B) \end{array}$$

commutes.

Of course, a similar result is true for f^* (and in the standard example of T, μ this is also true for $\otimes$).

From now on, we assume that $\mathfrak{S}$ is an o-minimal structure on the **ordered field** $(\mathbb{R}, +, \cdot)$ (i.e. $\mathfrak{S}$ contains the graph of **addition** and **multiplication** on $\mathbb{R}$, and contains therefore all **semialgebraic** sets). Then one has a **strong triangulation theorem** for $\mathfrak{S}$-sets ([vDr, thm. 8.2.9]), and such a triangulation is especially a "good cellular stratification" in the following sense:

We call a cellular decomposition a **good cellular stratification**, if

a. The decomposition is a **stratification** in the sense, that the **frontier** $\partial U := \bar{U} \backslash U$ of each cell $U = U_{i,j}$ is a union of lower dimensional cells.

b. Each cell U satisfies the following "goodness" condition: Let $j : U \to \bar{U}$ be the open inclusion into the closure of the cell. Then the natural morphism $\mathcal{F} \to Rj_* j^* \mathcal{F}$ is an isomorphism for each **constant sheaf** $\mathcal{F}$ on $\bar{U}$.

A **sufficient** condition for $(b.)$ is the following:

(con) Each point of ∂U has a fundamental system of open neighborhoods V_n $(n \in \mathbb{N})$ in $\bar{U}$, with $V_{n+1} \subset V_n$ such that each pair $(U \cap V_n, U \cap V_{n+1})$ is homeomorphic to a **pair of open convex** subsets of some affine space (since this implies

$$R\Gamma(U \cap V_n, \mathcal{F}) \xrightarrow{\sim} R\Gamma(U \cap V_{n+1}, \mathcal{F}) \xrightarrow{\sim} \mathcal{F}_x$$

for $x \in U \cap V_{n+1}$).

For example, $(\bar{U}, \partial U)$ is a topological manifold with boundary. Especially, a compact regular CW-complex (e.g. a finite simplicial complex) and a semilinear cell complex as in [Shiota, p. 59] is a "finite good cellular stratification".

Lemma 2.0.3. *Let $\mathcal{F}$ be a complex of sheaves of R-modules, constructible with respect to a finite good cellular stratification of X, with $\mathcal{F}_x \in ob(T)$ for all $x \in X$. Then $R\Gamma(X, \mathcal{F}) \in T$.*

This implies together with the previous results the following **finiteness theorem**:

Theorem 2.0.2. *Let $A \subset \mathbb{R}^n, B \subset \mathbb{R}^m$ and $f : A \to B$ be continuous and definable with respect to the o-minimal structure $\mathfrak{S}$, with B locally closed.*

1. *Rf_* maps $D_T(A, \mathfrak{S})$ into $D_T(B, \mathfrak{S})$.*
2. *If A is also locally closed, then $f^!$ maps $D_T(B, \mathfrak{S})$ into $D_T(A, \mathfrak{S})$.*
3. *Let R be commutative. Then $D(B, \mathfrak{S})$ is stable under $\otimes^L_R$, and*

$$Rhom_R(\cdot, \cdot) \quad maps \quad D^-(B, \mathfrak{S}) \times D^+(B, \mathfrak{S}) \quad into \quad D^+(B, \mathfrak{S}) \, .$$

Moreover, let T be equal to the the subcategory of **perfect complexes**, *i.e. complexes which are isomorphic to a bounded complex of finitely generated projective R-modules (compare [KS, exercise I.30]). Then $D_T(B, \mathfrak{S})$ is also stable under these functors and the* **biduality result**

$$\mathcal{F} \simeq D_B(D_B(\mathcal{F}))$$

holds for $\mathcal{F} \in ob(D_T(B, \mathfrak{S}))$. Here $D(\cdot)$ is the corresponding **duality functor** *(which we recall in section 2.2).*

If we apply (1.) for the standard example of T to the **constant sheaf** k_A, we get that the **cohomology** $H^*(A, k)$ is **finite dimensional** (which is a generalization of the well known fact that A has only a finite number of connected components ([vDr, prop. 3.2.18])).

Theorem 2.0.2 together with proposition 2.0.1 implies the following **Künneth theorem**:

Corollary 2.0.4. *Let $A_i \subset \mathbb{R}^{n_i}, B_i \subset \mathbb{R}^{m_i}$ be locally closed and $f_i : A_i \to B_i$ be continuous and definable with respect to the o-minimal structure $\mathfrak{S}$ $(i = 1, 2)$. Let $\mathcal{F}_i$ be $\mathfrak{S}$-constructible complexes on A_i $(i = 1, 2)$.*

1. *The natural morphism*

$$(Rf_{1*}\mathcal{F}_1) \boxtimes^L (Rf_{2*}\mathcal{F}_2) \to R(f_1 \times f_2)_*(\mathcal{F}_1 \boxtimes^L \mathcal{F}_2)$$

is an isomorphism.

2. *Assume moreover: R is commutative and the stalks of $\mathcal{F}_1$ are perfect complexes. Then one has a natural isomorphism*

$$(D_{A_1}\mathcal{F}_1) \boxtimes^L (D_{A_2}\mathcal{F}_2) \simeq D_{A_1 \times A_2}(\mathcal{F}_1 \boxtimes^L \mathcal{F}_2) \, ,$$

with $D(\cdot)$ the corresponding **duality functor**.

Consider for the standard example the subcategory T' of T, whose objects are the complexes with **vanishing Euler characteristic**. Then $D_{T'}(B)$ is just the "kernel" of the surjection

$$\chi_B : D_T(B) \to CF(B, \mathbb{Z}) ,$$

and this "kernel" is stable under the Grothendieck functors

$$Rf_*,\ f^*,\ Rf_!,\ f^! .$$

Similarly for $Rhom(\cdot, \cdot)$, $\otimes^L$:

$$\mathcal{F}, \mathcal{G} \in ob(D_T(B, \mathfrak{S})) \Longrightarrow Rhom(\mathcal{F}, \mathcal{G}),\ \mathcal{F} \otimes^L \mathcal{G} \in ob(D_{T'}(B, \mathfrak{S})) ,$$

if **at least** one of the complexes $\mathcal{F}$ or $\mathcal{G}$ belongs to $ob(D_{T'}(B, \mathfrak{S}))$.

Therefore one gets induced **Grothendieck functors** on the level of **constuctible functions** with a similar formalism and especially with a **biduality result** (compare with [Scha] and [KS, chapter 9.7]). Note that this approach is **simpler** than the one of Kashiwara and Schapira (we do not use [Scha, thm. 3.4], [KS, thm. 9.7.1]), and applies to other situations as well: valuation theory of convex **Polytopes**, constructible sheaves and functions for **stratified spaces**, in the **complex analytic** or **algebraic** context, in l-**adic cohomology** .... This will be explained in section 2.3 together with more precise references for this subject.

The final basic result about o-minimal structures is the **trivialization theorem** ([vDr, thm. 9.1.7]). It implies the following "generic base change theorem":

Theorem 2.0.3. *Let $A \subset \mathbb{R}^n$, $B \subset \mathbb{R}^m$ and $f : A \to B$ be continuous and definable with respect to the o-minimal structure $\mathfrak{S}$, with B locally closed. Fix an $\mathfrak{S}$-constructible complex $\mathcal{F}$ on A. Then there exists an* **open and dense** *$\mathfrak{S}$-subset U of B such that $\mathcal{F}|U$ and*

$$f_U := f : A_U := A \cap f^{-1}(U) \to U$$

have the following **base change property**: *for each cartesian diagram*

$$\begin{array}{ccc} A' & \xrightarrow{f'} & B' \\ \downarrow{\scriptstyle g'} & & \downarrow{\scriptstyle g} \\ A_U & \xrightarrow{f_U} & U, \end{array}$$

with g continuous definable and B' locally closed, the natural morphism

$$g^* Rf_{U*}(\mathcal{F}|A_U) \to Rf'_* g'^*(\mathcal{F}|A_U)$$

is an isomorphism. Especially, the natural morphism

$$(Rf_*\mathcal{F})_x \to R\Gamma(\{f = x\}, \mathcal{F})$$

is an isomorphism for all $x \in U$.

2.0.2 Definable spaces

Up till now we worked with definable subsets of an ambient space $\mathbb{R}^n$. But for many applications it is important to work with more general spaces. One possibility of generalization is to work with **definable spaces**, i.e. spaces with a finite definable atlas (see [vDr, chapter 10]), and to translate the results into this context. Especially for "$\mathfrak{S}$-manifolds", this corresponds to the "geometric categories" in [vDrMi]. The two basic results are then the **embedding theorem** for **regular** definable spaces [vDr, thm. 10.1.8], and the **quotient theorem** for **definable proper equivalence relations** [vDr, thm. 10.2.15].

The first implies for example that a **Nash-manifold** and the set of (real or) complex points of a **scheme** or an **algebraic space**, which is separated and of finite type over *spec*($\mathbb{C}$) (or *spec*($\mathbb{R}$)), can be realized as (i.e. is definable homeomorphic to) a **closed semialgebraic** subset of some $\mathbb{R}^n$. Especially, we can apply the previous results to **complex algebraically constructible** sheaves.

More generally, every **locally compact definable space** X is definably homeomorphic to a **closed definable** subset Y of some $\mathbb{R}^n$, by the following argument:

A locally compact space is regular so that X is definably homeomorphic to a definable subset Y' of some $\mathbb{R}^m$, where the topology of Y' is the induced subspace topology. Since X is locally compact, this implies that Y' is a locally closed subset. But then Y' is definably homeomorphic to a closed definable subset Y of $\mathbb{R}^{m+1}$, by the same proof as in [BCR, prop. 2.2.9], [vDrMi, C.3, p. 522] (or the beginning of the proof of theorem II.2.1' in [Shiota, p. 125]), since the function which measures the distance from a definable subset of $\mathbb{R}^m$ is continuous and definable (compare [BCR, prop. 2.2.8] or [Shiota, II.1.2]).

The **quotient theorem** implies, that the quotient X/K of a (locally) closed definable subset of some $\mathbb{R}^n$ by a **definable action** of a **compact definable** group (i.e. K is a compact definable subset of some $\mathbb{R}^m$ and the action and group operation are definable) can be realised as a closed definable subset Y of some $\mathbb{R}^N$ such that the projection map

$$p : X \to X/K \simeq Y$$

is definable (one can get Y as a closed subset, since X/K is locally compact and X/K exists as a definable proper quotient of X by [vDr, cor. 10.2.18]). Typical examples are:

1. The **real analytic** action of a **compact Lie group** K on a **compact** real analytic manifold, or a linear representation of such a group (with $\mathfrak{S} = \mathbb{R}_{an}$ the globally subanalytic sets).
2. We restrict an **algebraic** action of a **linear algebraic** group G (over $\mathbb{R}$ or $\mathbb{C}$) to a **maximal compact subgroup** K of the group of real or complex points (which corresponds to $\mathbb{R}_{an}$ in general, and to the semialgebraic context for linear reductive groups). Moreover, a maximal compact subgroup K of a complex linear reductive group G is Zariski-dense so that G and K have the same fixed point set.

Fix now a compact group K, a K-space X (i.e. we have a continuous action $a : K \times X \to X$) and a K-**monodromic** or **equivariant** complex $\mathcal{F} \in ob(D_T(X))$ (these notions will be explained and used in chapter 3 of this book). Let us denote the projection

$$\pi : X \to X/K =: Y$$

and the open inclusion $j : X \backslash X^K \to X$, with X^K the fixed point set of the action. Note that π is **proper** so that $(R\pi_* j_! j^* \mathcal{F})_y = 0$ for $y \in \pi(X^K)$, and

$$(R\pi_* j_! j^* \mathcal{F})_y = R\Gamma(\{\pi = y\}, \mathcal{F}) \quad \text{for} \quad y \notin \pi(X^K)\,.$$

Assume that $R\pi_* j_! j^* \mathcal{F}$ is **constructible** with respect to a **finite good cellular stratification** of Y. Note that we **don't** assume X to be compact or that the action extends to a compactification of X. Of course, the important example comes from the above discussion about definable proper quotients (X^K is then a $\mathfrak{S}$-set), if $\mathcal{F}$ is $\mathfrak{S}$-constructible (use theorem 2.0.1 and the triangulation theorem for $\mathfrak{S}$-sets [vDr, thm. 8.2.9]).

If we apply lemma 2.0.2 and lemma 2.0.3 to $R\pi_* j_! j^* \mathcal{F}$, we obtain various "localization" or "fixed point" theorems. This is the content of the next chapter 3 of this book. Note that from this point of view, the points $y \in Y$ correspond to the orbits $K(x) \simeq K/K_x$ (with K_x the stabilizer), which is quite natural.

Another abstraction of the theory of o-minimal structures on the real field to the more general context of **real analytic** manifolds is the theory of **analytic geometric categories**, as introduced and studied by van den Dries and Miller [vDrMi]. This is a generalization of the theory of **subanalytic** subsets of real analytic manifolds.

An **analytic geometric category** $\mathcal{S}$ is given if each real analytic manifold M is equipped with a collection $\mathcal{S}(M)$ of subsets of M such that the following conditions are satisfied (for each such manifold):

AG1. $\mathcal{S}(M)$ is a boolean algebra of subsets of M, with $M \in \mathcal{S}(M)$.

AG2. If $A \in \mathcal{S}(M)$, then $A \times \mathbb{R} \in \mathcal{S}(M \times \mathbb{R})$.

AG3. If $f : M \to N$ is a proper analytic map and $A \in \mathcal{S}(M)$, then $f(A) \in \mathcal{S}(N)$.

AG4. If $A \subseteq M$ and $(U_i)_{i \in I}$ is an open covering of M, then $A \in \mathcal{S}(M)$ if and only if $A \cap U_i \in \mathcal{S}(U_i)$ for all $i \in I$.

AG5. Every bounded set in $\mathcal{S}(\mathbb{R})$ has a finite boundary.

$\mathcal{S}$ corresponds **uniquely** ([vDrMi, D.10]) to an o-**minimal structure** $\mathfrak{S}$ **on** $\mathbb{R}_{an}$ (i.e. an o-minimal structure on the real field containing the structure $\mathbb{R}_{an}$ of globally subanalytic subsets). $\mathfrak{S}$ is defined by the subsets of $\mathbb{R}^n$, which belong to $\mathcal{S}$ as a subset of $\mathbb{P}^n(\mathbb{R})$ (for the standard inclusion $\mathbb{R}^n \hookrightarrow \mathbb{P}^n(\mathbb{R})$). Moreover, $\mathcal{S}$ can be recovered as the subsets of real analytic manifolds, which are **locally** (at each point of the ambient manifold) real analytic isomorphic to $\mathfrak{S}$-sets. Note that the last step is well defined, since the real analytic isomorphisms between **bounded** open subanalytic subsets of affine spaces are definable in $\mathfrak{S}$.

We introduce in this chapter a variant of this notion, by restricting ourselves to (analytic) **Nash** manifolds (so in the above constructions one should only look at Nash manifolds and semialgebraic maps). In this way, every o-minimal structure $\mathfrak{S}$ on the real field $(\mathbb{R}, +, \cdot)$ (and not only those on $\mathbb{R}_{an}$) corresponds uniquely to a category, which we call a **Nash geometric category**. This notion is a generalization of the theory of **locally semialgebraic** subsets of Nash manifolds.

A **Nash geometric category** $\mathcal{S}$ is given if each (real analytic) Nash manifold M is equipped with a collection $\mathcal{S}(M)$ of subsets of M such that the conditions NGi. :=AGi., $i = 1, 2, 5$ and the following conditions are satisfied (for each such manifold):

NG3. If $f : M \to N$ is a proper Nash map (i.e. analytic with a semialgebraic graph) and $A \in \mathcal{S}(M)$, then $f(A) \in \mathcal{S}(N)$.

NG4. If $A \subseteq M$ and $(U_i)_{i \in I}$ is a covering of M by open semialgebraic subsets, then $A \in \mathcal{S}(M)$ if and only if $A \cap U_i \in \mathcal{S}(U_i)$ for all $i \in I$.

Then all results of [vDrMi] extend to this context, if one makes the obvious modifications (which we state and explain in section 2.1). Especially, an "analytic geometric category" **induces** by restriction a "Nash geometric category", and this correspondence is **injective**, since the associated o-minimal structure is **unique**.

A Nash geometric category $\mathcal{S}$ induces an o-minimal structure $\mathfrak{S}(\mathcal{S})$ on the ordered field $(\mathbb{R}, +, \cdot)$, by the same definition as for analytic geometric categories ($\mathbb{R}^n$ is an open semialgebraic subset of the Nash manifold $\mathbb{P}^n(\mathbb{R})$). Moreover, $\mathcal{S}$ can be recovered as the subsets of Nash manifolds, which are **locally** (at each point of the ambient manifold) Nash isomorphic to $\mathfrak{S}$-sets. Note that the last step is well defined, since the Nash isomorphisms between **bounded** open semialgebraic subsets of affine spaces are definable in $\mathfrak{S}$.

The advantage of the notion of a Nash geometric category is the fact, that this last step is also well defined, if one starts with a **category of $\mathfrak{X}$-sets**, as defined by Shiota [Shiota], who uses a different axiomatic setting.

A family of subsets of all affine spaces $\mathbb{R}^n$ is called $\mathfrak{X}$ ([Shiota, p. viii, p. 95, 96]), if it satisfies the axioms:

$\mathfrak{X}$(i) Every algebraic set in any Euclidean space is an element of $\mathfrak{X}$.

$\mathfrak{X}$(ii) If $X_1 \subset \mathbb{R}^n$ and $X_2 \subset \mathbb{R}^n$ are elements of $\mathfrak{X}$, then $X_1 \cap X_2, X_1 \backslash X_2$ and $X_1 \times X_2$ are elements of $\mathfrak{X}$.

$\mathfrak{X}$(iii) If $X \subset \mathbb{R}^n$ is an element of $\mathfrak{X}$ and $p : \mathbb{R}^n \to \mathbb{R}^m$ is a linear map such that the restriction of p to $\bar{X}$ is proper, then $p(X)$ is an element of $\mathfrak{X}$.

$\mathfrak{X}$(iv) If $X \subset \mathbb{R}$ and $X \in \mathfrak{X}$, then each point of X has a neigborhood in X, which is a finite union of points and intervals.

Note that this is a generalization of the notion of an o-minimal structure on $(\mathbb{R}, +, \cdot)$ (which is the same as a family $\mathfrak{X}_0$ in the notation of Shiota). Moreover, a **Nash** isomorphism h between **bounded** open semialgebraic subsets U_i $(i = 1, 2)$ of affine spaces is **definable** in $\mathfrak{X}$, and $X \subset U_1$ belongs to $\mathfrak{X}$ if and only if $h(X) \subset U_2$ belongs to $\mathfrak{X}$ (by axiom $\mathfrak{X}$(iii) and the graph embedding, as in [Shiota, II1.6]).

Sometimes, Shiota assumes in addition the axiom ([Shiota, p. 97]):

$\mathfrak{X}$(v) If a subset X of $\mathbb{R}^n$ is an $\mathfrak{X}$-set locally at each point of $\mathbb{R}^n$, then X is an $\mathfrak{X}$-set.

Note that an o-minimal structure **never** satisfies this axiom. But if we **restrict** a Nash geometric category $\mathfrak{S}$ to all affine spaces $\mathbb{R}^n$, we get a family $\mathfrak{X}$ satisfying this axiom. Indeed, the notion of a Nash geometric category is **equivalent** to a family $\mathfrak{X}$, satisfying the axiom $\mathfrak{X}$(v). More precisely, the arguments of [vDrMi] imply the following (see the next section):

Theorem 2.0.4. 1. *A family $\mathfrak{X}$ induces a Nash geometric category $\mathcal{S}(\mathfrak{X})$, whose sets are the subsets of Nash manifolds, which are locally (at each point of the ambient manifold) Nash-isomorphic to $\mathfrak{X}$-sets.*

2. *A Nash geometric category $\mathcal{S}$ induces:*

   (a) *An o-minimal structure $\mathfrak{S}(\mathcal{S})$ on $(\mathbb{R}, +, \cdot)$, whose sets are the subsets of $\mathbb{R}^n$, which belong as a subset of $\mathbb{P}^n(\mathbb{R})$ (for the standard inclusion $\mathbb{R}^n \hookrightarrow \mathbb{P}^n(\mathbb{R})$) to $\mathcal{S}$.*

   (b) *By restriction to $\mathbb{R}^n$ $(n \in \mathbb{N})$ a family $\mathfrak{X}(\mathcal{S})$ satisfying the axiom $\mathfrak{X}(v)$.*

3. $\mathcal{S}(\mathfrak{S}(\mathcal{S})) = \mathcal{S} = \mathcal{S}(\mathfrak{X}(\mathcal{S}))$.

4. $\mathfrak{S}(\mathcal{S}(\mathfrak{X})) \subseteq \mathfrak{X} \subseteq \mathfrak{X}(\mathcal{S}(\mathfrak{X}))$, *with $\mathfrak{S}(\mathcal{S}(\mathfrak{X})) = \mathfrak{X}$, if $\mathfrak{X}$ is an o-minimal structure on $(\mathbb{R}, +, \cdot)$, and $\mathfrak{X} = \mathfrak{X}(\mathcal{S}(\mathfrak{X}))$, if $\mathfrak{X}$ satisfies the axiom $\mathfrak{X}(v)$.*

Remark, that this theorem allows an easy **comparison** (contrary to a remark in [vDrMi, p. 498]) between the **different** notions of "geometric categories".

We get a one to one correspondence between o-**minimal structures** $\mathfrak{S}$ on $(\mathbb{R}, +, \cdot)$, **Nash geometric categories** $\mathcal{S}$ and **families** $\mathfrak{X}$, **satisfying the axiom** $\mathfrak{X}$(v). Moreover, the analytic geometric categories (as a subset of the Nash geometric categories) correspond in this way to the o-minimal structures on $\mathbb{R}_{an}$, and to the families $\mathfrak{X}$, containing the subanalytic subsets of affine spaces and satisfying the axiom $\mathfrak{X}$(v) (since the above correspondence is compatible with the natural partial order on the families, which is induced by the inclusions of the subsets of the families).

Notice that $\mathfrak{S}(\mathcal{S}(\mathfrak{X}))$ is the **greatest** o-minimal structure **contained** in $\mathfrak{X}$. Every **bounded** (especially every compact) $\mathfrak{X}$-set belongs to it, and every $\mathfrak{X}$-map between such sets is **definable** in this structure. More precisely, the proof of theorem 2.0.4 gives the following characterization:

(*) For each $A \subseteq \mathbb{R}^n$, $A \in \mathfrak{S}(\mathcal{S}(\mathfrak{X}))$ if and only if $\tau_n(A)$ is an $\mathfrak{X}$-set locally at each point of $\mathbb{R}^n$.

Here we use the semialgebraic map $\tau_n : \mathbb{R}^n \to \mathbb{R}^n$, given by

$$\tau_n(x_1, \dots, x_n) := (x_1(1+x_1^2)^{-1/2}, \dots, x_n(1+x_n^2)^{-1/2}) .$$

Note that τ_n is a Nash isomorphism of $\mathbb{R}^n$ onto $]-1,1[^n$.

Similarly $\mathfrak{X}(\mathcal{S}(\mathfrak{X}))$ is the **smallest** family of this type, which **contains $\mathfrak{X}$ and satisfies the axiom** $\mathfrak{X}$(v). Moreover, it can be directly constructed as the family of sets, which are locally an $\mathfrak{X}$-set at each point of the ambient affine space (compare [Shiota, p. 268]).

For most of the results of [Shiota] one can therefore assume that one works with an o-minimal structure, or with a family $\mathfrak{X}$ satisfying the axiom $\mathfrak{X}$(v). Moreover, one can extend these results into the framework of analytic or Nash geometric categories, by using a suitable affine embedding of the ambient manifolds.

Let us finally remark, that we develop the theory of constructible sheaves not only in the context of o-minimal structures on $\mathbb{R}$, but also (as far as possible) in the context of definable spaces and analytic (Nash) geometric categories.

2.1 Geometric categories

In this section, we prove theorem 2.0.4 and explain the modifications compared to [vDrMi], which are necessary to translate their results into the context of Nash geometric categories. We assume that the reader is familar with [vDrMi]. We follow the notation and numbering of [vDrMi, sec. 1, app. D].

For example, the fact that the family of **subanalytic** subsets of a real analytic manifold is the "smallest" analytic geometric category ([vDrMi, p. 499]), translates of course into the fact that the family of **locally semialgebraic** subsets of Nash manifolds is the "smallest" Nash geometric category.

Fix a Nash geometric category $\mathcal{S}$, let M, N be (analytic) Nash manifolds, and let $A \in \mathcal{S}(M)$, $B \in \mathcal{S}(N)$.

1.1′ Every Nash map $f : M \to N$ (i.e. f is analytic with semialgebraic graph) is an $\mathcal{S}$-map.

1.2′ Given a covering (U_i) of M by open semialgebraic subsets, a map $f : A \to N$ is an $\mathcal{S}$-map if and only if each restriction $f|U_i \cap A : U_i \cap A \to N$ is an $\mathcal{S}$-map.

Moreover, the statements $1.i$ for $i = 3, \dots, 20$ of [vDrMi] remain true in the Nash geometric context without any modification. For example property 1.7 implies

$$cl(A) := \bar{A} \in \mathcal{S}(M) .$$

For their proof, we modify [vDrMi, app.D] in the following way.

D.1′ All **locally semialgebraic** sets are $\mathcal{S}$-sets, in particular $\mathbb{R}^n \in \mathcal{S}(\mathbb{P}^n(\mathbb{R}))$.

Since this is a local statement, it suffices to show that the sets

$$\{x \in M |\ f(x) = 0\} \quad \text{and} \quad \{x \in M |\ f(x) > 0\}$$

belong to $\mathcal{S}$(M), if M is an open semialgebraic subset of some affine space and $f : M \to \mathbb{R}$ is a polynomial map.

Then the proof of [vDrMi, p. 530] applies, because the embeddings

$$M \to M \times \mathbb{P}^1(\mathbb{R});\ x \mapsto (x, f(x)) \quad \text{and} \quad x \mapsto (x, 0)$$

and the projection $M \times \mathbb{P}^1(\mathbb{R}) \to M$ are proper Nash maps, and

$$\{(x, y) \in M \times \mathbb{P}^1(\mathbb{R}) |\ f(x) \cdot y_1^2 - y_2^2 = 0\}$$

is locally given by the vanishing of a polynomial.

This implies also D.2′ :=1.1′, since the graph of a Nash map is (locally) semialgebraic, and D.3′ :=1.2′ follows directly from the axiom NG4. Moreover, the statements $D.i$ for $i = 4, \dots, 9$ of [vDrMi] remain true in the Nash geometric context, with the same proof. For later applications let us just recall that D.4 implies the stability of $\mathcal{S}$-sets under **products**, und D.6 implies the stability under **proper images**: Let $A, A' \in \mathcal{S}(M)$ with $A' \subseteq A$, A **locally closed** and let $f : A \to N$ be a **proper** $\mathcal{S}$-map. Then $f(A') \in \mathcal{S}(N)$.

Now we come to the proof of theorem 2.0.4 (which is a generalization of [vDrMi, D.10]).

Proof. (1) The statement 1. corresponds to [vDrMi, D.10.3]. We already explained in the introduction, that $\mathcal{S}(\mathfrak{X})$ is well defined by the axioms $\mathfrak{X}$(i-iii). Moreover, NG4. follows from the definiton and NG.1,2 follow from $\mathfrak{X}$(i,ii). For NG3., we start with a proper Nash map

$$f : M \to N$$

and an $\mathcal{S}(\mathfrak{X})$-set A in M. Then the proof of [vDrMi, p. 533, 534] applies, if we take for $y \in N$ an open semialgebraic neigborhood V of y, with a Nash isomorphism

$$h : V \to h(V)$$

onto an open **bounded** (!) semialgebraic subset of $\mathbb{R}^n$ containing $[-1, 1]^n$. Similarly for

$$g_x : U_x \to g_x(U_x) , \quad \text{with} \quad x \in f^{-1}h^{-1}([-1, 1]^n) ,\ f(U_x) \subset V$$

and such that $g_x(A \cap U_x)$ belongs to $\mathfrak{X}$. Then the map

$$g_x(U_x) \to h(V), \ a \mapsto h(f(g_x^{-1}(a)))$$

is a Nash map between open **bounded** (!) semialgebraic subsets of affine spaces.

$$g_x(A \cap U_x) \cap [-1,1]^n \subset g_x(U_x)$$

belongs to $\mathfrak{X}$ by $\mathfrak{X}$(i,ii). Therefore its image under the above map belongs to $\mathfrak{X}$ by $\mathfrak{X}$(iii) (and the graph embedding). The rest of the proof works without changes and uses then only $\mathfrak{X}$(i,ii). Similarly, the proof of AG5. in [vDrMi, p. 534] applies also to the proof of NG5. in our situation, and uses $\mathfrak{X}$(iv) (instead of condition (6) for o-minimal structures).

(2) The statement 2.(a) corresponds to [vDrMi, D.10.2] and their proof applies without changes. The proof of 2.(b) goes as follows: $\mathfrak{X}$(i) follows from D.1′ and $\mathfrak{X}$(ii) follows from NG1. and D.4 ("stability under **products**"). Moreover, every bounded $\mathcal{S}$-set in $\mathbb{R}$ belongs by 2.(a) to an o-minimal structure and is therefore a finite union of intervals and points. This implies $\mathfrak{X}$(iv) (by D.1′ and NG1.). Note, that $X \in \mathcal{S}(M)$ implies $cl(X) \in \mathcal{S}(M)$ (this is a special case of property 1.7, which will follow from the first part of the statement 3. of the theorem). Then D.2′ :=1.1′ and D.6 imply the condition $\mathfrak{X}$(iii).

(3) The first equality in the statement 3. corresponds to the first part of [vDrMi, D.10.3]. The proof in [vDrMi, p. 534] applies (if one works with Nash isomorphisms instead of analytic isomorphisms). Let $A \in \mathcal{S}(\mathfrak{X}(\mathcal{S}))(M)$. By definition, there exists for all $x \in M$ a Nash isomophism $h_x : U_x \to V_x$, with U_x an open semialgebraic neighborhood of x in M and V_x an open semialgebraic subset of some affine space such that $h_x(A \cap U_x)$ belongs to $\mathcal{S}(V_x)$. But then $A \cap U_x$ belongs to $\mathcal{S}(U_x)$ (by NG3.) for all $x \in M$, and therefore $A \in \mathcal{S}(M)$ (by NG4.). Altogether we get

$$\mathcal{S}(M) \subseteq \mathcal{S}(\mathfrak{S}(\mathcal{S}))(M) \subseteq \mathcal{S}(\mathfrak{X}(\mathcal{S}))(M) \subseteq \mathcal{S}(M)\,.$$

(4) First note that [vDrMi, D.10.1] gives the characterization (*) of

$$\mathfrak{S}(\mathcal{S}(\mathfrak{X}))\text{-sets}\,.$$

The proof applies without changes, since τ_n has a semialgebraic graph and

$$N \times \mathbb{P}^n(\mathbb{R}) \to N$$

is a proper Nash map for each Nash manifold N. Here we also use the obvious fact, that $A \subset \mathbb{R}^n$ belongs to $\mathcal{S}(\mathfrak{X})$, if and only if A is an $\mathfrak{X}$-set locally at each point of $\mathbb{R}^n$. This fact implies already the second part of statement 4.

Let $A \subset \mathbb{R}^n$ belong to $\mathfrak{X}$. Then $\tau_n(A)$ belongs to $\mathfrak{X}$ and therefore A belongs to $\mathfrak{S}(\mathcal{S}(\mathfrak{X}))$ by condition (*), if $\mathfrak{X}$ is an o-minimal structure. This argument breaks down for more **general** $\mathfrak{X}$. Conversely, suppose $A \subset \mathbb{R}^n$ belongs to $\mathfrak{S}(\mathcal{S}(\mathfrak{X}))$ so that $\tau_n(A)$ is an $\mathfrak{X}$-set locally at each point of $\mathbb{R}^n$. For each $x \in \mathbb{R}^n$, there exists U_x open in $\mathbb{R}^n$ such that $\tau_n(A) \cap U_x$ belongs to $\mathfrak{X}$. Since $\tau_n(A)$ is bounded, finitely many of the U_x cover $\tau_n(A)$ so that $\tau_n(A)$ belongs to $\mathfrak{X}$ (by $\mathfrak{X}$(ii)). Hence, A belongs to $\mathfrak{X}$ by $\mathfrak{X}$(iii), since $\tau_n(A)$ is **bounded** (compare with [Shiota, II.1.6]). This proves the first part of statement 4., and the proof of theorem 2.0.4 is finished. □

We finally remark, that the statements D.i for $i = 11, \ldots, 17, 19$ of [vDrMi] remain true in the Nash geometric context, with the same proof, if one uses in the proof of D.11 (or D.19) an open covering $(U_i)_{i \in N}$ with U_i open semialgebraic subsets of M and

$$\phi_i : U_i \to \mathbb{R}^m \quad (\text{ or } h_i : U_i \to \mathbb{R}^m \ )$$

analytic Nash isomorphisms (with $h_i(U_i \cap A) \in \mathfrak{S}_m$).

For later applications, we just prove an interesting topological variant of D.16.1 and D.17. Let us first introduce some notions. Fix an analytic (Nash) manifold M and an analytic (Nash) geometric category $\mathcal{S}$. Let $A \in \mathcal{S}(N)$. We generalize our definition of a "cellular stratification" into this context in the following way:

1. A **definable stratification** of A is a partition of A into locally closed $\mathcal{S}$-subsets, called strata, which is **locally finite** at each point of M (and not only at each point of A) such that the **frontier** $\partial S := cl_A(S) \backslash S$ in A of each stratum S is a union of other strata T. Since this implies $dim\, T < dim\, S$ ([vDrMi, 1.16]), we get that the union

   $$Y_n \text{ of all strata } S \text{ with } \dim S \le n \quad (n \in \mathbb{N})$$

   is closed in A and moreover an $\mathcal{S}$-set, since the partition is locally finite at each point of M . Moreover, each stratum S of dimension n is open in $Y_n \backslash Y_{n-1}$.

2. A **cellular stratification** of a (locally compact) topological space X is a **locally finite** partition of X into **locally closed** subsets, called strata such that each stratum S is an "n-cell" for some n (i.e. is homeomorphic to some $\mathbb{R}^n$) and such that the **frontier** ∂S of each stratum S is a union of other cells T, with $dim\, T < dim\, S$.

3. A "cellular stratification" is called **good**, if each cell S satisfies the goodness condition (b.) (as on page 86).

4. A "cellular stratification" is **regular**, iff for each cell S the pair $(cl(S), \partial S)$ is homeomorphic to a closed ball with boundary

   $$(\{x \in \mathbb{R}^n \,|\, |x| \le 1\}, \{x \in \mathbb{R}^n \,|\, |x| = 1\}) \ .$$

A (definable) regular cellular stratification of a locally compact X (or closed definable A) is nothing else but the structure of a locally finite regular CW-complex (such that each cell is definable).

Proposition 2.1.1. *For every* **locally finite** *family $\mathfrak{F} \subseteq \mathcal{S}(M)$, there exists a definable (regular) cellular stratification of M, compatible with $\mathfrak{F}$.*

Here, a family $\mathfrak{A}$ of subsets is compatible with a similar family $\mathfrak{F}$, if for all $A \in \mathfrak{A}$ and $B \in \mathfrak{F}$, either $A \cap B = \emptyset$ or $A \subseteq B$. Fix $A \in \mathcal{S}(M)$ as above. Then also $\partial A \in \mathcal{S}(M)$ so that we can assume $A, \partial A \in \mathfrak{F}$. Then the union of all cells of a stratification as in the proposition, which are contained in A, is a definable (good) cellular stratification of A (and this is a regular stratification, if A is closed). The induced stratification of A is in the regular case a "good cellular stratification", since "regular" implies the condition (con) of page 86 with respect to the frontier $\partial_M U$ in M of a cell $U \subset A$, and this implies the same condition also for the frontier $\partial_A U$ in A.

Let us first consider the

Lemma 2.1.1. *For every locally finite family $\mathfrak{F} \subseteq \mathcal{S}(M)$, there exists a locally finite partition $\mathfrak{P} \subseteq \mathcal{S}(M)$ of M compatible with $\mathfrak{F}$, consisting of (regular) "cells" in the sense above.*

Proof. The proof is the same as in [vDrMi, proof of D.11], but we use for the "regular" case the **triangulation theorem** instead of the cell decomposition theorem. Since M has countable topology, there exists an open covering $(U_i)_{i \in N}$ of M (by semialgebraic subsets), analytic (Nash) isomorphism $\phi_i : U_i \to \mathbb{R}^m$, and a sequence V_i of compact $\mathcal{S}$-sets in M $(i \in \mathbb{N})$ such that

$$(int(V_i))_{i \in N} \text{ covers } M \text{ and } V_i \subseteq U_i \quad (i \in \mathbb{N}) .$$

For each $i \in \mathbb{N}$, take (by [vDr, thm. 3.2.11] or [vDr, thm. 8.2.9]) a finite partition $\mathfrak{D}_i \subset \mathfrak{S}_m$ (with $\mathfrak{S} := \mathfrak{S}(\mathcal{S})$) of $\phi_i(V_i)$ into (regular) cells, compatible with

$$\{\phi_i(A \cap (V_i \setminus \cup_{k<i} V_k) : A \in \mathfrak{F}\} .$$

Note that this is a finite collection of bounded $\mathfrak{S}$-sets. Since $\phi_i(V_i)$ is compact, a triangulation as in [vDr, thm. 8.2.9] corresponds to a closed complex, so that the cells are regular. Put $\mathfrak{P} := \cup_i \mathfrak{P}_i$, with

$$\mathfrak{P}_i := \{\phi_i^{-1}(D) : D \in \mathfrak{D}_i, D \in \phi_i(V_i \setminus \cup_{k<i} V_k)\} . \qquad \square$$

Proof of proposition 2.1.1. The proof is similar to [vDrMi, proof of D.12]. Take a locally finite partition $\mathfrak{P}_m \subseteq \mathcal{S}(M)$ as in lemma 2.1.1 (with $m := dim M$). The only problem is now the frontier condition. Assume that for a certain $k \in \{1, \dots, m\}$ we have constructed a locally finite partition $\mathfrak{P}_k \subseteq \mathcal{S}(M)$ of M, consisting of (regular) cells and compatible with

$$\mathfrak{F} \cup \{\partial S : S \in \mathfrak{P}_k, dim\, S > k\} .$$

By lemma 2.1.1 there is a locally finite partition $\mathfrak{H} \subseteq \mathcal{S}(M)$ of M, consisting of (regular) cells compatible with $\mathfrak{P}_k$ and with $\{\partial S : S \in \mathfrak{P}_k, dim\, S = k\}$. Let $\mathfrak{H}'$

denote those cells of $\mathfrak{H}$, which are contained in a cell $S \in \mathfrak{P}_k$, with $dim\ S < k$. Then

$$\mathfrak{P}_{k-1} := \{S : S \in \mathfrak{P}_k, dim\ S \geq k\} \cup \mathfrak{H}'$$

is a locally finite partition of M, consisting of (regular) cells and compatible with

$$\mathfrak{F} \cup \{\partial S : S \in \mathfrak{P}_{k-1}, dim\ S > k-1\} .$$

Note that we only changed the cells S with $dim\ S < k$. By induction, this leads to the required stratification $\mathfrak{P}_0$. □

Remark 2.1.1. One gets another proof of proposition 2.1.1, if one uses a proper analytic (semialgebraic) embedding of the ambient manifold N into some affine space. Then one is in the context of $\mathfrak{X}$-sets and can apply the (stronger) triangulation theorem [Shiota, thm. II.2.1′]. In a similar way, one can reduce results of this type about locally closed $\mathcal{S}$-sets to the case of closed sets (compare with the beginning of the proof of theorem II.2.1′ in [Shiota, p. 125]).

2.2 Constructible sheaves

In this section we give the proofs of the results for constructible sheaves. Moreover, we explain their abstraction in the context of definable spaces and analytic or Nash geometric categories. We follow the description of the results in the introduction.

For the abstract notions and results about sheaf theory in the framework of derived categories, see [KS] and also [IC]. All topological spaces are locally compact (or subsets of such spaces), with countable topology and of finite cohomological dimension (over $\mathbb{Z}$). For a ring R (associative with unit), let $D(X) := D(X, R)$ denote the derived category of complexes of sheaves of R-modules. Note that we allow unbounded complexes (see [Sp]).

Let us begin with the simple

Lemma 2.2.1. 1. *Let $\mathfrak{S}$ be an o-minimal structure on $\mathbb{R}$ (or on $(\mathbb{R}, +, \cdot)$). Consider a continuous definable map $f : A \to B$ between definable subsets of affine spaces (or between definable spaces). If $\mathcal{F}$ is a $\mathfrak{S}$-constructible complex of sheaves on B, then $f^*\mathcal{F}$ is also $\mathfrak{S}$-constructible.*

2. *Let $\mathcal{S}$ be a an analytic (Nash) geometric category and M, N be analytic (Nash) manifolds. Consider a continuous definable map $f : A \to B$ between definable subsets $A \subset M$, $B \subset N$, with A closed.*

   If $\mathcal{F}$ is an $\mathcal{S}$-constructible complex of sheaves on B, then $f^*\mathcal{F}$ is also $\mathcal{S}$-constructible.

The proof is obvious, since under the assumptions above the inverse image of a definable subset is definable (in 2. this corresponds to [vDrMi, D.7]). The statement of (2.) becomes false, if A is not closed. Take in the example of [vDrMi, rem. to D.7, p. 532] a constant sheaf on B.

Next, we prove the "cellular fibration lemma".

Proof of lemma 2.0.1. Let us denote the inclusions

$$i_k : X_{k-1} \to X_k \quad , \quad j_k : X_k \backslash X_{k-1} \to X_k$$

and the restrictions

$$f_k := f|X_k : X_k \to Y \quad (k = 1, \ldots, m) .$$

We prove by induction on k that $Rf_{k!}(\mathcal{F}|X_k)$ has locally constant cohomology sheaves on Y. Using

$$Rf_{k-1!}(\mathcal{F}|X_{k-1}) = Rf_{k!}Ri_{k*}i_k^*(\mathcal{F}|X_k)$$

and applying $Rf_{k!}$ to the standard distinguished triangle

$$Rj_{k!}j_k^*(\mathcal{F}|X_k) \longrightarrow \mathcal{F}|X_k \longrightarrow Ri_{k*}i_k^*(\mathcal{F}|X_k) \xrightarrow{[1]} ,$$

it is enough to show that $Rf_{k!}Rj_{k!}j_k^*(\mathcal{F}|X_k)$ has locally constant cohomology sheaves. Since $X_k \backslash X_{k-1}$ is the disjoint union of the finitely many open subsets $U_{k,j}$, we have to show this statement for $Rg_!\mathcal{G}$, with $\mathcal{G} := \mathcal{F}|U_{k,j}$ and

$$g := f|U_{k,j} : U_{k,j} \to Y .$$

But this is a local result on Y so that by condition (cf) we can assume:

$$g : Y \times \mathbb{R}^n \to Y$$

is the projection. $\mathcal{G}$ has by assumption locally constant cohomology sheaves so that the natural **adjunction** morphism

$$\mathcal{G} \to g^* Rg_* \mathcal{G}$$

is an isomorphism by the **Vietoris-Begle theorem** [KS, prop. 2.7.8] (this holds also for complexes which are not bounded from below, because Rg_* has finite cohomological dimension). Let $s : Y \to Y \times \mathbb{R}^n$ be a section of g, i.e. a continuous map with $g \circ s = id$, for example $y \mapsto (y, 0)$. Applying s^* to this isomorphism, we get

$$s^*\mathcal{G} \simeq s^* g^* Rg_* \mathcal{G} \simeq Rg_* \mathcal{G} .$$

Of course, $\mathcal{G}' := s^*\mathcal{G}$ has also locally constant cohomology sheaves. We now claim:

$$Rg_! g^* \mathcal{G}'[n] \simeq \mathcal{G}' . \tag{2.1}$$

By induction we can assume $n = 1$. Moreover, we identify $\mathbb{R} \simeq \,]0,1[$ so that g is the projection

$$Y\times \,]0,1[\, \to Y \, .$$

Consider the projection $h : Y \times [0,1] \to Y$ together with the inclusions

$$i : Y \times \{0,1\} \to Y \times [0,1] \quad \text{and} \quad j : Y\times \,]0,1[\, \to Y \times [0,1] \, .$$

Then we have

$$Rg_! g^* \mathcal{G}' \simeq Rh_* Rj_! j^* h^* \mathcal{G}' \, ,$$

and the distinguished triangle

$$Rj_! j^* h^* \mathcal{G}' \longrightarrow h^* \mathcal{G}' \xrightarrow{ad_i} i_* i^* h^* \mathcal{G}' \xrightarrow{[1]}$$

implies that $Rg_! g^* \mathcal{G}'[1]$ is isomorphic to

$$cone(\, Rh_* h^* \mathcal{G}' \xrightarrow{ad_i} Rh_* i_* i^* h^* \mathcal{G}' \,) \simeq cone(\, \mathcal{G}' \xrightarrow{(id,id)} \mathcal{G}' \oplus \mathcal{G}' \,) \, .$$

Here we use again the **Vietoris-Begle theorem** [KS, prop. 2.7.8], together with the fact that the composition

$$\mathcal{G}' \xrightarrow{ad_h} Rh_* h^* \mathcal{G}' \xrightarrow{ad_i} Rh_* i_* i^* h^* \mathcal{G}'$$

is isomorphic to the adjunction map for the projection $h \circ i : Y \times \{0,1\} \to Y$. Finally, from the comutative diagram

$$\begin{array}{ccccccc}
\mathcal{G}' & \xrightarrow{(id,0)} & \mathcal{G}' \oplus \mathcal{G}' & \xrightarrow{(0,id)} & \mathcal{G}' & \xrightarrow{[1]} & \\
\| & & \wr \downarrow (x,y) \mapsto (x,x-y) & & & & \\
\mathcal{G}' & \xrightarrow{(id,id)} & \mathcal{G}' \oplus \mathcal{G}' & & & &
\end{array}$$

we get

$$cone(\, \mathcal{G}' \xrightarrow{(id,id)} \mathcal{G}' \oplus \mathcal{G}' \,) \simeq \mathcal{G}' \, .$$

Note that this last argument works in any triangulated category. □

Remark 2.2.1. 1. The isomorphism (2.1) also follows easily from the **Künneth formula** (compare [KS, proof of prop. 3.2.3(iii)]). But the above argument has the advantage that it lifts into the equivariant context studied in chapter 3!

2. Especially for Y a point, we get from (2.1): $R\Gamma_c(\mathbb{R}^n, g^*\mathcal{G}')[n] \simeq \mathcal{G}'$.

With the help of lemma 2.0.1 and the **cell decomposition theorem**, we can give the

Proof of theorem 2.0.1. Let $i : B \to \mathbb{R}^m$ be the inclusion. Then $Rf_!\mathcal{F}$ is a $\mathfrak{S}$-constructible complex of sheaves on B if and only if $Ri_!Rf_!\mathcal{F}$ is a $\mathfrak{S}$-constructible complex of sheaves on $\mathbb{R}^m$: Let X_i $(i = 0, \ldots, n)$ be a filtration of B by definable subsets, adapted to $Rf_!\mathcal{F}$. Then

$$Y_0 := \partial B,\ Y_i := \partial B \cup X_{i-1}\ (i = 1, \ldots, n+1),\ Y_{n+2} := \mathbb{R}^m$$

is a similar filtration of $\mathbb{R}^m$, adapted to $Ri_!Rf_!\mathcal{F}$. Therefore we can assume $B = \mathbb{R}^m$. Consider the graph embedding

$$g := (f, id) : A \to \mathbb{R}^m \times A \subset \mathbb{R}^{m+n}\,.$$

Then $A' := \mathbb{R}^m \times A$ is a locally closed definable subset of $\mathbb{R}^{m+n}$ and $\mathcal{F}' := Rg_!\mathcal{F}$ is a $\mathfrak{S}$-constructible complex of sheaves on A' (since f has a definable graph). Denote by $\pi : \mathbb{R}^{m+n} \to \mathbb{R}^m$ the projection onto the first m coordinates. Then $f = \pi \circ g$ and therefore $Rf_! \simeq R\pi_!Rg_!$. So it is enough to treat the case $f = \pi$. By a similar factorization, we can also assume $A' = \mathbb{R}^{m+n}$ and $n = 1$. Let $\mathcal{F}$ be constructible with respect to the filtration

$$X_\bullet\ :\ \emptyset = X_{-1} \subset X_0 \subset \ldots \subset X_n = \mathbb{R}^{m+1}\,,$$

with X_i closed and definable. By the **cell decomposition theorem** ([vDr, thm. 3.2.11]), there is a "decomposition" of $\mathbb{R}^{m+1}$ into finitely many (definable) locally closed cells A such that $X_i \backslash X_{i-1}$ $(i = 0, \ldots, n)$ is a union of cells. By the definition ([vDr, p. 52]) of a "decomposition", $\pi^{-1}(\pi(A)) = \pi(A) \times \mathbb{R}$ has a partition

$$\{(-\infty, f_1), (f_1, f_2), \ldots, (f_n, \infty), \Gamma(f_1), \ldots, \Gamma(f_n)\}\,.$$

Here, $f_j : \pi(A) \to \mathbb{R}$ are continuous (definable) functions with $f_1 < \cdots < f_n$, $\Gamma(f_j)$ denotes the graph of f_j and

$$(f_{j-1}, f_j) := \{(x, r) \in \pi(A) \times \mathbb{R} |\ f_{j-1}(x) < r < f_j(x)\}$$

(with $f_0 \equiv -\infty$, $f_{n+1} \equiv \infty$). Therefore

$$\pi_A := \pi : \pi^{-1}(\pi(A)) \to \pi(A)$$

is a **cellular fibration** with respect to the filtration

$$\cup_j\ \Gamma(f_j) \subset \pi^{-1}(\pi(A))\,.$$

Moreover, this filtration is adapted to $\mathcal{F}|\pi^{-1}(\pi(A))$, since each $\Gamma(f_j)$ or (f_{j-1}, f_j) is contained in some $X_i \backslash X_{i-1}$. The **cellular fibration lemma** implies, that

$$(R\pi_!\mathcal{F})|\pi(A) \simeq R\pi_{A!}(\mathcal{F}|\pi^{-1}(\pi(A)))$$

has locally constant cohomology sheaves. Here we use the **base change isomorphism** for $R\pi_!$ (compare [KS, prop. 2.6.7]). But the partition of $\mathbb{R}^m$ given by the definable subsets $\pi(A)$ induces in general **not a filtration** by closed subsets. Therefore we have to take a "stratification" of $\mathbb{R}^m$, partitioning the $\pi(A)$ ([vDr, prop. 4.1.13]). □

Remark 2.2.2. 1. The proof above explains the analogy to the results and techniques used in l-adic cohomology. The "cellular fibration" corresponds to the "fibration élémentaire" in [SGA4$\frac{1}{2}$, (4.2), p. 36] or [SGA4, XI3.1, p. 68], and the "cell decomposition theorem" plays the role of the "bon voisinage" in [SGA4$\frac{1}{2}$, (4.2), p. 37] or [SGA4, XI3.2, p. 69].

2. By the "embedding theorem", theorem 2.0.1 holds also for a continuous definable map between locally compact definable spaces, if we work with an o-minimal structure on $(\mathbb{R}, +, \cdot)$.

Before we translate this result into the context of analytic (Nash) geometric categories, we prove the useful

Proposition 2.2.1. *Let $\mathcal{S}$ be an analytic (Nash) geometric category and N an analytic (Nash) manifold. Suppose that $\mathcal{F}$ is locally an $\mathcal{S}$-constructible complex of sheaves, i.e. each point of M has an open (semialgebraic) neighborhood U, with $\mathcal{F}|U$ $\mathcal{S}(U)$-constructible. Then $\mathcal{F}$ is $\mathcal{S}(N)$-constructible.*

Proof. Take a locally finite covering (U_i) of N by open subanalytic (semialgebraic) subsets, together with compact subanalytic (semialgebraic) subsets $V_i \subset U_i$ such that (V_i) is also a covering of N and such that $\mathcal{F}|U_i$ is $\mathcal{S}(U_i)$-constructible. Take an adapted filtration $X_{i,j}$ by $\mathcal{S}(U_i)$-sets. Then

$$((X_{i,j} \backslash X_{i,j-1}) \cap V_i)$$

is a locally finite family of $\mathcal{S}(N)$-sets. By proposition 2.1.1, there is a definable "stratification" of N, compatible with this family. Then $\mathcal{F}$ is constructible with respect to the filtration Y_i $(i \in \mathbb{N})$, where Y_i is the union of the strata S with $dim\ S \leq i$. Note that $Y_i = N$ for $i >> 0$, since we assume that N is of finite dimension. □

Corollary 2.2.1. *Let $\mathcal{S}$ be an analytic (Nash) geometric category. Denote by A, B closed $\mathcal{S}$-sets of the real analytic (Nash) manifolds M, N and let $f : A \to B$ be a proper $\mathcal{S}$-map. If $\mathcal{F}$ is an $\mathcal{S}$-constructible complex of sheaves on A, then $Rf_!\mathcal{F}$ is an $\mathcal{S}$-constructible complex of sheaves on B.*

Proof. Note that $Rf_! = Rf_*$, since f is proper. Let $i : B \to N$ be the inclusion. By using $i \circ f$ instead of f, we can assume $B = N$. Moreover, by proposition 2.2.1, we only have to show that $Rf_!\mathcal{F}$ is locally $\mathcal{S}$-constructible on N.

Take an open relative compact subset U of N, which is subanalytic (semialgebraic) and such that $cl(U)$ is contained in some open (semialgebraic) subset V, which is analytic (Nash) isomorphic to some $\mathbb{R}^n$. Then we can assume $N = V = \mathbb{R}^n$ and U is a bounded open subset, definable with respect to the o-minimal structure $\mathfrak{S}(\mathcal{S})$. Moreover, $f^{-1}(U)$ is a locally compact definable space with respect to this o-minimal structure, since the closure of $f^{-1}(U)$ is a compact $\mathcal{S}$-set. Of course, here we use the properness of f, together with the assumption, that A is a closed subset (compare [vDrMi, D.7]). Similarly,

$$f_U := f : f^{-1}(U) \to U$$

is definable with respect to $\mathfrak{S}(\mathcal{S})$ and $\mathcal{F}|f^{-1}(U)$ is $\mathfrak{S}(\mathcal{S})$-constructible. Therefore

$$(Rf_!\mathcal{F})|U \simeq Rf_{U!}(\mathcal{F}|f^{-1}(U))$$

is $\mathfrak{S}(\mathcal{S})$-constructible (especially locally $\mathcal{S}$-constructible) by remark 2.2.2(2). But this remark depends on the embedding theorem for regular definable spaces. A simpler argument goes as follows:

Take finitely many open "affine charts" V_i $(i = 1, \ldots, m)$ as above, which cover $f^{-1}(U)$. Let $W_i := f^{-1}(U) \cap V_i$, $U_i := \cup_{l \leq i} W_l$ and consider the inclusion

$$j_i : U_i \to f^{-1}(U)\,.$$

Then we show by induction on i, that

$$Rf_{U!}Rj_{i!}j_i^*\mathcal{F}'$$

is $\mathfrak{S}(\mathcal{S})$-constructible (with $\mathcal{F}' := \mathcal{F}|f^{-1}(U)$). Of course, the "affine case" $i = 1$ is a special case of theorem 2.0.1. For the induction step, consider the open inclusion $j : U_i \to U_{i+1}$, with closed complement $k : X := U_{i+1} \backslash U_i \to U_{i+1}$, and let f_X denote the restriction $f : X \to U$. Note that X is a closed definable subset of V_{i+1} so that

$$Rf_{U!}Rj_{i+1!}k_*k^*j_{i+1}^*\mathcal{F}' \simeq Rf_{X!}k^*j_{i+1}^*\mathcal{F}'$$

is also $\mathfrak{S}(\mathcal{S})$-constructible (by theorem 2.0.1). If we apply $Rf_{U!}Rj_{i+1!}$ to the standard distinguished triangle

$$Rj_!j^*j_{i+1}^*\mathcal{F}' \longrightarrow j_{i+1}^*\mathcal{F}' \longrightarrow k_*k^*j_{i+1}^*\mathcal{F}' \xrightarrow{[1]} ,$$

then the claim follows, since

$$Rf_{U!}Rj_{i+1!}Rj_!j^*j_{i+1}^*\mathcal{F}' \simeq Rf_{U!}Rj_{i!}j_i^*\mathcal{F}'$$

is $\mathfrak{S}(\mathcal{S})$-constructible by the induction hypothesis. □

Remark 2.2.3.
1. In the proof of the important corollary 2.2.1, we only used the **cell decomposition theorem** and the property: X definable implies ∂X definable, with $dim X > dim \partial X$ (so we don't use (!) the deeper triangulation and trivialization theorem, or results about Whitney stratifications).

2. Corollary 2.2.1 remains true, if one only assumes that A is a locally closed subset. In the analytic geometric context one only has to take $M' := M \backslash \partial A$ instead of M. But in the Nash geometric context, M' is in general not a Nash manifold (e.g. M' is not a semialgebraic subset of M). Therefore one has to take a proper semialgebraic embedding of M into some $\mathbb{R}^m$, and uses then a definable homeomorphism of A onto a closed $\mathcal{S}$-set in $\mathbb{R}^{m+1}$.

3. The statement of corollary 2.2.1 becomes false, if B is only locally closed. Take in the example of [vDrMi, rem. to D.6, p. 531] a constant sheaf on A'.

Proof of corollary 2.0.1. For the part (1.) we only use the following properties:

$$g := |f| : A \to [0, \epsilon]$$

is a proper map such that $(Rg_*\mathcal{F})|\,]0,\epsilon]$ has locally constant cohomology sheaves. Then one has

$$R\Gamma([0,\epsilon], Rg_*\mathcal{F}) \xrightarrow{\sim} R\Gamma([0,\epsilon[\,, Rg_*\mathcal{F}) \xrightarrow{\sim} (Rg_*\mathcal{F})_0\,,$$

which implies (1.a) by the **proper base change theorem** [KS, prop. 2.6.7]. Denote the inclusions $i : \{0\} \to [0,\epsilon[$ and $j :]0,\epsilon[\, \to [0,\epsilon[$. If we apply $R\Gamma_c([0,\epsilon[\,, \cdot)$ to the distinguished triangle

$$i_*i^!\mathcal{F}' \longrightarrow \mathcal{F}' \longrightarrow Rj_*j^*\mathcal{F}' \xrightarrow{[1]}$$

(with $\mathcal{F}' := (Rg_*\mathcal{F})|[0,\epsilon[\,)$, we get the isomorphism

$$R\Gamma_{\{0\}}(Rg_*\mathcal{F}) \xrightarrow{\sim} R\Gamma_c([0,\epsilon[\,, Rg_*\mathcal{F})\,,$$

since

$$R\Gamma_c([0,\epsilon[\,, Rj_*j^*\mathcal{F}') \simeq R\Gamma(\,]0,\epsilon], \mathcal{F}'') \simeq \mathcal{F}''_\epsilon \simeq 0\,,$$

with $\mathcal{F}'' := k_!j^*\mathcal{F}'$ and $k :]0,\epsilon[\, \to]0,\epsilon]$ the inclusion. This implies (a.2) by the commutative diagram

$$\begin{array}{ccc} R\Gamma_{\{f=0\}}(A,\mathcal{F}) & \longrightarrow & R\Gamma_c(\{|f|<\epsilon\},\mathcal{F}) \\ \downarrow\wr & & \downarrow\wr \\ R\Gamma_{\{0\}}(Rg_*\mathcal{F}) & \xrightarrow{\sim} & R\Gamma_c([0,\epsilon[\,, Rg_*\mathcal{F})\,. \end{array}$$

In the same way one proves part (2.), since $(Rg_*\mathcal{F})|\,]c,\infty[$ (with $g := |f|$ and $c >> 0$) has locally constant cohomology sheaves. □

Proof of corollary 2.0.2. By using $i \circ f$ instead of f (with $i : B \to \mathbb{R}$ the inclusion), we can assume $B = \mathbb{R}$. Moreover, it is enough to prove the case that A is a closed subset (otherwise use a definable homeomorphism onto a closed definable subset of $\mathbb{R}^{n+1}$). By using the graph embedding, we can then also assume $A = \mathbb{R}^n$. Let $g := \|\cdot\|_\infty : \mathbb{R}^n \to \mathbb{R}$ be the supremum norm. Note that this is a proper definable function. Then

$$h := (f, g) : \mathbb{R}^n \to \mathbb{R}^2$$

is a proper definable function, with

$$Rf_*\mathcal{F} \simeq R\pi_*Rh_*\mathcal{F}\,.$$

Here $\pi : \mathbb{R}^2 \to \mathbb{R}$ is the projection onto the first coordinate. Since h is proper, we have $Rh_* = Rh_!$ and we can apply theorem 2.0.1. So we can assume

$$f = \pi : \mathbb{R}^2 \to \mathbb{R}\,.$$

Let $\mathcal{F}$ be constructible with respect to the filtration

$$X_\bullet : \ \emptyset = X_{-1} \subset X_0 \subset \ldots \subset X_m = \mathbb{R}^2\,,$$

with X_i closed and definable. By the **cell decomposition theorem** ([vDr, thm. 3.2.11]), there is a "decomposition" of $\mathbb{R}^2$ into finitely many (definable) locally closed cells A such that $X_i \backslash X_{i-1}$ $(i = 0, \ldots, m)$ is a union of cells as in the proof of theorem 2.0.1. By the definition of such a decomposition, the set of projections $\pi(A)$ induces a decomposition of $\mathbb{R}$:

$$\big\{(-\infty, a_1), (a_1, a_2), \ldots, (a_k, \infty), \{a_1\}, \ldots, \{a_k\}\big\}\,.$$

We claim:

1. $R\pi_*\mathcal{F}$ is constructible with respect to the filtration $\{a_1, \ldots, a_k\} =: \Sigma \subset \mathbb{R}$.
2. The canonical map $(R\pi_*\mathcal{F})_x \to R\Gamma(\{\pi = x\}, \mathcal{F})$ is an isomorphism for all $x \in \mathbb{R}\backslash\Sigma$.

This is a local statement on $\mathbb{R}\backslash\Sigma$. Let I be one of the open intervals of this decomposition of $\mathbb{R}$. As in the proof of theorem 2.0.1, $\pi^{-1}(I) = I \times \mathbb{R}$ has a partition (compatible with $X_i\backslash X_{i-1}$):

$$\{(-\infty, f_1), (f_1, f_2), \ldots, (f_n, \infty), \Gamma(f_1), \ldots, \Gamma(f_n)\}\,.$$

Here, $f_j : I \to \mathbb{R}$ are continuous (definable) function with $f_1 < \cdots < f_n$, $\Gamma(f_j)$ denotes the graph of f_j and

$$(f_{j-1}, f_j) := \{(x, r) \in I \times \mathbb{R} |\ f_{j-1}(x) < r < f_j(x)\}$$

(with $f_0 \equiv -\infty, f_{n+1} \equiv \infty$). Let $I' \subset I$ be an open subinterval, whose closure is a compact subset of I. Then the $|f_j|$ are bounded on I' by some constant $M > 0$. By assumption on $\mathcal{F}$, we get that the natural morphism

$$R\Gamma(\{x\} \times [-(m+1), m+1], \mathcal{F}) \to R\Gamma(\{x\} \times [-m, m], \mathcal{F})$$

is an isomorphism for all $x \in I'$ and all $m \in \mathbb{N}, m > M$. Since

$$\pi_m := \pi : I' \times [-m, m] \to I'$$

is proper, this implies by the **proper base change theorem** ([KS, prop. 2.6.7]):

$$R\pi_{m+1*}(\mathcal{F}|I' \times [-(m+1), m+1]) \xrightarrow{\sim} R\pi_{m*}(\mathcal{F}|I' \times [-m, m])\,.$$

By a **Mittag-Leffler argument** ([KS, prop. 2.7.1(iii)]) we get the isomorphism

$$(R\pi_*\mathcal{F})|I' \xrightarrow{\sim} R\pi_{m*}(\mathcal{F}|I' \times [-m,m])\,,$$

together with a commutative diagram (for $x \in I'$):

$$\begin{array}{ccc} (R\pi_*\mathcal{F})_x & \xrightarrow{\sim} & (R\pi_{m*}(\mathcal{F}|I' \times [-m,m]))_x \\ \downarrow & & \downarrow \wr \\ \Gamma(\{x\} \times \mathbb{R}, \mathcal{F}) & \xrightarrow{\sim} & R\Gamma(\{x\} \times [-m,m], \mathcal{F})\,. \end{array}$$

Moreover, $R\pi_{m*}(\mathcal{F}|I' \times [-m,m])$ has locally constant cohomology sheaves by the **cellular fibration lemma** applied to the filtration

$$\cup_j \Gamma(f_j) \subset I' \times [-m,m]\,, \quad \text{with } f_0 \equiv -m,\ f_{n+1} \equiv m\,. \qquad \square$$

In a similar way, we show that a continuous definable function together with a $\mathfrak{S}$-constructible complex admits a **cohomological version of a Milnor fibration** (as in assumption 1.1.1 of chapter 1).

Consider $A \subset \mathbb{R}^n$ locally closed with $f : A \to \mathbb{R}$ continuous and definable with respect to the o-minimal structure $\mathfrak{S}$. Let $\mathcal{F}$ be a $\mathfrak{S}$-constructible complex of sheaves on A and $K \subset \{f = 0\}$ a compact definable subset. Let $g : \mathbb{R}^n \to [0,\infty[$ be the distance function $dist(\cdot, K)$ to K:

$$dist(x,K) := inf(\ \{\parallel x - y \parallel_\infty \mid y \in K\}\ )$$

with $\parallel \cdot \parallel_\infty$ the supremum norm. Note that g is a continuous definable function with $K = \{g = 0\}$. Moreover, there is a constant $c > 0$, with $dist(x,K) > c$ for all $x \in \partial A$ (since K is compact and ∂A is closed). Then

$$g' := g : A' := A \cap \{g \leq c\} \to [0,c]$$

is a proper definable map with $K = \{g' = 0\}$.

Lemma 2.2.2. *There exists sequences $\epsilon_n, \delta_n \searrow 0$, with $0 < \epsilon_n << \delta_n < c$ such that the family $K_n := \{|f| \leq \epsilon_n, g \leq \delta_n\}$ $(n \in \mathbb{N})$ has the following properties:*

1. *All $R^i f_*(\mathcal{F}|K_n^*)$ are locally constant for*

$$f : K_n^* := \{0 < |f| \leq \epsilon_n, g \leq \delta_n\} \to \{x \in \mathbb{R} |\ 0 < |x| \leq \epsilon_n\}\,.$$

2. *$\forall n\ \exists \epsilon$ with $0 < \epsilon \leq \epsilon_{n+1}$ such that*

$$R\Gamma(K_n \cap \{f = x\}, \mathcal{F}) \to R\Gamma(K_{n+1} \cap \{f = x\}, \mathcal{F})$$

is an isomorphism for $|x| \leq \epsilon$.

Proof. $h := (f, g') : A' \to \mathbb{R} \times [0, c]$ is a proper definable function, with

$$Rf_*\mathcal{F} \simeq R\pi_* Rh_* \mathcal{F}$$

and $\pi : \mathbb{R} \times [0, c] \to \mathbb{R}$ the projection onto the first coordinate. Since h is proper, we have $Rh_* = Rh_!$ and we can apply theorem 2.0.1. So we can assume

$$f = \pi : \mathbb{R} \times [0, c] \to \mathbb{R}\ ,\ K = \{(0, 0)\}\ ,$$

and g is the projection $\mathbb{R} \times [0, c] \to [0, c]$. Let $\mathcal{F}$ be constructible with respect to the filtration

$$X_\bullet\ :\ \emptyset = X_{-1} \subset X_0 \subset \ldots \subset X_m = \mathbb{R} \times [0, c]\ ,$$

with X_i closed and definable. By the **cell decomposition theorem** ([vDr, thm. 3.2.11]), there exists $a > 0$ and a "decomposition" of

$$B :=]-a, a[\ \times [0, c]$$

into finitely many (definable) locally closed cells such that

$$B \cap (X_i \backslash X_{i-1}) \quad (i = 0, \ldots, m)$$

is a union of cells A, with $\pi(A) \in \{\ ]-a, 0[\ ,\ ]0, a[\ , \{0\}\}$. Moreover

$$\pi^{-1}(\pi(A)) = \pi(A) \times [0, c]$$

has a partition

$$\{(0, f_1), (f_1, f_2), \ldots, (f_n, c), \Gamma(f_0), \ldots, \Gamma(f_{n+1})\}\ .$$

Here, $f_j : \pi(A) \to [0, c]$ are continuous (definable) functions, with

$$0 \equiv f_0 < f_1 < \cdots < f_n < f_{n+1} \equiv c\ ,$$

$\Gamma(f_j)$ denotes the graph of f_j and

$$(f_{j-1}, f_j) := \{(x, r) \in \pi(A) \times [0, c]|\ f_{j-1}(x) < r < f_j(x)\}\ .$$

Let $\Sigma \subset [0, c]$ be the **finite** set (compare [vDr, cor. 3.1.6])

$$\Sigma := \{f_j(0),\ lim_{x \nearrow 0}\ f_j(x),\ lim_{x \searrow 0}\ f_j(x)\},$$

and choose a δ, with $0 < \delta < c$ and $\Sigma \cap\ ]0, \delta] = \emptyset$. For all $\delta' \in\ ]0, \delta[$, there exists an $\epsilon > 0$, with

$$f_j(x) \notin [\delta', \delta] \text{ for all } x \in I \cap \{|x| \leq \epsilon\}\ \ (\ I \in \{\ ]-a, 0[\ ,\ ]0, a[\ , \{0\}\}\ )\ .$$

Then we can apply the **cellular fibration lemma** to the proper map

$$\pi_\delta := \pi : I' \times [0, \delta] \to I' ,$$

with $I' := I \cap \{|x| \leq \epsilon\}$ and $I =]-a, 0[\ ,\ ]0, a[$ so that

$$R\pi_{\delta *}(\mathcal{F}|I' \times [0, \delta])$$

has locally constant cohomology sheaves. Moreover, we have for $|x| \leq \epsilon$ the isomorphism:

$$R\Gamma(\{x\} \times [0, \delta], \mathcal{F}) \xrightarrow{\sim} R\Gamma(\{x\} \times [0, \delta'], \mathcal{F}) ,$$

since $\mathcal{F}|(\ \{x\} \times [\delta', \delta]\ )$ has locally constant cohomology sheaves. This implies the statement of the lemma. □

Remark 2.2.4.

1. For $K = \{pt\}$ a point, the statement of the lemma 2.2.2 is a local result so that it is also true for locally compact definable spaces, or in the context of analytic (Nash) geometric categories (this is also true for K a compact definable subset, but then one has to use a suitable affine embedding). Moreover, for an o-minimal structure on $(\mathbb{R}, +, \cdot)$ one can of course use the Euclidean norm instead of the supremum norm. Our proof of the lemma is similar to [GM2, proof of lem. 6.4].

2. The existence of a **cohomological version of a Milnor fibration** (in the sense above) for a definable function should be compared with the corresponding results about **triangulations** ([Shiota, II.3.3, II.3.1′, II.4.6]) or a_f-**stratifications** ([Ta1, Ta2]) of a definable function.

3. Lemma 2.2.2 allows us to use proposition 1.3.1 and corollary 1.4.1, which imply the Künneth isomorphisms of proposition 2.0.1 (and the same result for locally compact definable spaces, or in the context of analytic (Nash) geometric categories).

Now we come to the proof of lemma 2.0.2, which implies also the following version of this lemma:

Let $\mathcal{F}$ be a complex of sheaves of R-modules, constructible with respect to the **finite cellular decompositon** $X_\bullet$ of X, with $\mathcal{F}_x \in ob(T)$ for all $x \in X$. Denote by $U_{i,j} \simeq \mathbb{R}^{n_{i,j}}$ the open "cells" of $X_i \backslash X_{i-1}$. We only assume that T is a full subcategory of $D(R)$, with $0 \in ob(T)$ such that

(ext) T is stable under **extensions**: for a distinguished triangle

$$M' \to M \to M'' \xrightarrow{[1]}$$

in $D(R)$ with $M', M'' \in ob(T)$, we have $M \in ob(T)$.

Remark that this implies the following properties:

(a.) T is stable under **isomorphisms**: If $f : M' \to M$ is an isomorphism in $D(R)$, then $M' \in ob(T)$ implies $M \in ob(T)$ by the distinguished triangle

$$M' \xrightarrow{f} M \to 0 \xrightarrow{[1]} .$$

(b.) T is stable under **finite direct sums** by the distinguished triangle

$$M \to M \oplus M' \to M' \xrightarrow{[1]} .$$

Lemma 2.2.3. 1. *Suppose that $M \in ob(T)$ implies $M[-n_{i,j}] \in ob(T)$ (for all $n_{i,j}$ as above). Then $R\Gamma_c(X, \mathcal{F}) \in ob(T)$.*

2. *Suppose $M \in ob(T)$ implies $M[1] \in ob(T)$. Then $R\Gamma_c(X, \mathcal{F})[d] \in ob(T)$, with $d := max\{n_{i,j}\}$ the dimension of X.*

3. *Suppose $M \in ob(T) \Leftrightarrow M[1] \in ob(T)$. Let μ be an* **additive** *function on T and denote $\mu_{i,j} := \mu(\mathcal{F}_x)$, with $x \in U_{i,j}$. Then $R\Gamma_c(X, \mathcal{F}) \in ob(T)$, with*

$$\mu(R\Gamma_c(X, \mathcal{F})) = \sum_{i,j} (-1)^{n_{i,j}} \cdot \mu_{i,j} .$$

Proof. We follow the proof of lemma 2.0.1. Let us denote the inclusions

$$i_k : X_{k-1} \to X_k \quad \text{and} \quad j_k : X_k \backslash X_{k-1} \to X_k \quad (k = 1, \dots, m) .$$

We prove by induction on k that $R\Gamma_c(X_k, \mathcal{F})[s] \in ob(T)$, with $s := 0$ in the case (1.) and $s := d$ in the case (2.). Using

$$R\Gamma_c(X_{k-1}, \mathcal{F}) = R\Gamma_c(X_k, Ri_{k*}i_k^*\mathcal{F})$$

and applying $R\Gamma_c(X_k, \cdot)[s]$ to the standard distinguished triangle

$$Rj_{k!}j_k^*(\mathcal{F}|X_k) \longrightarrow \mathcal{F}|X_k \longrightarrow Ri_{k*}i_k^*(\mathcal{F}|X_k) \xrightarrow{[1]} ,$$

it is enough to show that

$$R\Gamma_c(X_k, Rj_{k!}\, j_k^*(\mathcal{F})[s] \simeq R\Gamma_c(X_k \backslash X_{k-1}, \mathcal{F})[s] \in ob(T) .$$

Since $X_k \backslash X_{k-1}$ is the disjoint union of the finitely many open subsets $U_{k,j}$, we have

$$R\Gamma_c(X_k \backslash X_{k-1}, \mathcal{F}) \simeq \oplus_j \; R\Gamma_c(U_{k,j}, \mathcal{F}) .$$

So it is enough to show $R\Gamma_c(U_{k,j}, \mathcal{F})[s] \in ob(T)$. But the natural morphism

$$R\Gamma_{\{x\}}(U_{k,j}, \mathcal{F}) \to R\Gamma_c(U_{k,j}, \mathcal{F})$$

is for $x \in U_{k,j}$ an isomorphism ([KS, prop.3.2.3]). This follows also from the proof of corollary 2.0.1: take a homeomorphism

$$h : (U_{k,j}, x) \xrightarrow{\sim} (\{z \in \mathbb{R}^n | \, |z| < 1\}, 0)$$

so that $Rr_* Rh_* \mathcal{F}$ has locally constant cohomology sheaves, where

$$r : \{z \in \mathbb{R}^n | \, |z| < 1\} \to [0, 1[\quad \text{is the function} \quad z \mapsto |z| \, .$$

Then we get from the isomorphism (2.1) and remark 2.2.1(2):

$$R\Gamma_c(U_{k,j}, \mathcal{F}) \simeq \mathcal{F}_x[-n_{i,j}] \quad (\, x \in U_{k,j} \,) \, .$$

This implies (1.), (2.) and the first part of (3.) (for the second case we use of course $d - n_{i,j} \geq 0$). Assume that $M \in T$ implies $M[n] \in T$ for all $n \in \mathbb{Z}$. Then we also get the second part of (3.) by

$$\mu(0) = 0 \, , \;\; \mu(M[1]) = -\mu(M) \quad \text{and} \quad \mu(M \oplus M') = \mu(M) + \mu(M') \, ,$$

since one always has distinguished triangles

$$M \to 0 \to M[1] \xrightarrow{[1]} \quad \text{and} \quad M \to M \oplus M' \to M' \xrightarrow{[1]} \, .$$

□

Remark 2.2.5. 1. The simplest case for (1.) is of course $F \in T \Rightarrow F[-1] \in T$. Another interesting case occurs, if all cells $U_{i,j}$ are of **even** dimension. For example in the complex algebraic context, with $U_{i,j} \simeq \mathbb{C}^{n_{i,j}}$ (e.g. X is a smooth projective variety with an algebraic $\mathbb{C}^*$-action, which has only a finite fixed point set. Compare [ChGi, thm. 2.4.3]). Another example is given by a compact smooth manifold with a Morse function, whose index at each critical point is even ([ChGi, thm. 2.4.21]). Then we can take for T the complexes of R-modules, whose cohomology sheaves **vanish in odd degrees**. If we apply (1.) to a single constructible sheaf $\mathcal{F}$ in degree 0 (e.g. a locally constant sheaf), we get that $H^*_c(X, \mathcal{F})$ vanishes in odd degrees. For a more interesting example, see [J].

2. The standard example for T as in (2.) is given by the complexes of R-modules, whose cohomology sheaves vanish in negative degrees. If we apply (2.) to a single constructible sheaf $\mathcal{F}$ in degree 0, we recover the well-known fact that $H^i_c(X, \mathcal{F}) = 0$ for $i > dim\ X$ (compare [KS, prop. 3.2.2(iii)]).

Lemma 2.0.2 implies proposition 2.0.2. Moreover, together with theorem 2.0.1 it implies corollary 2.0.3 by the **base change property**

$$(Rf_! \mathcal{F})_y \simeq R\Gamma_c(\, f = y\}, \mathcal{F})$$

([KS, prop. 2.6.7]). In the same way, we get by theorem 2.0.1, corollary 2.2.1 and lemma 2.2.3 the following

Corollary 2.2.2. *Let T be a full subcategory of $D(R)$, stable under extensions, with $0 \in ob(T)$. Assume*

$$M \in ob(T) \Rightarrow M[-1] \in ob(T) \quad (\, or \quad M \in ob(T) \Rightarrow M[1] \in ob(T)\,)\,.$$

Let $D_T(A)$ denote the full subcategory of the derived category $D(A) := D(A,R)$ of sheaves of R-modules on A, whose objects are the constructible complexes, with $\mathcal{F}_x \in ob(T)$ for all $x \in A$. Consider a continuous map $f : A \to B$, whose maximal fiber dimension is d. Then $Rf_!$ (or $Rf_![d]$) maps $D_T(A)$ into $D_T(B)$ in any one of the following cases:

1. *$\mathfrak{S}$ is an o-minimal structure on $\mathbb{R}$, and $f : A \to B$ is a definable map between definable locally closed subsets of affine spaces.*
2. *$\mathfrak{S}$ is an o-minimal structure on $(\mathbb{R}, +, \cdot)$, and $f : A \to B$ is a definable map between locally compact definable spaces.*
3. *$\mathcal{S}$ is an analytic (Nash) geometric category, and $f : A \to B$ is a* **proper** *definable map between definable closed subsets $A \subset M$, $B \subset N$ of the analytic (Nash) manifolds M, N.*

The final stability result, which already holds for any o-minimal structure on $\mathbb{R}$, deals with the derived tensor product.

Suppose we have rings R_i $(i = 1, 2, 3)$ with a tensor product:

$$\otimes^L : D(X, R_1) \times D(X, R_2) \to D(X, R_3)\,,$$

which commutes with the functor f^* for a continuous map f. For example R is commutative and we take $R_i := R$ $(i = 1, 2, 3)$. Let $X \simeq \mathbb{R}^n$ be a cell and suppose that $\mathcal{F}_i \in ob(D(X, R_i))$ has locally constant cohomology sheaves $(i = 1, 2)$. Let s be a constant map on X. By the **Vietoris-Begle theorem**, we have as in the proof of lemma 2.0.1: $\mathcal{F}_i \simeq s^*\mathcal{F}'_i$ for some $\mathcal{F}'_i \in ob(D(R_i))$. Then we get

$$\mathcal{F}_1 \otimes^L \mathcal{F}_2 \simeq s^*(\mathcal{F}'_1 \otimes^L \mathcal{F}'_2)\,,$$

so that $\mathcal{F}_1 \otimes^L \mathcal{F}_2$ has locally constant cohomology sheaves. By the **cell decomposition theorem** [vDr, 3.2.11] (or the **embedding theorem** [vDr, 10.1.8] for regular definable spaces) and proposition 2.1.1, this implies the

Lemma 2.2.4. *Consider one of the following cases:*

1. *A is a topological space with a cellular decomposition or cellular stratification.*
2. *$\mathfrak{S}$ is an o-minimal structure on $\mathbb{R}$ (or on $(\mathbb{R}, +, \cdot)$) and A is a definable subset of an affine space (or a regular definable space).*
3. *$\mathcal{S}$ is an analytic (Nash) geometric category and A is a definable subset of the analytic (Nash) manifold M.*

If $\mathcal{F}_i \in ob(D(A, R_i))$ $(i = 1, 2)$ are constructible with respect to this structure, then the same is true for $\mathcal{F}_1 \otimes^L \mathcal{F}_2$.

Remark 2.2.6. Let T_i be full subcategories of $D(R_i)$ $(i = 1, 2, 3)$ such that $\otimes^L$ maps $T_1 \times T_2$ into T_3. Then $\otimes^L$ maps

$$D_{T_1}(X, R_1) \times D_{T_2}(X, R_2) \quad \text{into} \quad D_{T_3}(X, R_3) .$$

For example, R is commutative, $R_i := R$ $(i = 1, 2, 3)$ and T_i $(i = 1, 2, 3)$ denotes the **perfect** complexes of R-modules ([KS, exercise I.30]).

We consider now a more general version of lemma 2.0.3. Let X be a topological space with a **locally finite good cellular stratification**. Denote by $X_\bullet$ the induced filtration, with X_i the union of all cells of dimension $\leq i$. Let d be the dimension of X (i.e. $X_{d-1} \neq X_d = X$), and recall that we assumed all spaces to be of finite dimension.

As in lemma 2.2.3, we fix a full subcategory T of $D(R)$ stable under extensions, with $0 \in ob(T)$. Define $D_T(X)$ as the full subcategory of $D(X, R)$, consisting of the $X_\bullet$-constructible complexes of sheaves $\mathcal{F}$, with $\mathcal{F}_x \in ob(T)$ for all $x \in X$ (and similarly for X' a union of cells). Let us consider the inclusions

$$i_k : X_{k-1} \to X_k \ , \quad l_k : X_k \to X \ , \quad j_k : X_k \backslash X_{k-1} \to X_k \quad \text{and} \quad j'_k : U_k \to X_k \ ,$$

with U_k a union of k-dimensional cells.

Lemma 2.2.5. 1. *Rj'_{k*} maps $D_T(U_k)$ into $D_T(X_k)$.*
Assume $M \in ob(T) \Rightarrow M[-1] \in ob(T)$ (or $M \in ob(T) \Rightarrow M[1] \in ob(T)$).

2. *$l^!_{d-k}$ (or $l^!_{d-k}[k]$) maps $D_T(X)$ into $D_T(X_{d-k})$.*

3. *Suppose that the stratification has only* **finitely** *many cells. Then $R\Gamma(X, \cdot)$ (or $R\Gamma(X, \cdot)[d]$) maps $D_T(X)$ into T.*

Proof. 1. Suppose first that U_k is the disjoint union of finitely many cells $U_{k,i}$. Then

$$Rj'_{k*}\mathcal{F} \simeq \oplus \, Rj'_{k,i*}(\mathcal{F}|U_{k,i}) \ ,$$

with $j'_{k,i} : U_{k,i} \to X_k$ the inclusion. Therefore, we can assume $U_k =: U$ is one cell. Since ∂U_k is a union of lower-dimensional cells, we can also assume $X_k = cl(U)$ (with $j := j'_k$). Let $\mathcal{F} \in ob(D_T(U))$ be given. Let s be a constant map on $cl(U)$. By the **Vietoris-Begle theorem** (as in the proof of lemma 2.0.1) we have $\mathcal{F} \simeq j^* s^* \mathcal{F}'$ for some $\mathcal{F}' \in ob(T)$. Since we deal with a **good** stratification, this implies

$$s^*\mathcal{F}' \simeq Rj_* j^* s^* \mathcal{F}' \simeq Rj_* \mathcal{F} \ .$$

The first isomorphism holds not only for a constant sheaf, but also for a complex with constant cohomology sheaves, since Rj_* has finite cohomological dimension. This implies (1.) for U_k the disjoint union of **finitely** many cells. Since the statement of (1.) is a local result on X_k, we get also the general case, because we assumed the stratification to be **locally finite**.

2. We prove (2.) by induction on k, where the case $k = 0$ is obvious, since l_d is the identity on X. For the induction step, consider the distinguished triangle

$$Ri_{d-k*}i^!_{d-k}l^!_{d-k}\mathcal{F} \longrightarrow l^!_{d-k}\mathcal{F} \longrightarrow Rj_{d-k*}j^*_{d-k}l^!_{d-k}\mathcal{F} \xrightarrow{[1]} ,$$

with

$$i^*_{d-k}Ri_{d-k*}i^!_{d-k}l^!_{d-k}\mathcal{F} \simeq l^!_{d-k-1}\mathcal{F} .$$

Then the claim follows, if we apply the induction hypothesis to $l^!_{d-k}\mathcal{F}$ and part (1.) to $j^*_k l^!_{d-k}\mathcal{F}$.

3. In the same way we prove

$$R\Gamma(X_k, \mathcal{F}) \in ob(T) \quad (\text{ or } R\Gamma(X_k, \mathcal{F})[k] \in ob(T) \,)$$

by induction on k (starting with $k = -1$). Consider the distinguished triangle

$$Ri_{k*}i^!_k l^*_k\mathcal{F} \longrightarrow l^*_k\mathcal{F} \longrightarrow Rj_{k*}j^*_k l^*_k\mathcal{F} \xrightarrow{[1]} ,$$

with

$$R\Gamma(X_k, Ri_{k*}i^!_k l^*_k\mathcal{F}) \simeq R\Gamma(X_{k-1}, l^!_{k-1}l_{k*}l^*_k\mathcal{F}) .$$

By (2.) we have $l^!_{k-1}l_{k*}l^*_k\mathcal{F}[s] \in D_T(X_{k-1})$ (with $s = 0$ or $s = k-1$) and we can apply the induction hypothesis to this complex. Let $X_k \backslash X_{k-1}$ be the disjoint union of the finitely many cells $U_{k,j}$. Then

$$R\Gamma(X_k, Rj_{k*}j^*_k l^*_k\mathcal{F}) \simeq \oplus_j \; R\Gamma(U_{k,j}, \mathcal{F}) .$$

But for $x \in U_{k,j}$, the natural morphism

$$R\Gamma(U_{k,j}, \mathcal{F}) \to \mathcal{F}_x$$

is an isomorphism by the **Vietoris-Begle theorem** (compare [KS, prop. 3.2.3(i)]), and this implies the induction step. □

Remark 2.2.7. Let $R = k$ be a field and let T denote the cohomological bounded complexes with finite dimensional cohomology, concentrated in nonnegative degrees. Fix an o-minimal structure on $(\mathbb{R}, +, \cdot)$. If we apply (3.) to the constant sheaf k_A, with A a definable subset of some affine space, we get that $H^*(A, k)$ is finite dimensional, with $H^i(A, k) = 0$ for $i > dim\ A$.

Corollary 2.2.3. *Consider one of the following cases:*

(a.) *X is locally compact and has a* **locally finite good cellular stratification**.

(b.) *X is a locally closed subspace of some affine space, definable with respect to an o-minimal structure on $(\mathbb{R}, +, \cdot)$.*

(c.) *X is a locally closed subspace of an analytic (Nash) manifold, definable with respect to an analytic (Nash) geometric category.*

Let R be a commutative ring, and fix strict full **triangulated** *subcategories T_1 of $D^-(R)$, T_2, T_3 of $D^+(R)$ such that $Rhom_R(\cdot,\cdot)$ maps $T_1 \times T_2$ into T_3. Let $D_{T_i}(X)$ $(i = 1, 2, 3)$ be defined as before, with the obvious constructibility notion in the cases above. Then $Rhom_R(\cdot,\cdot)$ maps $D_{T_1}(X) \times D_{T_2}(X)$ into $D_{T_3}(X)$.*

Proof. First notice, that $Rhom_R(\cdot,\cdot)$ is defined by [Sp] on the whole derived category (with the usual properties), since R is commutative. By the **triangulation theorem** [vDr, thm. 8.2.9] and proposition 2.1.1, we can reduce the cases (b.),(c.) to the case (a.). For the proof of the case (a.) we use the notation of lemma 2.2.5 and follow [IC, p. 136, proof of thm. 8.6].

(1.) Suppose first that X is just one "cell" and let $\mathcal{F}_i \in ob(D_{T_i}(X))$ be given $(i = 1, 2)$. Let s be a constant map on X. By the proof of lemma 2.0.1, we have $\mathcal{F}_i \simeq s^*\mathcal{F}'_i$ for some $\mathcal{F}'_i \in ob(T_i)$. Then we get by [IC, p. 171, prop. 10.21]:

$$s^* Rhom_R(\mathcal{F}'_1, \mathcal{F}'_2) \simeq Rhom_R(\mathcal{F}_1, \mathcal{F}_2)\,,$$

and the assertion follows. Note that [IC, p.171, prop.10.21] and its proof applies, since for X a cell our assumptions on $\mathcal{F}_i$ imply

$$\mathcal{F}_1 \in ob(D^-(X)) \quad \text{and} \quad \mathcal{F}_2 \in ob(D^+(X))\,.$$

(2.) Let $Rhom_R(\mathcal{F}_1, \mathcal{F}_2) =: \mathcal{H}$. We prove by induction on k :

$$l_k^! \mathcal{H} \in ob(D_{T_3}(X_i))\,,$$

where the case $k = -1$ with $X_{-1} = \emptyset$ is obvious. For the induction step, consider the distinguished triangle

$$Ri_{k*}i_k^! l_k^! \mathcal{H} \longrightarrow l_k^! \mathcal{H} \longrightarrow Rj_{k*}j_k^* l_k^! \mathcal{H} \xrightarrow{[1]}\,,$$

with $Ri_{k*}i_k^! l_k^! \mathcal{H} \simeq Ri_{k*}l_{k-1}^! \mathcal{H}$. Then it is enough to show

$$Rj_{k*}j_k^* l_k^! \mathcal{H} \in ob(D_{T_3}(X_k))\,.$$

But [Sp, prop.6.19] implies

$$Rj_{k*}j_k^* l_k^! \mathcal{H} \simeq Rj_{k*}Rhom_R\big((l_k \circ j_k)^* \mathcal{F}_1, (l_k \circ j_k)^! \mathcal{F}_2\big)\,,$$

with

$$(l_k \circ j_k)^* \mathcal{F}_1 \in ob(D_{T_1}(X_k))\,, \text{ and } (l_k \circ j_k)^! \mathcal{F}_2 \simeq j_k^* l_k^! \mathcal{F}_2 \in ob(D_{T_2}(X_k))$$

by lemma 2.2.5(2). So by (1.) we get (since this is a local result)

$$Rhom_R\big((j_k \circ l_k)^* \mathcal{F}_1, (j_k \circ l_k)^! \mathcal{F}_2\big) \in ob(D_{T_3}(X_k \backslash X_{k-1}))\,,$$

and the claim follows from lemma 2.2.5(1). □

Note that corollary 2.2.3 and lemma 2.2.4 imply part (3.) of theorem 2.0.2, up to the biduality statement.

Corollary 2.2.4. *Consider an inclusion $l : B \to A$, with A, B subsets of X in one of the following cases:*

(a.) *X has a* **locally finite regular cellular stratification** *and A, B are unions of cells.*

(b.) *X is an affine space and A, B are definable with respect to the o-minimal structure $\mathfrak{S}$ on $(\mathbb{R}, +, \cdot)$.*

(c.) *X is an analytic (Nash) manifold and A, B are definable with respect to the analytic (Nash) geometric category $\mathcal{S}$.*

Fix a full subcategory T of $D(R)$ stable under extensions, with $0 \in ob(T)$. Assume $M \in ob(T) \Rightarrow M[-1] \in ob(T)$. Let $D_T(Y)$ $(Y = A, B)$ be defined as before, with the obvious constructibility notion in the cases above (in the last case we mean constructible with respect to $\mathcal{S}(X)$). Then one has:

1. *Rl_* maps $D_T(B)$ into $D_T(A)$.*
2. *Assume that B is a locally closed subset of A. Then $l^!$ maps $D_T(A)$ into $D_T(B)$.*

Proof. The case (c.) follows from the case (a.) by proposition 2.1.1. In the case (b.), we first have to take a Nash isomorphism of X onto a bounded open semialgebraic subset. Then we can work with $cl(X)$ instead of X and we use the **triangulation theorem** [vDr, thm. 8.2.9] to reduce this case to (a.). Since $cl(X)$ is compact, such a triangulation is indeed a **regular** cell stratification. So we only have to consider the case (a.). Note that the induced stratifications of A and B are **good** cellular stratifications.

(1.) We prove the result by induction on $k = dim\, B$ (and start with $k = -1$, i.e. $B = \emptyset$). For the induction step, consider the usual filtration $B_\bullet$ of B indexed by the dimension of the cells. Let us denote the inclusions

$$i : B_{k-1} \to B_k = B\ , \quad l : B \to A \ \text{ and } \ j : B \backslash B_{k-1} \to B\ .$$

Consider the distinguished triangle

$$Ri_* i^! \mathcal{F} \longrightarrow \mathcal{F} \longrightarrow Rj_* j^* \mathcal{F} \xrightarrow{[1]} \ .$$

Then we can use lemma 2.2.5(2) and the induction hypothesis for $i^! \mathcal{F}$, with

$$Rl_* Ri_* \simeq R(l \circ i)_*\ .$$

Similarly

$$Rl_* Rj_* \simeq R(l \circ j)_*\ ,$$

and we can use lemma 2.2.5(1) for the inclusion $l \circ j : X_k \backslash X_{k-1} \to A$.

(2.) For the second result, look at the inclusions

$$k : B \to cl_A(B) \quad \text{and} \quad i : cl_A(B) \to A\,.$$

Then $cl_A(B)$ is also a union of strata with $l^! \simeq k^! i^!$. Since B is open in $cl_A(B)$, we have $k^! \simeq k^*$. Therefore, we can assume that B is a closed subset of A. Denote the open inclusion $j : A\backslash B \to A$. Then the claim follows from (1.) (applied to j) and the distinguished triangle

$$Ri_* i^! \mathcal{F} \longrightarrow \mathcal{F} \longrightarrow Rj_* j^* \mathcal{F} \xrightarrow{[1]} .$$

□

Remark 2.2.8. An important example for (1.) in the case (c.) is the following: $A = X$, B is the open complement of a closed definable subset, and we look at $l^*\mathcal{F}$ for some $\mathcal{F} \in ob(D_T(X))$. Then of course $l^*\mathcal{F}$ is constructible with respect to $\mathcal{S}(X)$. But the above results in case (c.) **do not** apply, if we **only assume** that $\mathcal{F} \in ob(D(B))$ is constructible with respect to $\mathcal{S}(B)$!

Corollary 2.2.5. *Consider one of the following cases:*

(a.) *M, N are affine spaces and $A \subset M, B \subset N$ are definable with respect to the o-minimal structure $\mathfrak{S}$ on $(\mathbb{R}, +, \cdot)$.*

(b.) *M, N are analytic (Nash) manifolds and $A \subset M, B \subset N$ are definable with respect to the analytic (Nash) geometric category $\mathcal{S}$.*

Let $f : A \to B$ be continuous and definable, with B locally closed. Fix a full subcategory T of $D(R)$ stable under extensions, with $0 \in ob(T)$. Let $D_T(Y)$ $(Y = A, B)$ be defined as before (in the last case, we mean constructibility with respect to the $\mathcal{S}$-structure of the ambient manifold).

1. *Consider the case* (a.) *and assume $M \in ob(T) \Rightarrow M[-1] \in ob(T)$. Then Rf_* maps $D_T(A)$ into $D_T(B)$.*

2. *Let also A be locally closed in the case* (a.)*, or closed in the case* (b.)*. Assume $M \in ob(T) \Leftrightarrow M[-1] \in ob(T)$. Then $f^!$ maps $D_T(B)$ into $D_T(A)$.*

Proof. (1.) Let $i : B \to N$ be the inclusion. Since B is locally closed, we have

$$i^* R(i \circ f)_* \simeq i^* Ri_* Rf_* \simeq Rf_*\,.$$

So it is enough to consider $i \circ f$ instead of f, and we can assume $B = N$. By using a Nash isomorphism of M onto a bounded open semialgebraic subset of some affine space, we can assume that A is bounded. Consider the graph isomorphism

$$h := (id, f) : A \to h(A) \subset M \times N\,,$$

and the inclusion

$$i : h(A) \to A' := cl(h(A))\,.$$

Then $f = \pi \circ i \circ h$, with $\pi : A' \to N$ the restriction of the projection onto N. Here π is definable and moreover **proper** (!), since A is bounded. So the claim follows, if we apply corollary 2.2.4(1) to i and corollary 2.2.2(1) to π.

(2.) Consider the graph isomorphism

$$h := (id, f) : A \to h(A)$$

and the (locally) closed inclusion

$$i : h(A) \to M \times B \, .$$

Then $f = \pi \circ i \circ h$, with $\pi : M \times B \to B$ the projection, and we get

$$f^! \simeq h^! i^! \pi^! \, , \quad \text{with } h^! \simeq h^* \, .$$

Moreover, π is a topological submersion with fiber M so that

$$\pi^! \simeq \pi^*[dim\ M] \quad (\text{ locally on } M \times B\ )$$

(compare [KS, prop. 3.3.2]). So the claim follows, if we apply lemma 2.2.1 to h (here we use in case (b.) that A and therefore also $h(A)$ is closed), corollary 2.2.4(2) to i and lemma 2.2.1 to π. Note that $(\pi^*\mathcal{F})_x \simeq \mathcal{F}_{\pi(x)}$ so that all stalk properties are preserved by π^*. □

Remark 2.2.9. Since we consider in the case (2.) above only locally compact spaces of bounded dimension, we have the right adjoint functor $f^!$ of $Rf_!$ with its usual properties even for **unbounded** complexes. The case of a commutative base ring R is already worked out by Spaltenstein [Sp]. But his arguments also work in the general case by the following modifications: suppose we consider complexes of sheaves of left R-modules. Use in [Sp, sec. 6.13] a right resolution $\mathcal{K}$ of the constant sheaf R_X by a bounded below complex of c-soft sheaves in the category of sheaves of R-**bimodules**, and define $f^! := Rf^!_{\mathcal{K}}$ as in [Sp, p. 148/149]. Then one gets a natural morphism between complexes of sheaves of abelian groups

$$Hom_R(f_!(\mathcal{K} \otimes_R \mathcal{A}), \mathcal{B}) \to f_* Hom_R(\mathcal{A}, f^!_{\mathcal{K}}\ \mathcal{B}) \, ,$$

and the arguments of Spaltenstein [Sp] work without further modifications.

Remark 2.2.10. 1. One should compare the results above and their proof with the corresponding results in l-adic cohomology ([SGA4$\frac{1}{2}$, thm. 1.1, cor. 1.5]).

2. Corollary 2.2.5 implies the parts (1.) and (2.) of theorem 2.0.2.

3. By the embedding theorem, the statements of corollary 2.2.5 are also true for definable spaces, if we assume: B is locally compact and A is regular in the case (1.), and A, B are locally compact in the case (2.).

In the same way one gets the

Proof of corollary 2.0.4(1). 1. Let us first consider the case, that $f_i : A_i \to B_i$ are open inclusions. Then the claim follows from proposition 2.0.1 applied to $Rf_{i*}\mathcal{F}_i$, which are constructible by corollary 2.2.5(1).

2. The **Künneth morphism** is compatible with **composition** of maps (compare section 1.4). So by composition with the inclusions $B_i \to cl(B_i)$, we can assume that B_i is a closed subset. As before, we can also assume that A_i is bounded. Then we consider again the graph isomorphism

$$h_i := (id, f_i) : A_i \to h_i(A_i) \subset M_i \times N_i$$

so that the restriction of the projection to the closure

$$\pi_i : cl(h_i(A_i)) =: A'_i \to B_i$$

is **proper** ($i = 1, 2$). Then we apply the case 1. to the inclusions $h_i(A_i) \to A'_i$ (note that A_i and therefore also $h_i(A_i)$ is locally closed), and the general **Künneth isomorphism** (1.16) for **proper** maps (compare also with [KS, Ex. II.18], [IC, p. 171, 10.20(b)]) to the projections π_i. □

An interesting application of the previous results is the following (compare [KS, prop. 8.4.9])

Corollary 2.2.6. *Let $A \subset \mathbb{R}^n$ be locally closed and definable with respect to the o-minimal structure $\mathfrak{S}$ on $(\mathbb{R}, +, \cdot)$. Let $\mathcal{F}$ be $\mathfrak{S}$-constructible on A. Fix a strict full* **triangulated** *subcategory T of $D(R)$. For $x \in A$, denote by i_x the inclusion $\{x\} \to A$. Then the following conditions are equivalent:*

1. *For all $x \in A$ we have $i_x^*\mathcal{F} \in ob(T)$.*
2. *For all $x \in A$ we have $i_x^!\mathcal{F} \in ob(T)$.*

Proof. As in lemma 2.2.2, let $g : \mathbb{R}^n \to \mathbb{R}$ be the distance function to $K = \{x\}$ and choose $c > 0$, with

$$g' := g : A' := A \cap \{g \leq c\} \to [0, c] \quad \text{proper}\,.$$

1. $\Rightarrow$ 2.: Since $Rg'_*(\mathcal{F}|A') =: \mathcal{G}$ is constructible, there exits $\epsilon > 0$, such that $\mathcal{G}|\,]0, \epsilon]$ has locally constant cohomology sheaves. Then we get

$$i_x^!\mathcal{F} \simeq R\Gamma_c(\{g' < \epsilon\}, \mathcal{F})\,,$$

and the result follows from the distinguished triangle

$$R\Gamma_c(\{g' < \epsilon\}, \mathcal{F}) \longrightarrow R\Gamma(\{g' \leq \epsilon\}, \mathcal{F}) \longrightarrow R\Gamma(\{g' = \epsilon\}, \mathcal{F}) \xrightarrow{[1]} .$$

2. $\Rightarrow$ 1.: We prove the claim by induction on $d := dim\ A$ (where the case $d = 0$ is obvious, since in this case, $i_x^* \simeq i_x^!$). With ϵ as before, we also get

$$i_x^*\mathcal{F} \simeq R\Gamma(\{g' \leq \epsilon\}, \mathcal{F})\,.$$

By the distinguished triangle above, it is enough to show

$$\mathcal{G}_\epsilon \simeq R\Gamma(\{g' = \epsilon\}, \mathcal{F}) \in ob(T) .$$

But $\mathcal{G}_\epsilon \simeq \mathcal{G}_{\epsilon/2} \simeq i^!_{\epsilon/2}\mathcal{G}[-1]$ (compare with the proof of lemma 2.2.3(1), or with [KS, prop. 3.2.3]). Let us denote the inclusion $k : \{g' = \epsilon/2\} \to A$. Then one has the **base change isomorphism** ([KS, Ex. III.9] or [IC, p. 159, 10.7(4)]):

$$i^!_{\epsilon/2}\mathcal{G} \simeq R\Gamma(\{g' = \epsilon/2\}, k^!\mathcal{F}) .$$

Moreover, for ϵ small enough, one has

$$dim\ \{g' = \epsilon/2\} < dim\ A$$

(by [vDr, prop. 4.1.5]). So we can apply the induction hypotheses to $k^!\mathcal{F}$ and the result follows. □

Especially, for T equal to the the subcategory of **perfect** complexes (see [KS, exercise I.30]) and the base ring R commutative, this together with corollary 2.0.1 implies that

$$\mathcal{F} \in D_T(A, \mathfrak{S})$$

is **cohomologically constructible** in the sense of [KS, def.3.4.1]. Therefore, the **biduality result** [KS, prop. 3.4.3(ii)] holds for $\mathcal{F} \in ob(D_T(A, \mathfrak{S}))$ (compare also with [SHS, Exp. 10, thm. 2.1]):

$$\mathcal{F} \xrightarrow{\sim} D_A(D_A(\mathcal{F})) , \tag{2.2}$$

with $D_A\mathcal{F} := Rhom_R(\mathcal{F}, c^!R)$ and c a constant map on A. Since the above result is a local result, it is also true for locally compact definable spaces, and in the context of analytic (Nash) geometric categories.

As an application one gets (compare [SGA5, prop. 1.12, p. 12]) the

Corollary 2.2.7. *Consider one of the following cases:*

(a.) *M, N are affine spaces and $A \subset M, B \subset N$ are definable with respect to the o-minimal structure $\mathfrak{S}$ on $(\mathbb{R}, +, \cdot)$.*

(b.) *M, N are analytic (Nash) manifolds and $A \subset M, B \subset N$ are definable with respect to the analytic (Nash) geometric category $\mathcal{S}$.*

Let $f : A \to B$ be continuous and definable, with A, B locally closed in the case (a.), *or closed in the case* (b.). *Let R be commutative and denote by T the full subcategory of* **perfect** *complexes in $D(R)$.*

1. *$f^!(D_B\mathcal{F}) \simeq D_A(f^*\mathcal{F})$ for $\mathcal{F} \in ob(D(A))$, and $Rf_*(D_A\mathcal{F}) \simeq D_B(Rf_!\mathcal{F})$ for $\mathcal{F} \in ob(D(A))$.*

2. *In the case* (b) *we assume in addition that f is proper. Then one has an isomorphism $Rf_!(D_A\mathcal{F}) \simeq D_B(Rf_*\mathcal{F})$ for $\mathcal{F} \in ob(D_T(A))$.*

3. *For $\mathcal{F} \in ob(D_T(B))$ one has an isomorphism $f^*(D_B\mathcal{F}) \simeq D_A(f^!\mathcal{F})$.*

Proof. (1.) follows from

$$f^! Rhom_R(\mathcal{F}, c^! R) \simeq Rhom_R(f^*\mathcal{F}, f^! c^! R)$$

([Sp, prop. 6.19]) and

$$Rf_* Rhom_R(\mathcal{F}, f^! c^! R) \simeq Rhom_R(Rf_! \mathcal{F}, c^! R)$$

([Sp, prop. 6.15.(1)]), together with $f^! c^! \simeq (c \circ f)^!$ ([Sp, prop. 6.17(b)]).

(2.) We use 1. and apply the **biduality isomorphism** to $\mathcal{F}$ and $Rf_!(D_A\mathcal{F})$ (which is allowed, since $Rf_!(D_A\mathcal{F}) \in ob(D_T(B))$ by previous results):

$$D_B(Rf_*\mathcal{F}) \simeq D_B(Rf_*(D_A D_A \mathcal{F})) \simeq D_B D_B(Rf_!(D_A\mathcal{F})) \simeq Rf_!(D_A\mathcal{F}) .$$

3. In the same way one gets

$$D_A(f^!\mathcal{F}) \simeq D_A(f^!(D_B D_B \mathcal{F})) \simeq D_A D_A(f^*(D_B\mathcal{F})) \simeq f^*(D_B\mathcal{F}) . \qquad \square$$

Corollary 2.2.8. *Consider one of the following cases:*

(a) *M_i are affine spaces and $A_i \subset M_i$ are locally closed and definable with respect to the o-minimal structure $\mathfrak{S}$ on $(\mathbb{R}, +, \cdot)$.*

(b) *M_i are analytic (Nash) manifolds and $A_i \subset M_i$ are closed and definable with respect to the analytic (Nash) geometric category $\mathcal{S}$ $(i = 1, 2)$.*

Let R be commutative and denote by T the full subcategory of perfect complexes in $D(R)$. Then one has a natural isomorphism

$$(D_{A_1}\mathcal{F}_1) \boxtimes^L (D_{A_2}\mathcal{F}_2) \simeq D_{A_1 \times A_2}(\mathcal{F}_1 \boxtimes^L \mathcal{F}_2) ,$$

for $\mathcal{F}_1 \in ob(D_T(A_1))$ and $\mathcal{F}_2 \in ob(D(A_2))$.

Proof. As we explained before, $\mathcal{F}_1 \in D_T(A_1)$ is **cohomologically constructible** in the sense of [KS, Def. 3.4.1]. Then the natural morphism

$$(D_{A_1}\mathcal{F}_1) \boxtimes^L (D_{A_2}\mathcal{F}_2) \to Rhom_R(q_1^*\mathcal{F}_1, q_2^!(D_{A_2}\mathcal{F}_2))$$

is an isomorphism ($q_i : A_1 \times A_2 \to A_i$ denotes the projection). Since this a local result, we can assume that $\mathcal{F}_1$ is bounded and of finite tor-dimension (i.e. isomorphic to a **bounded complex of flat** sheaves). Then we can by truncation also assume that $\mathcal{F}_2$ is bounded from below and the claim follows from [KS, prop. 3.4.4] (or we change in the following argument the numbering of the $\mathcal{F}_i$ and apply [SHS, theorem 4.1, Exp. 9]). Then the statement of the corollary follows from corollary 2.2.7(1):

$$q_2^!(D_{A_2}\mathcal{F}_2) \simeq D_{A_1 \times A_2}(q_2^*\mathcal{F}_2) ,$$

and [Sp, prop. 6.7(2)]:

$$Rhom_R(q_1^*\mathcal{F}_1, D_{A_1 \times A_2}(q_2^*\mathcal{F}_2)) \simeq D_{A_1 \times A_2}(q_1^*\mathcal{F}_1 \otimes^L q_2^*\mathcal{F}_2) . \qquad \square$$

Finally we prove the "generic base change theorem":

Proof of theorem 2.0.3. Let $\mathcal{F}$ be constructible with respect to the filtration X_i $(0 \le i \le N)$ of A by definable subsets. By [vDr, thm. 9.1.7], there is a partition of B into definable subsets B_j $(1 \le j \le M)$ such that $(f; X_i)$ is **definable trivial** over each B_j, i.e.

$$f : f^{-1}(B_j) \to B_j$$

is (definable) isomorphic to a projection

$$B_j \times F \to B_j$$

such that $f^{-1}(B_j) \cap X_i$ corresponds under this isomorphism to $B_j \times F_i$, for some (definable) subset F_i of F.

By taking a definable **cell stratification** partitioning each of the B_j ([vDr, prop. 4.1.13]), we can assume that the B_j define such a cell stratification. Let U be the union of the open cells of this stratification (i.e. of the cells S, with $S \not\subset \bar{T}$ for a cell $T \neq S$). Then U is an **open and dense definable** subset of B. Moreover each such cell S is open and closed in U. Since the statement of the theorem is local on U, we can assume that U is just one cell, with

$$f_U : A_U = U \times F \to U$$

the projection. Moreover, $\mathcal{F}$ is a complex of sheaves, whose restriction to the fibers of the projection $U \times F \to F$ has locally constant cohomology sheaves. Consider a cartesian diagram as in the theorem

$$\begin{array}{ccc} A' & \xrightarrow{f'} & B' \\ \downarrow{\scriptstyle g'} & & \downarrow{\scriptstyle g} \\ A_U & \xrightarrow{f_U} & U. \end{array}$$

Then f' is (isomorphic to) the projection $A' = B' \times F \to B'$ and $g'^*\mathcal{F}$ is a complex of sheaves, whose restriction to the fibers of the projection $B' \times F \to F$ has locally constant cohomology sheaves. Fix a point

$$p' \in B' \quad \text{with} \quad p := g'(p') \in U\,.$$

By the **conic structure theorem** [vDr, thm. 9.2.3] (which is an easy application of the trivialization theorem), we can assume (after shrinking B'), that B' is the **open cone** with base

$$B' \cap \{x \in \mathbb{R}^k : ||x - p'|| = \epsilon\}$$

and vertex p'. Here the open cone with base the empty set is by definition the vertex. Since B' was assumed to be locally closed, we get that

$$B'_\epsilon := B' \cap \{x \in \mathbb{R}^k : \; ||x - p'|| \le \epsilon\}$$

is a **compact** set for ϵ small enough. Then

$$B'_\epsilon \times F \to F$$

is a **proper** map in the sense of [KS, Def. on p. 103]. Therefore we can apply [KS, prop. 2.7.8], which implies the isomorphism

$$R\Gamma(B' \times F, g'^*\mathcal{F}) \xrightarrow{\sim} R\Gamma(\{p'\} \times F, g'^*\mathcal{F}) .$$

The same argument works also for $f_U : A_U = U \times F \to U$ and $\mathcal{F}$. Since we have these isomorphisms for a fundamental system of open neighborhoods of p' or p, we finally get the commutative diagram

$$\begin{array}{ccccc} (g^* Rf_{U*}\mathcal{F})_{p'} & \xleftarrow{\sim} & (Rf_{U*}\mathcal{F})_p & \xrightarrow{\sim} & R\Gamma(\{f = p\}, \mathcal{F}) \\ \downarrow & & \downarrow & & \downarrow \wr \\ (Rf'_* g'^* \mathcal{F})_{p'} & \xrightarrow{\sim} & R\Gamma(\{f' = p'\}, g'^*\mathcal{F}) & = & R\Gamma(\{f' = p'\}, g'^*\mathcal{F}) . \end{array}$$

□

Remark 2.2.11. In the proof above we only used that

$$f_U := f : \ A_U := A \cap f^{-1}(U) \to U$$

is a locally trivial fibration (in a filtration preserving way), and that A, B have locally a conic structure (over a compact base). Therefore, it implies also the similar statement in the **complex algebraic** context. The **generic fibration result** follows from stratification theory, see for example [Ver], especially [Ver, cor. 5.1].

2.3 Constructible functions

We explain in this section the relation between the standard Grothendieck functors for **constructible sheaves** and their counterparts in the theory of **constructible functions** (compare [Scha, Viro] and [KS, sec. 9.7]).

First we extend [PK, prop. 1, sec. 3] into the context of a space with a **finite cellular stratification**. Let X be a locally compact space and Z a subset of X. Suppose X has a finite cellular decomposition such that Z is the disjoint union of cells U_j $(j = 1, \ldots, N)$, and define

$$\chi(Z) := \sum_j (-1)^{dim\ U_j} \quad (\text{ with } \chi(\emptyset) := 0).$$

Lemma 2.3.1. *Suppose X has a finite cellular stratification such that $Z \subset X$ is the disjoint union of cells. The Euler characteristic $\chi(Z)$ of Z does* **not** *depend on the choice of the cellular stratification of X (with Z a union of cells).*

Proof. If Z is (locally) closed, this follows from the **Euler-Poincaré formula**, i.e. lemma 2.0.2 applied to the constant sheaf k_Z with $R = k$ a field and μ the usual Euler charateristic:

$$\chi(Z) = \sum_i (-1)^i \cdot dim_k H^i_c(Z, k).$$

In general, define by induction

$$Z_{i+1} := \partial Z_i = cl(Z_i) \setminus Z_i \text{ with } Z_0 := Z .$$

Then Z_i and $cl(Z_i)$ is the union of cells of such a cellular stratification, with $Z_i = \emptyset$ for $i >> 0$ (since the frontier of a cell is a union of cells of smaller dimension). Then the claim follows from

$$\chi(Z) = \sum_{i \geq 0} (-1)^i \cdot \chi(cl(Z_i)) . \qquad \square$$

Remark 2.3.1. By [vDr, prop. 4.1.13] (or proposition 2.1.1), this lemma applies especially to a subset Z of an affine space X (or a real analytic (Nash) manifold X), which is definable with respect to some o-minimal structure on $\mathbb{R}$ (which is relative compact and definable with respect to some analytic (Nash) geometric category). Therefore $\chi(Z)$ does not depend on the choice of this structure !

Example 2.3.1. *For a* **compact convex** *subset $Z \neq \emptyset$ of an affine space, we get*

$$\chi(Z) = \sum (-1)^i \cdot dim_k \ H^i(Z, k) = 1 .$$

Example 2.3.2. *Let $Z \neq \emptyset$ be a* **convex** *subset of an affine space V, which is definable with respect to an o-minimal structure on $\mathbb{R}$. Suppose $Z = ri(Z)$, where $ri(Z)$ denotes the interior of Z in its affine hull $aff(Z)$ (see [Br, sec. 1.3]). Then we can take Z as a cell of a cellular stratification of $cl(Z)$. This implies*

$$\chi(Z) = (-1)^{dim(Z)} .$$

Moreover, $dim(Z)$ is in this case the dimension of the affine space aff(Z).

Let $f : A \to B$ be a continuous map. We fix a commutative ring R, a strict full **triangulated** subcategory T of $D(R)$ and an **additive** map μ on T with values in the abelian group G. We consider from now on always one of the following cases:

1. A, B are locally compact spaces with a fixed cellular decomposition or cellular stratification, and f is a **definable** or **cellular** map in the sense that $f^{-1}(U)$ is a union of cells of A, for each cell U of B.

2. We fix an o-minimal structure $\mathfrak{S}$ on $\mathbb{R}$ (on $(\mathbb{R}, +, \cdot)$), A, B are locally closed definable subsets of affine spaces (are locally compact definable spaces), and f is a definable map.

3. We fix an analytic (Nash) geometric category $\mathcal{S}$, A, B are **closed** definable subsets of analytic (Nash) manifolds, and f is a definable map.

Note that all cases are stable under **products**: $A_1 \times A_2$, the projections

$$A_1 \times A_2 \to A_i$$

and the product of definable maps are definable.

Consider for example the case (1.): A cellular decomposition (stratification) of A_i, with associated filtration by dimension $A_{i,j}$ $(i = 1, 2)$, induces a cellular decomposition (stratification) of $A_1 \times A_2$, with associated filtration by dimension defined as

$$(A_1 \times A_2)_k := \cup_{i+j \leq k} A_{1,i} \times A_{2,j} \,.$$

The corresponding cells are the products of the cells of A_1 and A_2.

In each of these cases, we denote by $D_T(A)$ the strict full triangulated subcategory of $D(A, R)$, whose objects are the **constructible** complexes $\mathcal{F}$, with $\mathcal{F}_x \in ob(T)$ for all $x \in A$. Similarly, we denote by $CF(A, G)$ the abelian group of G-valued **constructible functions**. Here we call

$$\phi : A \to G \quad \text{constructible}\,,$$

if all the sets $\{\phi = g\}$ are definable $(g \in G)$, and this family of subsets is **finite** in the case (2.), or **locally finite** in the cases (1.) and (3.). Moreover, in case (1.) definable means by definition a union of cells.

We introduce also the following notation: $D_{T,c}(A)$ denotes the strict full triangulated subcategory of $D_T(A)$, whose objects are the constructible complexes $\mathcal{F}$ with **compact support** in the sense, that there exists a compact definable subset K of A (depending on $\mathcal{F}$), with

$$\mathcal{F}|(A \backslash K) \simeq 0 \,.$$

In the same way, one defines the abelian group $CF_c(A, G)$ of G-valued constructible functions with compact support. If we use this notation in case (1.), then we assume that we consider a cell **stratification** (so that the closure of a definable subset is definable).

Let us for example consider the case (2.) (with $G = \mathbb{Z}$) for the simplest o-minimal structure on $\mathbb{R}$, the **semilinear** sets ([vDr, sec. 1.7]). In this simple but important example, $CF_c(V, \mathbb{Z})$ is generated by the indicator functions of **convex polytopes** $1_P \equiv [P]$ in the affine space V, subject to the relations ([Mor, thm. 2.(2)]):

(val) $[P] + [Q] = [P \cup Q] + [P \cap Q]$ for each pair of convex polytopes P, Q, whose union is also a convex polytope.

This group is related to **valuation** theory for convex polytopes (see [Mor] and [PK]). Moreover, $CF(V,\mathbb{Z})$ is generated by the indicator functions of **polyhedra**, i.e. subsets which are the intersection of finitely many closed halfspaces ([Br, p. 51]), subject to the same relations as in 2.3, with polyhedrons instead of convex polytopes ([Mor, cor. 1]). $CF(V,\mathbb{Z})$ is also generated ([vDr, cor. 1.7.6]) by the indicator functions of the **basic semilinear** sets:

$$\{x \in V|\ f_i = 0,\ g_j > 0\}\ ,$$

with f_i $(i = 0,\dots,m)$ and g_j $(j = 0,\dots,k)$ **affine** functions. Note that such a basic semilinear set Z is convex with $Z = ri(Z)$.

Let us come back to the general case. As in the introduction, we have the additive map

$$\mu_A : D_T(A) \to CF(A,G)\ ,\ \mathcal{F} \mapsto \mu(\mathcal{F}) := (x \mapsto \mu(\mathcal{F}_x))\ ,$$

which induces also an additive map

$$\mu_A : D_{T,c}(A) \to CF_c(A,G)\ .$$

Since the inverse image under f of a definable subset is definable, we have $f^* : D_T(B) \to D_T(A)$ and a natural group homomorphism

$$f^* : CF(B,G) \to CF(A,G),\ f^*(\phi) := \phi \circ f$$

such that the diagram

$$\begin{array}{ccc} D_T(B) & \xrightarrow{f^*} & D_T(A) \\ \downarrow{\scriptstyle \mu_B} & & \downarrow{\scriptstyle \mu_A} \\ CF(B,G) & \xrightarrow{f^*} & CF(A,G) \end{array}$$

commutes. The same is true for the sheaves/functions with **compact** support, if f is **proper**. Moreover, the pullback f^* for constructible functions is **functorial** in f, i.e. the identity map is definable with $id^* = id$, and the composition of definable maps is definable, with $(f \circ g)^* = g^* \circ f^*$.

Similarly, if T is stable under the tensor product (i.e. $\otimes^L$ maps $T \times T$ into T), then also $D_T(A)$ is stable under $\otimes^L$ (by lemma 2.2.4 and remark 2.2.6). Suppose that μ is a **multiplicative** map, i.e. G is a commutative ring and

$$\mu(M \otimes^L M') = \mu(M) \cdot \mu(M') \quad \text{for } M, M' \in ob(T)\ .$$

Then we also have a commutative diagram (and similarly for the compact support case):

$$\begin{array}{ccc} D_T(A) \times D_T(A) & \xrightarrow{\otimes^L} & D_T(A) \\ \downarrow{\scriptstyle \mu_A \times \mu_A} & & \downarrow{\scriptstyle \mu_A} \\ CF(A,G) \times CF(A,G) & \xrightarrow{m} & CF(A,G)\ , \end{array}$$

where the map m is just pointwise multiplication. Moreover, this multiplication **commutes** with pullback:

$$m(f^*(\phi_1), f^*(\phi_2)) = f^*(m(\phi_1, \phi_2)) .$$

Especially, μ commutes with the exterior product $\boxtimes^L$, if we define $\phi_1 \boxtimes \phi_2$ by

$$\phi_1 \boxtimes \phi_2(x_1, x_2) := \phi_1(x_1) \cdot \phi_2(x_2) .$$

Then one gets for the product of two definable maps the formula

$$(f_1 \times f_2)^*(\phi_1 \boxtimes \phi_2) = f_1^*(\phi_1) \boxtimes f_2^*(\phi_2) .$$

More interesting is a similar result for the functor $Rf_!$. Note that $Rf_!$ maps $D_T(A)$ into $D_T(B)$ in case (2.) and for proper f also in case (3.) (by corollary 2.2.2). Similarly, $Rf_!$ respects in both cases the subcategories of complexes with compact support. By lemma 2.0.1, lemma 2.0.2 and the **base change isomorphism** for $Rf_!$ [KS, prop. 2.6.7]), the same is true in case (1.), if we make the following assumption (compare [PK, Def. 4.(C), sec. 3]):

(ccf) For each cell U of B, $f_U := f : U' := f^{-1}(U) \to U$ is a **cellular fibration** with respect to a filtration U'_i of U' such that $U'_i \backslash U'_{i-1}$ is a disjoint union of finitely many relative open subsets, which are cells of A.

Assume now that μ is surjective (otherwise consider the image of μ, which is always a subgroup of G). We claim that μ_A is also surjective. Consider first one of the following cases:

(a.) $\mu_A : D_T(A) \to CF(A, G)$ in the case (1.) with a finite cell decomposition, or in the case (2.).

(b.) $\mu_A : D_{T,c}(A) \to CF_c(A, G)$ in all cases.

Then $CF(A, G)$ (or $CF_c(A, G)$) is **generated** by functions ϕ of the following type: take a locally closed (and relatively compact) definable subset X of A, a $g \in G$ and define $\phi := g \cdot 1_X$ as

$$\phi(x) := \begin{cases} g & \text{for } x \in X , \\ 0 & \text{for } x \notin X . \end{cases}$$

Take a complex $M \in ob(T)$, with $\mu(M) = g$. Then $\phi = \mu(\mathcal{F})$ for $\mathcal{F}$ the trivial extension of the sheaf complex c^*M, with c a constant map on X. Let us consider the remaining cases. Then there exists for a given constructible function ϕ an **adapted definable stratification**, i.e. a finite filtration

$$A_\bullet : \emptyset = A_{-1} \subset A_0 \subset \ldots \subset A_m = A$$

by definable closed subsets such that the connected components $U_{i,j}$ of $A_i \backslash A_{i-1}$ are open in $A_i \backslash A_{i-1}$, with $\phi | U_{i,j}$ constant (in case (1.) use the given cell stratification and in case (3.) use proposition 2.1.1). Then

$$\phi = \mu(\mathcal{F}) \text{ for } \mathcal{F} := \oplus_i \mathcal{F}_i ,$$

with $\mathcal{F}_i$ the trivial extension of a sheaf complex on $A_i \backslash A_{i-1}$, whose restriction to $U_{i,j}$ is defined as before. Moreover, this argument shows that ϕ has a representation as a **locally finite** sum

$$\phi = \sum_k g_k \cdot 1_{X_k} ,$$

with X_k a locally finite family of locally closed definable subsets (which in case (1.) for a regular cellular complex or in case (3.) can be chosen as relatively compact cells). Conversely, each such function belongs to $CF(A, G)$.

Lemma 2.3.2. *Under the assumptions as before, there exists a (unique) group homomorphism $f_! : CF(A, G) \to CF(B, G)$ (and similarly for the compact support case) such that the diagram*

$$\begin{array}{ccc} D_T(A) & \xrightarrow{Rf_!} & D_T(B) \\ \downarrow{\scriptstyle \mu_A} & & \downarrow{\scriptstyle \mu_B} \\ CF(A,G) & \xrightarrow{f_!} & CF(B,G) \end{array}$$

commutes.

Denote by T' the full subcategory of T, whose objects are the $M \in ob(T)$ such that $\mu(M) = 0$. Then T' and therefore also $D_{T'}(A)$ is a strict full triangulated subcategory of T (of $D_T(A)$). By the same references as before, $Rf_!$ maps $D_{T'}(A)$ in $D_{T'}(B)$. Then lemma 2.3.2 follows from the following simple abstract argument.

Lemma 2.3.3. *Let the triangulated categories T_i and additive functions μ_i on T_i with values in the abelian groups G_i be given ($i = 0, \ldots, n$). Suppose μ_i is surjective for $i = 1, \ldots, n$ and let $F : \times_{i=1}^n T_i \to T_0$ be a functor, which is exact in each variable. Then there exists a unique $\mathbb{Z}$-multilinear map f making the diagram*

$$\begin{array}{ccc} \times_{i=1}^n ob(T_i) & \xrightarrow{F} & ob(T_0) \\ \downarrow{\scriptstyle \times_{i=1}^n \mu_i} & & \downarrow{\scriptstyle \mu_0} \\ \times_{i=1}^n G_i & \xrightarrow{f} & G_0 \end{array}$$

commutative, if and only if $\mu_0(F(M_1, \ldots, M_n)) = 0$ for each family $M_i \in ob(T_i)$ such that $\mu_i(M_i) = 0$ for **at least** *one i.*

Proof. The implication $\Rightarrow$ is obvious. For the other implication, consider $M_i, M_i' \in ob(T_i)$ such that $\mu_i(M_i) = \mu_i(M_i')$ for $i = 1, \ldots, n$. We claim

$$\mu_0(F(M_1, \ldots, M_n)) = \mu_0(F(M_1', \ldots, M_n')) .$$

By fixing the other variables, it is enough to consider the case $i = 1$. Since μ_1 is additive, we get

$$\mu_1(M_1 \oplus M_1'[1]) = \mu_1(M_1) - \mu_1(M_1') = 0 .$$

By assumption we get $\mu_0(F(M_1 \oplus M_1'[1])) = 0$. But F is an exact functor. This implies

$$F(M_1 \oplus M_1'[1]) = F(M_1) \oplus F(M_1')[1] ,$$

and

$$0 = \mu_0(F(M_1 \oplus M_1'[1])) = \mu_0(F(M_1) - \mu_0(F(M_1') .$$

Since the μ_i for $i = 1, \ldots, n$ are surjective, we get a unique map f, making the diagram commutative. Moreover, f is multilinear. As before, we can assume $i = 1$. Since μ_i is additive and F is exact, we get for $M_1, M_1' \in ob(T_1)$:

$$f(-\mu_1(M_1)) = f(\mu_1(M_1[1])) = \mu_0(F(M_1)[1]) = -\mu_0(F(M_1)) = -f(\mu_1(M_1)) ,$$

and

$$\begin{aligned} f(\mu_1(M_1) + \mu_1(M_1')) &= f(\mu_1(M_1 \oplus M_1')) = \mu_0(F(M_1 \oplus M_1')) \\ &= \mu_0(F(M_1) \oplus F(M_1')) = \mu_0(F(M_1)) + \mu_0(F(M_1')) \\ &= f(\mu_1(M_1)) + f(\mu_1(M_1')) . \end{aligned}$$

□

Remark 2.3.2. 1. Lemma 2.3.3 and its proof can also be reformulated in the language of **Grothendieck groups** of triangulated categories (compare [KS, Ex. I.27] and [SGA5, exp. VIII]).

2. Suppose we want to show, that a (multi-)functor between constructible sheaves induces (or factors over) a similar (multi-)linear map between constructible functions. By this lemma, we only have to show that it preserves the corresponding **stalkwise vanishing condition**. This approach is simpler and more general than the one of Kashiwara and Schapira (we don't use [Scha, thm. 3.4], [KS, thm. 9.7.1]), and applies to other situations as well. For example to the complex analytic/algebraic context, or in l-adic cohomology. For a beautiful example compare with [KL, thm. 3.1.2, thm. 4.1, thm. 7.3.1].

When one knows, that a functor between constructible sheaves induces a similar map between constructible functions, then one can ask for an **intrinsic** description of this map. In the case above, this can be done for $f_!$ in the following way. Consider first the case B a point. Under the previous assumptions, such a function ϕ in $CF(A,G)$ (or $CF_c(A,G)$) has a presentation

$$\phi = \sum_{k=1}^{m} g_k \cdot 1_{X_k} ,$$

with $g_k \in G$ and X_k definable subsets (which in case (3.) are also relatively compact). Then one gets

$$f_!(\phi) = \sum_{k=1}^{m} g_k \cdot \chi(X_k) ,$$

with χ the **Euler characteristic** as defined at the beginning of this section. It follows, that the right value is **independent** of this presentation of ϕ, and this is often denoted as the integral of ϕ with respect to the Euler characteristic (compare [Scha] or [PK]):

$$\int_A \phi \; d\chi := \sum_{k=1}^{m} g_k \cdot \chi(X_k) .$$

Our χ corresponds also to $(-1)^{dimV} \cdot \chi^{loc}$ in [Mor, lem. 5,p. 24]). This notation is motivated by the fact, that χ is a **valuation** or **finitely additive measure** on the algebra of definable subsets, i.e. for each pair of definable subsets one has:

$$\chi(P) + \chi(Q) = \chi(P \cup Q) + \chi(P \cap Q). \tag{2.3}$$

Moreover, this value of $\int_A \phi \; d\chi$ is independent of which T and surjective μ we have used for its sheaf theoretic definition, and in the case (2.) and (3.) it can be **directly** defined and studied (compare [vDr, chapter 4]). By lemma 2.3.1 and remark 2.3.1, it is in case (2.) or (3.) even independent of the used o-minimal structure (or analytic (Nash) geometric category).

For general B, one can apply the above arguments to the fibers of f and get

$$f_!(\phi)(x) = \int_{\{f=x\}} \phi|\{f = x\} \; d\chi .$$

Note that we get in case (1.) by the assumptions on f an induced cell decomposition or stratification of $\{f = x\}$ such that the inclusion of the fiber is a cellular map. In the cases (2.) and (3.) we can use the constructible function $\phi \cdot 1_{\{f=x\}}$ on A and get

$$f_!(\phi)(x) = \int_A \phi \cdot 1_{\{f=x\}} \; d\chi .$$

Example 2.3.3. *Let $f : V \to W$ be a linear map of vector spaces and Z be a convex definable subset. Then $f_!(1_Z) = 1_{f(Z)}$, if Z is compact. Similarly,*

$$f_!(1_Z) = (-1)^d \cdot 1_{f(Z)} , \quad \text{with } d := dim(Z) - dim(f(Z))$$

in the case (2.), if $Z = ri(Z)$ is equal to the interior in its affine hull.

As we remarked before, the map $f_!$ can in our context also directly be studied **without** the use of constructible sheaves. The same remark applies to the following properties:

(a.) The pushdown map is **functorial**: $(f \circ g)_! = f_! \circ g_!$, if each of the maps f, g and $f \circ g$ satisfies our assumptions, e.g. in case (1.) we have to assume that all three maps are cellular fibrations over the cells of their targets. In special cases, the functoriality is sometimes called a "Fubini theorem" ([PK, Mor]).

(b.) Consider a cartesian diagram of definable maps

$$\begin{array}{ccc} A' & \xrightarrow{f'} & B' \\ \downarrow{\scriptstyle g'} & & \downarrow{\scriptstyle g} \\ A & \xrightarrow{f} & B, \end{array}$$

with f a cellular fibration over the cells of B in case (1.) or f proper in case (3.). Then f' has also this property and one has

$$g^* f_! = f'_! g'^* : CF(A, G) \to CF(B', G) .$$

Similarly, g proper implies g' proper and in this case one also has

$$g^* f_! = f'_! g'^* : CF_c(A, G) \to CF_c(B', G) .$$

(c.) In the case of a commutative ring G, we have for f as before a **projection formula**

$$f_!(m(\phi_1, f^*(\phi_2)) = m(f_!(\phi_1), \phi_2) .$$

Moreover, in the "compact support case" one only needs that ϕ_1 has a compact support.

(d.) If f_i satisfies the assumptions above ($i = 1, 2$), then the same is true for $f_1 \times f_2$, with

$$(f_1 \times f_2)_!(\phi_1 \boxtimes \phi_2) = f_{1!}(\phi_1) \boxtimes f_{2!}(\phi_2) .$$

As a special case of (a.) let us just notice the following formula for the Euler characteristic of (locally closed) definable subsets with respect to an o-minimal structure ([vDr, cor. 4.2.11]): Let $f : A \to B$ be a **surjective** definable map, whose fibers all have the same Euler characteristic e. Then

$$\chi(A) = e \cdot \chi(B). \tag{2.4}$$

This is a generalization of $\chi(A \times B) = \chi(A) \cdot \chi(B)$. This formula for the usual topological interpretation of the Euler characteristic is one of its basic properties. For a nice application (in the case of complex algebraic varieties) see for example [Rie]. Another application is given by the

Lemma 2.3.4. *Let Z be a closed convex subset of the affine space V, which is not compact, but proper in the sense that it contains no lines. Suppose that Z is definable with respect to an o-minimal structure on $\mathbb{R}$. Then $\chi(Z) = 0$.*

Proof. Since Z is a proper closed convex subset, there exists a linear function $f : V \to \mathbb{R}$, whose restriction to Z is proper and bounded from below (see for example [HNP, prop. 1.11, cor. 1.13]). So the non-empty fibers of f are compact convex and have therefore the same Euler characteristic 1. This implies

$$\chi(Z) = \chi(f(Z)) .$$

$f(Z) \subset \mathbb{R}$ is closed convex, bounded from below but not from above (because Z is by assumption non-compact). So we get $f(Z) = [a, \infty[$ (for some $a \in \mathbb{R}$) and $\chi([a, \infty[\,) = 0$, since $[a, \infty[$ is the disjoint union of the cells $\{a\}$, $]a, \infty[$. □

Remark 2.3.3. Lemma 2.3.4 remains true without the assumptions that Z is definable, if we use the topological definition, because the same argument as before implies $Rf_! k_Z \simeq k_{f(Z)}$ ([KS, prop. 2.7.8]).

For simplicity, we consider from now on only the cases (2.) and (3.).

A typical application of the above formalism deals with **convolution** (compare [Scha, sec. 4], [KS, Ex. IX.10] or [PK, 2.prop.-def.2]).

Let A be a definable (commutative) **monoid**, i.e. we have a continuous definable multiplication map

$$a : A \times A \to A ,$$

with $a(a(x, y), z) = a(x, a(y, z))$ (and $a(x, y) = a(y, x)$ in the commutative case), together with a **unit** $e \in A$ (i.e. $a(e, x) = x$ for all x).

Then one defines for a commutative ring G the convolution product of two constructible functions $\phi_i \in CF(A, G)$ in case (2.) (or $\phi_i \in CF_c(A, G)$ in case (3.)) by

$$\phi_1 * \phi_2 := a_!(\phi_1 \boxtimes \phi_2) . \tag{2.5}$$

Then $(CF(A, G), *)$ or $(CF_c(A, G), *)$ is an assocative (commutative) ring with unit $1_{\{e\}}$. The associativity for example follows from the following commutative diagram (whose left square is cartesian):

$$\begin{array}{ccccc} A \times A & \xleftarrow{p_{1,2}} & A \times A \times A & \xrightarrow{id \times a} & A \times A \\ {\scriptstyle a}\downarrow & & \downarrow{\scriptstyle a \times id} & & \downarrow{\scriptstyle a} \\ A & \xleftarrow[p_1]{} & A \times A & \xrightarrow[a]{} & A , \end{array}$$

where the left horizontal maps are the projections on the indicated factors. Then one gets by the properties (a.)-(d.) above:

$$\begin{aligned} (\alpha * \beta) * \gamma &= a_!(\, a_!(\alpha \times \beta) \ \times \gamma\,) \\ &= a_!(\, (a \times id)_!(\, \alpha \times \beta \times \gamma\,)\,) = q_!(\, \alpha \times \beta \times \gamma\,) , \end{aligned}$$

with $q := a \circ (a \times id) = a \circ (id \times a)$. But a similar calculation shows that the last term is also equal to

$$\alpha * (\beta * \gamma) .$$

Consider a continuous definable map $f : A \to B$ of two such (commutative) monoids (A, a, e) and (B, b, e'), with $f(e) = e'$ and $f(a(x, y)) = b(f(x), f(y))$ so that the following diagram commutes:

$$\begin{array}{ccc} A \times A & \xrightarrow{a} & A \\ {\scriptstyle f\times f}\downarrow & & \downarrow{\scriptstyle f} \\ B \times B & \xrightarrow[b]{} & B . \end{array}$$

Then we get from the properties (a.) and (d.) that $f_!$ is a ring homomorphism

$$(CF(A, G), *, 1_{\{e\}}) \to (CF(A, G), *, 1_{\{e'\}})$$

in case (2.), or

$$(CF_c(A, G), *, 1_{\{e\}}) \to (CF_c(A, G), *, 1_{\{e'\}})$$

in case (3.). Especially, for $f : A \to \{e'\}$ a constant function one gets (compare [Scha, thm.4.2.(ii)] and [PK, 2.prop.-def.3]):

$$\int_A (\phi_1 * \phi_2)\, d\chi = (\int_A \phi_1\, d\chi) \cdot (\int_A \phi_2\, d\chi) . \tag{2.6}$$

The standard example for a commutative monoid (A, a) is a vector space V with a the addition map. Then one can take for f as above a linear map $f : V \to W$ of vector spaces . Then one gets for two convex definable subsets A, B:

$$1_A * 1_B = \begin{cases} 1_{A+B} & \text{in the case (3.), with } A, B \text{ compact.} \\ (-1)^d \cdot 1_{A+B} & \text{in the case (2.), with } C = ri(C) \text{ for } C = A, B. \end{cases}$$

Here $A + B := \{a + b |\ a \in A,\ b \in B\}$ is the **Minkowski sum** of A and B, with $d := dim(A) + dim(B) - dim(A + B)$ in the case (2.). For the compact case, compare with [Scha, ex.4.3] and [PK, 4. proof of prop.-def. 2.3]).

Moreover, one has in this example the following **inversion formula** ([Scha, ex. 4.4], [PK, 2. thm.2] and [Mor, (6) on p. 28, sec. 5.3]): let A be a **compact convex** (definable) subset of V. Then

$$1_A * (-1)^{dim(A)} \cdot 1_{-ri(A)} = 1_{\{0\}} . \tag{2.7}$$

Let us sketch a proof of this formula. Fix $x \in V$. Then

$$(1_A * 1_{-ri(A)})(x) = \chi(\ A \cap (x + ri(A))\ ) ,$$

and this is $(-1)^{dim(A)}$ for $x = 0$. Suppose $x \neq 0$ and $A' := A \cap (x + ri(A)) \neq \emptyset$. Fix a decomposition $V = W \oplus \mathbb{R} \cdot x$ of V and consider the linear projection

$$\pi : V = W \oplus \mathbb{R} \cdot x \to W \, .$$

Then $\pi_!(1_{A'}) = 0$, since any non-empty fiber of $\pi : A' \to W$ is an interval of type $[a, b[$ (compare [Br, thm. 3.3]), with $\chi([a, b[\,) = 0$. By functoriality, we get

$$\chi(A') = \int_W \pi_!(1_{A'}) \, d\chi = 0 \, .$$

Of course, a similar formalism also works in the context of constructible sheaves (one uses for example the Künneth isomorphism for $Rf_!$, and compare with [KS, Ex. II.20]).

The approach through constructible sheaves is more important, if one can deduce results about constructible functions, which cannot (so easily) be proved directly. This applies for example to the relation between constructible functions and **Lagrangian cycles** (compare [KS, thm. 9.7.11] and also with chapter 5 of this book), to the **duality** transformation and the **pushdown** f_* in the o-minimal context.

Another example comes from constructible sheaves in the **complex analytic** or **algebraic** context. If A is a complex analytic (algebraic) space, then one can consider A as a subanalytic (semialgebraic) space as in our case:

- In the algebraic context, it is just a locally compact definable space with respect to the o-minimal structure given by the **semialgebraic** sets.
- In the complex analytic context, one first has to use a **topological embedding** as (i.e. a real analytic homeomorphism onto) a closed real analytic subset of a real analytic manifold (e.g. of some $\mathbb{R}^N$, compare [GMT, ch. VI, thm. 1.2, p. 111]). Then one works with the analytic geometric category given by the **subanalytic** sets.

Then the category of **complex analytically (algebraically)** constructible complexes of sheaves of R-modules (with a "stalk property" T)

$$D_C(A, R) \quad \text{or} \quad D_{C-T}(A, R)$$

is a **full subcategory** of the corresponding category of subanalytically (semialgebraically) constructible complexes of sheaves. Similarly, one has the subgroup

$$CF_C(A, G) \subset CF(A, G)$$

of G-valued **complex analytically (algebraically)** constructible functions.

Consider a surjective **additive** map $\mu : T \to G$ as before. Then the induced map

$$\mu_A : D_{C-T}(A, R) \to CF_C(A, G)$$

is also surjective. This can be proved as before, with the help of a corresponding **stratification** result (compare [Ver]). Let $f : A \to B$ be a complex algebraic or a **proper** complex analytic map. Then $Rf_!$ maps $D_C(A, R)$ into $D_C(B, R)$. This can be deduced by stratification techniques (compare [Ver]). Another sheaf theoretic approach is given in chapter 4 of this book (see theorem 4.0.2). For a "micro-local" approach compare with [KS, sec. 8.5].

So we get an induced pushdown map

$$f_! : CF_C(A, G) \to CF_C(B, G) .$$

This goes back to MacPherson [MP] in his work on Chern classes of singular spaces. Compare also with [CEP, Exp. 1, prop. 1.1] and [CEP, Exp. 7, (3.1)], where resolution of singularities is used.

Remark 2.3.4. By the description in chapter 1 and 5 of the **nearby cycle** functor ψ_f for a complex analytic (algebraic) function, one gets in the same way a **nearby cycle** and **vanishing cycle** functor ψ_f, ϕ_f for the group of **complex analytically (algebraically)** constructible functions (compare with [CEP, Exp. 7, prop. 3.4], where resolution of singularities is used).

In the following, we consider only the case (2.) for an o-minimal structure on $(\mathbb{R}, +, \cdot)$, or the case (3.), and introduce a **duality** transformation for constructible functions.

Assume that T is a strict full **triangulated** subcategory of $D^b(R)$, which is also **stable** under the duality functor $D_{pt} := Rhom(\cdot, R)$ (with R commutative). Consider $\mathcal{F} \in ob(D_T(A))$. Then

$$D_A(\mathcal{F}) := Rhom(\mathcal{F}, c^! R) \quad (\text{ with } c : A \to \{pt\} \,)$$

is **constructible** by corollary 2.2.5(2) and corollary 2.2.3. Moreover

$$D_A(\mathcal{F}) \in ob(D_T(A)) .$$

This follows from corollary 2.2.6 and

$$i_x^! D_A(\mathcal{F}) \simeq D_{pt}(i_x^* \mathcal{F})$$

for each inclusion $i_x : \{x\} \to A$ (compare corollary 2.2.7(1)).

Let μ be an additive map on T with values in the abelian group G, which is surjective and **invariant** under D_{pt}. The standard example is $R = k$ a field, T denotes the (cohomological) bounded complexes with **finite dimensional cohomology** and μ is the usual **Euler characteristic** in $G = \mathbb{Z}$.

Denote by T' the full subcategory of T, whose objects are the $M \in ob(T)$ such that $\mu(M) = 0$. Then $D_{T'}(A)$ is stable under D_A by the same argument and references as above. So we can apply lemma 2.3.3 and get an induced duality transformation

$$D_A : CF(A, G) \to CF(A, G)$$

(and similarly for the compact support case) such that the following diagram commutes:

$$\begin{array}{ccc} D_T(A) & \xrightarrow{D_A} & D_T(A) \\ \big\downarrow{\scriptstyle \mu_A} & & \big\downarrow{\scriptstyle \mu_A} \\ CF(A,G) & \xrightarrow{D_A} & CF(A,G)\,. \end{array}$$

We now explain, how one **computes** the dual of a constructible function for the case that T is a subcategory of the **perfect** complexes, as in the case of the standard example (compare [Scha, lem.-def. 2.4], [KS, (9.7.10)] and [PK, def. of "star", p. 353]).

Fix a point $x \in A$ and a definable open neighborhood U, together with a proper definable function

$$r : U \to [0, c[\quad (\text{ for some } c > 0\,)$$

such that $\{x\} = \{r = 0\}$. For example the distance function to x in some definable coordinates. Then one gets by corollary 2.2.7(3):

$$\mu(i_x^* D_A(\mathcal{F})) = \mu(D_{pt}(i_x^! \mathcal{F})) = \mu(i_x^! \mathcal{F})\,,$$

and by corollary 2.0.1(1.b):

$$\mu(i_x^* D_A(\mathcal{F})) = \mu(\; R\Gamma_c(\{r < \epsilon\}, \mathcal{F})\;)\,,$$

for $\epsilon > 0$ small enough. Hence one gets for $\phi \in CF(A, G)$ the formula

$$(D_A\phi)(x) = \int_{\{r<\epsilon\}} \phi|_{\{r<\epsilon\}}\, d\chi = \int_A 1_{\{r<\epsilon\}} \cdot \phi\, d\chi\,, \tag{2.8}$$

for $\epsilon > 0$ small enough. Note that the left side is independent of the choice of r and the right side is independent of the choice of T and μ ! Moreover, the biduality result (2.2) implies (for T a subcategory of the perfect complexes) the **biduality formula**

$$\phi \in CF(A, G) \Rightarrow D_A(D_A\phi) = \phi\,. \tag{2.9}$$

Similarly one defines the **link operator**

$$\Lambda : CF(A, G) \to CF(A, G),\ \Lambda := id - D_A\,.$$

By the previous calculation and corollary 2.0.1, one gets for $\phi \in CF(A, G)$ the formula

$$(\Lambda\phi)(x) = \int_{\{r=\epsilon\}} \phi|_{\{r=\epsilon\}}\, d\chi = \int_A 1_{\{r=\epsilon\}} \cdot \phi\, d\chi\,, \tag{2.10}$$

with $\epsilon > 0$ small enough. This operator is very important in the study of **real algebraic** sets (compare [McPa2]).

Example 2.3.4. *Let X be a definable locally closed subset of A, which is a* **good** *cell (in the sense of condition (b.) as on page 86). For example, A is an affine space and X is a convex definable subset with $X = ri(X)$, or X is a cell of a definable good cellular stratification. Then*

$$D_A(g \cdot 1_X) = (-1)^{dim(X)} \cdot g \cdot 1_{cl(X)} . \tag{2.11}$$

This follows from the next argument (compare [KS, ex. 9.7.3]). Let us denote the inclusions

$$j : X \to cl(X) \, , \; k : cl(X) \to A$$

and consider the complex

$$\mathcal{F} := Rk_! Rj_! j^* c^* M \, ,$$

with $M \in ob(T)$ and c a constant map on $cl(X)$. Then

$$\mu_A(\mathcal{F}) = g \cdot 1_X \, , \quad \text{with } g := \mu(M) \, .$$

But

$$D_A(Rk_! Rj_! j^* c^* M) \simeq Rk_! Rj_*(D_X(j^* c^* M)) \, ,$$

and $D_X(j^* c^* M)$ has locally constant cohomology sheaves so that

$$D_X(j^* c^* M) \simeq j^* c^* M' \quad \text{for some } M' \in ob(D(R)) \, .$$

Moreover, by the proof of lemma 2.2.3 (or [KS, prop. 3.2.3]), we have for the inclusion $i_x : \{x\} \to X$:

$$M' \simeq i_x^* j^* c^* M' \simeq i_x^! j^* c^* M'[-dim(X)] \, .$$

Together with $i_x^!(D_X(j^* c^* M)) \simeq D_{pt}(i^* j^* c^* M)$, we get

$$M' \simeq D_{pt}(M)[-dim(X)] \, .$$

Especially $M' \in ob(T)$, with $\mu(M') = (-1)^{dim(X)} \cdot \mu(M)$. By the **goodness** condition one has $Rj_* j^* c^* M' \simeq c^* M'$, and this implies $D_A(\mathcal{F}) \simeq Rk_! c^* M'$, together with

$$\mu_A(D_A(\mathcal{F})) = (-1)^{dim(X)} \cdot g \cdot 1_{cl(X)} \, .$$

This example implies another description of the **duality** transformation D_A for constructible functions. For a given $\phi \in CF(A, G)$, choose a presentation of ϕ as a (locally) finite sum

$$\phi = \sum_k g_k \cdot 1_{X_k} \, ,$$

with X_k a definable **good** cell. Use for example an adapted (locally) finite good cellular stratification. Then one gets

$$D_A(\phi) = \sum_k (-1)^{dim(X_k)} \cdot g_k \cdot 1_{cl(X_k)} \, .$$

This applies especially to the simple example of the o-minimal structure given by the **semilinear** sets of vector spaces. This is not a structure on $(\mathbb{R}, +, \cdot)$, but we can consider it as a substructure of the semialgebraic subsets. Then the reader may check that the **semilinear** constructible sheaves are **stable** under D_V, by the same proof as in section 2.2.

Then $CF(V;\mathbb{Z})$ is generated by the indicator functions of basic semilinear sets (or polyhedra) and we can apply this example to these sets (or to the relative interior of a polyhedron). Especially, the duality involution D_V for semialgebraic constructible functions restricts to a **duality involution** D_V for **semilinear** constructible functions (and this corresponds to the transformation "star" in [PK], or to the duality involution studied in [Mor, sec. 3.3]). This duality involution on $CF_c(V;\mathbb{Z})$ induces for each abelian group G' a well-known involution $*$ on

$$Hom_Z(CF_c(V;\mathbb{Z}), G')$$

related to **valuation theory of convex polytopes** (compare [Klain, thm. 3.1]).

If for example X is such a **polyhedron** (i.e. the intersection of finitely many closed halfspaces), then we can apply this description of the duality to the relative interior of the closed faces F of X (see [Br, cor. 8.4, 8.5]):

$$1_X = \sum_F 1_{ri(F)}$$

implies under **duality** the well known formula (compare [Mor, p. 27] and [PK, 2.prop.-def. 6]):

$$(-1)^{dim(X)} \cdot 1_{ri(X)} = \sum_F (-1)^{dim(F)} \cdot 1_F \,.$$

The same formula holds for a simplicial triangulation of a convex polytope X (by the same argument). This implies for valuations on convex polytopes a formula of Macdonald (see for example [Klain, thm. 0.1]).

Let us come back to the general context. Corollary 2.2.8 implies the

Lemma 2.3.5. *Let ϕ_i be constructible functions with values in the commutative ring G $(i = 1, 2)$. Then*

$$(D_{A_1}\phi_1) \boxtimes (D_{A_2}\phi_2) = D_{A_1 \times A_2}(\phi_1 \boxtimes \phi_2) \,.$$

If one has a **duality involution** for constructible functions, then one can introduce the transformations

$$f^! := D_A \circ f^* \circ D_B \quad \text{and} \quad hom(\cdot, \cdot) := D(\, m(\,\cdot, D(\cdot))\,)$$

of such functions as in [KS, (9.7.14,15)].

We consider here the more interesting case of the **direct image** functor f_*. As before, assume that T is a strict full **triangulated** subcategory of the **perfect** complexes, **stable** under D_{pt}, and that μ is an **additive** function on T with values in G, which is surjective and **invariant** under D_{pt}.

Lemma 2.3.6. *Assume $f : A \to B$ is a (proper) definable map. Then one has for all $\phi \in CF_c(A, G)$ (or $\phi \in CF(A, G)$) the formula*

$$D_B(f_!\phi) = f_! D_A(\phi) .$$

This follows directly from the counterpart for constructible sheaves, corollary 2.2.7(2) and the fact that $Rf_* = Rf_!$ in the corresponding context, since f is proper (on the support).

Lemma 2.3.7. *Assume (A, a, e) is a definable monoid (such that $a : A \times A \to A$ is proper). Then one has for $\phi_i \in CF_c(A, G)$ (or $\phi_i \in CF(A, G)$) for $i = 1, 2$ the formula*

$$(D_A\phi_1) * (D_A\phi_2) = D_A(\phi_1 * \phi_2) .$$

This follows from lemma 2.3.5 and lemma 2.3.6.

Let us now consider the case a continuous definable map in the case (2.) of an o-minimal structure on $(\mathbb{R}, +, \cdot)$ (which is not studied in [Scha] or [KS]). Then we can define

$$f_* : CF(A, G) \to CF(A, G)$$

by the formula

$$f_* := D_B \circ f_! \circ D_A \tag{2.12}$$

so that the duality transformation **exchanges** f_* and $f_!$.

Another more direct definition follows from lemma 2.3.3. Assume only T is a strict full **triangulated** subcategory of $D(R)$ and μ is a surjective **additive** function. Denote again by T' the full subcategory of T, whose objects are the $M \in ob(T)$ such that $\mu(M) = 0$. Then Rf_* maps $D_{T'}(A)$ into $D_{T'}(B)$ by theorem 2.0.2(1). So by lemma 2.3.3, we can define f_* by the commutative diagram

$$\begin{array}{ccc} D_T(A) & \xrightarrow{Rf_*} & D_T(B) \\ \downarrow{\scriptstyle \mu_A} & & \downarrow{\scriptstyle \mu_B} \\ CF(A, G) & \xrightarrow{f_*} & CF(B, G) . \end{array}$$

This is an extension of the above definition by corollary 2.2.7(2). Of course, this definition is functorial in the map f.

Let us translate this definition into the language of **constructible functions**. Consider first the case B a point. Choose a proper definable function

$$r : A \to [0, \infty[\ .$$

Assume for example that A is a **closed** subset of a vector space and use the restriction of a definable norm function. If A is only **locally closed**, choose first a definable homeomorphism onto a closed definable subset of another vector space. Then one gets by corollary 2.0.1 (2.a):

$$\mu(\ R\Gamma(A,\mathcal{F})\ ) = \mu(\ R\Gamma(\{r \leq c\},\mathcal{F})\ )$$

for $c >> 0$ big enough. So one gets for f a constant function

$$f_*(\phi) = \int_A 1_{\{r\leq c\}} \cdot \phi \; d\chi \quad \text{for } c >> 0\,.$$

Let us introduce the **new Euler characteristic** χ' in this way by

$$\chi'(X) := f_*(1_X) = \chi(X \cap \{r \leq c\}) \quad \text{for } c >> 0\,.$$

Here we consider for T, μ the standard example. This is **independent** of the choice of r (since this is true for the left term), and defines a **valuation** on the algebra of definable sets of A. Hence one gets for $\phi \in CF(A,G)$ the formula

$$f_*(\phi) = \int_A \phi \; d\chi'\,. \tag{2.13}$$

Example 2.3.5. *Suppose $A = V$ is a vector space and $X \neq \emptyset$ is a definable* **closed** *and* **convex** *subset. Take for r a definable norm function. Then $X \cap \{r \leq c\}$ is a compact convex set and we get $\chi'(X) = 1$. By duality, we also get*

$$\chi'(ri(X)) = (-1)^{dim(X)} \cdot \chi(X)\,.$$

Especially $$\chi'(ri(X)) = \begin{cases} (-1)^{dim(X)} & \textit{for } X \textit{ compact.} \\ 0 & \textit{for } X \textit{ noncompact but proper.} \end{cases}$$

This applies also to the **semilinear** context. Especially, χ' corresponds in that case (up to a sign) to the Euler characteristic used in [Mor] (compare [Mor, p. 16]).

Consider now the case of a continuous definable map $f : A \to B$. Fix $x \in B$ and a definable open neighborhood U, together with a proper definable function

$$r' : U \to [0, c'[\quad \text{for some } c' > 0 \text{ such that } \{x\} = \{r' = 0\}\,.$$

Corollary 2.0.1(1.a) implies for $\phi \in CF(A,G)$ the formula (with $r : A \to [0,\infty[$ as before):

$$(f_*\phi)(x) = \int_A 1_{\{r'\circ f\leq c'\}} \cdot \phi \; d\chi' = \int_A 1_{\{r'\circ f \leq c',\, r\leq c\}} \cdot \phi \; d\chi\,, \tag{2.14}$$

where we first choose $c' > 0$ small enough, and then $c >> 0$ big enough.

Example 2.3.6. *Suppose $f : V \to W$ is a linear map of vector spaces and X is a definable subset of V, which is* **closed** *and* **convex**. *Assume that $f(X)$ is* **closed**. *Then one has by duality:*

$$\begin{aligned} f_*(1_X) &= f_*(\, D_V((-1)^{dim(X)} \cdot 1_{ri(X)} \,) = D_W(\, f_!((-1)^{dim(X)} \cdot 1_{ri(X)}) \,) \\ &= D_W(\, (-1)^{dim(f(X))} \cdot 1_{ri(f(X))} \,) = 1_{f(X)} \,. \end{aligned}$$

Especially, one gets in this case for $\phi = 1_X$:

$$f_*(\phi)(x) = \int_{\{f=x\}} \phi|\{f = x\} \, d\chi' \,,$$

which is **not** *true in general.*

Note that this example can be applied to a **polyhedron**. But the indicator functions of polyhedra **generate** the group of $\mathbb{Z}$-valued **semilinear** constructible functions. So in this special case, one can calculate $f_*(\phi)$ for all such ϕ as a "fiber integral" with respect to χ' (and this is the definition of $f_*(\phi)$ used in [Mor, def. 4]).

Let us finally remark that one has in general for such a function

$$f_*(D_V(\phi)) \neq D_W(f_*(\phi)) \,.$$

So formula (1) in [Mor, p. 28] is in general **not** true for constructible functions **without compact support**, contrary to a remark in [Mor, p. 28]. Of course, this is true for functions with compact support.

Take for example $W = \{0\}$ and $\phi = 1_X$ for a closed polyhedron X, which is not compact, e.g.

$$X := [0, \infty[\, \subset V := \mathbb{R} \,.$$

Then

$$D_W(f_*(\phi)) = \chi'(X) = 1 \,, \text{ but } f_*(D_V(\phi)) = D_W(f_!(\phi)) = \chi(X) = 0 \,.$$

One gets a more interesting example, if one takes the **exterior product** of this example with respect to a linear map $f' : V' \to W'$ and a constructible function $\phi' \in CF_c(V', G)$, because in general one has the formula

$$(f_1 \times f_2)_*(\phi_1 \boxtimes \phi_2) = f_{1*}(\phi_1) \boxtimes f_{2*}(\phi_2) \,. \tag{2.15}$$

This follows by duality from the same formula for the map $f_!$, or from the counterpart for constructible sheaves (see corollary 2.0.4(1)).

Chapter 3

Localization results for equivariant constructible sheaves

Introduction

Most results of this chapter are applications of the two basic theorems, which we explain in this introduction. They give a unifying approach to generalizations of many different well-known results.

We consider only locally compact spaces with countable topology, and of finite cohomological dimension. For simplicity we assume in this introduction that the base ring R is **commutative**, and that:

- G is a countable **discrete** group, or
- G is (topologically isomorphic to) a real analytic **Lie group** with **finitely** many connected components.

Then one has for a G-space X the **equivariant derived category** $D^b_G(X)$, together with the **functors**

$$Rf_!, Rf_* : D^b_G(X) \rightleftarrows D^b_G(Y) \ : \ f^*$$

for a continuous G-equivariant map $f : X \to Y$ of G-spaces. If moreover $i := f$ is a **closed** inclusion, then one also has the functor

$$i^! : D^b_G(Y) \to D^b_G(X) \,.$$

In the case of a **discrete** group, this is just the bounded derived category of the abelian category of "equivariant sheaves" in the sense of Grothendieck [Groth2], with the corresponding standard functors. But in the case of a **Lie group**, one has to use the **triangulated category** $D^b_G(X)$ and the **functors** of Bernstein-Lunts [BL]. This is in general **not** the derived category of the abelian category of "equivariant sheaves". But in the case of a **finite** group, both possible definitions of the categories and functors are **equivalent**.

We recall in section 3.1 the relevant properties of [BL]. For example, one has a natural **t-structure** and an **exact forgetful functor**

$$For : D^b_G(X) \to D^b(X) ,$$

commuting with the standard functors (especially with the corresponding cohomology functor h). Moreover, we explain in section 3.1 also some other properties, which are important for our applications.

Consider $\mathcal{F} \in ob(D^b_G(X))$, and denote the **quotient map**

$$q : X \to Y := X/G .$$

Assume that Y is a "good" topological space in the sense above (e.g. Y is locally compact). Then one has

$$R\Gamma_?(X, \mathcal{F}) \simeq R\Gamma_?(Y, Rq_?\mathcal{F}) \qquad (? = ! \text{ or } *) .$$

Our approach to localization results applies, if Y is a "nice" topological space, and $Rq_?For(\mathcal{F})$ is a **constructible** complex of sheaves.

Let us give a more precise definition. Assume Y is locally compact.

1. A **cellular decomposition** of Y is a filtration

$$Y_\bullet : \emptyset = Y_{-1} \subset Y_0 \subset \ldots \subset Y_m = Y$$

   by closed subsets such that $Y_i \backslash Y_{i-1}$ (for $i = 1, \ldots, m$) is the disjoint union of open subsets $U_{i,j}$ homeomorphic to some affine space $\mathbb{R}^{n_{i,j}}$. We call $U_{i,j}$ the corresponding **cells** of Y.

2. A cellular decomposition of Y is **(locally) finite**, if the family of cells $\{U_{i,j}\}$ is (locally) finite on Y.

3. A **cellular stratification** is a cellular decomposition such that the **frontier** $\partial U := \bar{U} \backslash U$ of each cell $U = U_{i,j}$ is a union of lower dimensional cells.

4. We call a cellular stratification **good**, if each cell U satisfies the following "goodness" condition: Let $j : U \to \bar{U}$ be the open inclusion into the **closure** of the cell. Then, the natural morphism $\mathcal{F} \to Rj_*j^*\mathcal{F}$ is an isomorphism for each **constant** sheaf $\mathcal{F}$ on $\bar{U}$.

5. $\mathcal{G} \in ob(D^b(Y))$ is called **constructible** with respect to such a filtration $Y_\bullet$ by closed subsets (or the filtration is called **adapted** to $\mathcal{G}$), if the restriction of the cohomology sheaves of $\mathcal{G}$ to the $Y_i \backslash Y_{i-1}$ are locally constant.

For example, a locally compact CW-complex Y has a canonical locally finite cellular decomposition (which is finite for Y compact). Assume moreover that Y is a **regular** CW-complex. Then this decomposition is also a **good cellular stratification**.

Before we can formulate the first main result, we have to recall some more notions (compare for example with [SGA5, exp. VIII]).

Let D be a triangulated category (e.g. $D^b_G(X)$).

- A **strict full triangulated** subcategory T of D is a full subcategory of D, with

$$(1.)\ 0 \in ob(T)\ , \quad (2.)\ M \in ob(T) \Leftrightarrow M[1] \in ob(T)\ ,$$

  and (3.) T is stable under **extensions**: for a distinguished triangle

$$M' \to M \to M'' \xrightarrow{[1]}$$

  in D, with $M', M'' \in ob(T)$, we have $M \in ob(T)$. This implies that T is **stable** under **isomorphisms** and **finite direct sums**.

- An **additive** function μ on T is a function from the (isomorphism-classes of) objects of T to an abelian group A, with

$$\mu(M) = \mu(M') + \mu(M'')$$

  for each distinguished triangle $M' \to M \to M'' \xrightarrow{[1]}$ in T. This implies especially:

$$M' \simeq M \Rightarrow \mu(M) = \mu(M')\ ,\ \mu(M[1]) = -\mu(M)\ ,$$
$$\mu(0) = 0 \quad \text{and} \quad \mu(M \oplus M') = \mu(M) + \mu(M')\ .$$

The **universal** additive function is given by $M \to \{M\}$, with $\{M\}$ the class of M in the **Grothendieck group** $K(T)$ of T, i.e. any other additive function is induced from this class map by a group homomorphism $K(T) \to A$.

Here $K(T)$ is the quotient of the free abelian group on the (isomorphism-classes of) objects of T, modulo the subgroup generated by $M - M' - M''$ for each distinguished triangle as above. If the reader wants to be careful about set-theoretical difficulties, then he should always assume that T is "essentially small", i.e. has a **set** of isomorphism-classes of objects.

Similarly for the **Grothendieck group** $K(Ab)$ of an **abelian category** Ab, which is defined in the same way by the "relation" $M = M' + M''$ for each **short exact sequence**

$$0 \to M' \to M \to M'' \to 0\ .$$

Let us apply this to $D^b_G(X)$. Then one has (compare lemma 3.3.1):

$$K(D^b_G(X)) \simeq K(Sh_G(X))\ ,$$

with $Sh_G(X)$ the **abelian category** of equivariant sheaves.

If G acts trivially on X', then the category $Sh_G(X', R)$ of equivariant sheaves of R-modules is **equivalent** to the category

$$Sh(X', R[G/G_0])$$

of sheaves of $R[G/G_0]$-modules (by example 3.1.2), with $R[G/G_0]$ the corresponding group ring of the discrete group G/G_0 of components. This applies especially to $Y = X/G$ and X' a point.

Now we can formulate the first main result:

Theorem 3.0.1. *Fix a strict full triangulated subcategory T of $D^b_G(\{pt\})$. Consider an equivariant complex $\mathcal{F} \in ob(D^b_G(X))$ satisfying the* **stalk property**

$$(Rq_? \mathcal{F})_y := i_y^*(Rq_? \mathcal{F}) \in ob(T) \quad \text{for all } y \in Y \quad (? = ! \text{ or } *)\,,$$

with $i_y : \{y\} \to Y$ the inclusion of the point $y \in Y$.

1. *Assume $Rq_? For(\mathcal{F})$ $(? = ! \text{ or } *)$ is* **constructible** *with respect to a* **finite cellular decomposition** *of $Y = X/G$. In the case $? = *$, let Y also be* **compact**. *Then*
$$R\Gamma_?(X, \mathcal{F}) \in ob(T) \qquad (? = ! \text{ or } *)\,.$$
Consider also an additive function μ on T, and denote by U_j $(j \in J)$ the cells of the decomposition, with n_j the dimension of U_j. Then one gets
$$\mu(R\Gamma_?(X, \mathcal{F})) = \sum_j (-1)^{n_j} \cdot \mu(\, (Rq_? \mathcal{F})_{y_j}\,) \quad (\, y_j \in U_j\,)\,.$$

2. *Assume $Rq_* For(\mathcal{F})$ is* **constructible** *with respect to a* **finite good cellular stratification** *of $Y = X/G$, and has for $? = *$ the stalk property above. Then $R\Gamma(X, \mathcal{F}) \in ob(T)$.*

Remark 3.0.1. In the case $G = \{e\}$ (i.e. in the corresponding context without a group action), this theorem follows from lemma 2.2.3 and lemma 2.2.5(3). So for theorem 3.0.1 one only has to check, that the simple proof of chapter 2 "lifts up" into the **equivariant context**.

If one wants to **apply** this theorem, one first has to check, that $Rq_? For(\mathcal{F})$ is **constructible** with respect to an appropriate **cellular decomposition or stratification**. In section 3.1, we explain this for the following important cases (in the last two examples we use freely our notions from chapter 2. A reader unfamiliar with them may have first a look at the introduction of chapter 2):

- X is a locally compact G-space with the structure of an equivariant CW-decomposition as in [tDieck, p. 98] (e.g. G is a discrete group acting cellulary on a locally compact CW-complex as in [tDieck, p. 101]). Then X/G is a locally compact CW-complex with respect to the induced filtration. Assume that $\mathcal{F}$ is constructible with respect to the corresponding filtration on X. Then $Rq_!\mathcal{F}$ is constructible with respect to the induced filtration (compare example 3.1.4). For X/G **compact**, $Rq_!\mathcal{F}$ is constructible with respect to a **finite cellular decomposition** (or stratification, if X/G is also a **regular** CW-complex).

- We fix an o-minimal structure on $(\mathbb{R}, +, \cdot)$ (compare [vDr]). For example the structure of **semialgebraic** sets. Assume G is a compact definable group, with a definable action on the locally compact definable space X. Then X/G can be realized as a closed definable subset of some $\mathbb{R}^N$ such that

$$q : X \to X/G$$

is a **proper** definable map ([vDr, thm. 11.1.8, cor. 11.2.18]). If $\mathcal{F}$ is constructible with respect to this structure (i.e. with respect to a finite filtration by definable closed subsets), then the same is true for $Rq_!\mathcal{F}$ (by theorem 2.0.1). By [vDr, prop. 4.1.13, thm. 8.2.9], $Rq_!\mathcal{F}$ is then constructible with respect to a **finite good cellular stratification** (compare chapter 2).

- G acts real analytically and **properly** on a real analytic manifold M. Fix an analytic geometric category (compare [vDrMi]). For example the structure of **subanalytic** sets. Let X be a definable closed G-invariant subset of M, and let $\mathcal{F}$ be constructible on X with respect to this analytic geometric category. Then X/G can be realized as a closed definable subset of a real analytic manifold (see lemma 3.1.3), and $Rq_!\mathcal{F}$ is also constructible with respect to this structure (example 3.1.7, corollary 3.3.1). If G has only **finitely** many components, then $Rq_*\mathcal{F}$ is also constructible with respect to this structure (example 3.1.7). Let X/G be **compact**. Then $Rq_?\mathcal{F}$ is constructible with respect to a **finite good cellular stratification** (compare chapter 2).

In the following, we assume that we are in the context of one of these examples. Depending on suitable "exact or cohomology functors", we get different strict full triangulated subcategories T of $D^b_G(\{pt\})$, and therefore different results, that can be deduced from theorem 3.0.1.

Let us first apply the **forgetful** functor For. So we only consider **stalk information** of $For(\mathcal{F})$. From the equivariance of $\mathcal{F}$ we only use the properties:

1. $For(\mathcal{F})$ has **locally constant** cohomology sheaves on the **orbits** of G.

2. $For(\mathcal{F})_x \simeq For(\mathcal{F})_{g(x)}$ for all $g \in G$, $x \in X$.

A typical application is the following (this implies results of [EvMi, LawYau, J3]. See section 3.2 for more general results and examples):

Example 3.0.1. *Let G be a compact torus (or a finite p-group for some prime number p), and denote by X^G the fixed point set. Assume that the cohomology* *of the stalks of* $For(\mathcal{F})$ *are* **finite-dimensional** *vector spaces, and denote by* χ *the usual* **Euler characteristic**. *Then*

$$H^*_?(X, For(\mathcal{F})) \quad and \quad H^*_?(X^G, For(\mathcal{F})) \quad (? = ! \ or \ *)$$

are also finite dimensional with

$$\begin{aligned} &\chi(\ R\Gamma_?(X, For(\mathcal{F}))\ ) = \chi(\ R\Gamma_?(X^G, For(\mathcal{F}))\ )\,,\ or \\ &\chi(\ R\Gamma_?(X, For(\mathcal{F}))\ ) \equiv \chi(\ R\Gamma_?(X^G, For(\mathcal{F}))\ ) \quad mod\, p\,. \end{aligned} \tag{3.1}$$

Next we apply the **cohomology functor**

$$h : D^b_G(\{pt\}) \to Sh_G(\{pt\}, R) \simeq mod(R[G/G_0])\,.$$

The following result is explained in section 3.1.

Example 3.0.2. *Assume* G *is* **finite** *and* R *is a noetherian ring (or* G *is a* **discrete** *group acting* **properly** *on* X *and* $R = k$ *is a field of characteristic* 0*). Let* T *be the strict full triangulated subcategory of* $D^b(mod(R[G]))$*, given by the complexes with* **finitely generated** *cohomology (by the* **perfect** *complexes). Assume that the cohomology of the stalks of* $For(\mathcal{F})$ *are* **finitely generated** R*-modules. Then*

$$R\Gamma_c(X, \mathcal{F}) \in ob(T)\,.$$

Moreover, the class of $R\Gamma_c(X, R)$ *for the equivariant constant sheaf* $\mathcal{F} = R_X$ *is given by the formula*

$$\{R\Gamma_c(X, R)\} = \sum_j (-1)^{n_j} \cdot \{R[G] \otimes_{R[G_{x_j}]} R\} \quad \in K(T)\,, \tag{3.2}$$

with $G_x := \{g \in G|\ g(x) = x\}$ *the stabilizer of* $x \in X$ *and* $q(x_j)$ *a point in the cell* U_j *of dimension* n_j *(for a finite cellular decomposition of* X/G *adapted to* $Rq_! R_X$*).*

This formula is a generalization of a result of Verdier [Ver2]. One gets a similar formula for any other **additive** function on T.

Other applications deal with "trace formulae" (see section 3.3 for more general versions of the next two examples).

Example 3.0.3. *Let* G *be a* **finite cyclic** *group generated by* g*, and let* R *be a noetherian ring of finite homological dimension. Then one can define for a finitely generated* G*-module* M *the* **trace** $tr(g; M)$ *of* g *on* M*. Assume that the cohomology of the stalks of* $For(\mathcal{F})$ *are* **finitely generated** R*-modules. Then this is also true for*

$$H^*_?(X, \mathcal{F}) \quad and \quad H^*_?(X^g, \mathcal{F}) \quad (? = ! \ or \ *)\,.$$

Moreover

$$\sum (-1)^j \cdot tr(g; H^j_?(X, \mathcal{F})) = \sum (-1)^j \cdot tr(g; H^j_?(X^g, \mathcal{F}))\,. \tag{3.3}$$

This **trace formula** also holds for a G-**equivariant** constructible complex of sheaves, if the **compact** Lie group G is **topologically** generated by g, i.e. G is the closure of the cyclic group generated by g (compare corollary 3.3.2).

Consider now the case of a **Lie group** with **finitely** many connected components. Let again $R = k$ be a field of characteristic 0. Let T be the strict full triangulated subcategory of $D^b_G(\{pt\})$, given by the complexes $\mathcal{F}$ such that $For(\mathcal{F})$ has **finite-dimensional** cohomology over k. Then $K(T)$ is isomorphic to the **Grothendieck group** $Rep(G/G_0, k)$ of the **abelian category** of finite-dimensional $k[G/G_0]$-modules.

Example 3.0.4. *Assume that the cohomology of the stalks of* $For(\mathcal{F})$ *are* **finite-dimensional**. *Then*

$$R\Gamma_?(X, \mathcal{F}) \in ob(T) \quad (? = !\ or\ *)\,.$$

Let moreover K *be a* **compact torus** *in the* **center** *of* G *(which in the o-minimal context is also definable). Then one gets for* $\mathcal{F}$ *as before:*

$$R\Gamma_?(X^K, \mathcal{F}) \in ob(T) \quad (? = !\ or\ *)\,,$$

together with the "localization formula"

$$\{R\Gamma_?(X, \mathcal{F})\} = \{R\Gamma_?(X^K, \mathcal{F})\} \quad \in Rep(G/G_0, k)\,. \tag{3.4}$$

In the case of a **Lie group** with **finitely** many connected components, one can also consider another **cohomology functor** on $D^b_G(\{pt\})$.

The category $D^b_G(\{pt\}, R)$ is by [BL, prop. 2.7] equivalent to the full subcategory of $D^b(BG, R)$, which consists of complexes with locally constant cohomology sheaves. Here BG is a **classifying space** of G. We denote by $H_G := \Gamma(BG, \cdot) \circ h$ the corresponding cohomology functor

$$D^b_G(\{pt\}, R) \xrightarrow{\ \ h\ \ } Sh(BG, R) \xrightarrow{\Gamma(BG,\cdot)} mod(R)\,.$$

Then one gets for $N \in ob(D^b_G(\{pt\}, R))$ the **graded module**

$$H^*_G(N) := \oplus_i\, H_G(N[i])$$

over the skew-commutative **graded ring**

$$A_G := H^*_G(R) = H^*(BG, R)\,.$$

This applies especially to

$$H^*_{G,?}(X, \mathcal{F}) := H^*_G(R\Gamma_?(X, \mathcal{F})) \quad (? = !\ \text{or}\ *)\,.$$

A homomorphism $\phi : G' \to G$ of such Lie groups induces a continuous map $\bar{\phi} : BG' \to BG$ and hence a homomorphism

$$A_G = H^*(BG, R) \to H^*(BG', R) = A_{G'}\,.$$

Let $S \subset H^*(BG, R)$ be a **multiplicatively closed** subset of homogeneous elements in the center of $H^*(BG, R)$. For a graded $H^*(BG, R)$-module M^* one can then define the localized module $S^{-1}M^*$. This localization functor is **exact**. Especially, if

$$M'^* \to M^* \to M''^*$$

is an exact sequence of $H^*(BG, R)$-modules, then

$$S^{-1}M'^* = 0 = S^{-1}M''^* \quad \text{implies} \quad S^{-1}M^* = 0\,.$$

Similarly

$$S^{-1}H^*_G(N) = 0 \Leftrightarrow S^{-1}(H^*_G(N[1])) = 0 \quad \text{for } N \in ob(D^b_G(\{pt\}, R))\,,$$

since only the grading is changed, up to a sign. Therefore, the subcategory T of $D^b_G(\{pt\}, R)$ given by the complexes N with $S^{-1}H^*_G(N) = 0$, is a **strict full triangulated subcategory**.

A variant of (the proof of) theorem 3.0.1 implies the following localization result for the **equivariant cohomology** (see section 3.4, and compare with [tDieck, GKM, EvMi2, AtBo]):

Example 3.0.5. *Let Y be a closed definable G-invariant subset of X (in the case of a CW-complex X, a "definable" subset is by definition a union of cells). Assume*

$$S \cap ker(H^*(BG, R) \to H^*(BG_x, R)) \neq \emptyset$$

for all $x \in U := X\backslash Y$. Then one has for $\mathcal{F} \in ob(D^b_G(X, R))$:

$$S^{-1}(H^*_{G,?}(X, \mathcal{F})) = S^{-1}(H^*_{G,?}(Y, \mathcal{F})) \quad (? = !\ or\ *)\,. \tag{3.5}$$

Our second main theorem is a similar **localization result for the Grothendieck group of equivariant constructible sheaves** (see theorem 3.3.1 for a more general version).

Let the ground ring R be a noetherian ring of finite homological dimension. Let T be the strict full triangulated subcategory of $D^b_G(\{pt\})$, given by the complexes $\mathcal{F}$ such that $For(\mathcal{F})$ has **finitely generated** cohomology modules over R. By the assumption on R, the last condition is stable under $\otimes^L_R$. The Grothendieck group $K(T)$ is therefore a **commutative ring**, with unit given by the class of the trivial G-module R. This ring $K(T)$ is by lemma 3.3.1 isomorphic to the Grothendieck group $Rep(G/G_0, R)$ of the **abelian category** of $R[G/G_0]$-modules, which are finitely generated over R. A homomorphism $\phi : G' \to G$ of Lie groups induces a homomorphism

$$G'/G'_0 \to G/G_0\,,$$

and therefore a restriction map

$$res : Rep(G/G_0, R) \to Rep(G'/G'_0, R)\,.$$

Let us denote by $K^G_{perf}(X)$ the **Grothendieck group** of the strict full triangulated subcategory $T(X)$ of $D^b_G(X)$, given by the complexes $\mathcal{F}$ with the following properties:

1. $For(\mathcal{F})$ is constructible (in the appropriate sense).

2. All stalk complexes of $\mathcal{F}$ belong to T.

This is in a natural way (by pullback and $\otimes^L_R$) a **module** over $Rep(G/G_0, R)$.

Let $i : Y \to X$ be the inclusion of a closed definable G-invariant subset. Then the functor i^* induces a $Rep(G/G_0, R)$-**linear** map

$$i^* : K^G_{perf}(X) \to K^G_{perf}(Y) \,.$$

Theorem 3.0.2. *Let $S \subset Rep(G/G_0, R)$ be a multiplicatively closed subset. Let Y be a closed definable G-invariant subset of X. Assume*

$$S \cap ker(res : Rep(G/G_0, R) \to Rep(G_x/(G_x \cap G_0), R)) \neq \emptyset$$

for all $x \in X \backslash Y$. Then i^* induces an isomorphism

$$S^{-1}K^G_{perf}(X) \xrightarrow[\sim]{i^*} S^{-1}K^G_{perf}(Y) \,.$$

By functoriality this implies for $\mathcal{F} \in ob(T(X))$ especially the formula

$$\{R\Gamma_c(X, \mathcal{F})\} = \{R\Gamma_c(Y, \mathcal{F})\} \quad \in S^{-1}Rep(G/G_0, R) \,. \tag{3.6}$$

In the o-minimal context, one has this formula also for $R\Gamma(\cdot, \mathcal{F})$!

More explicit examples are given in section 3.3. Compare also with [J2, J3, Saito] for similar results in the **complex algebraic** context of (holonomic) D-modules or perverse sheaves.

3.1 Equivariant sheaves

In this section we recall or prove the results about equivariant (constructible) complexes of sheaves, that we use in the following sections for the (proof of the) different localization results.

3.1.1 Equivariant sheaves and monodromic complexes

At the beginning, we make **no** assumptions about our topological spaces (e.g. properties like Hausdorff, locally compact or finite cohomological dimension).

Let us fix a topological group G with identity e, multiplication m and a topological space X with a **continuous** (left) action

$$a : G \times X \to X .$$

A complex $\mathcal{F} \in ob(D(X))$ is called a **monodromic** complex (with respect to this action), if there exists an **isomorphism**

$$\alpha : a^*\mathcal{F} \xrightarrow{\sim} p^*\mathcal{F}, \tag{3.7}$$

with $p : G \times X \to X$ the projection. Pulling back this isomorphism by the map

$$s_g : X \to G \times X \ , \ x \mapsto (g, x) \quad \text{for a given } g \in G \, ,$$

one gets an isomorphism

$$\alpha_g : g^*\mathcal{F} \xrightarrow{\sim} \mathcal{F} .$$

We call the pair $(\mathcal{F}, \alpha)$ a **naive equivariant complex** (compare [ChGi, Def. 5.1.6 and p. 438],[Saito]), if moreover

1. The pullbacks by $id_G \times a$ and $m \times id$ of α are related by the equation

   $$q^*\alpha \circ (id_G \times a)^*\alpha = (m \times id_X)^*\alpha \, ,$$

   where $q : G \times G \times X \to G \times X$ is the projection along the first factor G:

   $$\begin{array}{ccc} (a \circ (id_G \times a))^*\mathcal{F} & \xrightarrow{(id_G \times a)^*\alpha} & (p \circ (id_G \times a))^*\mathcal{F} = (a \circ q)^*\mathcal{F} \\ \| & & \downarrow{\scriptstyle q^*\alpha} \\ (a \circ (m \times id_X))^*\mathcal{F} & \xrightarrow{(m \times id_X)^*\alpha} & (p \circ (m \times id_X))^*\mathcal{F} = (p \circ q)^*\mathcal{F} . \end{array}$$

2. $\alpha_e : \mathcal{F} = e^*\mathcal{F} \xrightarrow{\sim} \mathcal{F}$ is the identity.

   Pulling back the first identities via the map

   $$X \to G \times G \times X \ , \ x \mapsto (g, h, x)$$

gives the **cocycle condition**

$$\alpha_{g \circ h} = \alpha_h \circ h^*(\alpha_g) .$$

So one gets especially a **right** action of G on $\mathcal{F}$ (or a left action of G, if one starts with the inverse isomorphism $p^*\mathcal{F} \xrightarrow{\sim} a^*\mathcal{F}$).

A **morphism** of monodromic (or naive equivariant complexes) $(\mathcal{F}_i, \alpha_i)$ $(i = 1, 2)$ is just a morphism $\phi : \mathcal{F}_1 \to \mathcal{F}_2$ such that the following diagram commutes:

$$\begin{array}{ccc} a^*\mathcal{F}_1 & \xrightarrow[\sim]{\alpha_1} & p^*\mathcal{F}_1 \\ \downarrow{\scriptstyle a^*\phi} & & \downarrow{\scriptstyle p^*\phi} \\ a^*\mathcal{F}_2 & \xrightarrow[\sim]{\alpha_2} & p^*\mathcal{F}_2 . \end{array}$$

If $\mathcal{F}$ is a sheaf (concentrated in degree zero) and α is a sheaf morphism, then $(\mathcal{F}, \alpha)$ is called a G-**equivariant sheaf** on X. For G a **discrete** group, this is just a sheaf on X, with a compatible action of G (as in [Groth2]). Especially for $G = \{e\}$ it is just a sheaf on X. In general an **equivariant sheaf** on X corresponds to a sheaf on X with a **continuous** action of G on the corresponding "espace étalé", **covering** the action of G on X (compare [Schei, Def. 8.2.1, p. 89/90]).

For example, a **constant** sheaf "is" always an equivariant sheaf.

Since the topological pullback of sheaves is an **exact** functor, one gets in this way the **abelian category** $Sh_G(X)$ of equivariant sheaves. Moreover, the **cohomology sheaves** of a naive equivariant (or monodromic) complex are equivariant (or monodromic) sheaves with respect to the induced isomorphism

$$\alpha^i : a^*h^i(\mathcal{F}) \simeq h^i(a^*\mathcal{F}) \xrightarrow{\sim} h^i(p^*\mathcal{F}) \simeq p^*h^i(\mathcal{F}) .$$

But the category of naive equivariant complexes is in general **not a triangulated** category. If $(\mathcal{F}, \alpha)$ is a naive equivariant complex, then this is also true for $(\mathcal{F}[1], \alpha[1])$. A distinguished triangle of naive equivariant complexes $(\mathcal{F}_i, \alpha_i)$ $(i = 1, 2, 3)$

$$\mathcal{F}_1 \xrightarrow{\phi_1} \mathcal{F}_2 \xrightarrow{\phi_2} \mathcal{F}_3 \xrightarrow{\phi_3} \mathcal{F}_1[1],$$

with ϕ_i $(i = 1, 2, 3)$ a morphism of naive equivariant complexes in the sense above, induces also a **long exact sequence of cohomology sheaves** in the abelian category of equivariant sheaves (compare [Saito, def. 3.1]).

But in general, one **cannot** extend a morphism of naive equivariant complexes to such a triangle (because the induced isomorphism of a cone is not unique). Similarly, the equivariant derived category $D_G(X)$ introduced in [BL] is **not** (!) the derived category of the abelian category of equivariant sheaves. Only for a **discrete** group, these categories are **equivalent** [BL, thm. 8.3.1].

Lemma 3.1.1. *Let X_i be a topological space with an action of the topological group G_i, and consider a G_i-monodromic (or naive equivariant) complex $(\mathcal{F}_i, \alpha_i)$ on X_i $(i = 1, 2)$. Suppose that we have a continuous group homomorphism $\phi : G_1 \to G_2$ and a continuous map $f : X_1 \to X_2$ such that the diagram*

$$\begin{array}{ccc} G_1 \times X_1 & \xrightarrow{a_1} & X_1 \\ \downarrow{\scriptstyle \phi\times f} & & \downarrow{\scriptstyle f} \\ G_2 \times X_2 & \xrightarrow{a_2} & X_2 \end{array}$$

commutes.

1. *Then $f^*\mathcal{F}_2$ is a G_1-monodromic (or naive equivariant) complex with respect to*

$$a_1^* f^*\mathcal{F}_2 = (\phi \times f)^*(a_2^*\mathcal{F}_2) \xrightarrow{(\phi\times f)^*\alpha_2} (\phi \times f)^*(p_2^*\mathcal{F}_2) = p_1^* f^*\mathcal{F}_2 .$$

2. *Suppose that* $X_1 = X_2$, $G_1 = G_2$ *and that we have a bifunctor*

$$\otimes^L : D(X_1) \times D(X_1) \to D(X_1) ,$$

commuting with the topological pullback of sheaf complexes (e.g. we consider sheaves of R*-modules with* R *a commutative ring and* $\otimes^L = \otimes^L_R$*). Then*

$$\mathcal{F}_1 \otimes^L \mathcal{F}_2$$

is a G_1*-monodromic (or naive equivariant) complex with respect to* $\alpha_1 \otimes^L \alpha_2$.

3. *Suppose that* X_i, G_i *are locally compact of finite cohomological dimension, and* $G_1 = G_2$, *with* $\phi = id$. *Then* $Rf_!\mathcal{F}_1$ *is a* G_1*-monodromic (or naive equivariant) complex with respect to*

$$a_2^*(Rf_!\mathcal{F}_1) \simeq R(id \times f)_!(a_1^*\mathcal{F}_1) \xrightarrow{R(id\times f)_!\alpha_1} R(id \times f)_!(p_1^*\mathcal{F}_1) \simeq p_2^*(Rf_!\mathcal{F}_1) .$$

The proof of (1.) and (2.) is obvious. The isomorphism used in (3.) is the **base change isomorphism** for $Rf_!$ ([KS, prop. 2.6.6]). Note that in the case (3.) the commutative "action diagram" is also cartesian. Then (3.) follows from the fact, that this "base change isomorphism" commutes with pullbacks.

Let $p_i : X := X_1 \times X_2 \to X_i$ denote the projections ($i = 1, 2$). Then (1.) and (2.) imply that

$$\mathcal{F}_1 \boxtimes^L \mathcal{F}_2 := (p_1^*\mathcal{F}_1) \otimes^L (p_2^*\mathcal{F}_2)$$

is a $G_1 \times G_2$-monodromic (or naive equivariant) complex with respect to $\alpha_1 \boxtimes^L \alpha_2$.

For a general group G, the advantage of this definition of an equivariant sheaf is given by the following example ([BL, lem. on p. 3]):

Example 3.1.1. *Let* X *be a* **free** G*-space. This means that the stabilizer* G_x *of every* $x \in X$ *is* **trivial**, *and that the quotient map*

$$q : X \to \bar{X} := G\backslash X$$

is a **locally trivial fibration**. *Then* $q^* : Sh(\bar{X}) \to Sh_G(X)$ *is an equivalence of categories (with* q^* *as in lemma* 3.1.1(1), *and* $\phi : G \to \{e\}$*). The inverse functor is*

$$q_*^G : Sh_G(X) \to Sh(\bar{X}) ,$$

where q_*^G *denotes the subsheaf of* G*-invariants of* q_*.

Lemma 3.1.2. *Consider the situation of lemma* 3.1.1, *with* $f = id_{X_1}$. *Assume that the map* ϕ *is* **0-acyclic** *in the sense of [BL, def.* 1.9.1*]. Then the functor*

$$f^* : Sh_{G_2}(X_1) \to Sh_{G_1}(X_1)$$

is an **equivalence** *of categories* .

Let us first recall the definition of an n-**acyclic** map ($n \in \mathbb{N}_0$, $n = \infty$) in the sense of [BL, def. 1.9.1]. A continuous map $g : X \to Y$ is called n-**acyclic**, if

1. For any $\mathcal{F} \in ob(Sh(Y))$ the adjunction morphism $\mathcal{F} \to g_*g^*\mathcal{F}$ is an isomorphism and $R^ig_*g^*\mathcal{F} = 0$ for $i = 1, \ldots, n$.

2. For any base change $Y' \to Y$ the induced map $X \times_Y Y' \to Y'$ satisfies also the first property.

f is ∞-**acyclic** if it is n-acyclic for all n. Similarly, a topological **space** M is n-**acyclic** in the sense of [BL, 1.9.4], if it is non-empty, connected, locally connected and satisfies (1.) above for the constant map $g : M \to \{pt\}$.

Proof of the lemma. By adjunction of (g^*, g_*) one gets by the condition (1.) above the isomorphism

$$Hom_{Sh(X)}(g^*\mathcal{F}, g^*\mathcal{F}') \simeq Hom_{Sh(Y)}(\mathcal{F}, g_*g^*\mathcal{F}') \simeq Hom_{Sh(Y)}(\mathcal{F}, \mathcal{F}') , \qquad (3.8)$$

for $\mathcal{F}, \mathcal{F}' \in ob(Sh(Y))$. Note, that the **composition** and therefore also the **product** of 0-acyclic maps is also 0-acyclic. Then the claim follows, if one applies the bijection (3.8) to $\phi \times id$ and $\phi \times \phi \times id$. □

One has the following simple criterion for a 0-**acyclic** map ([BL, 1.9.4]). Assume that one of the following condition holds:

(a.) $g : X \to Y$ is a local trivial fibration with all fibers non-empty and connected.

(b.) $g : X \to Y$ is locally fibered (i.e. for each $x \in X, y \in Y$ there are neigbourhoods U of x and V of y such that $g : U \to V$ is a fibration) with all fibers non-empty, connected and locally connected.

Then g is 0-acyclic.

Example 3.1.2. 1. *Suppose the* **connected** *group G operates* **trivially** *on X (for example X is a point). Then $Sh_G(X)$ is* **equivalent** *to $Sh(X)$ (apply (a.) to $\phi : G \to \{e\}$).*

2. *Suppose the* **identity component** *G_0 of G is* **open**, *and G_0 operates* **trivially** *on X so that one has an induced action of the* **discrete** *group*

$$\pi_0(G) = G/G_0 \ \ on \ X .$$

Then $Sh_G(X)$ is **equivalent** *to $Sh_{G/G_0}(X)$ (apply (a.) to the canonical projection $\phi : G \to G/G_0$). If already G operates* **trivially** *on X (and we consider sheaves of R-modules), then $Sh_G(X, R)$ is* **equivalent** *to*

$$Sh(X, R[\pi_0(G)]) ,$$

with $R[\pi_0(G)]$ the group ring of $\pi_0(G)$ over R. Especially, $Sh_G(\{pt\}, R)$ is **equivalent** *to the category*

$$mod(R[\pi_0(G)])$$

of R-modules with an action of $\pi_0(G)$.

3. *Let G be (topologically isomorphic to) a* **Lie group** *and H a* **closed connected normal** *subgroup. Suppose H operates* **trivially** *on X. Then $Sh_G(X)$ is* **equivalent** *to $Sh_{G/H}(X)$ (apply (b.) to the canonical projection $\phi : G \to G/H$).*

Remark 3.1.1. Let the group G be definable with respect to an o-minimal structure on $(\mathbb{R}, +, \cdot)$. Then G has only finitely many connected components ([vDr, prop. 2.18]). Moreover, by a result of Pillay [Pi, prop. 2.5, rem. 2.6], G is topologically isomorphic to a **Lie group**.

Another useful property of a **monodromic** sheaf $(\mathcal{F}; \alpha)$ is the following: Consider for $x \in X$ the **orbit** $G(x) := a(G \times \{x\})$, and assume that

$$a : G \times \{x\} \to G(x)$$

has **locally a continuous section**, i.e. for each $g \in G$, $y \in G(x)$, there are neigbourhoods U of (g, x) in $G \times \{x\}$ and V of y, together with a continuous map $s : V \to U$ such that $a \circ s = id_V$. Then $\mathcal{F}|G(x)$ is **locally constant**, since

$$\mathcal{F}|V \simeq s^*(a^*\mathcal{F}|U) \simeq s^*(p^*\mathcal{F}|U) \simeq c^*\mathcal{F}_x \quad (\, c : V \to \{x\} \,)\,.$$

Example 3.1.3. 1. *G is a* **Lie group** *and $x \in X$ has the following properties:*

(a) *The* **stabilizer** *G_x of x is a* **closed** *subgroup.*

(b) *The natural map $G/G_x \to G(x)$ is a homeomorphism.*

Then G_x is a **closed Lie subgroup** *of G ([Hoch, VIII. thm.1.1]) and the projection $G \to G/G_x$ is a submersion. Both conditions are satisfied, if G operates* **properly** *on X, i.e. the map $G \times X \to X \times X$, $(g, x) \mapsto (x, g(x))$ is proper ([tDieck, prop. 3.19, p. 28]). For example G is a compact group and X is Hausdorff.*

2. *G, X and the action $a : G \times X \to X$ is* **definable** *with respect to an o-minimal structure on $(\mathbb{R}, +, \cdot)$. Then the same is true for*

$$a : G \times \{x\} \to G(x)\,,$$

and this has **generically** *(i.e. over an open dense definable subset of $G(x)$) a continuous section (e.g. by the trivialization theorem [vDr, thm. 9.1.7]). By* **translation** *with the elements of G, we get such a section on an open neighbourhood of each point of $G(x)$.*

From now on, we consider only locally compact spaces with countable topology of finite cohomological dimension (unless we have to work with ∞**-acyclic** maps or spaces).

3.1.2 Equivariant derived categories

Next we recall some of the properties of the **equivariant derived category** $D_G^*(X)$ (with $* = b, +$) of Bernstein-Lunts.

In this context we **always** assume that the base ring R is commutative (as in [BL]), and that G is topologically isomorphic to a **Lie group** with the following properties ([BL, 3.1, 2.9, 12.4.1]):

(*+) For every $n \in \mathbb{N}$ there exists an **n-acyclic free** G-space M_n which is a **manifold**.

(**) If we use the **unbounded** equivariant derived category $D_G^+(X)$, then we assume **in addition** that there exists an **∞-acyclic free** "smooth" G-space M of the type

$$M \simeq \varinjlim M_n\ , \quad M_1 \hookrightarrow M_2 \hookrightarrow M_3 \ldots\ ,$$

with M_n as in (*+) and $M_n \hookrightarrow M_{n+1}$ closed embeddings of submanifolds ($n \in \mathbb{N}$).

By [BL, lem. 3.1, lem. 12.4.2] we know the conditions (*+) and (**) for a **Lie group** with one of the following properties:

a) G is a linear group, i.e. a closed subgroup of $GL(k, \mathbb{R})$ for some k.

b) G has a finite number of connected components.

Similarly, if G satisfies the condition (*+) or (**), then the same is true for any closed subgroup (by the induced action).

Remark 3.1.2. Many of the following results of Bernstein-Lunts [BL] hold under weaker assumptions than (*+) and (**) above. But one has to be careful with [BL, criterion 1.9.4] ! As remarked by one of the referees, the proof of this "criterion" in [BL, A.4, p. 14/15] contains an **error** in [BL, step 4, p.15] (the functor $R^i f_*$ do not commute in general with infinite products for $i > 0$). But we only need this "criterion" in the following **special** form, where the proof of [BL, step 4, p. 15] can easily be modified: Suppose M_n is an **n-acyclic manifold** ($n \in \mathbb{N}$, or M is an **∞-acyclic manifold** as in (**) above). Then the **constant map** $M_n \to \{pt\}$ (or $M \to \{pt\}$) is **n-acyclic (or ∞-acyclic)** !

Bernstein-Lunts introduce in [BL] an **equivariant derived category**

- $D_G^b(X)$ under the condition (*+) , and
- $D_G^+(X)$ under the condition (**) ,

with the following properties:

1. $D_G^*(X)$ is a **triangulated** category with a t-**structure**, whose heart is naturally equivalent to $Sh_G(X)$ ([BL, p. 25, 2.9.7]). We denote the corresponding **cohomology functor** just by

$$h : D_G^*(X) \to Sh_G(X) .$$

2. One has an **exact forgetful functor** ([BL, p. 19, sec. 2.9])

$$For : D_G^*(X) \to D^*(X)$$

such that $For(h(\mathcal{F})) \simeq h^0(For(\mathcal{F}))$, with h^0 the usual cohomology sheaf in degree zero. Moreover $\mathcal{F} \simeq 0$ if and only if $For(\mathcal{F}) \simeq 0$.

3. For a closed subgroup H of G one has a **restriction functor**

$$Res_{H,G} : D_G^*(X) \to D_H^*(X) ,$$

with $Res_{\{e\},G} \simeq For$ ([BL, 2.6.1, 2.9.8]).

4. For $q : X \to \bar{X} = X/G$ the quotient map one has a **pullback functor**

$$q^* : D^*(\bar{X}) \to D_G^*(X) .$$

5. Moreover, one has standard functors **commuting** with the restriction functor, which under For correspond to the classical ones ([BL, thm. 3.4.1, sec. 2.9]):

   (a) If the ground ring R is of finite homological dimension, one has

   $$\otimes^L , \; Rhom : D_G^b(X) \times D_G^b(X) \to D_G^b(X) .$$

   (b) For a continuous map of G-spaces $f : X \to Y$ (i.e. f commutes with the G-action), one has

   $$f^* : D_G^*(Y) \to D_G^*(X)$$

   and

   $$Rf_*, Rf_! : D_G^b(X) \rightleftarrows D_G^b(Y) \; : f^! .$$

   Moreover, one has the usual **adjunction** properties for

   $$(f^*, Rf_*) \quad \text{and} \quad (Rf_!, f^!) .$$

Another useful property is the fact that the **forgetful functor**

$$For \; : \; D_G^*(X) \to D^*(X)$$

induces on $For(\mathcal{F})$ an isomorphism $\alpha : a^*\mathcal{F} \xrightarrow{\sim} p^*\mathcal{F}$ such that $(For(\mathcal{F}), \alpha)$ is a **naive equivariant** complex. This follows from the description of $D_G^*(X)$ in terms of

acyclic resolutions of X ([BL, 2.1.3]). Consider a resolution $p' : P \to X$. This is a G-equivariant map with P a **free** G-space. Denote by q the projection $P \to P/G$. Then $\mathcal{F} \in ob(D^*_G(X))$ is given by a triple

$$(\mathcal{F}_X, \overline{\mathcal{F}}, \beta) , \quad \text{with } \mathcal{F}_X \in ob(D^*(X)) , \ \overline{\mathcal{F}} \in ob(D^*(P/G))$$

and β an isomorphism

$$\beta : p'^*(\mathcal{F}_X) \simeq q^*\overline{\mathcal{F}} ,$$

with p' an n-**acyclic** map for $n >> 0$ or $n = \infty$.

By lemma 3.1.1(1), $q^*\overline{\mathcal{F}}$ has an induced structure of a naive equivariant complex (apply the lemma to q and the homomorphism $G \to \{e\}$). Then $p'^*(\mathcal{F}_X)$ gets an induced structure of a naive equivariant complex by the isomorphism β. But this structure has to come from such a structure on $\mathcal{F}_X$, since the map p' is n-acyclic for $n >> 0$ or $n = \infty$ (this argument is similar to lemma 3.1.2, and one uses [BL, prop.1.9.2]). But $\mathcal{F}_X$ is just $For(\mathcal{F})$ ([BL, def. 2.1.4]).

Let us finally remark that for a **discrete** group G one can also work with $D^*(Sh_G(X))$ (with $* = b, +$). Then one can apply the classical results of Grothendieck [Groth2]. In this context, the base ring R is not necessarily commutative, and one has similar functors as before. The only exception is the functor $f^!$, which maybe is not defined in general but exists at least for $f : X \to Y$ a **closed inclusion**.

Now we are ready for the

Proof of theorem 3.0.1. Since $R\Gamma_?(X, \mathcal{F}) = R\Gamma_?(Y, Rq_?\mathcal{F})$, we can assume $X = Y$, i.e. G acts trivially. Moreover, for Y compact we have $R\Gamma_c(Y, \cdot) = R\Gamma(Y, \cdot)$. Let us denote the corresponding filtration

$$Y_\bullet \ : \emptyset = Y_{-1} \subset Y_0 \subset \ldots \subset Y_m = Y ,$$

and the inclusions

$$i : Y_{m-1} \to Y \ , \ j : U := Y \backslash Y_{m-1} \to Y .$$

We can assume $U \neq \emptyset$.

For the proof of (1.), we apply $R\Gamma_c(Y, \cdot)$ to the standard distinguished triangle ([BL, 1.4.5, 2.9.8])

$$Rj_!j^*(\mathcal{F}) \longrightarrow \mathcal{F} \longrightarrow Ri_*i^*(\mathcal{F}) \xrightarrow{[1]} .$$

Of course, one has this also in the case of $D^b(Sh_G(Y, R))$ for a **discrete** group. Then

$$R\Gamma_c(Y, Ri_*i^*(\mathcal{F})) \simeq R\Gamma_c(Y_{m-1}, \mathcal{F}) .$$

By induction on m, it is enough to consider

$$R\Gamma_c(Y, Rj_!j^*(\mathcal{F})) \simeq R\Gamma_c(U, \mathcal{F}) .$$

But U is the topological sum of finitely many cells U_j, and this implies

$$R\Gamma_c(U, \mathcal{F}) \simeq \oplus_j R\Gamma_c(U_j, \mathcal{F}) .$$

So we can also assume that $U = U_j$ is just one cell of dimension n_j. By the proof of lemma 2.0.1 and remark 2.2.1 on page 100 one gets

$$R\Gamma_c(U_j, \mathcal{F})[n_j] \simeq \mathcal{F}_{y_j} \quad (y_j \in U_j) .$$

The proof of lemma 2.0.1 "lifts up" into the equivariant context, because it uses only abstract properties of triangulated categories together with the fact that the **adjunction** map

$$\mathcal{F} \to g^* Rg_* \mathcal{F}$$

for the projection

$$g : U = \mathbb{R}^n \simeq \mathbb{R}^{n-1} \times]0,1[\to \mathbb{R}^{n-1}$$

is an isomorphism. But this adjunction map is an isomorphism if and only if its cone is zero ([Nee, cor. 1.2.6]), which can be checked after applying For. But in the non-equivariant context it follows from the **Vietoris-Begle theorem** [KS, prop. 2.7.8].

For the proof of (2.), we apply $R\Gamma(Y, \cdot)$ to the standard distinguished triangle ([BL, 1.4.5, 2.9.8])

$$Ri_*i^!(\mathcal{F}) \longrightarrow \mathcal{F} \longrightarrow Rj_*j^*(\mathcal{F}) \xrightarrow{[1]} .$$

Again, one has this also in the case of $D^b(Sh_G(Y, R))$ for a **discrete** group. We claim, that $Rj_*j^*(\mathcal{F})$ is constructible with respect to $Y_\bullet$, and that its stalks belong to T. Then the triangle above implies the same property also for $Ri_*i^!(\mathcal{F})$, with

$$R\Gamma(Y, Ri_*i^!(\mathcal{F})) \simeq R\Gamma(Y_{m-1}, i^!(\mathcal{F})) .$$

Then we can argue by induction on m, and we only have to show

$$R\Gamma(Y, Rj_*j^*(\mathcal{F})) \simeq R\Gamma(U, \mathcal{F}) \in ob(T) .$$

It is enough to consider the case $U = U_j$ is just one cell, with $Y = cl(U)$ the closure of U. Denote by f a constant map on U. Then the adjunction morphism

$$j^*\mathcal{F} \to f^*Rf_*j^*\mathcal{F}$$

is (as above) an isomorphism in $D^b_G(U)$. This implies for $y \in U$:

$$\mathcal{F}_y \simeq (f^*Rf_*j^*\mathcal{F})_y \simeq Rf_*j^*\mathcal{F} \quad \text{in } D^b_G(\{pt\}) .$$

Especially

$$R\Gamma(U, \mathcal{F}) \simeq \mathcal{F}_y \in ob(T) .$$

Let c be a constant map on the closure $cl(U)$ of U so that $f = c \circ j$. Then the adjunction morphism

$$c^* Rf_* j^* \mathcal{F} \to Rj_* j^* c^* Rf_* j^* \mathcal{F} \simeq Rj_* j^* \mathcal{F}$$

is an isomorphism, since this can be checked after applying For. Then it follows from the **goodness assumption** on the stratification. This implies our claim. □

3.1.3 Examples and stalk formulae

As explained in the introduction, our approach to localization results as in (the proof of) theorem 3.0.1 applies to the **equivariant** or **monodromic** complex $\mathcal{F}$ on the G-space X, if the quotient space $\bar{X} = X/G$ has a **(good) cellular stratification or decomposition** such that $Rq_*\mathcal{F}$ or $Rq_!\mathcal{F}$ is **constructible** with respect to the corresponding filtration (with q the projection map).

Let us explain the most important examples.

Example 3.1.4. *X is a locally compact G-space with the structure of an* **equivariant CW-decomposition** *as in [tDieck, p. 98] (e.g. G is a discrete group acting cellulary on a locally compact CW-complex [tDieck, p. 101]). Let X_n ($n \in \mathbb{N}$) denote the corresponding filtration Then $\bar{X}$ is a locally compact CW-complex with respect to the induced filtration $\bar{X}_n := X_n/G$ ([tDieck, Ex. 1.17.2, p. 103]). Assume that $\mathcal{F}$ is constructible with respect to this filtration on X. Then $Rq_!\mathcal{F}$ is constructible with respect to the induced filtration.*

For the constructibility of $Rq_!\mathcal{F}$ it is enough to work out the case, that $X = Y \times \mathbb{R}^n$ is just one "equivariant open cell", i.e. G is only acting on the factor Y with $Y/G = pt$ a point. Consider the cartesian diagram

$$\begin{array}{ccc} Y \times \mathbb{R}^n & \xrightarrow{q} & Y/G \times \mathbb{R}^n \\ \downarrow{\scriptstyle g} & & \downarrow{\scriptstyle g'} \\ Y & \xrightarrow{q'} & \{pt\} = Y/G\,, \end{array}$$

and a complex $\mathcal{F}$ on $Y \times \mathbb{R}^n$, with **locally constant** cohomology sheaves. Then

$$\mathcal{F} \simeq g^* Rg_* \mathcal{F}$$

by the **Vietoris-Begle theorem** ([KS, prop. 2.7.8]). By the **base change isomorphism** for $Rq_!$ ([KS, prop. 2.6.7]) one gets

$$Rq_!\mathcal{F} \simeq Rq_!(g^* Rg_* \mathcal{F}) \simeq g'^*(Rq'_!(Rg_*\mathcal{F}))\,.$$

The same argument applies to the

Example 3.1.5. *Let the space $\bar{X}$ have a (good) cellular stratification or decomposition and let $\mathcal{F}$ be constructible with respect to the corresponding filtration. Assume X is a* **free** *G-space with*

$$q : X \to \bar{X} \quad \textit{the quotient map}\,,$$

e.g. G is a discrete group and X is a covering space of $\bar{X}$, with covering group G. Then $Rq_!q^*\mathcal{F}$ is also **constructible** *with respect to this filtration. As we will later see, if G is a Lie group with* **finitely** *many connected components, this is also true for $Rq_*q^*\mathcal{F}$.*

Example 3.1.6. *G, X and the action $a : G \times X \to X$ are* **definable** *with respect to an o-minimal structure on $(\mathbb{R}, +, \cdot)$, and G is a* **compact** *group. Then $\bar{X}$ is a* **locally compact definable space** *such that*

$$q : X \to \bar{X} \quad \textit{is a proper definable map}$$

([vDr, 10. cor.2.18]). If $\mathcal{F}$ is constructible with respect to this o-minimal structure, then the same is true for $Rq_!\mathcal{F} = Rq_*\mathcal{F}$ (theorem 2.0.1). By [vDr, prop. 4.1.13, thm. 8.2.9], $Rq_!\mathcal{F}$ is then constructible with respect to a finite good cellular stratification (see chapter 2 for more on this).

In the following, we will explain the

Example 3.1.7. *Let G be a real analytic Lie group acting real analytically and* **properly** *on a real analytic manifold M. Then there exists a continuous subanalytic and G-invariant map*

$$f : M \to M' := \mathbb{R}^p \quad (\textit{ for some } p\,)$$

such that the induced mapping $\bar{f} : M/G \to f(M)$ is a homeomorphism, with $f(M)$ a **closed** *subanalytic subset of M' ([Kan, thm. 7.3]). Fix an analytic geometric category and a definable closed G-invariant subset X of M (compare with [vDrMi] and chapter 2). Let $\mathcal{F}$ be constructible on X with respect to this analytic geometric category. Then we have:*

1. *Assume G is* **compact** *so that f is a proper map. Then $Rf_*\mathcal{F} = Rf_!\mathcal{F}$ is constructible on $f(X)$.*

2. *Assume G has only* **finitely** *many connected components, and the cohomology sheaves of $\mathcal{F}$ are locally constant an the G-orbits (e.g. they are monodromic). Then $Rf_*\mathcal{F}$ and $Rf_!\mathcal{F}$ are constructible on $f(X)$.*

3. *Assume all stabilizers G_x for $x \in X$ are* **finite** *(e.g. G is a discrete group), the ground ring R is a* **field of characteristic** *0 and $\mathcal{F}$ underlies an* **equivariant** *complex in $D^*_G(X)$ (i.e. is in the image of the forgetful functor For). Then $Rf_!\mathcal{F}$ is constructible on $f(X)$.*

Fix an analytic geometric category. We first explain, that the induced map $f : X \to f(X)$ is a **quotient** in the category of **closed definable** subsets of real analytic manifolds and continuous definable maps (compare [Kan]).

Lemma 3.1.3. 1. *Let A be a (closed) G-invariant definable subset of X. Then $f(A)$ is a (closed) definable subset of M'. Especially, $f(X)$ is a closed definable subset and $f : X \to f(X)$ is definable.*

2. *Let B be a closed definable subset of a real analytic manifold N, and $g : X \to B$ be a continuous definable and G-invariant map. Then*

$$\bar{g} \circ \bar{f}^{-1} : f(X) \to B$$

is continuous and definable, with $\bar{(\cdot)}$ the induced map on X/G.

Proof. Take compact subanalytic subsets L'_i of M ($i \in \mathbb{N}$) such that $q(L'_i)$ is a locally finite cover of M/G (compare [Kan, p. 38, proof of thm. 7.3]). Then $f(L'_i)$ is locally finite on M'. Fix a definable (closed) subset A of X. Then $A \cap L'_i$ is a (closed) relatively compact definable subset of M. Therefore $f(A \cap L'_i)$ is a (closed) definable subset of M' ([vDrMi, D.6]). Assume that A is also G-invariant. Then

$$f(A) = \cup_i \, f(A \cap L'_i)\,,$$

and this union is locally finite on M'. Therefore $f(A)$ is a (closed) definable subset of M' and $f(X)$. Moreover, $f : X \to f(X)$ is definable by [vDrMi, D.5], and $\bar{f} : X/G \to f(X)$ is a homeomorphism. Then

$$h := \bar{g} \circ \bar{f}^{-1} : f(X) \to B$$

is continuous (with g as in the second statement), and we only have to show that the graph $graph(h)$ of h is a definable subset of $M' \times N$. But

$$graph(h) = (f,g)(X) \quad \text{and} \quad (f,g) : X \to M' \times N$$

is definable ([vDrMi, 1.6, D.9]). So the claim follows from the same argument as before, since $(f,g)(L'_i)$ is locally finite on $M' \times N$. □

Remark 3.1.3. Let U be an open subset of M' as above. Then lemma 3.1.3 can also be applied to the induced map

$$f : (f^{-1}(U), X \cap f^{-1}(U)) \to (U, U \cap f(X))\,.$$

Let us now explain the statements about the constructibility of $Rf_*\mathcal{F}$ and $Rf_!\mathcal{F}$. The case (1.) follows from corollary 2.2.1. By proposition 2.2.1, the other cases are **local** results on M'. Moreover, by the above lemma and remark, we can use any "local subanalytic realization" of the quotient in the sense above. Fix the point $x \in X$ and denote by H the stabilizer G_x of x. This is a compact Lie subgroup, since the action is proper. By [Kan, thm. 2.5], there exists a **real analytic slice** S at x: S is an H-invariant analytic submanifold of M containing x, GS is an open subset of M and there is a G-equivariant analytic mapping

$$f : GS \to G/H \text{ such that } f^{-1}(eH) = S\,.$$

Therefore GS is G-equivariant analytic isomorphic to the induced space $G \times_H S$ ([Illma, lem. 5.2]), and we only have to consider this induced space.

Then M/G is homeomorphic to S/H in such a way that the canonical projection corresponds to the map

$$G \times_H S \to S/H$$

induced by the H-equivariant projection $G \times S \to S$. Here one has the usual twisted action of H on $G \times S$, given by

$$h((g,s)) := (g \cdot h^{-1}, h(s))\,.$$

Consider now the case (2.) that G has only **finitely** many connected components. Since H is a compact subgroup, there exists ([Hoch, XV. thm.3.1]) a compact subgroup K of G containing H, and a K equivariant analytic isomorphism of manifolds (for the right K-action given by multiplication on K and G):

$$\mathbb{R}^p \times K \simeq G\,.$$

So we have to study the map $q : \mathbb{R}^p \times (K \times_H S) \to S/H$ induced by the H-equivariant projection $K \times S \to S$. Here we consider S/H as a closed subanalytic subset in some real analytic manifold such that

$$K \times_H S \to S/H$$

is subanalytic. But the cohomology sheaves of $\mathcal{F}$ are locally constant on the fibers of q (since the fibers correspond to the G-orbits). Especially, the cohomology sheaves are locally constant on the fibers of the projection

$$\mathbb{R}^p \times (K \times_H S) \to K \times_H S\,.$$

Fix a radius $r > 0$ and denote the inclusion

$$i : M(r) := \{x \in \mathbb{R}^p : ||x|| \leq r\} \times (K \times_H S) \to \mathbb{R}^p \times (K \times_H S)\,.$$

Using a **Mittag-Leffler argument** ([KS, prop. 2.7.1(iii)]), one gets in the same way as in the proof of corollary 2.0.2 on page 106 :

Corollary 3.1.1. *The natural morphism $Rq_*\mathcal{F} \to Rq_*Ri_*i^*\mathcal{F}$ (induced by the adjunction map for i) is an isomorphism. Similarly, the natural morphism*

$$(Rq_*\mathcal{F})_x \to R\Gamma(\{q = x\}, \mathcal{F}) \quad \textit{is an isomorphism.}$$

But $i^*\mathcal{F}$ is construcible on $M(r)$ and the definable map $q \circ i$ is proper. So

$$Rq_*\mathcal{F} \simeq Rq_*Ri_*i^*\mathcal{F}$$

is constructible (corollary 2.2.1). Note that the same argument applies also in the example 3.1.5. Then the projection q is locally isomorphic to a projection

$$q : \mathbb{R}^p \times K \times S \to S ,$$

with S open in $\bar{X}$ and $H = \{e\}$. The **projection formula** implies

$$Rq_*q^*\mathcal{F} \simeq Rq_*Ri_*i^*q^*\mathcal{F} \simeq (R(q \circ i)_*R_{M(r)}) \otimes^L \mathcal{F} ,$$

and $R(q \circ i)_*R_{M(r)}$ has locally constant cohomology sheaves. Let us come back to the case (2.) of the example. Denote the inclusion

$$j : \{x \in \mathbb{R}^p : ||x|| < r\} \times (K \times_H S) \to \{x \in \mathbb{R}^p : ||x|| \leq r\} \times (K \times_H S) .$$

Then one gets in the same way the isomorphism

$$Rq_!\mathcal{F} \simeq Rq_*Ri_*Rj_!(i \circ j)^*\mathcal{F} , \tag{3.9}$$

with $Rj_!(i \circ j)^*\mathcal{F}$ constructible on $M(r)$.

We consider now the case (3.) of the example. Let us denote the maps in the commutative diagram

$$\begin{array}{ccc} G \times S & \xrightarrow{q'} & S \\ \downarrow{\scriptstyle p} & & \downarrow{\scriptstyle p'} \\ G \times_H S & \xrightarrow{q} & S/H. \end{array}$$

Since $G \times_H S$ is the induced space $ind(S)$, we have by [BL, thm. 2.6.3, 2.9.8] an equivalence of categories (for $* = b, +$)

$$\nu^* : D^*_G(ind(S)) \to D^*_H(S) .$$

The inverse functor can be described in the following way. Define the action of $G \times H$ on $G \times S$ by left multiplication of G and the twisted action of H as above. Then q' is ϕ-equivariant with respect to the projection homomorphism $\phi : G \times H \to H$, and one has the pullback functor

$$Q^*_{q'} \ : \ D^*_H(S) \to D^*_{G \times H}(G \times S)$$

(compare [BL, 6.5-6.7]). Since the action of H on $G \times S$ is free, $D^*_{G \times H}(G \times S)$ is equivalent to $D^*_G(ind(S))$ (compare [BL, thm. 2.6.2, 2.9.8]). This equivalence is also given by the functor

$$Q_{p*} \ : \ D^*_{G \times H}(G \times S) \to D^*_G(ind(S))$$

([BL, 6.12.6]), where we use the projection homomorphism $G \times H \to G$. Then our inverse functor is given by the composition of these two functors.

Let us forget the action of G. Then we get a commutative diagram

$$\begin{array}{ccccc} D_H^*(S) & \xrightarrow{Q_{q'}^*} & D_{G\times H}^*(G \times S) & \xrightarrow{Q_{p*}} & D_G^*(ind(S)) \\ \downarrow{\scriptstyle id} & & \downarrow{\scriptstyle Res_{H,G\times H}} & & \downarrow{\scriptstyle Res_{\{e\},G}} \\ D_H^*(S) & \xrightarrow{q'^*} & D_H^*(G \times S) & \xrightarrow{Q_{p*}} & D^*(ind(S)) \,. \end{array}$$

The left square is commutative by [BL, 6.6.2-4, 6.7], and the right one by [BL, thm. 7.3]. But in the case (3.) of the example, we assume that H is a finite group. Therefore we also have a commutative diagram

$$\begin{array}{ccccc} D_H^*(S) & \xrightarrow{q'^*} & D_H^*(G \times S) & \xrightarrow{Q_{p*}} & D^*(ind(S)) \\ {\scriptstyle i}\downarrow\wr & & {\scriptstyle i}\downarrow\wr & & \downarrow{\scriptstyle id} \\ D^*(Sh_H(S)) & \xrightarrow{q'^*} & D^*(Sh_H(G \times S)) & \xrightarrow{Rp_*^H} & D^*(ind(S)) \,. \end{array}$$

Here i is the equivalence of [BL, thm. 8.3.1], q'^* is the pullback functor induced by the exact functor of lemma 3.1.1(1), and Rp_*^H is the derived functor of the left exact functor p_*^H which describes the H-invariant subsheaf of the direct image sheaf p_*. The commutativity of the diagram follows from [BL, 8.4, cor. 8.4.2].

So in the situation of the case (3.) of the example, we have

$$\mathcal{F} \simeq Rp_*^H(q'^*\mathcal{G})$$

for some $\mathcal{G} \in ob(D^*(Sh_H(S))$. Since $\mathcal{F}$ is constructible, the same is true for $For(\mathcal{G})$, because by the description of ν^* in [BL, 2.6.3], $\mathcal{G}$ is isomorphic to the pullback of $Res_{H,G}(\mathcal{F})$ by the H-equivariant analytic map

$$S \to G \times_H S \;,\; s \mapsto (e,s) \,.$$

Assume now that the ground ring is a field k of characteristic 0. Then the functor p_*^H is exact for a **finite** map (i.e. proper with finite fibers). Moreover

$$Rq_! p_*^H \simeq p'^H_* Rq'_! \;,$$

since both functors can be calculated by an injective resolution (compare [Groth2, prop. 5.1.1]), and p_*^H applied to an injective sheaf in $Sh_H(S)$ is a flabby sheaf ([Groth2, cor. on p. 198]. A flabby resolution can be used for calculating the functor $Rq_!$). Altogether we get the isomorphism

$$Rq_!\mathcal{F} \simeq Rq_! p_*^H(q'^*\mathcal{G}) \simeq p'^H_* Rq'_!(q'^*\mathcal{G}) \,.$$

But the **projection formula** implies

$$Rq'_!(q'^*\mathcal{G}) \simeq (Rq'_!k_{G\times S}) \otimes_k \mathcal{G} .$$

Moreover, $Rq'_!k_{G\times S}$ has locally constant cohomology sheaves. Therefore $p'_*Rq'_!q'^*\mathcal{G}$ and also its direct summand $p'^H_*Rq'_!q'^*\mathcal{G}$ is constructible on S.

This finishes the discussion of our examples. For later use let us remark, that the above discussion implies for an equivariant sheaf $\mathcal{F} \in ob(Sh_G(X))$ also an important **stalk formula** (i.e. for $S = \{x\}$), in the general case of a **proper** action.

Assume that G is (topologically isomorphic to) a Lie group satisfying the condition (*+), and let H be a **compact** subgroup. Consider the following commutative diagram of equivariant maps and group homomorphisms:

$$\begin{array}{ccc} (G\times\{x\}, G\times H) & \xrightarrow{(q',\phi')} & (\{x\}, H) \\ \Big\downarrow{\scriptstyle (p',\psi)} & & \Big\downarrow{\scriptstyle (id,\psi')} \\ (G\times_{H_0}\{x\}, G\times H/H_0) & \xrightarrow{(q,\phi)} & (\{x\}, H/H_0) . \end{array}$$

The group homomorphisms are the obvious one and the action of $G \times H$ is given as before by the left multiplication of G and the twisted action of H. By [BL, 6.6] we get a commutative diagram:

$$\begin{array}{ccc} Sh_{G\times H}(G\times\{x\}) & \xleftarrow{Q^*_{q'}} & Sh_H(\{x\}) \\ \Big\uparrow{\scriptstyle Q^*_{p'}} & & \Big\uparrow{\scriptstyle Q^*_{id}} \\ Sh_{G\times H/H_0}(G\times_{H_0}\{x\}) & \xleftarrow{Q^*_q} & Sh_{H/H_0}(\{x\}) \end{array}$$

(and similarly for the bounded equivariant derived categories).

But if we apply the definition [BL, def. 6.5] of Q^*_{id} to the trivial resolutions

$$H\times\{x\} \to \{x\} \quad \text{and} \quad H/H_0\times\{x\} \to \{x\} ,$$

we get the equality of Q^*_{id} and id^* as defined in lemma 3.1.1(1). This is therefore an equivalence by example 3.1.2(2), and the inverse of the induction equivalence

$$\nu^* : Sh_G(ind(\{x\})) \to Sh_H(\{x\}) \simeq Sh_{H/H_0}(\{x\})$$

is given by the composition

$$Sh_{H/H_0}(\{x\}) \xrightarrow{Q^*_q} Sh_{G\times H/H_0}(G\times_{H_0}\{x\}) \xrightarrow{Q_{p*}} Sh_G(ind(\{x\})) ,$$

with $p : G\times_{H_0}\{x\} \to ind(\{x\})$ the quotient with respect to the free action of the finite group H/H_0 (which is equivariant with respect to the projection homomorphism $G\times H/H_0 \to G$).

So we can apply to this situation the same arguments as before. Assume that the ground ring $R = k$ is a field of characteristic 0. Then we get for

$$\mathcal{F} \in ob(Sh_G(X)),\ i \in \mathbb{N} \text{ and the projection } q : X \to X/G$$

the following **stalk formula** in $Sh(\{q(x)\}, k)$:

$$(R^i q_! \mathcal{F})_{q(x)} \simeq \Gamma^{H/H_0}\left(H^i_c(G/H_0, k) \otimes_k Res_{H,G}(\mathcal{F})_x\right), \tag{3.10}$$

with

$$H^i_c(G/H_0, k) \in ob(mod(k[H/H_0]))\ ,\ Res_{H,G}(\mathcal{F})_x \in ob(Sh_{H/H_0}(\{x\}, k))\ ,$$

and Γ^{H/H_0} the functor given by the H/H_0-invariants (where we consider on $\otimes_k$ the diagonal action of H/H_0). But $\otimes_k$ and Γ^{H/H_0} commute with (arbitrary) direct sums. By using the partition

$$G/H_0 = \cup_{g \in G/G_0}\ gG_0/H_0\ ,$$

one gets

$$H^i_c(G/H_0, k) \simeq H^i_c(G_0/H_0, k) \otimes_k k[G/G_0]\ ,$$

with the action of H/H_0 on the right side given by the natural homomorphism

$$k[H/H_0] \to k[G/G_0]\ .$$

Altogether, we get the **stalk formula**

$$(R^i q_! \mathcal{F})_{q(x)} \simeq H^i_c(G_0/H_0, k) \otimes_k \Gamma^{H/H_0}\left(\ k[G/G_0] \otimes_k Res_{H,G}(\mathcal{F})_x\ \right). \tag{3.11}$$

And this equation holds in $Sh_G(\{x\}, k) \simeq mod(k[G/G_0])$.

In the o-minimal context, H_0 and G_0/H_0 are also definable so that the k vector space $H^i_c(G_0/H_0, k)$ is **finite dimensional**. Assume only that G is a Lie group. Since G_0 is a connected Lie group, it is homeomorphic to a product $\mathbb{R}^p \times K$. Here K is a maximal compact subgroup of G_0 containing H_0. Therefore

$$H^i_c(G_0/H_0, k) \simeq H^{i-p}(K/H_0, k)$$

is also in this case a **finite-dimensional** k vector space.

If we assume that G has only **finitely** many connected components, then also G is homeomorphic to a product $\mathbb{R}^p \times K$, with K as a maximal compact subgroup of G containing H_0. In this case one also has for

$$q' : G/H_0 \simeq G \times_{H_0} \{x\} \to \{x\}$$

the **projection formula**

$$R^i q'_*(q'^* \mathcal{G}) \simeq (R^i q'_* k_{G/H_0}) \otimes_k \mathcal{G}\ ,$$

because these direct image sheaves are isomorphic to the corresponding direct image sheaves of the **proper** map $K/H_0 \to \{x\}$. But the isomorphisms of corollary 3.1.1 imply also the same isomorphisms in the **equivariant** context, because these morphisms of functors are defined by natural **adjunction** properties, which "lift up" into the equivariant context. As before, one gets therefore in the situation of example 3.1.7(2) for $\mathcal{F} \in ob(Sh_G(X))$ the **stalk formula**:

$$(R^i q_* \mathcal{F})_{q(x)} \simeq H^i(G_0/H_0, k) \otimes_k \Gamma^{H/H_0}\big(k[G/G_0] \otimes_k Res_{H,G}(\mathcal{F})_x \big) . \tag{3.12}$$

And this equation holds again in $Sh_G(\{x\}, k) \simeq mod(k[G/G_0])$, with

$$H^i(G_0/H_0, k) \simeq H^i(K_0/H_0, k)$$

a **finite-dimensional** k vector space (for k a field of characteristic 0).

Let us finally consider the case of a **discrete** group G, with commutative ground ring R and $\mathcal{F} \in ob(Sh_G(X, R))$ (as in [Groth2]). Let H be a compact and therefore **finite** subgroup. If one applies the **projection formula** directly to the map

$$q : G/H \times \{x\} \simeq G(x) \to \{x\} = \{x\}/H ,$$

one gets the **stalk formula** in $Sh_G(\{x\}, R) \simeq mod(R[G])$:

$$(q_! \mathcal{F})_{q(x)} \simeq \Gamma_c(G/H, R) \otimes_R Res_{H,G}(\mathcal{F})_x \simeq R[G] \otimes_{R[H]} Res_{H,G}(\mathcal{F})_x . \tag{3.13}$$

Note that $q_!$ is in this case exact.

Consider especially a **proper** G-space X and a complex

$$\mathcal{F}' \in ob(D^b(X/G)) .$$

Then we can apply this formula to

$$q^* \mathcal{F}' \in ob(D^b_G(X)) ,$$

and one has

$$Res_{H,G}(q^* \mathcal{F}')_x \simeq \mathcal{F}'_{q(x)}$$

with the **trivial** action of $H := G_x$.

Let us now explain theorem 3.0.1(1) in the case of a **discrete** group by some examples based on the **stalk formula** (3.13).

Fix an o-minimal structure on $(\mathbb{R}, +, \cdot)$, and consider the action of a **finite** group G by **definable automorphisms** on a definable space X. Then X/G is a definable space and the projection $q : X \to X/G$ is a **finite** definable map. Let R be a commutative noetherian ring, denote by fin the **abelian category** of finitely generated $R[G]$-modules, and by T the **strict full triangulated** subcategory of bounded complexes of $R[G]$-modules with finitely generated cohomology. Let μ be the universal **additive** function on fin (or T), which associates to such a module (or complex) M its class $\{M\}$ in the **Grothendieck group**

$$K(fin) = K(T) =: Rep(G, R) \,.$$

Consider an equivariant complex $\mathcal{F} \in ob(D^b_G(X))$, whose stalks have **finitely generated** cohomology as R-modules. Then $Rq_!\mathcal{F}$ is constructible, and the cohomology modules of its stalks belong to fin. Choose an **adapted finite definable cell decomposition** of X/G. Then theorem 3.0.1(1) and equation (3.13) imply the formula:

$$\{R\Gamma_c(X, \mathcal{F})\} = \sum_j (-1)^{n_j} \cdot \{R[G] \otimes_{R[G_{x_j}]} Res_{G_{x_j},G}(\mathcal{F})_{x_j}\} \in Rep(G, R) \,, \quad (3.14)$$

with $q(x_j)$ a point in the cell U_j of dimension n_j.

Example 3.1.8. *Take especially for $\mathcal{F}$ the (equivariant) constant sheaf R_X. For a* **conjugacy class** *(H) of a subgroup H of G, let $X_{(H)}$ denote the set of points in X, whose stabilizer belongs to (H) (we leave it as an exercise for the reader, to check that $X_{(H)}$ is a locally closed definable G-invariant subset of X). Choose a finite definable cell decomposition of X/G such that each $q(X_{(H)})$ is a union of cells. Then one gets the formula*

$$\{R\Gamma_c(X, R)\} = \sum_{(H)} \chi_c(X_{(H)}/G) \cdot \{R[G] \otimes_{R[H]} R\} \in Rep(G, R) \,, \qquad (3.15)$$

with $\chi_c(\cdot)$ the Euler characteristic of $H^*_c(\cdot, \mathbb{Q})$.

Remark 3.1.4. For $R = \mathbb{Q}$ (or any field of characteristic 0), this formula is true in a much more general situation (compare [Ver]). But the above generalization to a commutative noetherian ring needs **not** to be true.

Fix an analytic geometric category, and consider the **proper** action of a **discrete** group G by **real analytic automorphisms** on a real analytic manifold M. Let X be a closed definable invariant subset. Then X/G can be realized as a closed definable subset of a real analytic manifold. Let $R = k$ be a field of characteristic 0. Consider an equivariant constructible complex $\mathcal{F} \in ob(D^b_G(X))$, whose stalks have **finite-dimensional** cohomology as k-vector spaces. Then $Rq_!\mathcal{F}$ is constructible by example 3.1.7(3).

Denote by T the **triangulated** category of **perfect** complexes of $k[G]$-modules, i.e. of complexes isomorphic to a bounded complex of finitely generated projective $k[G]$-modules. Let μ be the universal **additive** function on T, which associates to such a complex P its class $\{P\}$ in the Grothendieck group $K(T)$. If H is a **finite** group, then each complex in $ob(D^b(k[H]))$, whose underlying cohomology spaces are **finite dimensional** over k, is **perfect** as a complex of $k[H]$-modules (since k a field of characteristic 0). The **stalk formula** (3.13) implies, that the **cohomology sheaves**

$$h^i(Rq_!\mathcal{F}) \simeq Rq_!(h^i\mathcal{F}) \qquad (i \in \mathbb{Z})$$

have **perfect** stalks as $k[G]$-modules. Therefore

$$(Rq_!\mathcal{F})_y \in ob(T) \quad \text{for all } y \in X/G\,.$$

Assume now that X/G is **compact**, and choose an **adapted finite definable cell decomposition** of X/G. Then theorem 3.0.1(1) and equation (3.13) imply the formula:

$$\{R\Gamma_c(X,\mathcal{F})\} = \sum_j (-1)^{n_j} \cdot \{k[G] \otimes_{k[G_{x_j}]} Res_{G_{x_j},G}(\mathcal{F})_{x_j}\} \in K(T)\,, \qquad (3.16)$$

with $q(x_j)$ a point in the cell U_j of dimension n_j.

Let us finish this section with the following "real analytic counterpart" of [BL, thm. 9.1]:

Proposition 3.1.1. *Consider the situation of example 3.1.7(3), i.e. all stabilizers G_x for $x \in X$ are finite, and the ground ring R is a field of characteristic 0. The quotient map $f : X \to f(X)$ is a ϕ-map in the sense of [BL, def. 6.0] with respect to group homomorphism $G \to \{e\}$. Therefore one has the* **general inverse and direct image functors** *([BL, sec. 6]):*

$$Q_f^* : D^+(f(X)) \to D_G^+(X) \text{ and } Q_{f*} : D_G^+(X) \to D^+(f(X))\,.$$

1. *The functor Q_{f*} preserves the natural t-structure, i.e. it maps $Sh_G(X)$ into $Sh(f(X))$. In particular Q_{f*} preserves the bounded category D^b.*
2. *$Q_{f*}Q_f^* \simeq id$.*
3. *Suppose $For(\mathcal{F})$ is constructible on X ($\mathcal{F} \in ob(D_G^+(X))$). Then $Q_{f*}\mathcal{F}$ is* **constructible** *on $f(X)$.*

Proof. All results are local on $f(X)$. So we can use a real analytic slice S (compare also with [BL, proof of thm. 9.1]). Consider the commutative diagram

$$\begin{array}{ccc} S & \xrightarrow{\nu} & X = G \times_H S \\ \downarrow{\scriptstyle p'} & & \downarrow{\scriptstyle f} \\ S/H & \xrightarrow{\sim} & f(X), \end{array}$$

with H a finite group and ν the H-equivariant embedding

$$S \to G \times_H S\,,\ s \mapsto (e,s)\,.$$

Then $Q_{\nu*}$ is the inverse to the induction equivalence

$$Q_\nu^* = \nu^* \ : \ D_G^+(X) \to D_H^+(S) \ .$$

But $Q_{p'*} \simeq Q_{f*} \circ Q_{\nu*}$ ([BL, 6.12.2]). So one gets $Q_{f*} \simeq Q_{p'*} \circ \nu^*$, and it is enough to prove the theorem for H and $Q_{p'*}$. The statements (1.) and (2.) follow from [BL, thm.8.6.1]. The statement (3.) follows from the identification

$$Q_{p'*} \simeq Rp'^H_* \simeq p'^H_* \ . \qquad \square$$

Remark 3.1.5. 1. By a result of Illmann [Illma], a proper **smooth** action of a Lie group G on a **smooth** manifold M is isomorphic to a proper real analytic action on a real analytic manifold. Therefore all results of the real analytic context apply also to the **smooth context**, if we only consider complexes of sheaves, whose cohomology sheaves are locally constant on the submanifolds

$$M_{(H)} := \{x \in M |\ G_x = gHg^{-1} \text{ for some } g \in G\}$$

corresponding to the **conjugacy classes** (H) of the stabilizers (e.g. a locally constant sheaf). Note that in the real analytic context the $M_{(H)}$ are subanalytic subsets of M ([Kan, lem. 7.6]).

2. Suppose in addition that the action is **almost free**, i.e. all stabilizers G_x for $x \in M$ are **finite**. Then M/G is an "orbifold" and the category $Sh_G(M)$ is equivalent to the category of sheaves on this "orbifold" ([MoePr]). Moreover, $R\Gamma(M/G, Q_{f*}\mathcal{F})$ calculates the corresponding cohomology. We leave it to the reader to prove similar "localization results" for

$$R\Gamma(X/G, Q_{f*}\mathcal{F}) \quad (\ \mathcal{F} \in ob(D_G^+(X))\ )\ .$$

3.2 Localization results for additive functions

Let us now explain some localization results for **additive** functions. Fix a topological group G and topological space X with a continuous action

$$a : G \times X \to X\ .$$

Denote by X^G the closed fixed point set of the action of G on X, and by j the open inclusion $X\backslash X^G \to X$.

Theorem 3.2.1. 1. *Let $\mathcal{F}$ be a* **monodromic** *complex on X. Assume that X/G has a* **finite cellular decomposition** *and $Rq_!j_!j^*\mathcal{F}$ is* **constructible** *with respect to this decomposition. Consider a strict full triangulated subcategory T of $D(R)$, and an additive function μ on T (with values in the abelian group A). Assume the* **stalk condition** *$\mathcal{F}_x \in ob(T)$ for all $x \in X\backslash X^G$.*

(a) *Let* $G \simeq (S^1)^n$ *be a compact torus. Then*

$$R\Gamma_c(X, X^G; \mathcal{F}) \in ob(T) \text{ and } \mu(R\Gamma_c(X, X^G; \mathcal{F})) = 0 .$$

(b) *Let* G *be finite. Then* $R\Gamma_c(X, X^G; \mathcal{F}) \in ob(T)$. *Let the index of all stabilizers* G_x $(x \in X \backslash X^G)$ *be divisible by* $d \in \mathbb{N}$. *Then*

$$\mu(R\Gamma_c(X, X^G; \mathcal{F})) \text{ is also divisible by } d ,$$

i.e. lies in the image of the map $A \to A$, $a \mapsto d \cdot a$.

2. *Make the same assumption as before, with a* **finite good cellular stratification** *instead of a decomposition.*

(a) *Let* $G \simeq (S^1)^n$ *be a compact torus. Then*

$$R\Gamma(X, X^G; \mathcal{F}) \in ob(T) \text{ and } \mu(R\Gamma(X, X^G; \mathcal{F})) = 0 .$$

(b) *Let* G *be finite. Then* $R\Gamma(X, X^G; \mathcal{F}) \in ob(T)$. *Let the index of all stabilizers* G_x $(x \in X \backslash X^G)$ *be divisible by* $d \in \mathbb{N}$. *Then*

$$\mu(R\Gamma(X, X^G; \mathcal{F})) \text{ is also divisible by } d .$$

3. *Do not assume that* $\mathcal{F}$ *is monodromic. Then* (1.b) *and* (2.b) *remain true, if we only assume that the* **function**

$$X \backslash X^G \to A , \; x \mapsto \mu(\mathcal{F}_x)$$

is G-**invariant**. *Similarly in the case* (1.a) *and* (2.a), *if we work in the* **o-minimal context** *of example 3.1.6.*

Proof. In the theorem G is especially a compact group. Therefore the projection $q : X \to X/G$ is proper. This implies

$$R\Gamma_?(X, X^G; \mathcal{F}) \simeq R\Gamma_?(X/G, Rq_! Rj_! j^* \mathcal{F}) \quad (? = ! \text{ or } *) .$$

Moreover, $(Rq_! j_! j^* \mathcal{F})_x = 0$ for $y \in q(X^G)$ and

$$(Rq_! j_! j^* \mathcal{F})_y = R\Gamma(\{q = y\}, \mathcal{F}) \text{ for } y \notin q(X^G) .$$

Consider a point $y = q(x) \notin q(X^G)$.

(a.) Assume G is a compact torus $(S^1)^n$. Then a monodromic complex has **locally constant** cohomology sheaves on the orbit $G(x) \simeq G/G_x$, and the function

$$x' \mapsto \mu(\mathcal{F}_{x'}) \text{ is constant on } G(x) .$$

Let us denote this constant by μ. Now G/G_x is a connected abelian Lie group. So it is a torus of positive dimension. Especially, it is (homeomorphic to) a compact semialgebraic set with $\chi(G(x)) = 0$. Therefore

$$R\Gamma(G(x), \mathcal{F}) \in ob(T) \quad \text{and} \quad \mu(R\Gamma(G(x), \mathcal{F})) = \mu \cdot \chi(G(x)) = 0 \, .$$

This follows from theorem 3.0.1(1) applied to the case of the trivial group $\{e\}$.

The same conclusion can also be applied in the case (3.) of the theorem, because in that case $\mathcal{F}|G(x)$ is constructible on the definable set $G(x)$ (compare chapter 2). So the claim of the theorem follows from theorem 3.0.1 (for the trivial group $\{e\}$). Note that the full subcategory T' of complexes M in T, with $\mu(M) \in A'$ for a subgroup A' of A (e.g. $\{0\}$ or $d \cdot A$), is also a strict full triangulated subcategory !

(b.) Assume G is a finite group. Then $G(x)$ is a finite set, especially

$$R\Gamma(G(x), \mathcal{F}) \simeq \oplus_{x' \in G(x)} \mathcal{F}_{x'} \in ob(T) \, .$$

If $\mathcal{F}$ is a monodromic complex, we have

$$\mathcal{F}_{x'} \simeq \mathcal{F}_x \text{ for all } x' \in G(x) \, .$$

The function $x' \mapsto \mu(\mathcal{F}_{x'})$ is therefore constant on $G(x)$. By assumption this is also true in the case (3.). So we get

$$\mu(R\Gamma(G(x), \mathcal{F})) \;=\; \sharp G(x) \cdot \mu(\mathcal{F}_x) \, .$$

By assumption, $\sharp G(x)$ is divisible by d and the claim follows as before from theorem 3.0.1 (for the trivial group $\{e\}$). □

We can apply this theorem to our examples in the following cases:

- In the case (1.) and (3.b) to the example 3.1.4, if the quotient complex X/G is **compact**.
- To the example 3.1.5, if the base space has an **appropriate cell decomposition or stratification**.
- **Always** in the o-minimal context of example 3.1.6.
- To the example 3.1.7, if the quotient X/G (or equivalently X) is **compact**.

A more interesting result in the case of example 3.1.7 is the following

Corollary 3.2.1. *Let G' be a* **closed normal** *subgroup of the Lie group G. Assume G has only* **finitely** *many connected components, and consider the situation of example* 3.1.7, *with $X/G \simeq f(X)$* **compact**. *Let $\mathcal{F}$ be a* **constructible G-monodromic** *complex on X, with*

$$\mathcal{F}_x \in ob(T) \text{ for all } x \in X \backslash X^{G'}$$

(and T, μ as in theorem 3.2.1*).*

1. *$R\Gamma_c(X, X^{G'}, \mathcal{F})$ and $R\Gamma(X, X^{G'}; \mathcal{F}) \in ob(T)$.*
2. *Let $G' \simeq (S^1)^n$ be a compact torus. Then*

$$\mu(\ R\Gamma_c(X, X^{G'}; \mathcal{F})\ ) \ = \ 0 \ = \ \mu(\ R\Gamma(X, X^{G'}; \mathcal{F})\ ) \, .$$

3. *Let G' be finite. Let the index of all stabilizers G'_x ($x \in X \backslash X^{G'}$) be divisible by $d \in \mathbb{N}$. Then*

$$\mu(\ R\Gamma_c(X, X^{G'}; \mathcal{F})\ ) \text{ and } \mu(\ R\Gamma(X, X^{G'}; \mathcal{F})\ )$$

are also divisible by d.

Proof. $X^{G'}$ is G-invariant, since G' is a normal subgroup. Denote the open inclusion $j : X \backslash X^{G'} \to X$. Then $Rj_! j^* \mathcal{F}$ is also constructible and G-monodromic. By example 3.1.7, we already know that

$$Rf_? Rj_! j^* \mathcal{F} \quad (? = ! \text{ or } *)$$

is constructible on $f(X)$. But $f(X)$ is a compact definable set. So we only have to show, that the stalks of this complex belong to $ob(T)$, and have the properties stated above.

That the stalks belong to $ob(T)$, follows already from corollary 3.1.1 and the proof of the example 3.1.7: they are isomorphic to the cohomology (with compact support) over a relatively compact definable subset of the fiber. For the other properties, we factor the projection $f : G(x) \to \{f(x)\}$ as the composition of

$$q : G(x) \to G(x)/G' \text{ and } p : G(x)/G' \to \{f(x)\} \, .$$

Since

$$\mathcal{H} := Rq_! \big((Rj_! j^* \mathcal{F}) | G(x) \big) = Rq_* \big((Rj_! j^* \mathcal{F}) | G(x) \big)$$

is G-monodromic on $G(x)/G'$ (by lemma 3.1.1(3)), it has locally constant cohomology sheaves. This implies as above

$$R\Gamma_?(G(x)/G', \mathcal{H}) \simeq R\Gamma_?(K, \mathcal{H}) \quad (? = ! \text{ or } *)$$

for some relatively compact definable subset K of $G(x)/G'$. Then we only have to apply theorem 3.2.1 to $q^{-1}(K)$. □

Let us now illustrate theorem 3.2.1 and corollary 3.2.1 by some examples.

We begin with the simplest case of a **constant** sheaf k_X (which is always monodromic), with $R = k$ a field. Denote by T the complexes M with **finite-dimensional** cohomology $H^*(M)$ and choose for μ the usual **Euler characteristic** χ. Moreover, we assume that in the case of a finite group G is a p-**group** for some prime number p (so that one can take $d = p$ in the theorem).

In this case we get in our examples, that the relative cohomology

$$H^*_?(X, X^G, k) \quad \text{or} \quad H^*_?(X, X^{G'}, k)$$

with closed or compact support is **finite dimensional** and has **vanishing** Euler characteristic (mod p).

This result is for compact G-complexes (and often also in a more general situation) classical and well known (see [tDieck, thm. 6.7, p. 220] or [Bre, thm. 4.3, 7.10, 10.9]). Other references are [EvMi, prop. 1.2] and [LawYau, Thm. 4.1, 4.7], where also very interesting applications are given.

Of course, the power of theorem 3.2.1 and corollary 3.2.1 lies in the following facts:

- It applies to constructible sheaf complexes.
- We can use any additive function $\mu : T \to A$.
- In the o-minimal context, we need no compactness or finite orbit-type assumptions.

Especially interesting is the **complex algebraic** context. In this case we can deduce complex algebraic results by using "real methods" (e.g. the o-minimal structure of semialgebraic sets).

Consider a complex algebraic action of a finite group or a complex torus G on a complex algebraic variety X (i.e. on the set of complex points of a seperated scheme of finite type over $spec(\mathbb{C})$). In the case of a complex torus consider a maximal compact subgroup K of G. This is a compact real torus, which is **Zariski dense** in G. Especially, we have $X^G = X^K$. For a finite group let $K := G$. Then X, the group K and the induced action is definable with respect to the o-minimal structure of **semialgebraic** sets.

Consider the derived category $D^b_{G,c}(X)$ of G-equivariant **complex algebraically constructible** complexes, i.e. $For(\mathcal{F})$ is constructible with respect to a finite filtration by closed algebraic subsets (compare [BL, 4.2]). Here we consider as a ground ring $R = k$ a field, and constructibility includes the assumption, that the stalk cohomology is finite dimensional. The underlying complex $For(\mathcal{F})$ of $\mathcal{F} \in ob(D^b_{G,c}(X))$ is then also semialgebraically constructible and monodromic. So we can apply theorem 3.2.1 (with $\mu = \chi$ the usual Euler characteristic):

Corollary 3.2.2. *Let $\mathcal{F} \in ob(D^b_{G,c}(X))$ be given.*

1. *Let G be a complex torus. Then*

$$\chi(R\Gamma_c(X, X^G; \mathcal{F})) = 0 = \chi(R\Gamma(X, X^G; \mathcal{F})) .$$

2. *Let G be a finite group. Let the index of all stabilizers G_x ($x \in X \backslash X^G$) be divisible by $d \in \mathbb{N}$. Then*

$$\chi(R\Gamma_c(X, X^G; \mathcal{F})) \text{ and } \chi(R\Gamma(X, X^G; \mathcal{F}))$$

are divisible by d.

As an example for $\mathcal{F}$, one can take any G-equivariant perverse sheaf (e.g. the intersection cohomology complex [BL, 5.2]). The first part of the corollary is a generalization of [J2, thm. 3.6, cor. 3.7]. Assume X is smooth and we take as a ground ring $k = \mathbb{C}$. Then the **equivariant Riemann-Hilbert correspondence** ([BL, 4.2])

$$DR : D^b_{G,rh}(X) \to D^b_{G,c}(X)$$

translates these results into formulae for the deRham cohomology of **equivariant** (complexes of) **regular holonomic D-modules**. Especially, we get a (real) "topological" proof of a generalization of [J2, thm. 2.14] (we can allow complexes and do not need the assumption X projective. Of course, Joshua argues just in the opposite way and gave an "analytic" proof of these topological results).

Our proof of part (2.) of the corollary uses only the fact, that the finite group G operates **semialgebraically** on X. So it applies in the case of a complex algebraic variety X, which is already defined over $spec(\mathbb{R})$, also to the group G generated by the **antiholomorphic involution** σ corresponding to this **real structure**. In this case, $X^G = X(\mathbb{R})$ is just the set of **real points** and for the constant sheaf k_X one gets the well-known **congruences** modulo 2:

$$\chi(R\Gamma_c(X(\mathbb{C}), k)) \equiv \chi(R\Gamma_c(X(\mathbb{R}), k)) \quad (mod\ 2)\,,$$

and

$$\chi(R\Gamma(X(\mathbb{C}), k)) \equiv \chi(R\Gamma(X(\mathbb{R}), k)) \quad (mod\ 2)\,.$$

Another important example of the same type deals with the **Milnor fibration** of a **real analytic function**. Consider a real analytic subset X of an open subset of $\mathbb{R}^n$ and a real analytic function $f : X \to \mathbb{R}$.

Denote by $f_C : X_C \to \mathbb{C}$ a complexification of X and f. Fix a point $x \in X$ and assume for simplicity $f_C(x) = 0$. Then a **complex Milnor fiber** of f_C in x is given as

$$M_{f,C} := \{x' \in X_C |\ f_C(x') = z,\ ||x|| \leq \delta\} \quad (0 < |z| << \delta << 1)\,.$$

Compare for example with lemma 1.1.1 and example 1.1.3 for the fact, that the **cohomology** of $M_{f,C}$ is independent of all choices. By the same reference, one has for the real function f a **right (or left) real Milnor fiber** in x:

$$M^*_f := \{x' \in X |\ f(x') = *z,\ ||x|| \leq \delta\} \quad (0 < z << \delta << 1)\,,$$

with $* = +, -$ such that the **cohomology** of M^*_f is also independent of all choices (for $* = +, -$).

Of course, the cohomology of the **left** Milnor fiber is in general **different** from the cohomology of the **right** Milnor fiber. But **both are the fixed point set** of the complex conjugation acting on a corresponding **complex** Milnor fiber. So one gets in both cases $* = +, -$ the following congruences:

$$\chi(R\Gamma(M_{f,C}, k)) \equiv \chi(R\Gamma(M_f^*, k)) \quad (mod\ 2)\ . \tag{3.17}$$

In particular

$$\chi(R\Gamma(M_f^+, k)) \equiv \chi(R\Gamma(M_f^-, k)) \quad (mod\ 2)\ .$$

This is from the "topological viewpoint" surprising and depends on the existence of a "complexification". If one applies this to the function $f = r$ which is given as the distance to the point x, one gets the classical result of Sullivan [Su], that the Euler characteristic

$$\chi(R\Gamma(M_r^+, k))$$

of a **link** M_r^+ of X at x is **even** (since $M_r^- = \emptyset$). Compare also with [McPa] for more important relations of this type.

3.3 Localization results for Grothendieck groups and trace formulae

3.3.1 Grothendieck groups

Next we deal with a "localization theorem" for the **Grothendieck group** of equivariant constructible sheaves. For the definition and basic results about the Grothendieck group $K(Ab)$ of an abelian category Ab, or $K(T)$ of a triangulated category T compare with [SGA5, exp. VIII].

For simplicity, we assume that:

- G is a countable **discrete** group, or
- G is (topologically isomorphic to) a real analytic **Lie group** satisfying the condition (*+).

In the first case we use $D^b_G(X) := D^b(Sh_G(X))$ and the corresponding functors as defined by [Groth2]. In the second case we use the equivariant derived category $D^b_G(X)$ and the corresponding functors of Bernstein-Lunts [BL].

Moreover, we assume in this section that the base ring R is a **commutative noetherian ring of finite homological dimension** so that we have in both cases above the bifunctor

$$\otimes^L_R : D^b_G(X) \times D^b_G(X) \to D^b_G(X)\ .$$

Recall, that in the second case the triangulated category $D^b_G(X)$ is in general **not** the derived category of the abelian category $Sh_G(X)$. But it has a t-**structure** with a corresponding **cohomology functor**

$$h : D^b_G(X) \to Sh_G(X) .$$

This induces an isomorphism of **Grothendieck groups**

$$K(D^b_G(X)) \simeq K(Sh_G(X))$$

by the following simple abstract

Lemma 3.3.1. *Let T be a triangulated category with a* **nondegenerate** t-**structure** *and denote by $h : T \to Ab$ the corresponding* **cohomology functor**, *with Ab the* **heart** *of the t-structure. Denote by T^b the full subcategory of T, whose objects M are defined by the condition*

$$h^i(M) := h(M[i]) = 0 \quad \textit{for } |i| >> 0 .$$

Then T^b is a **strict full triangulated subcategry** *and one has a natural isomorphism of abelian groups $K(T^b) \simeq K(Ab)$.*

Proof. T^b is a strict full triangulated subcategory, since h is a cohomology functor, i.e. a distinguished triangle induces under h a long exact cohomology sequence. So one gets a well-defined group homomorphism

$$\alpha : K(T^b) \to K(Ab) , \ \{M\} \mapsto \sum (-1)^i \{h^i(M)\} .$$

The sum is finite by the definition of T^b. But Ab is an abelian **admissible** subcategory of T and therefore also of T^b by [BBD, thm. 1.3.6]. Recall that admissible means, that any short exact sequence in Ab is induced from a distinguished triangle in T by forgetting the map of degree 1. Therefore one also has a well-defined group homomorphism

$$\beta : K(Ab) \to K(T^b) , \ \{M\} \mapsto \{M\}.$$

Then one has $\alpha \circ \beta = id$, since $M \in ob(Ab)$ implies $h^i(M) = 0$ for $i \neq 0$ by [BBD, 1.3.4], or in our case [BBD, prop. 1.3.7]. Since the t-structure is nondegenerate, one has by definition for $M \in ob(T)$ the implication $h^i(M) = 0$ for all i implies $M = 0$ ([BBD, prop. 1.3.7]). Then one shows $\beta \circ \alpha(M) = M$ by induction over

$$l(M) := \sharp\{i \in \mathbb{Z} |\, h^i(M) \neq 0\} .$$

For the induction step one uses the distinguished triangle

$$h^i(M)[-i] \longrightarrow \tau^{\geq i} M \longrightarrow \tau^{>i} M \xrightarrow{[1]} ,$$

with $\tau^{\geq i}$ and $\tau^{>i}$ the corresponding truncation functors of the t-structure (compare [BBD, prop. 1.3.3(ii)] and [KS, proof of prop. 10.1.10]). This finishes the proof. For the classical counterpart of this lemma see [SGA5, cor. on p. 359]. □

The natural t-structure on $D^b_G(X)$ is of course nondegenerate. Moreover, in the **complex algebraic context** this applies also to the perverse t-structure [BL, 5.1] (with respect to middle perversity, where the ground ring is a field of characteristic zero, and constructibility includes the assumption, that all stalks are finite dimensional). Especially, the Grothendieck groups of the abelian categories of **equivariant perverse sheaves** $Perv_G(X)$ and **equivariant algebraically constructible sheaves** $Sh_{G,c}(X)$ are equivalent:

$$K(Perv_G(X)) \simeq K(D^b_{G,c}(X)) \simeq K(Sh_{G,c}(X)) .$$

See [ABV] for a very interesting application of this isomorphism of Grothendieck groups.

Recall that a complex $M \in ob(D(R))$ is called **perfect**, if it is isomorphic to a bounded complex of finitely generated projective R-modules. Since we assume that R is a noetherian ring of finite homological dimension (e.g. a field or $\mathbb{Z}$), this is equivalent to the condition $M \in ob(D^b(R))$ and the cohomology of M is finitely generated. The strict full triangulated subcategory $perf$ of perfect complexes in $D(R)$ is stable under the tensor product $\otimes^L_R$ ([SGA5, p. 368]). Its Grothendieck group

$$K^0(R) := K(perf)$$

is therefore a **commutative ring** (with unit given by the class of R).

Consider a second strict full triangulated subcategory T of $D^b(R)$ such that $\otimes^L_R$ maps $T \times perf$ into T. This is for example always the case if T is stable by direct summands or R is a field ([SGA5, p. 378, prop. 3.6]). For example, we can take $T = perf$. Then $\otimes^L_R$ induces on $K(T)$ the structure of a $K^0(R)$-module.

We extend these definitions now into the context of equivariant constructible sheaves, where the **constructibility** of $forget(\cdot)$ is defined with respect to

1. a filtration by closed subsets,
2. an o-minimal structure, or
3. an analytic geometric category.

Denote by $D^b_{G,T}(X)$ the strict full triangulated subcategory of $D^b_G(X)$, whose objects $\mathcal{F}$ are defined by the condition:

- $For(\mathcal{F})$ is constructible.
- $For(\mathcal{F})_x \in ob(T)$ for all $x \in X$.

We denote the corresponding **Grothendieck group** by $K^G_T(X)$. Since R is a noetherian ring, we can define in the same way $Sh_{G,fin}(X)$ as the **abelian category** of **equivariant constructible sheaves**, whose stalks are finitely generated. By assumption, R is also of finite homological dimension so that lemma 3.3.1 implies the equality

$$K^G_{perf}(X) \simeq K(Sh_{G,fin}(X)) .$$

Especially for X a point, this is the Grothendieck group

$$Rep(G/G_0, R) := K^G_{perf}(\{pt\}) = K_{fin}(mod(R[G/G_0]))$$

of the abelian category of $R[G/G_0]$-modules, whose underlying R-module is finitely generated. This is the **representation ring** of G/G_0 over R.

We recall now some **functorial properties** of these groups in the **o-minimal or analytic geometric context**, which are similar to other equivariant theories. For example equivariant algebraic K-theory [ChGi, chapter 5], equivariant topological K-theory [Seg, AtSeg] or equivariant D-modules [J2, J3, Saito]).

- The bifunctor

$$\otimes^L_R : D^b_{G,T}(X) \times D^b_{G,perf}(X) \to D^b_{G,T}(X)$$

  induces on $K^G_T(X)$ the structure of a unitary module over the commutative ring with unit $K^G_{perf}(X)$.

- A definable continuous G-equivariant map $f : X \to Y$ induces a pullback

$$f^* : D^b_{G,T}(Y) \to D^b_{G,T}(X) ,$$

  commuting with $\otimes^L_R$ and functorial in f (i.e. $(f \circ g)^* \simeq g^* \circ f^*$). This induces a ring (or module-) homomorphism

$$f^* : K^G_{perf}(Y) \to K^G_{perf}(X) \quad \text{or} \quad f^* : K^G_T(Y) \to K^G_T(X) ,$$

  which is also **functorial** in f. Especially, $K^G_T(X)$ is a $Rep(G/G_0, R)$-module (by the constant map to a point).

- A (proper) definable continuous G-equivariant map f in the o-minimal (analytic geometric) context induces a pushdown

$$Rf_! : D^b_{G,T}(X) \to D^b_{G,T}(Y) ,$$

  functorial in f (i.e. $R(f \circ g)_! \simeq Rf_! \circ Rg_!$). For the following properties one only uses, that $Rf_!$ maps $D^b_{G,T}(X)$ into $D^b_{G,T}(Y)$ (e.g. in the context of example 3.1.7(2)). Then one has a **projection formula**

$$Rf_!(\cdot \otimes^L_R f^*(\cdot)) \simeq Rf_!(\cdot) \otimes^L_R (\cdot) : D^b_{G,T}(X) \times D^b_{G,perf}(Y) \to D^b_{G,T}(Y) .$$

  This induces a homomorphism

$$f_! : K^G_T(X) \to K^G_T(Y) ,$$

  which is a **module homomorphism** over $K^G_{perf}(Y)$, and therefore also over $Rep(G/G_0, R)$.

- In the o-minimal context one has for a group G and such a map f also the pushdown
$$Rf_* : D^b_{G,T}(X) \to D^b_{G,T}(Y),$$
functorial in f. This induces a homomorphism
$$f_* : K^G_T(X) \to K^G_T(Y),$$
which is a **module homomorphism** over $Rep(G/G_0, R)$.

- The restriction functor
$$Res^G_H : D^b_{G,T}(X) \to D^b_{H,T}(X)$$
induces a **restiction map**
$$Res^G_H : K^G_T(X) \to K^H_T(X),$$
and all the above morphisms (or homomorphisms) **commute** with this restriction functor (map).

Remark 3.3.1. 1. In most equivariant theories one has a **pushdown** only for a **proper** morphism. So the general pushdown $f_!, f_*$ in the o-minmal context is not typical (but they exist in the algebraic context for equivariant holonomic D-modules).

2. The **projection formula** holds in the case of a discrete group by the same proof as in the non-equivariant situation. In the case of the equivariant derived category of [BL], it follows from the definition of the functors in terms of acyclic resolutions, and the fact that the isomorphism of the projection formula (in the non-equivariant situation) commutes with pullbacks (as explained in section 1.4).

3. That in the o-minimal context f_* is **linear** over $Rep(G/G_0, R)$, follows from the **Künneth isomorphism** of corollary 2.0.4(1) applied to the map $f \times id$, with id the identity on a point. This isomorphism lifts up into the **equivariant** context as before. The Künneth morphism is defined by some **adjunction** properties (see section 1.4), which also hold in the equivariant context. Then one shows that the cone of this morphism is zero, which can be checked after applying the forgetful functor For.

4. That these morphisms commute with the restriction functor, follows from [BL, thm. 3.4.1] (or from the methods of [Groth2] in the case of a discrete group).

Another standard property of an equivariant theory is an **induction equivalence**. Let us explain this in our context.

Let $H \subset G$ be a **closed definable** subgroup. Consider an H-space X and the **induced** G-space

$$ind(X) = G \times_H X = (G \times X)/H \ ,$$

with the usual twisted action of H on $G \times X$. By [BL, thm2.6.2] one has an **equivalence**

$$\nu^* : D^b_G(ind(X)) \to D^b_H(X) \ ,$$

with ν the H-equivariant embedding

$$X \to G \times_H X \ , \ x \mapsto (e, x) \ .$$

More precisely, this is given by the composition of For^G_H and

$$\nu^* : D^b_H(ind(X)) \to D^b_H(X) \ .$$

In the o-minimal context, this induced space $ind(X)$ is also definable for definable H. In the analytic geometric context, we consider only Lie groups and a real analytic action of H on a real analytic manifold M, leaving the closed definable subset X invariant. But then $ind(X)$ is a closed definable subset of the real analytic manifold $ind(M)$. Moreover, ν is in both cases a **definable map**. So we get an induced exact functor

$$\nu^* : D^b_{G,T}(ind(X)) \to D^b_{H,T}(X) \ ,$$

and this is also an **equivalence** of categories. For this we need the following useful

Lemma 3.3.2. *Let $\mathcal{F} \in ob(D^b_{H,T}(X))$ be given. Then there exists a finite H-* **invariant** *filtration*

$$X_0 \subset \ldots \subset X_n = X$$

by closed definable subsets X_i such that $For(\mathcal{F})$ is constructible with respect to this filtration.

This is easy for an H-space X with only finitely many H-orbits, or for a discrete H-space X with a trivial action. In general, we use some special properties of cs-**stratifications** (see chapter 4 for more on this notion).

Proof. Since $For(\mathcal{F})$ is constructible, it is by definition constructible with respect to a finite filtration

$$X'_0 \subset \ldots \subset X'_m = X$$

by closed definable subsets. Let us first consider the o-minimal case. Then one can choose this filtration in such a way, that it is a cs-stratification in the sense of [HS, def. 1.1]. Use either an **adapted triangulation** as in [vDr, thm. 8.2.9], or an

adapted Whitney stratification as in [Ta1] or [Shiota, II.1.14]. By [HS, cor. 3.4], there is a **unique coarsest** cs**-stratification**

$$X_0 \subset \ldots \subset X_n = X$$

such that $For(\mathcal{F})$ is constructible with respect to this filtration. Each connected component of $X_i \backslash X_{i-1}$ is then a union of connected components of the $X'_j \backslash X'_{j-1}$ ([HS, def. 3.1]). Especially, the X_i are **definable**. Fix an element $g \in G$. Then

$$g^* For(\mathcal{F}) \simeq For(\mathcal{F}) .$$

By the definition of the unique coarsest cs-stratification, this implies that X_i is **invariant** under g. In the analytic geometric context, one can use this argument in local coordinates, and the **locally** defined filtration X_i defines a **globally well defined** filtration ([HS, cor. 3.9]). Or one starts with a **definable Whitney stratification** in the analytic geometric context ([vDrMi, D.16(1)]), and applies the argument above. □

Let us now explain the **induction equivalence** in the **constructible** context. Start with an $\mathcal{F} \in ob(D^b_{H,T}(X))$, and recall the maps

$$G \times_H X \xleftarrow{\ p\ } G \times X \xrightarrow{\ q'\ } X ,$$

where one has the twisted action of $G \times H$ on $G \times X$. Consider

$$\mathcal{G} \in ob(D^b_G(G \times_H X)) ,$$

with $\nu^* \mathcal{G} \simeq \mathcal{F}$. Then $Q_p^* \mathcal{G} \simeq Q_{q'}^* \mathcal{F}$. By applying For, one gets

$$p^* For(\mathcal{G}) \simeq q'^* For(\mathcal{F}) .$$

But $q'^* For(\mathcal{F})$ is constructible with respect to an H**-invariant** filtration $q'^{-1}(X_i)$ by closed definable subsets. Then $For(\mathcal{G})$ is constructible with respect to the definable filtration

$$p(q'^{-1}(X_i)) \qquad (i = 1, \ldots, n) ,$$

because $G \times X$ is a **free** H-space. Especially, p' is a **locally trivial fibration**. Choose any (local) section s of p'. Then

$$For(\mathcal{G}) \simeq s^* q'^* For(\mathcal{F}) .$$

Moreover

$$(For(\mathcal{G}))_{p(g,x)} \simeq (q'^* For(\mathcal{F}))_{(g,x)} \simeq (For(\mathcal{F}))_x \in ob(T) .$$

This implies the **induction equivalence**, and one has an induced **group isomorphism**

$$\nu^* : K_T^G(ind(X)) \xrightarrow{\sim} K_T^H(X) .$$

Remark 3.3.2. 1. We will **use** this group isomorphism **only** in the easy case X a point, or a discrete H-set with **trivial** action, where lemma 3.3.2 is trivial. Note that the proof of this lemma is one of the few places in this book, where we use explicitly **geometric** properties (i.e. the existence of a unique coarsest cs-stratification).

2. The same argument as before implies this induction equivalence also for equivariant constructible complexes in the **complex analytic or algebraic** context (compare [BL, prop. 5.1(iii)]).

For the case of a point, the induced **forget homomorphism**

$$For_H^G : K_{perf}^G(\{pt\}) \to K_{perf}^H(\{pt\})$$

is the usual **restriction map** (induced by the homomorphism $H/H_0 \to G/G_0$):

$$res = res_{H/H_0}^{G/G_0} : Rep(G/G_0, R) \to Rep(H/H_0, R) .$$

Since the forgetful functor **commutes** with all functors (and by the fact, that ν^* commutes with $\otimes_R^L$), one gets the following commutative diagram:

$$\begin{array}{ccc} K_T^G(ind(X)) \times Rep(G/G_0, R) & \xrightarrow{\otimes} & K_T^G(ind(X)) \\ \downarrow{\scriptstyle \nu^* \times res} & & \downarrow{\scriptstyle \nu^*} \\ K_T^H(X) \times Rep(H/H_0, R) & \xrightarrow{\otimes} & K_T^H(X) . \end{array} \tag{3.18}$$

Example 3.3.1. *Consider the isomorphism $\nu^* : K_T^G(ind(X)) \to K_T^H(X)$. Then the action of*

$$Rep(G/G_0, R) \quad on \quad K_T^G(ind(X))$$

is isomorphic by ν^* to the **induced** *action on the $Rep(H/H_0, R)$-module $K_T^H(X)$.*

We finally need the following **splitting result**:

Example 3.3.2. *Consider one of the following situations:*

1. *X has a finite increasing filtration X_i $(i = 0, \ldots, n)$ by closed G-invariant subsets, with $Y := X_0$ and $X = X_n$.*

2. *Y is a closed definable G-invariant subspace of the definable G-space X with respect to an o-minimal structure on $(\mathbb{R}, +, \cdot)$.*

3. *We fix an analytic geometric category, a real analytic G-manifold M and a closed definable G-invariant subset X of M such that Y is a closed definable G-invariant subspace of X.*

Denote by $i : Y \to X$ (or $j : U := X \backslash Y \to X$) the inclusion (of its open complement). In case (3.), *constructibility on U is* **defined** *with respect to filtrations by closed subsets of U, which are definable in M ! Then one has a split short exact sequence*

$$0 \longrightarrow K_T^G(U) \xrightarrow{j_!} K_T^G(X) \xrightarrow{i^*} K_T^G(Y) \longrightarrow 0\,. \tag{3.19}$$

*This is a split short exact sequence of $Rep(G/G_0, R)$-modules, since $j_!, i^*$ are $Rep(G/G_0, R)$-***linear**.

Note that $Rj_! = j_!$ maps in all cases

$$D^b_{G,T}(U) \quad \text{into} \quad D^b_{G,T}(X)\,.$$

In the case (1.), constructibility is defined with respect to the (induced) filtration. In the case (3.), this follows from our special definition of constructibility. Then the splitting sequence (3.19) is a special case of the following abstract

Lemma 3.3.3. *Consider three triangulated categories D, D_U, D_F and exact functors*

$$D_F \xrightarrow{i_!} D \xrightarrow{j^*} D_U\,.$$

Suppose $i_!$ and j^* have an exact left adjoint i^* and $j_!$ such that the adjunction morphisms $i^*i_! \to id$, $j^*j_! \to id$ are **isomorphisms**, *and $i^*j_! \simeq 0$. By* **adjunction** *this implies also $0 \simeq j^*i_!$. Finally assume that for each $K \in ob(D)$ there exists a distinguished triangle*

$$j_!j^*K \longrightarrow K \longrightarrow i_!i^*K \xrightarrow{[1]} \,.$$

Then one has a natural **isomorphism** *of abelian groups*

$$K(D) \simeq K(D_F) \oplus K(D_U)\,.$$

Proof. Since all functors are exact, one gets induced group homomorphisms of the corresponding Grothendieck groups, which we denote by the same symbol. Consider the homomorphism

$$(i^*, j^*) : K(D) \to K(D_F) \oplus K(D_U)\,,$$

and

$$i_! \oplus j_! : K(D_F) \oplus K(D_U) \to K(D)\,.$$

These are inverse homomorphisms, since our assumptions imply on the level of Grothendieck groups the identities

$$id = i^*i_!\,, \quad id = j^*j_!\,, \quad i^*j_! = 0 = j^*i_!$$

and $id = i_!i^* + j_!j^*$ (by the existence of the distinguished triangles above). □

Remark 3.3.3. 1. The **dual** statement of the lemma is also true. The assumptions in the lemma used for the "splitting of Grothendieck groups" is just "half" of the structure used in [BBD, 1.4.3] for the "gluing of t-structures" [BBD, thm. 1.4.10].

2. In most of the other equivariant theories, one has a **long exact sequence** instead of a split short exact sequence. But the abstract splitting result of lemma 3.3.3 applies also in the algebraic context of equivariant holonomic D-modules (compare [J3, lem. 2.5] and [Saito, thm. 3.14]).

With these results, we can now easily prove a very general localization result for our equivariant Grothendieck groups (we formulate them in a style similar to [tDieck, chap. III.3]).

Let us fix a **multiplicatively closed** subset (=: multiplicative subset) S of the commutative representation ring $Rep(G/G_0, R)$, i.e.

$$1 \in S \quad \text{and} \quad a, b \in S \Rightarrow ab \in S\,.$$

For a $Rep(G/G_0, R)$-module M, we denote by $S^{-1}M$ the corresponding **localization** of M. Let us first explain the

Proposition 3.3.1. *Let G be a* **discrete** *group. Consider a G-space X with quotient map $q : X \to X/G =: Y$. Assume*

1. *Y has a* **finite cellular decomposition** *Y_i $(i = 0, \ldots, n)$ such that*

$$q : q^{-1}(C) \to C$$

is a **covering** *map for each cell C of Y.*

2. *For each cell C there is a point $x_c \in q^{-1}(C)$, with stabilizer G_{x_c} such that*

$$S \cap ker(res : Rep(G, R) \to Rep(G_{x_c}, R)) \neq \emptyset\,.$$

Consider the G-equivariant sheaves on X, which are constructible with respect to the induced filtration $X_i := q^{-1}(Y_i)$ $(i = 0, \ldots, n)$. Then one has

$$S^{-1}K^G_{perf}(X) = 0\,.$$

Proof. Consider the filtration $X_0 \subset \ldots \subset X_n = X$ by G-invariant closed subsets. By example 3.3.2, it is enough to treat the case $X_{n-1} = \emptyset$ (localization is an exact functor). Then Y is the disjoint union of **finitely** many cells C, which are open and closed in Y. Therefore, X is also a disjoint union of the subsets $q^{-1}(C)$, which are open and closed in X. Then $K^G_{perf}(X)$ is the **product** of the

$$K^G_{perf}(q^{-1}(C))\,.$$

Therefore it is enough to show that for each cell C, there is an element in S which **annihilates** $K^G_{perf}(q^{-1}(C))$. So we can assume:

$$X = q^{-1}(C)\,, \quad \text{with } q : X \to C \quad \text{a covering}\,.$$

But C is contractible. Therefore one has a homeomorphism

$$h : X \to C \times \pi\,,$$

with π a discrete set and $q = p \circ \pi$ for the projection

$$p : C \times \pi \to C\,.$$

One can identify π with the set $\pi_0(X)$ of connected components of X. This has then a natural G-action, and h becomes a G-equivariant map. So we can assume $X = C \times \pi$ and $q = p$. Consider the second projection $f : C \times \pi \to \pi$, and the section

$$s : \pi \to C \times \pi\,,\ u \mapsto (q(x_c), u)\,.$$

Then we have the functor

$$f^* : D^b_{G,perf}(C) \to D^b_{G,perf}(X)\,,$$

and we claim that Rf_* restricts to a functor

$$Rf_* : D^b_{G,perf}(X) \to D^b_{G,perf}(C)\,.$$

Let $\mathcal{F} \in ob(D^b_{G,perf}(X))$ be given. Then $For(\mathcal{F})$ has by definition **locally constant** cohomology sheaves. By the **Vietoris-Begle theorem** [KS, prop. 2.7.8] we get as in the proof of lemma 2.0.1 on page 99 :

$$s^* For(\mathcal{F}) \simeq Rf_* For(\mathcal{F})\,,$$

which implies our claim. Similarly, this **Vietoris-Begle theorem** implies that the **adjunction** morphisms

$$id \to Rf_* f^* \quad \text{and} \quad f^* Rf_* \to id$$

for these **equivariant** functors are isomorphisms, since this can be checked after applying For. This implies

$$f^* : K^G_{perf}(\pi) \xrightarrow{\sim} K^G_{perf}(q^{-1}(C))\,,$$

as $Rep(G, R)$-modules. But π is G-isomorphic to the **induced space** G/G_{x_c}, and by assumption we have

$$S \cap ker(res : Rep(G, R) \to Rep(G_{x_c}, R)) \neq \emptyset\,.$$

Then the result follows from example 3.3.1. □

Instead of a discrete group, let us consider a **Lie group** G. Let X be a G-space. Then we would like to apply the proposition to the induced action of the **discrete** group G/G_0 on the quotient X/G_0. Let $H := G_x$ be the stabilizer of $x \in X$. Then the **stabilizer** of its image $[x] \in X/G_0$ is given by

$$(G/G_0)_{[x]} = (H \cdot G_0)/G_0 \simeq H/(H \cap G_0)\,. \tag{3.20}$$

Moreover, the natural homomorphism

$$H/H_0 \to G/G_0$$

has the kernel $(H \cap G_0)/H_0$. So it factors over

$$(H/H_0)/((H \cap G_0)/H_0) \simeq H/(H \cap G_0)\,.$$

The main technical problem is the question if the quotient map

$$q:\ X/G_0 \to Y = (X/G_0)/(G/G_0) = X/G$$

satisfies the **assumption (1.)** of proposition 3.3.1.

We explain now, why this is **always** the case in the situation of

- example 3.1.4 with X/G **compact**.
- example 3.1.6.
- example 3.1.7 with X/G **compact**.

Let us first consider the example 3.1.4. We assume moreover that all stabilizers G_x $(x \in X)$ are **compact** (e.g. the action is proper). Then one has to consider a G-space of the type

$$G/H \times S\,,$$

with S a cell with trivial action (and the left action of G on G/H). By assumption, H is a compact subgroup. For later applications we consider the following more general situation:

H is a **compact** subgroup, S is an H-space and we consider the **induced** G-space $G \times_H S$. Denote $H^0 := H \cap G_0$. This is a closed normal subgrup of finite index in H. We assume in addition, that the finite map

$$S/(H \cap G_0) \to S/H$$

is a **covering** (for a trivial H-space this is the identity map). Then one gets a commutative diagram

$$\begin{array}{ccccc} G_0\backslash(G \times_{H^0} S) & \xrightarrow{\sim} & G/G_0 \times G_0\backslash(G_0 \times_{H^0} S) & \xrightarrow{\sim} & G/G_0 \times S/H^0 \\ \downarrow{\scriptstyle\alpha} & & \downarrow & & \downarrow{\scriptstyle\beta} \\ G_0\backslash(G \times_H S) & \xrightarrow{q} & G\backslash(G \times_H S) & \xrightarrow{\sim} & S/H\,. \end{array} \tag{3.21}$$

Here β is the covering

$$G/G_0 \times S/H^0 \to S/H^0 \to S/H \, .$$

α is the quotient map with respect to the induced action of the finite group H/H^0. But this finite group operates **freely** so that α is also covering (compare [tDieck, cor. 3.24, p. 29]). Then q must also be a covering.

Consider now the example 3.1.6. Then

$$q : X/G_0 \to X/G$$

is a **finite definable** map. By the **trivialization theorem** [vDr, thm. 9.1.7], there are finitely many definable subsets of X/G, such that q is a **trivial fibration** over these subsets. Choose a finite stratification by cells, partitioning each of these sets [vDr, prop. 1.13]. Then also

$$q : q^{-1}(C) \to C$$

is a trivial fibration for each cell C. But q is a finite map. Especially, each fiber is finite. Therefore $q : q^{-1}(C) \to C$ is a finite covering for each cell C.

Let us come to the example 3.1.7.

Proposition 3.3.2. *Consider the situation of example* 3.1.7, *and fix a family* X_j *of* G*-invariant definable subsets of* X *such that the family* $f(X_j)$ *is locally finite on* $f(X)$. *Denote the quotient maps*

$$p : X \to X/G_0 \quad \textit{and} \quad q : X/G_0 \to X/G \simeq f(X) =: Y \, .$$

Then there exists a **definable regular cellular stratification**

$$Y_0 \subset \ldots \subset Y_m = Y$$

with the following properties:

1. *The stratification is compatible with the family* $f(X_j)$.

2. *For each cell* C *of this stratification,* $q : q^{-1}(C) \to C$ *is a covering.*

3. *The induced filtration* $q^{-1}(Y_i)$ $(i = 0, \ldots, n)$ *is a definable cellular decomposition of* Y *by relatively compact cells.*

4. *The family of closures of these cells is locally finite on* X/G_0.

Proof. Fix $x \in X$ and take a real analytic slice S' at x. Then GS' is open in M and real analytic G-isomorphic to $G \times_H S'$, with H the compact stabilizer of x. As before, we can identify

$$f : GS' \to f(GS')$$

with the canonical projection

$$G \times_H S' \to S'/H \, .$$

Take now a compact definable neigborhood S of x in $X \cap S'$, which is invariant under H, e.g. the intersection of X with a small ball in S with respect to an H-invariant real analytic metric (compare [Kan, proof of thm. 1.17]). Consider for this S the diagram (3.21). Then

$$q' : S/(H \cap G_0) \to S/H$$

can be realized as a finite definable map between compact definable subsets of suitable affine spaces. But these sets and map are then also definable with respect to the o-minimal structure corresponding to the given analytic geometric structure, as in [vDrMi, D.10] (compare with the introduction of chapter 2 for more on this).

As explained before, we can find a finite stratification of S/H by definable cells such that

$$q' : q'^{-1}(C) \to C$$

is a finite covering for each cell C. This is then also true for any connected definable subset C of such a cell.

$$q : q^{-1}(C) \to C$$

is then also a covering, by the diagram (3.21). Each connected component of $q^{-1}(C)$ is definable and relatively compact. Moreover, the family of the closures of these connected components is locally finite on $q'^{-1}(S/H)$. This follows from the diagram (3.21) and the fact, that q' and α are finite (i.e. proper with finite fibers).

We apply the construction of the proof of proposition 2.1.1 and lemma 2.1.1 on page 97. Recall that $f(X_i)$ is a locally finite family of subsets, which are definable by lemma 3.1.3. We choose in the proof of lemma 2.1.1 the V_i of the type S_i/H_i as in the discussion before. Add the corresponding cells $C_{i,k}$ to our given locally finite family of definable sets, and apply the proof of proposition 2.1.1 and lemma 2.1.1. Since $f(X)$ is a closed definable subset of the ambient manifold, we end up with a **definable regular cellular stratification** Y_i $(i = 0, \ldots, n)$ of Y. By construction, each cell C of this stratification is contained in a cell of type $C_{i,k}$. This implies the proposition. □

This proposition implies also the following sharpening of example 3.1.7.

Corollary 3.3.1. *In the situation of example 3.1.7, consider an equivariant constructible complex $\mathcal{F} \in ob(D^*_G(X))$. Then $Rf_! For(\mathcal{F})$ is constructible on $f(X)$.*

Proof. By example 3.1.7.2 we already know, that

$$Rp_! For(\mathcal{F})$$

is constructible on $p(X)$. By lemma 3.3.2, this is constructible with respect to a G/G_0-invariant filtration by definable closed sets

$$Z_i \quad (i = 0, \ldots, m) .$$

Consider the finite family of G-invariant definable subsets of X, given by

$$p^{-1}(Z_i \backslash Z_{i-1}) \quad (i = 0, \ldots, m) .$$

If we apply proposition 3.3.2, we get a definable cell stratification, such that for each cell C the restriction of the cohomology sheaves of

$$Rp_! For(\mathcal{F}) \quad \text{to} \quad q^{-1}(C)$$

are locally constant. Since

$$q : q^{-1}(C) \to C$$

is also a covering, we get as in example 3.1.4 the fact, that

$$Rq_! Rp_! For(\mathcal{F})$$

is constructible with respect to this definable cell stratification. □

Now we can prove the

Theorem 3.3.1. *Consider one of the following situations:*

1. *X is a* **proper** *G-space with the structure of an equivariant CW-decomposition for the group G, with X/G* **compact**.

2. *We fix an o-minimal structure on $(\mathbb{R}, +, \cdot)$ and consider a definable action of a* **compact** *definable group on the definable space X.*

3. *We fix an analytic geometric category and consider a real analytic* **proper** *G-manifold M for some (real analytic) Lie group G. Let X be a closed definable G-invariant subset of M such that X/G is* **compact**.

Denote by $i : Y \to X$ the inclusion of a closed definable G-invariant subset of X (in case (1.) this is by definition a union of equivariant cells). Fix a **multiplicatively closed** *subset S of $Rep(G/G_0, R)$. Assume*

$$S \cap ker(res : Rep(G/G_0, R) \to Rep(G_x/(G_x \cap G_0), R)) \neq \emptyset$$

for all $x \in U := X \backslash Y$. Then the **pullback** *i^* induces an* **isomorphism** *of* **localized** *$Rep(G/G_0, R)$-modules:*

$$S^{-1} K_T^G(X) \xrightarrow[\sim]{i^*} S^{-1} K_T^G(Y) .$$

Proof. As explained before, X/G has in all cases a finite (definable) cellular filtration, with Y/G a closed union of cells such that we can apply proposition 3.3.1 to the discrete group G/G_0 and the map

$$(X \backslash Y)/G_0 \to (X \backslash Y)/G\,,$$

with its induced cellular filtration. So we have

$$S^{-1}K_{perf}^{G/G_0}((X\backslash Y)/G_0) = 0\,,$$

where this group is defined by those sheaves, which are constructible with respect to the induced filtration of $(X\backslash Y)/G_0$. By example 3.3.2, we only have to show that

$$S^{-1}K_T^G(X\backslash Y) = 0\,.$$

But $K_T^G(X\backslash Y)$ is a module over

$$K_{perf}^{G/G_0}((X\backslash Y)/G_0)\,,$$

and the $Rep(G/G_0, R)$-module structure factors by definition over the ring homomorphism

$$Rep(G/G_0, R) \to K_{perf}^{G/G_0}((X\backslash Y)/G_0)\,.$$

Remark that in case (3.) the used induced filtration on

$$(X\backslash Y)/G_0$$

is the restriction of a definable filtration on X/G_0. Therefore, the trivial extension by 0 of the considered constructible sheaves on $(X\backslash Y)/G_0$ (or their pullbacks to $X\backslash Y$) are constructible. This implies the theorem. □

By the definition of the isomorphism i^* for the localized Grothendieck groups, the **inverse isomorphism** is induced by $i_* = i_!$ (compare example 3.3.2). This implies the following **functorial property** of this isomorphism. Let

$$f : (X, Y) \to (X', Y')$$

be a definable map such that

$$Rf_! \text{ maps } D^b_{G,T}(X) \text{ into } D^b_{G,T}(X')\,.$$

Its restriction $f : Y \to Y'$ has then also the corresponding property. Assume that we can apply theorem 3.3.1 to (X, Y) and (X', Y'). Then the following diagram commutes:

$$\begin{array}{ccc} S^{-1}K_T^G(X) & \xrightarrow[\sim]{i^*} & S^{-1}K_T^G(Y) \\ \downarrow f_! & & \downarrow f_! \\ S^{-1}K_T^G(X') & \xrightarrow[\sim]{i'^*} & S^{-1}K_T^G(Y')\,. \end{array} \tag{3.22}$$

Moreover, in the o-minimal context this is also true for f_*.

Let us now illustrate this result in the **o-minimal context** for a **finite** group G and the map f to a point. So we consider a definable space X with an action of the **finite** group G by **definable automorphisms**.

Example 3.3.3. *Let G be a finite abelian group, and consider the ground ring $R = \mathbb{C}$. Then*

$$Rep(G, \mathbb{C}) \text{ is the group ring } \mathbb{Z}[M] ,$$

with M the abelian group of characters $\chi : G \to \mathbb{C}^*$. We denote the corresponding element in $\mathbb{Z}[M]$ by e^{χ}. Let H_i $(i \in I)$ represent the stabilizers of the points $x \in X \backslash Y$, with $Y := X^G$. Then for each $i \in I$, there exists a **non-trivial** *character χ_i, with $\chi_i | H_i = 1$. This implies of course*

$$1 - e^{\chi_i} \in ker(\ res : Rep(G, \mathbb{C}) \to Rep(H_i, \mathbb{C})\ ) .$$

Denote by S the multiplicative subset generated by the elements $1 - e^{\chi_i}$ $(i \in I)$. Consider an equivariant constructible complex $\mathcal{F} \in ob(D^b_{G,perf}(X))$. Then we get the equality of the following **virtual cohomology representations** *(as elements in the localized ring, for $? = c, *$):*

$$\sum (-1)^k \{H^k_?(X, \mathcal{F})\} = \sum (-1)^k \{H^k_?(X^G, \mathcal{F})\} \in S^{-1} Rep(G, \mathbb{C}) .$$

Remark 3.3.4. This can be applied for example to the **complex algebraic** context, and is then a generalization of [J3, thm. 4.3] for the case of a **finite abelian group of complex algebraic transformations**. The more general case of a linear complex diagonalizable group (considered in [J3]) cannot be directly deduced in this way. This will be obtained later on from another localization result.

3.3.2 Trace formulae

Other applications are related to trace formulae.

Example 3.3.4. *Let G be a finite group, and consider the ground ring $R = \mathbb{C}$. Then*

$$Rep(G, \mathbb{C}) = K_{fin}(mod(\mathbb{C}[G])) .$$

Each finite dimensional complex representation A induces in the usual way a **character function**

$$\chi_A : G \to \mathbb{C} ,\ g \mapsto tr(g : A \to A) =: tr(g, A) ,$$

with $\chi_{A \oplus B} = \chi_A + \chi_B$ and $\chi_{A \otimes B} = \chi_A \cdot \chi_B$. By additivity, this induces a character function for each virtual representation $\chi \in Rep(G, \mathbb{C})$, which we just denote by the same symbol χ. Fix a conjugacy class γ in G, and denote by Y the union of *the fixed point sets X^g, for $g \in \gamma$. Let H_i $(i \in I)$ represent the stabilizers of the points $x \in X \backslash Y$. Then for each $i \in I$, there exists an element*

$$\chi_i \in Rep(G, \mathbb{C}) \ , \quad \text{with } \chi_i(\gamma) \neq 0 \ \text{ and } \chi_i | H_i = 0$$

(compare [AtSeg, lem. 1.2]). Let S be the multiplicative subset of $Rep(G, \mathbb{C})$, given by the elements $\chi \in Rep(G, \mathbb{C})$ with $\chi(\gamma) \neq 0$. This implies of course

$$\chi_i \in S \cap ker(\ res : Rep(G, \mathbb{C}) \to Rep(H_i, \mathbb{C})\ ) \ .$$

Consider an equivariant constructible complex $\mathcal{F} \in ob(D^b_{G,perf}(X))$. Then we get the equality of the following **virtual cohomology representations** *(as elements in the localized ring, for $? = c, *$):*

$$\sum (-1)^k \{H^k_?(X, \mathcal{F})\} = \sum (-1)^k \{H^k_?(X^G, \mathcal{F})\} \ \in S^{-1} Rep(G, \mathbb{C}) \ .$$

We get especially the following **trace formula** *for $g \in \gamma$:*

$$tr(g, H^*_?(X, \mathcal{F})) = tr(g, H^*_?(Y, \mathcal{F})) \ ,$$

with $tr(g, \cdot)$ the corresponding **Lefschetz number** *(i.e. the alternating sum of the traces of the induced endomorphisms on the cohomology spaces).*

Our formulation follows [AtSeg, thm. 1.1], which describes the corresponding localization result for **compact groups in topological equivariant K-theory**.

For the trace formula, it is enough to consider the finite cyclic group generated by g. If we consider this finite cyclic group as our G, then Y is just the fixed point set X^g of g, and one gets for an equivariant constructible complex

$$\mathcal{F} \in ob(D^b_{G,perf}(X))$$

the **trace formula** (for $? = c, *$):

$$tr(g, H^*_?(X, \mathcal{F})) = tr(g, H^*_?(X^g, \mathcal{F})) \ . \tag{3.23}$$

Since G acts in this case trivially on X^g, the restriction of the cohomology sheaves of $\mathcal{F}$ to X^g are just constructible sheaves of $\mathbb{C}[G]$-modules with finite dimensional stalks. If we associate to such a complex of sheaves the function

$$x \mapsto tr(g, \mathcal{F}_x) \ ,$$

we get a complex valued **constructible function** $tr(g, \mathcal{F})$. As in section 2.3, one can calculate the trace

$$tr(g, H^*_?(X^g, \mathcal{F}))$$

as a corresponding **integral of the constructible function** $tr(g, \mathcal{F})$ (compare [GM2, prop. 11]).

The advantage of our result is the fact, that it applies in the o-minimal context to any equivariant constructible sheaves without any compactness assumptions. We will see later on, how one can extend the formula (3.23) from the case of a definable automorphism of **finite order** to the case of an element of a **compact definable group**. But even for a **finite automorphism**, it has important applications. Of course, the approach through the **localization result for Grothendieck groups** is quite heavy. But it has the advantage that it gives **functorial results** with respect to appropriate maps. For the special case of a map to a point, i.e. for results about the cohomology (with compact support), we will also give a very easy direct proof.

For the moment, let us give some examples.

Example 3.3.5. *Consider the constant sheaf $\mathbb{C}_X$, which is always an equivariant sheaf. Let g be a definable automorphism of X of* **finite order**. *Assume X is cohomologically* **acyclic**, *in the sense that*

$$H^*(X, \mathbb{C}) = \mathbb{C} \quad \textit{placed in degree } 0\,.$$

The condition $H^0(X, \mathbb{C}) = \mathbb{C}$ is equivalent to X connected. Therefore

$$tr(g, H^*(X, \mathbb{C})) = 1\,,$$

and the trace formula (3.23) *implies $X^g \neq \emptyset$. Of course, this conclusion follows from the fact, that X and the action of g is* **definable** *with respect to an o-minmal structure. This is for example* **not** *true for a* **smooth** *action on $\mathbb{R}^n$. For another proof of the existence of a fixed point, in the special case of a* **real algebraic action on a real algebraic variety**, *see [PeRa].*

Other applications of the trace formula (3.23) are:

- The **Lefschetz fixed point formula** [Ro, prop. 1.6] for a finite order automorphism σ of a semisimple connected algebraic group G defined over $\mathbb{Q}$. One considers a σ-stable arithmetic subgroup Γ of $G(\mathbb{Q})$ and a finite-dimensional complex representation V of $G(\mathbb{R})$, with a compatible σ-action. Let K be a maximal compact subgroup of $G(\mathbb{R})$. Then the **discrete** group Γ operates **properly** on $X = G(\mathbb{R})/K$, and V induces a **σ-equivariant constructible sheaf** $\mathcal{V}$ on the "orbifold" X/Γ. But X/Γ is (in general) **not** compact or definable. But the action of σ **extends to the Borel-Serre compactification** $\bar{X}/\Gamma$, which is a **compact** "orbifold with corners". Moreover, $\mathcal{V}$ **extends** to an σ-equivariant constructible sheaf on this compactification. Then one can deduce the trace formula (3.23) by using a suitable triangulation of $\bar{X}/\Gamma$.
- The "micro-local trace formula" [ABV, thm. 25.8]. Its proof is reduced in [ABV] to the trace formula (3.23) in the **subanalytic context for a compact subanalytic space**. In [ABV] the **deep** and general Lefschetz formula of [GM2] is used.

Moreover, the authors of [ABV] ask for a **simple** proof of the trace formula (3.23) ([ABV, p. 274]). We will now give such a simple proof.

Let G be a **finite cyclic group** generated by the element g. Consider a G-space X together with the quotient map

$$q : X \to Y := X/G$$

such that Y has a **finite cellular decomposition or stratification** as in theorem 3.0.1. Denote the inclusion $j : U := X \setminus X^g \to X$, and suppose that $q(X^g)$ is a union of cells of Y (as in the situation of the theorem).

Consider the strict full triangulated subcategory T of complexes of $R[G]$-modules, which are **perfect** as R-modules. For such a complex A, one can define the **trace**

$$tr(g, A) \text{ of } g : A \to A\,,$$

and this induces an **additive** function $tr(g, \cdot)$ on T ([SGA5, p.376]). For $R = k$ a field, this is the usual alternating sum of the traces of g on the finite-dimensional cohomology spaces.

Now $q : X \to Y = X/G$ is a **finite** map (i.e. proper with finite fibers), since G is finite. This implies $Rq_! = Rq_* = q_*$. Consider a complex

$$\mathcal{F} \in ob(D^b_G(X))$$

such that all stalks $For(\mathcal{F})_x$ $(x \in X)$ are **perfect** complexes of R-modules. Since q is finite, we have

$$(q_*For(\mathcal{F}))_{q(x)} = \sum_{x' :\, q(x')=q(x)} For(\mathcal{F})_{x'}\,.$$

Especially, the stalk $\mathcal{F}_{q(x)}$ belongs to $ob(T)$ for all $x \in X$. If $x \notin X^g$, then g acts on the fiber

$$\{x' : q(x') = q(x)\}$$

as a **non-trivial cyclic permutation**. This implies automatically (by the definition of the trace):

$$tr(g, (q_*For(\mathcal{F}))_{q(x)}) = 0 \quad (x \notin X^g)\,.$$

Assume that $q_*For(\mathcal{F})$ is constructible on Y. Since $q(Y)$ is a union of cells, this is also true for

$$q_*For(\mathcal{F}|Y) \text{ and } q_*Rj_!j^*For(\mathcal{F})\,,$$

and in the case of a finite cellular **stratification** also for

$$q_*Rj_*j^*For(\mathcal{F})\,.$$

By theorem 3.0.1 we get for $? = !$ or $*$:

$$R\Gamma_?(X, \mathcal{F}) \text{ and } R\Gamma_?(X^g, \mathcal{F}) \in ob(T)\,.$$

If we apply this theorem to $q : U = X \backslash X^g \to q(U)$, we get for $? = !$ or $*$:

$$R\Gamma_?(U, \mathcal{F}) \in ob(T) \text{ and } tr(g, R\Gamma_?(U, \mathcal{F})) = 0 .$$

Therefore, we get in this general context the **trace formula**

$$tr(g, R\Gamma_?(X, \mathcal{F})) = tr(g, R\Gamma_?(X^g, \mathcal{F})) \quad (? = ! \text{ or } *) . \tag{3.24}$$

Note that the argument above would also work for a commutative ring, which is not necessarily noetherian of finite homological dimension. Let us now extend the trace formula to the case of a **compact definable group**.

Corollary 3.3.2. *Fix an o-minimal structure on* $(\mathbb{R}, +, \cdot)$*, and a commutative noetherian ground ring* R *of finite homological dimension. Let* T *be the triangulated category of perfect complexes. Consider the definable action of a* **compact** *definable group* G *on the definable space* X*. Fix* $g \in G$*, and let* X^g *be the fixed point set of* g*. Suppose* G *is* **topologically generated** *by* g *(i.e.* G *is the closure of the cyclic group generated by* g*). Then one has for a constructible complex*

$$\mathcal{F} \in ob(D^b_{G,T}(X))$$

the trace formula (3.24).

Proof. Since the compact definable group G is topologically generated by g, it is **abelian** and the product of a compact torus G_0 with a finite cyclic group G' (recall that G is homeomorphic to a Lie group [Pi]). Since G_0 is then the connected component of the identitity, it is also definable. If G_0 is trivial, we are just in the case of a finite cylic group as above. We therefore can assume $G_0 \neq \{e\}$. Let g' be the generator of G' corresponding to

$$g/G_0 \in G/G_0 \simeq G' .$$

Then

$$tr(g, R\Gamma_?(*, \mathcal{F})) = tr(g', R\Gamma_?(*, \mathcal{F})) \text{ for } * = X, X^g ,$$

since this depends only on the image of g in G/G_0. Moreover

$$X^g = (X^{g'})^{G_0} .$$

By the trace formula for the group G', we can assume $X = X^{g'}$. Then we apply the **restriction functor** $res_{G',G}$ and consider

$$res_{G',G}(\mathcal{F}) \in ob(D^b_{G',T}(X)) .$$

But G' is a **finite group acting trivially** on $X = X^{g'}$. We can therefore consider $res_{G',G}(\mathcal{F})$ as an element in

$$ob(D^b(X, R[G'])) .$$

Since G is abelian, each $g_0 \in G_0$ induces for all $i \in \mathbb{Z}$ an isomorphism of cohomology sheaves in $Sh(X, R[G'])$:

$$g_0^*(h^i(res_{G',G}(\mathcal{F}))) \simeq h^i(res_{G',G}(\mathcal{F})) .$$

Just look at the induced action of the elements of G on $h^i(For(\mathcal{F}))$. Consider the projection

$$q : X \to Y := X/G = X/G_0 .$$

By theorem 3.0.1 we only have to show that

$$tr\Big(g', \big(Rq_! h^i(res_{G',G}(\mathcal{F}))\big)_{q(x)}\Big) = 0 \text{ for all } x \notin X^{G_0} .$$

Recall that by the assumption on R, a complex of R-modules is perfect iff it is cohomologically bounded with finitely generated cohomology, and that the trace is an **additive** function. But the above isomorphisms for $g_0 \in G_0$ imply that this trace function is **constant** on the fibers of q. Then theorem 3.0.1(1) for the case of a trivial group $\{e\}$ implies:

$$tr\Big(g', \big(Rq_! h^i(res_{G',G}(\mathcal{F}))\big)_{q(x)}\Big) \simeq \chi(q^{-1}(\{q(x)\})) \cdot tr\Big(g', h^i(res_{G',G}(\mathcal{F}))_x\Big) .$$

But $q^{-1}(\{q(x)\})$ is for $x \notin X^{G_0}$ homeomorphic to a compact torus of positive dimension, and has therefore a **vanishing Euler characteristic**. □

Remark 3.3.5. Let G' be a **compact real analytic Lie group**, and consider an o-minimal structure on $\mathbb{R}_{an}$, i.e. containing all globally subanalytic sets. Consider a definable action of G' on a definable space X (with respect to this structure). Fix an element $g \in G'$. The **closure** G of the the cyclic group generated by g is a closed subgroup of G'. Then G is a **closed real analytic** Lie subgroup of G'. Especially, it is definable and the inclusion $G \to G'$ and therefore also the induced action of G on X is definable. If we consider

$$\mathcal{F} \in ob(D^b_{G',T}(X)) ,$$

then we can apply the corollary to $res_{G,G'}(\mathcal{F})$. The trace formula (3.24) holds therefore in this context.

Example 3.3.6. *Let X be a* **compact** *real analytic manifold with a real analytic metric. Then the group $I(X)$ of* **isometries** *of X is a compact real analytic Lie group, acting real analytically on X (compare [tDieck, p. 15] and the references given there). If we apply the trace formula to the* **constant sheaf** *R_X, we get for an* **isometry** *g the equality*

$$tr(g, R\Gamma(X, R)) = \chi(X^g) .$$

Of course, this holds also in the smooth context and for $R = \mathbb{C}$ it is a very special case of the results of [AtSeg]. Moreover, in this form it is a classical result of Kobayashi [Ko].

Of course, one can also deduce the examples 3.3.3 and 3.3.4 from the theorem 3.0.1. More interesting in this direction is the following application:

Corollary 3.3.3. *Suppose we are in the context of*

- *example* 3.1.6, *or*
- *example* 3.1.7(2) *with* X/G **compact**.

Let $R = k$ *be a field of characteristic* 0, *and denote by* C *(or* T*) the abelian category of finite dimensional* k*-vector spaces (or the triangulated category of cohomologically bounded complexes with finite dimensional cohomology). Let* Y *be a closed definable and* G*-invariant subset of* X*. Assume that for all* $x \in X\backslash Y$ *the group*

$$W(G_x) := N(G_x)/G_x$$

(with $N(G_x)$ *the normalizer of* G_x *in* G*) contains a* **circle subgroup**,*i.e. a closed subgroup homeomorphic to* S^1*. Consider a complex*

$$\mathcal{F} \in ob(D^b_{G,T}(X)) .$$

Then one has in $Rep(G/G_0, k)$ *the following equality (for* $? = c, *$*):*

$$\{R\Gamma_?(X, \mathcal{F})\} = \{R\Gamma_?(Y, \mathcal{F})\} \in Rep(G/G_0, k) . \tag{3.25}$$

Proof. Consider the projection $q : U := X\backslash Y \to U/G$. By theorem 3.0.1, we only have to show:

$$\{(Rq_?\mathcal{F}))_{q(x)}\} = \{0\} \in Rep(G/G_0, k) \text{ for all } x \in U .$$

By a truncation argument (with respect to the t-structure on $D^b_{G,T}(X)$) and induction, it is enough to show this for a single sheaf

$$\mathcal{F} \in ob(Sh_{G,C}(X)) .$$

Now we apply the stalk formulae (3.11) and (3.12):

$$(R^i q_? \mathcal{F})_{q(x)} \simeq H^i_?(G_0/H_0, k) \otimes_k \Gamma^{H/H_0}(k[G/G_0] \otimes_k Res_{H,G}(\mathcal{F})_x) ,$$

with $H = G_x$ the stabilizer of x. But

$$A := \Gamma^{H/H_0}(k[G/G_0] \otimes_k Res_{H,G}(\mathcal{F})_x)$$

is as a k-vector space a direct summand of

$$k[G/G_0] \otimes_k Res_{H,G}(\mathcal{F})_x$$

Therefore it is finite dimensional, since $\mathcal{F}_x \in ob(C)$ and G/G_0 is by assumption finite. By additivity we get the equality:

$$\{(Rq_?\mathcal{F})_{q(x)}\} = \chi(R\Gamma_?(G_0/H_0, k)) \cdot \{A\} \ \in Rep(G/G_0, k) .$$

But

$$\chi(R\Gamma_?(G_0/H_0,k)) = (|H/H_0|/|G/G_0|) \cdot \chi(R\Gamma_?(G/H,k)) ,$$

and G/H has a **free** right action of $W(H)$ ([tDieck, prop. 1.15, p. 6]). Moreover, this group contains a circle group S. We claim now that this implies

$$\chi(R\Gamma_?(G/H,k)) = 0 .$$

Consider first the case of a compact Lie group G. Then S is also a Lie group, and the action of S on G/H is real analytic. The claim follows then from theorem 3.2.1(1.a).

Consider now the case of a Lie group with **finitely** many connected components. Let S' be the inverse image of S in $N(H)$ under the natural projection

$$N(H) \to N(H)/H .$$

This is a compact subgroup of $N(H)$, containing H. By [Hoch, XV. thm. 3.1], there exists a compact subgroup K of G containing S', and a K equivariant analytic isomorphism of manifolds

$$\mathbb{R}^p \times K \simeq G$$

for the right K-action given by multiplication on K and G. But K/H is by construction invariant under the right action of S, and

$$\chi(R\Gamma_?(G/H,k)) = \begin{cases} \chi(R\Gamma(K/H,k)) & \text{for } ? = * , \\ (-1)^p \cdot \chi(R\Gamma(K/H,k)) & \text{for } ? = c . \end{cases}$$

So we are back in the case of a compact group K with a free S-action on K/H, and the claim follows. □

Remark 3.3.6. The idea to consider the condition " $W(G_x) := N(G_x)/G_x$ contains a circle subgroup" is taken from (the proof of) [tDieck, prop. 2.2, p. 240].

Example 3.3.7. *Let $G, X, R = k$ and C, T be as in corollary 3.3.3. Let $K \subset G$ be a* **compact torus**, *which is a definable subgroup contained in the* **center** *of G. Consider the fixed point set $Y := X^K$ of K, and a point $x \in X \backslash Y$. Since K is in the center of G, we get the inclusions*

$$K \hookrightarrow N(G_x) \text{ and } K/(K \cap G_x) \hookrightarrow W(G_x) .$$

But $K/(K \cap G_x)$ is a non-trivial compact connected abelian Lie group, and therefore $W(G_x)$ contains a **circle subgroup**. *So we get for $\mathcal{F} \in ob(D^b_{G,T}(X))$:*

$$\{R\Gamma_?(X,\mathcal{F})\} = \{R\Gamma_?(X^K,\mathcal{F})\} \in Rep(G/G_0,k) .$$

Assume moreover, $k = \mathbb{C}$ and G is the product $G' \times K$, with G' a finite (abelian) group. Then

$$X^G = (X^K)^{G'} , \; G/G_0 \simeq G' \text{ and } Rep(G/G_0,\mathbb{C}) \simeq Rep(G',\mathbb{C}) .$$

Let S be the **multiplicative subset** *of $Rep(G/G_0, \mathbb{C})$ as in example* 3.3.4 *(or* 3.3.3*), corresponding to the finite (abelian) group G'. These examples together with the above equation imply the "localization formula":*

$$\{R\Gamma_?(X, \mathcal{F})\} = \{R\Gamma_?(X^G, \mathcal{F})\} \in S^{-1} Rep(G/G_0, \mathbb{C}) . \tag{3.26}$$

Remark 3.3.7.

1. For a very special case of this kind of "fixed point reduction" for a central compact torus, compare with [ChGi, prop. 2.5.1].
2. Consider the action of a **compact Lie group** K in the context of an o-minimal structure on $\mathbb{R}_{an}$ (i.e. containing the globally subanalytic sets). Fix an element $g \in K$. Then one can apply the equation (3.26) to the **closure** G of the cyclic subgroup generated by g. In this case one has $X^G = X^g$.
3. Consider the **algebraic action of a linear complex diagonalizable group** on a complex algebraic variety. Then we can apply the equation (3.26) to a maximal compact subgroup K of G. But
$$K/K_0 = G/G_0 \quad \text{and} \quad X^K = X^G ,$$
since K is **Zariski dense** in G. Therefore, this equation is also true in this complex algebraic context for the linear complex diagonalizable group G. This is a generalization of [J3, thm. 4.3].

3.4 Equivariant cohomology

For simplicity we assume in this section, that G is (topologically isomorphic to) a Lie group with **finitely** many connected components, and the ground ring R is a commutative noetherian ring of finite homological dimension. So we can apply all functors of [BL], especially $\otimes^L_R$.

The category $D^b_G(\{pt\}, R)$ is by [BL, prop. 2.7] **equivalent** to the full subcategory of $D^b(BG, R)$, which consists of complexes with **locally constant cohomology** sheaves. Here BG is a **classifying space** of G. By looking at complexes in degree 0, one gets that

$$Sh_G(\{pt\}, R)$$

is equivalent to the category of locally constant sheaves of R-modules on BG. But this corresponds exactly to $R[\pi_1(BG)]$-modules (after the choice of a base point), and the isomorphism

$$\pi_1(BG) \simeq \pi_0(G) = G/G_0$$

gives us back our old equivalence between $Sh_G(\{pt\}, R)$ and $mod(R[\pi_0(G)])$. For a **connected** Lie group G, we consider especially complexes with **constant** cohomology sheaves.

We denote by $H_G := \Gamma(BG, \cdot) \circ h$ the corresponding cohomology functor

$$D^b_G(\{pt\}, R) \xrightarrow{h} Sh(BG, R) \xrightarrow{\Gamma(BG,\cdot)} mod(R) .$$

Then one gets for $N \in ob(D^b_G(\{pt\}, R))$ the **graded module**

$$H^*_G(N) := \oplus_i H_G(N[i])$$

over the skew-commutative **graded ring**

$$A_G := H^*_G(R) = H^*(BG, R) .$$

This applies especially to

$$H^*_{G,?}(X, \mathcal{F}) := H^*_G(R\Gamma_?(X, \mathcal{F})) \quad (? = ! \text{ or } *) .$$

A homomorphism $\phi : G' \to G$ of such Lie groups induces a continuous map $\bar{\phi} : BG' \to BG$ and hence a homomorphism

$$A_G = H^*(BG, R) \to H^*(BG', R) = A_{G'} .$$

For an $A_{G'}$-module M let ${}_{A_G}M$ denote the induced A_G-module (with respect to the homomorphism $A_G \to A_{G'}$). Let especially G' be a closed subgroup of G, and consider the homogenous space

$$Z := G/G' \simeq G \times_{G'} \{pt\} ,$$

together with the ϕ-map $\nu : \{pt\} \to G/G'$, $pt \mapsto e/G'$. The functor

$$\nu^* : D^b_G(Z) \to D^b_{G'}(\{pt\})$$

induces by [BL, thm. 2.6.3] an **equivalence** of categories, and the inverse functor is given by the general direct image functor Q_* corresponding to ν ([BL, 6.12.6]). Especially

$$\mathcal{F} \simeq Q_* \nu^* \mathcal{F} \quad \text{for } \mathcal{F} \in ob(D^b_G(Z)) .$$

This implies by [BL, 13.5] the following isomorphism of A_G-modules:

$$H^*_G(Z, \mathcal{F}) \simeq {}_{A_G}H^*_{G'}(\nu^* \mathcal{F}) . \tag{3.27}$$

Let $S \subset H^*(BG, R)$ be a **multiplicatively closed** subset of homogeneous elements in the center of $H^*(BG, R)$. For a graded $H^*(BG, R)$-module M^* one can then define the localized module $S^{-1}M^*$. This localization functor is **exact**. We already explained in the introduction, that the subcategory T of $D^b_G(\{pt\}, R)$ given by the complexes N with $S^{-1}H^*_G(N) = 0$, is a **strict full triangulated subcategory**.

Consider now a **proper** G-space X and the projection $q : X \to Y := X/G$. Fix a point $y \in Y$ and denote the inclusions

$$i_y : \{y\} \to Y \quad \text{and} \quad i : q^{-1}(\{y\}) \to X .$$

Then one has for $\mathcal{F} \in ob(D^b_G(X))$ isomorphisms ([BL, 1.4.6, thm. 3.4.3]):

$$Rq_! i^* \mathcal{F} \simeq i_y^* Rq_! \mathcal{F} \quad \text{and} \quad Rq_* i^! \mathcal{F} \simeq i_y^! Rq_* \mathcal{F} \, .$$

Equation (3.27) implies therefore

$$H^*_G(i_y^! Rq_* \mathcal{F}) \simeq H^*_G(q^{-1}(\{y\}), i^! \mathcal{F}) \simeq \, _{A_G}H^*_{G_x}(\nu^* i^! \mathcal{F}) \, ,$$

with G_x the stabilizer of $x \in q^{-1}(\{y\})$.

Especially for **compact** G we have $Rq_! = Rq_*$, and equation (3.27) implies in this case also

$$H^*_G(i_y^* Rq_! \mathcal{F}) \simeq H^*_G(q^{-1}(\{y\}), i^* \mathcal{F}) \simeq \, _{A_G}H^*_{G_x}(\nu^* i^* \mathcal{F}) \, ,$$

with G_x the stabilizer of $x \in q^{-1}(\{y\})$.

We would like to have in general, that the **equivariant cohomology** $H^*_G(\cdot)$ of

$$i_y^* Rq_! \mathcal{F} \, , \; i_y^! Rq_! \mathcal{F} \, , \; i_y^* Rq_* \mathcal{F} \text{ and } i_y^! Rq_* \mathcal{F}$$

is (isomorphic to) a module **induced** from an $H^*_{G_x}(\cdot)$-module.

We consider for this a **maximal compact** subgroup K of G containing G_x. The homomorphism $A_G \to A_{G_x}$ factors then as

$$A_G \to A_K \to A_{G_x} \, .$$

But one has topologically $G \simeq \mathbb{R}^p \times K$. The restriction functor $Res_{K,G}$ is then fully faithful ([BL, thm. 3.7.3]). This implies $A_G \simeq A_K$ and the A_G-module

$$H^*_G(i_y^* Rq_! \mathcal{F}) \simeq H^*_G(R\Gamma_c(q^{-1}(\{y\}), i^* \mathcal{F}))$$

is isomorphic to the A_K-module

$$H^*_K(R\Gamma_c(q^{-1}(\{y\}), i^*(Res_{K,G}\mathcal{F}))) \simeq H^*_K(K/G_x, k^! i^*(Res_{K,G}\mathcal{F})) \, ,$$

with k the inclusion

$$\{0\} \times K/G_x \to \mathbb{R}^p \times K/G_x \simeq q^{-1}(\{y\}) \, .$$

Compare with the description given before corollary 3.1.1 on page 162. Therefore it is **in general** an A_G-module, which is (isomorphic to) a module induced from an A_{G_x}-module (by this argument, we do not need any duality result).

By corollary 3.1.1, the stalks of $Rq_* \mathcal{F}$ have in our examples the same property:

$$H^*_G(i_y^* Rq_* \mathcal{F}) \simeq H^*_G(R\Gamma(q^{-1}(\{y\}), i^* \mathcal{F}))$$

is isomorphic to the A_K-module (with k as above)

$$H^*_K(R\Gamma(q^{-1}(\{y\}), i^*(Res_{K,G}\mathcal{F}))) \simeq H^*_K(K/G_x, k^* i^*(Res_{K,G}\mathcal{F})) \, .$$

We leave it to the reader to show that $H^*_G(i_y^! Rq_! \mathcal{F})$ is isomorphic to a module induced from an A_{G_x}-module.

This allows us to apply the following "variation" of theorem 3.0.1:

Theorem 3.4.1. *Fix a strict full triangulated subcategory T of $D^b_G(\{pt\})$. Consider an equivariant complex $\mathcal{F} \in ob(D^b_G(X))$.*

1. *Assume $Rq_?For(\mathcal{F})$ $(? = !$ or $*)$ is* **constructible** *with respect to a* **finite cellular decomposition** *$\emptyset = Y_{-1} \subset Y_0 \subset \ldots \subset Y_m = Y = X/G$. In the case $? = *$, let Y also be* **compact**. *Denote the closed inclusion $i_k : Y_k \to Y$ $(k = 0, \ldots, m)$. Assume the* **costalk condition**

$$i_y^!(i_k^*(Rq_?\mathcal{F})) \in ob(T) \quad (? = ! \text{ or } *)$$

for all $k = 0, \ldots, m$ and $y \in Y_k \backslash Y_{k-1}$, with i_y the inclusion $\{y\} \to Y_k$. Then

$$R\Gamma_?(X, \mathcal{F}) \in ob(T) \qquad (? = ! \text{ or } *) .$$

2. *Assume $Rq_*For(\mathcal{F})$ is* **constructible** *with respect to a* **finite good cellular stratification** *$\emptyset = Y_{-1} \subset Y_0 \subset \ldots \subset Y_m = Y = X/G$. Denote the closed inclusion $i_k : Y_k \to Y$ $(k = 0, \ldots, m)$. Assume the* **stalk condition**

$$i_y^*(i_k^!(Rq_*\mathcal{F})) \in ob(T)$$

for all $k = 0, \ldots, m$ and $y \in Y_k \backslash Y_{k-1}$, with i_y the inclusion $\{y\} \to Y_k$. Then

$$R\Gamma(X, \mathcal{F}) \in ob(T) .$$

Proof. The proof is almost identical to the proof of theorem 3.0.1 (e.g. we can asssume $X = Y$). In the case (1.) one uses in addition, that one has for a point y of a cell U a natural morphism

$$i_y^!(\mathcal{F}|U) \to R\Gamma_c(U, \mathcal{F})$$

induced from the **adjunction** map $Ri_{y!}i_y^! \to id$. Moreover, this is an isomorphism, if $For(\mathcal{F}|U)$ has locally constant cohomology sheaves. This can be checked after applying For, in which case it follows from [KS, prop. 3.2.3] (or from our proof of lemma 2.2.3 on page 109). The inductive proof of the case (2.) works also in the context of theorem 3.4.1(2), since we explicitly assume that the corresponding stalks $i_y^*(i_k^!(Rq_?\mathcal{F}))$ belong to T. □

By [BL, 1.4.6, thm. 3.4.3] and the preceding discusion, theorem 3.4.1 implies therefore the following localization result for the equivariant cohomology:

Corollary 3.4.1. *Consider a G-space X and a complex $\mathcal{F} \in ob(D^b_G(X))$ as in one of the examples 1.-3. of the introduction. In the case of example 1. or 3. let X/G be compact. Let $i : Y \to X$ be the inclusion of a closed definable G-invariant subset of X (in the case of CW-complex X, a "definable" subset is by definition a union of cells). Assume*

$$S \cap ker(H^*(BG, R) \to H^*(BG_x, R)) \neq \emptyset$$

for all $x \in U := X \backslash Y$. *Then one has for* $\mathcal{F} \in ob(D^b_G(X,R))$:

$$S^{-1}(H^*_{G,?}(X,Y,\mathcal{F})) = 0 \qquad (? = !\ or\ *)\,,$$

and i^* *induces an isomorphism*

$$S^{-1}(H^*_{G,?}(X,\mathcal{F})) \xrightarrow[\sim]{i^*} S^{-1}(H^*_{G,?}(Y,\mathcal{F}))\,. \tag{3.28}$$

Our formulation of the corollary follows [tDieck, ch. III.3], where he studies this localization result for the equivariant cohomology $H^*_G(\cdot,R)$ (corresponding to the constant equivariant sheaf).

Example 3.4.1. *Consider an* **algebraic action of a complex algebraic group** G *on a seperated complex algebraic variety* X*, together with a complex*

$$\mathcal{F} \in ob(D^b_G(X))$$

such that $For(\mathcal{F})$ *is* **complex algebraically constructible**. *If* Y *is a closed* G*-invariant algebraic subvariety, then the statement of corollary* 3.4.1 *is also true in this context: Let* K *be a* **maximal compact** *subgroup of* G*. This is a compact real analytic subgroup of* G*. Therefore* K *and the induced action of* K *on* X *are definable in the o-minimal category* $\mathbb{R}_{an}$ *of* **globally subanalytic** *sets. Moreover*

$$For(Res_{K,G}(\mathcal{F})) \simeq For(\mathcal{F})$$

is constructible with respect to $\mathbb{R}_{an}$ *and* $Res_{K,G}$ *is fully faithfull ([BL, thm.* 3.7.3*])! Especially*

$$H^*_{G,?}(X,Y,\mathcal{F}) \simeq {}_{A_G}H^*_{K,?}(X,Y,Res_{K,G}(\mathcal{F})) \quad (? = !\ or\ *)\,,$$

and the following diagram commutes

$$\begin{array}{ccc} H^*(BG,R) & \longrightarrow & H^*(BG_x,R) \\ \downarrow\wr & & \downarrow \\ H^*(BK,R) & \longrightarrow & H^*(BK_x,R)\,. \end{array}$$

Let us make this more explicit for the case of a compact torus $K = (S^1)^n$ and $R = \mathbb{C}$. Then

$$S := H^*_K(\mathbb{C}) = Sym^*((Lie(K) \otimes_R \mathbb{C})^*) = \mathbb{C}[Lie(K) \otimes_R \mathbb{C}]$$

can by identified with the **complex polynomials on the complexified Lie algebra** $Lie(K) \otimes_R \mathbb{C}$ of K. For $x \in X$, with stabilizer K_x one gets similarly

$$H^*_{K_x}(\mathbb{C}) = H^*_{K^0_x}(\mathbb{C}) = \mathbb{C}[Lie(K^0_x) \otimes_R \mathbb{C}]\,,$$

with K_x^0 the identity component of K_x. Here the first equality is true, since K_x/K_x^0 is finite and we use as a ground ring a field of characteristic 0. For $x \notin X^G =: Y$, there are **linear polynomials** in the kernel of the homomorphism

$$H^*(BK, \mathbb{C}) \to H^*(BK_x^0, \mathbb{C}) ,$$

which vanish on $Lie(K_x^0) \otimes_R \mathbb{C} \subset Lie(K) \otimes_R \mathbb{C}$ (compare [tDieck, proof of thm.3.13, p.196]). Let us denote by

$$spt(M) := \bigcap_{f \in S,\ f \cdot M = 0} \{x \in Lie(K) \otimes_R \mathbb{C} \mid f(x) = 0\}$$

the **support** of an S-module M. Then one has for an S-module M **induced** from an $H^*_{K_x}(\mathbb{C})$-module the estimate

$$spt(M) \subset Lie(K_x^0) \otimes_R \mathbb{C} .$$

Especially, corollary 3.4.1 implies in the case of a **compact torus** the "localization theorem" [GKM, thm. 6.2] for equivariant constructible sheaves (and the same result for the equivariant cohomology with compact support. Compare also with [AtBo, EvMi2]). Moreover, in the o-minimal context, we **do not** have to assume that the action of the torus is compactifiable (as in [GKM]). More precisely, we get the

Corollary 3.4.2. *Let $G = K$ be a compact torus. Consider a K-space X and a complex $\mathcal{F} \in ob(D^b_K(X))$ as in one of the examples 1.-3. of the introduction. In the case of example 1. or 3. let X/G be compact. Let $i : Y \to X$ be the inclusion of a closed definable K-invariant subset of X, with $X^G \subset Y$. Then*

$$H^*_{K,?}(X, Y, \mathcal{F}) \qquad (? = !\ or\ *)$$

is a **torsion module** *over S, and there are finitely many $x_i \in X \backslash Y$ $(i \in I)$ such that*

$$spt(H^*_{K,?}(X, Y, \mathcal{F})) \subset \bigcup_{i \in I} Lie(K^0_{x_i}) \otimes_R \mathbb{C} \quad (? = !\ or\ *) .$$

Chapter 4

Stratification theory and constructible sheaves

Introduction

In this chapter we prove **cohomological** versions of some of the basic results of stratification theory. Here we lay down the foundations for our "Morse theory for constructible sheaves", which we develop in the next chapter. Many of these results are already interesting in themselves, independently of their use in this "deeper" Morse theory.

For nice surveys about stratified spaces compare with chapter I.1 of the work [GM] of Goresky-MacPherson, the articles [Tr1, Tr2] of Trotman and the recent monograph [Pflaum].

There are many different notions of "stratified spaces" in the literature. The following fits nicely with our notion of "constructible sheaves" as used in this book.

Definition 4.0.1. *Let X be a topological space.*

1. *A* **filtration** $X_\bullet$ *of X is a filtration $\emptyset = X_{-1} \subset X_0 \subset \cdots \subset X_n = X$ by closed subsets. We call $depth(X_\bullet) := \sharp\{i|\ X^i \neq \emptyset\} - 1$ the* **depth** *of $X_\bullet$, with $X^i := X_i \backslash X_{i-1}$ $(i = 0, \dots, n)$.*

2. *A complex of sheaves $\mathcal{F} \in ob(D(X))$ is called* **constructible** *with respect to $X_\bullet$ (or $X_\bullet$-constructible), if the restriction of $\mathcal{F}$ to all $X^i := X_i \backslash X_{i-1}$ $(i = 0, \dots, n)$ has locally constant cohomology sheaves. Let us denote by $D(X_\bullet)$ the corresponding full subcategory of all $X_\bullet$-constructible complexes.*

3. *A* **topological stratification** *of X is a filtration $X_\bullet$ of X such that X^i $(i = 0, \dots, n)$ is a topological manifold of pure dimension i (or empty). We call $dim(X) := sup\{i|X^i \neq \emptyset\}$ the* **dimension** *of X (with $dim(\emptyset) := -\infty$). The connected components of the X^i are called the* **strata** *(of this stratification).*

4. *Let X be a closed subset of the smooth manifold N, where we fix a degree $k = 1, \ldots, \infty, \omega$ of smoothness (with $\omega =$ real analytic). A* **smooth stratification** *of X is a topological stratification $X_\bullet$ of X such that X^i $(i = 0, \ldots, n)$ is a smooth submanifold of N.*

Remark 4.0.1. Usually, a stratification of X is defined as a **locally finite** partition of X into **locally closed** (sub-)manifolds S (of N) such that S is pure-dimensional (or connected), and $S \cap \overline{T} \neq \emptyset$ for two different "pieces" S, T implies $dim(S) < dim(T)$. Then $X_i := \bigcup_{dim(S) \leq i} S$ defines a stratification of X as above, whose strata are the connected components of the "pieces" S.

Conversely, we show later on, that in "good cases" the strata of a stratification in our sense are **locally finite** and satisfy the **frontier condition**:

$$S \cap \overline{T} \neq \emptyset \Rightarrow S \subset \overline{T} . \tag{4.1}$$

Let $f : Y \to X$ be a continuous map. Then a filtration $X_\bullet$ of X induces a filtration $Y_\bullet$ of Y by $Y_i := f^{-1}(X_i)$. Suppose f is the inclusion of an open subset, or of a closed subset, which is a union of strata. If $X_\bullet$ is a topological (smooth) stratification, then the same is true for $Y_\bullet$. Similarly, let $f : U \to N$ be the inclusion of an open subset of the smooth manifold N. Then the induced filtration $Y_\bullet$ of $Y := U \cap X \subset U$ is also a smooth stratification.

Nowadays there are many different "regularity conditions", which impose conditions on how the strata of a stratification fit together, and some of them imply suitable topological properties like the "frontier condition" above. The most famous "regularity conditions" are the properties **b** of Whitney and **w** of Verdier. We will recall in section 4.1 the definitions of these (and many other) conditions.

These stratifications are "locally topological trivial" along their strata by the famous "first isotopy lemma of Thom", which is one of the most important results about stratified spaces. Its proof is quite involved and uses a theory of "integrating vector fields":

- **roguse** vector fields in the context of **w-regular** stratifications [Ver, TrWi].
- **controlled** vector fields in the context of **b-regular** stratifications. This theory was first worked out by Mather in his unpublished notes [Mather]. Other presentations can be found in [GWPL, Verona, Shiota, Pflaum].

Later on, Bekka showed that this theory of "controlled vector fields" also applies to his new **C-regular** stratified spaces [Bekka1, Bekka4], which in general is a weaker regularity condition than **b-regularity** [Bekka1, Bekka3].

Even if one is only interested in applications on Whitney **b-regular** stratified spaces, our results apply (with the same proof) to a more general class of stratified spaces, which we call "**d-regular**" or "distance-regular". This includes especially the stratifications which are only "**locally C-regular**" in the sense of Bekka [Bekka2, Bekka3].

Definition 4.0.2. *Let $X_\bullet$ be a smooth stratification of the closed subset X of the smooth manifold N. We call the stratification $X_\bullet$* **d^0-regular** *or "distance-regular", if for* **each** *germ of a* **smooth retraction**

$$\pi : (U, x) \to (X^i, x)$$

(i.e. $x \in X^i$, U is an open neighborhood of x in N and $\pi : U \to U \cap X^i$ is a smooth map with $\pi|(U \cap X^i) = id$), there **exists** *an open neighborhood U' of x in $U \cap X$ and a continuous function*

$$\rho : U' \to \{y \in \mathbb{R} |\ y \geq 0\}$$

with the following properties:

1. *$\{\rho = 0\} = U' \cap X^i$, and $\rho|(U' \backslash X^i)$ is* **smooth** *(i.e. ρ is locally at each point of $U' \backslash X^i$ induced from a smooth function on an open neighborhood in N).*
2. *$(\pi, \rho) : U' \cap X^j \to X^i \times \{y \in \mathbb{R} |\ y > 0\}$ is a* **submersion** *for all $j > i$.*

If one can choose the above function ρ always as a **smooth** *function on U', then we call $X_\bullet$* **d-regular**.

One can formulate this definition also in the following way: For **each** germ of a smooth retraction

$$\pi : (U, x) \to (X^i, x) ,$$

there **exists** a distance function

$$\rho : U' \to \{y \in \mathbb{R} |\ y \geq 0\}$$

such that (π, ρ) is the germ of a "good tube" for X^i at x (in the sense of [Bekka1, p. 45]), and ρ or $\rho|(U' \backslash X^i)$ is **smooth** in the sense above.

Note that this is a **local** condition which is invariant under diffeomorphisms of the ambient manifolds. Moreover, the trivial stratification $\emptyset \subset X$ of a closed smooth submanifold X of N is always **d-regular** (since the second condition is empty in that case). If we would like to indicate the degree k of smoothness, we also use the notion of **d^0_k**- and **d_k-regularity**.

Similarly, any Whitney **b-regular** stratification is also **d-regular**: Given **any** C^k-chart for X^i as a submanifold of N at x:

$$\phi : (U, U \cap X^i, x) \to (\mathbb{R}^n, \mathbb{R}^i \times \{0\}, 0) ,$$

there is a neighborhood of x in which

$$(\pi_\phi, \rho_\phi) := (\phi^{-1} \circ \pi_i \circ \phi, \rho_i \circ \phi)$$

is a **submersion** on each $U \cap X^j$ (for $j > i$), with

$$\pi_i(x_1, \ldots, x_n) := (x_1, \ldots, x_i, 0, \ldots, 0) \quad \text{and} \quad \rho_i(x_1, \ldots, x_n) := \sum_{j=i+1}^{n} x_j^2 .$$

This last condition is called **b^k-regularity** [Tr3], and the implication

$$\mathbf{b} - \text{regularity} \quad \Rightarrow \quad \mathbf{b^k} - \text{regularity}$$

is an easy exercise (which we solve in the next section). Note, that by the **inverse function theorem** any germ of a smooth retraction

$$\pi : (U, x) \to (X^i, x)$$

is of the form $\pi = \pi_\phi$ for a suitable C^k-chart ϕ.

Usually, this "geometric version" of **b-regularity** is just the starting point on a long way towards a proof of the "first isotopy lemma of Thom". But it is already sufficient for our quick proof of a **cohomological version** of the "first isotopy lemma".

All our results are based on the **non-characteristic deformation lemma** of Kashiwara ([KS, prop. 2.7.2], [KS2, thm. 1.4.3]). This implies for a C^k-manifold M directly the following "micro-local" characterization of the (bounded below) complexes of sheaves with locally constant cohomology sheaves:

Proposition 4.0.1. *For $\mathcal{F} \in ob(D^+(M))$ fixed, the following conditions are equivalent:*

1. *All cohomology sheaves of $\mathcal{F}$ are locally constant.*
2. $(R\Gamma_{\{f \geq 0\}}(\mathcal{F}))_x = 0$ *for* **each** *germ of a C^k*-**submersion** $f : (M, x) \to (\mathbb{R}, 0)$.

To simplify the notation, we introduce also the following

Definition 4.0.3. *Let $X_\bullet$ be a smooth stratification of the closed subset X of the smooth manifold N, where we fix a degree $k = 1, \ldots, \infty, \omega$ of smoothness. The stratification $X_\bullet$ satisfies the* **local stratified acyclicity** *property (abbreviated* **LSA**), *if*

$$(R\Gamma_{\{f|X \geq 0\}}(\mathcal{F}))_x = 0 \quad \textit{for all } x \in X^i \ \ (i = 1, \ldots, n),$$

for all $\mathcal{F} \in ob(D(X_\bullet))$ and for **each** *germ of a C^k-function $f : (N, x) \to (\mathbb{R}, 0)$ such that $f|X^i$ is a* **submersion** *in x.*

Since X has finite (cohomological) dimension, it would be enough to ask this "vanishing condition" only for all $\mathcal{F} \in ob(D^+(X_\bullet))$, i.e. for all bounded from below $X_\bullet$-constructible complexes, or even only for all $X_\bullet$-constructible sheaves.

With this definition we get the following

Corollary 4.0.1 (cohomological version of the first isotopy lemma). *Let X be a closed subset of the C^k-manifold N, endowed with a smooth stratification $X_\bullet$. Let also $g : N \to M$ be a C^k-map of C^k-manifolds such that $g|X$ is proper and a stratified submersion, i.e. $g : X^i \to M$ a submersion for all i. Assume that the stratification $X_\bullet$ satisfies the property* **LSA**. *Then all cohomology sheaves of $Rg_*\mathcal{F}$ are* **locally constant** *for* **any** $\mathcal{F} \in ob(D(X_\bullet))$.

Using this corollary, we get by induction on the depth of the stratification an easy proof of the fact, that a $\mathbf{d^0}$**-regular** stratification satisfies the property **LSA** (compare proposition 4.1.2). Especially, we can apply the **cohomological version of the first isotopy lemma** to a Whitney **b-regular** stratification ! Here we only use the following simple implications of regularity conditions:

$$\mathbf{LSA} \longleftarrow \mathbf{d^0} \longleftarrow \mathbf{b^k} \longleftarrow \mathbf{b}\,.$$

The property **LSA** of a stratification $X_\bullet$ is also closely related to a "micro-local" characterization of the $X_\bullet$-constructible complexes in terms of the **micro-support** $\mu supp(\cdot)$ of Kashiwara-Schapira [KS, KS2].

Let us first recall the definition of the **micro-support** ([KS, def.5.1.2, p.221], [KS2, def. 0.1, p. 5; def. 3.1.2, p. 53]):

Definition 4.0.4. *Let M be a smooth manifold for a fixed degree of smoothness $k = 1, \ldots, \infty, \omega$. The* **micro-support** *$\mu supp(\mathcal{F})$ of the complex of sheaves $\mathcal{F} \in ob(D^+(M))$ is the subset of the cotangent bundle T^*M defined by $p \notin \mu supp(\mathcal{F}) :\Leftrightarrow$ there exists an open neighborhood U of p in T^*M such that we have*

$$(R\Gamma_{\{f\geq 0\}}(\mathcal{F}))_{x'} = 0 \tag{4.2}$$

for **any smooth** *function germ $f : (M, x') \to (\mathbb{R}, 0)$ with $df_{x'} \in U$.*

Therefore $\mu supp(\mathcal{F})$ is a **closed conic** subset of T^*M, i.e. a closed subset invariant under the natural $\mathbb{R}^+$-action, given by multiplication on the fibers of the vector bundle $\pi : T^*M \to M$. Moreover

$$\pi(\ \mu supp(\mathcal{F})\ ) = \mu supp(\mathcal{F}) \cap T^*_M M = supp(\mathcal{F})\,, \tag{4.3}$$

where we identify M with the zero-section $T^*_M M$ (for the last inclusion "$\subset$" look at a constant function germ f). Here we use the notation $T^*_N M$ for the **conormal bundle** of a (locally closed) submanifold N of M, i.e. for the kernel of the natural vector bundle epimorphism $T^*M|_N \to T^*N$ dual to the inclusion $TN \hookrightarrow TM|_N$.

Remark 4.0.2. One can show that the micro-support $\mu supp(\mathcal{F})$ depends only on the underlying C^1-structure of the smooth manifold M, i.e. it doesn't matter if we use smooth- or C^1-function germs in the defining "vanishing condition" (4.2). Compare with [KS, prop. 5.1.1, p. 218] and [KS2, thm. 3.1.1, p. 49].

Consider now a **smooth stratification** $X_\bullet$ of the closed subset $X \subset M$, and a smooth function germ $f : (M, x) \to (\mathbb{R}, 0)$, with $x \in X^i$. Then

$$f : (X^i, x) \to (\mathbb{R}, 0)$$

is a not submersion in x if and only if $df_x \in T^*_{X^i} M$. Moreover, if we denote by $k : X \to M$ the inclusion, then one has for $\mathcal{F} \in ob(D^+(X))$ the isomorphism ([KS, prop. 3.1.9(ii)]):

$$(R\Gamma_{\{f|X\geq 0\}}(\mathcal{F}))_x \simeq (R\Gamma_{\{f\geq 0\}}(Rk_*\mathcal{F}))_x\,.$$

Therefore one gets by definition the

Corollary 4.0.2. *Assume that* $X_\bullet$ *satisfies the property* **LSA**. *Consider an* $X_\bullet$*-constructible complex* $\mathcal{F} \in ob(D^+(X))$. *Then*

$$\mu supp(Rk_*\mathcal{F}) \subset \bigcup_i cl(T^*_{X^i}M) \subset T^*M\,, \tag{4.4}$$

with $cl(\cdot)$ *the closure of the corresponding conormal bundle.*

Recall that the smooth stratification $X_\bullet$ is Whitney **a-regular** at $x \in X^i$:$\Leftrightarrow$ if $x_n \in X^j$ (for $i < j$) is a sequence converging to x and the sequence of tangent planes $T_{x_n}X^j$ converges to some limiting plane τ, then $T_xX^i \subset \tau$. Therefore $X_\bullet$ is Whitney **a-regular** (i.e. **a**-regular in all $x \in X$) if and only if

$$\bigcup_i T^*_{X^i}M \subset T^*M$$

is a **closed** subset. This implies by definition the

Corollary 4.0.3. *Assume that* $X_\bullet$ *is Whitney* **a-regular**. *Then the following conditions are equivalent:*

1. $X_\bullet$ *satisfies the property* **LSA**.
2. *One has for any constructible complex* $\mathcal{F} \in ob(D^+(X_\bullet))$ *the estimate*

$$\mu supp(Rk_*\mathcal{F}) \subset \bigcup_i T^*_{X^i}M\,. \tag{4.5}$$

As we will see in the following chapters, this corollary becomes very interesting in the framework of the theory of **Lagrangian cycles** (compare [KS, chap. IX] and [GrM, SchVi, Sch2]).

Moreover, this estimate characterizes the constructibility:

Lemma 4.0.1. *Let* $X_\bullet$ *be a Whitney* **a-regular** *stratification of the closed subset* $X \subset M$, *satisfying the property* **LSA**. *Consider any* $\mathcal{G} \in ob(D^+(M))$ *with*

$$\mu supp(\mathcal{G}) \subset \bigcup_i T^*_{X^i}M\,.$$

Then $supp(\mathcal{G}) \subset X$ *and* $\mathcal{G}|X$ *is* $X_\bullet$*-constructible.*

As an application, we get a "micro-local" characterization of the complexes $\mathcal{F} \in ob(D^+(M))$, which are constructible with respect to a "geometric category" (compare with [KS2, thm. 8.2.6] and [KS, thm. 8.4.2] for the case, where the "geometric category" is given by the **subanalytic** subsets of real analytic manifolds). Here we consider one of the following "geometric categories" (compare with the introduction of chapter 2 for more details):

- M is an affine space $\mathbb{R}^n$, and $\mathcal{S}$ is an **o-minimal structure** on the real field $(\mathbb{R}, +, \cdot)$ ([vDr, vDrMi]), e.g. the structure of **semialgebraic** subsets of real affine spaces. We could also assume, that M is a real analytic $\mathcal{S}$-manifold as in [vDrMi, p. 507, 508] (e.g. a real analytic Nash manifold).

- M is a real analytic manifold, and $\mathcal{S}$ is an **analytic geometric category** ([vDrMi, SchVi]), e.g. the structure of **subanalytic** subsets of real analytic manifolds.

- M is a real analytic Nash manifold, and $\mathcal{S}$ is a **Nash geometric category** (as in chapter 2), e.g. the structure of **locally semialgebraic** subsets of real analytic Nash manifolds.

Definition 4.0.5. *Suppose $\mathcal{S}$ is one of these "geometric categories".*

1. *$\mathcal{F} \in ob(D(M))$ is called $\mathcal{S}$-**constructible**, if $\mathcal{F}$ is $X_\bullet$-constructible for a finite filtration $X_\bullet$ by closed subsets $X_i \subset N$, which are definable with respect to $\mathcal{S}$, i.e. with $X_i \in \mathcal{S}(M)$.*

2. *Consider a **locally closed conic** subset $\Lambda \subset T^*M$, which is **definable** with respect to $\mathcal{S}$. Λ is called **isotropic**, if there is a dense open $\Lambda' \subset \Lambda$ such that Λ' is a C^1-submanifold with $\omega|\Lambda' \equiv 0$, where ω is the canonical 1-form on T^*M (i.e. $\omega|T_p\Lambda' = 0$ for all $p \in \Lambda'$).*

The same definition applies also to the complex analytic (algebraic) context, i.e. M is a complex analytic (algebraic) manifold, if we use complex analytic (algebraic) instead of definable subsets. Note, that for a **locally conic** C^1-submanifold Λ', this agrees with the usual definition of an isotropic submanifold of the symplectic manifold T^*M (compare [KS, app. A.2, p. 481-483]).

Theorem 4.0.1. *Fix the manifold M and a "geometric category" $\mathcal{S}$ as before. Let the complex of sheaves $\mathcal{F} \in ob(D^+(M))$ be given. Then the following conditions are equivalent:*

1. *$\mathcal{F}$ is $\mathcal{S}$-**constructible**.*

2. *$\mu supp(\mathcal{F})$ is contained in a **closed conic definable isotropic** subset of T^*M.*

The proof of this theorem depends on the fact, that a locally closed definable set and a proper definable map (i.e. with a definable graph) can be stratified by suitable Whitney **b-regular** stratifications !

Especially, the theorem and (a variant of) its proof also applies to the complex analytic (algebraic) context. Here of course "conic" means invariance under the natural $\mathbb{C}^*$-action, given by multiplication on the fibers of the holomorphic cotangent bundle. A beautiful application is the theorem of Kashiwara [KS, thm. 11.3.7], that the **solution complex** $Rhom_{D_X}(\mathcal{M}, \mathcal{O}_X)$ of a **holonomic** D_X-module on the complex (algebraic) manifold X is complex analytically (algebraically) constructible. Here one uses the estimate ([KS, thm. 11.3.3]):

$$\mu supp\big(Rhom_{D_X}(\mathcal{M}, \mathcal{O}_X)\big) \subset char(\mathcal{M}) \quad \text{for a coherent } D_X\text{-module } \mathcal{M}.$$

Note that for a **holonomic** D_X-module $\mathcal{M}$ the characteristic variety $char(\mathcal{M})$ is **Lagrangian** (i.e. isotropic of pure dimension equal to dim(X)).

In the second section of this chapter we extend some well-known results about constructible sheaves on **stratified pseudomanifolds** [IC] and **cone-like stratified spaces** (or a **cs**-stratification) [HS] to a more general class of topological stratifications, which we call a **cohomologically cone-like** (or **ccs**-)stratification. This includes especially our **d^0-regular** stratified spaces, i.e. one has the following implications between these notions of stratifications:

$$\text{stratified pseudomanifolds} \longrightarrow \mathbf{cs} \longrightarrow \mathbf{ccs} \longleftarrow \mathbf{d^0}\,.$$

Definition 4.0.6. *Let $X_\bullet$ be a topological stratification. We call $X_\bullet$ a* **cohomologically cone-like** *or* **ccs**-*stratification, if for all $x \in X^i$ there* **exists** *a* **proper** *map*

$$(\pi, \rho) : (U, U \cap X^i) \to (B \times [0, \epsilon[\,, B \times \{0\})\,,$$

with U an open neighborhood of x in $X \backslash X_{i-1}$, B an open ball of dimension i and $\epsilon > 0$ such that

1. *$(\pi, \rho) : U \cap X^i \to B \times \{0\}$ is a* **homeomorphism**, *with $\rho(U \backslash X^i) \subset\,]0, \epsilon[$.*
2. *Denote by $(\pi, \rho)'$ the induced map $(\pi, \rho) : U' := U \backslash X^i \to B \times\,]0, \epsilon[$. Then all cohomology sheaves of $R(\pi, \rho)'_*(\mathcal{G}|U')$ are* **locally constant** *for* **any** *$\mathcal{G} \in ob(D(X \backslash X_i))$, which is constructible with respect to the induced stratification of $X \backslash X_i$.*

A neighborhood U of x as above is called a **distinguished neighborhood** *(with respect to $X_\bullet$).*

The results of [IC, HS] depend on the "local cone-like normal triviality" of their stratified spaces. Since we have in our context no local topological triviality, we develop new proofs of these results. For this we use as a substitute some properties of suitable **local direct image sheaves** as in the definition above (which would also follow from the "local cone-like normal triviality"). Then the arguments are close to the "usual geometric picture", but they work in a purely sheaf theoretic context.

We get especially the

Proposition 4.0.2. *Suppose $X_\bullet$ is a* **ccs**-*stratification.*

1. *The strata of $X_\bullet$ (i.e. the connected components of the X^i) satisfy the* **frontier condition** *(4.1), and the stratification is* **locally finite** *(i.e. a compact subset of X meets only finitely many strata).*

2. *Fix a* **distinguished neighborhood** *of the point* $x \in X^i$. *Then the natural morphisms* $R\Gamma(U, \mathcal{F}) \to \mathcal{F}_x$ *and* $R\Gamma_{\{x\}}(\mathcal{F}) \to R\Gamma_c(U, \mathcal{F})$ *are isomorphisms for all constructible complexes* $\mathcal{F} \in ob(D(X_\bullet))$.

3. *Let* $j : U \to V$ *be the inclusion of* **locally closed** *subsets of* X, *which are unions of strata of* $X_\bullet$. *Then the corresponding categories of constructible complexes of sheaves (for the induced stratifications)*

$$D(U_\bullet) \quad \text{and} \quad D(V_\bullet)$$

are **stable** *under the functors*

$$j^*,\ j^!,\ j_! \quad \text{and}\ Rj_*\,.$$

Assume in addition that we consider complexes of sheaves of R*-modules over a* **commutative** *ring* R. *Let* $D_{perf}(X_\bullet)$ *be the full subcategory of* $X_\bullet$*-constructible complexes* $\mathcal{F}$, *whose stalk complexes* $\mathcal{F}_x$ *are* **perfect** *complexes for all* $x \in X$, *i.e. which are isomorphic to a bounded complex of finitely generated projective* R*-modules. Then also*

$$D_{perf}(U_\bullet) \quad \text{and} \quad D_{perf}(V_\bullet)$$

are stable under these four functors. Moreover, $D_{perf}(X_\bullet)$ *is stable under* $\otimes^L$, $Rhom(\cdot,\cdot)$ *and the* **duality functor** D_X *on* X, *with*

$$\mathcal{F} \xrightarrow{\sim} D_X(D_X\mathcal{F}) \quad \text{for} \quad \mathcal{F} \in ob(D_{perf}(X_\bullet))\,. \tag{4.6}$$

For more general versions of these results in the context of **stratified** maps, compare with proposition 4.2.1, corollary 4.2.1 and corollary 4.2.2. In particular, we explain the following application to the complex analytic (algebraic) context:

Theorem 4.0.2. *Consider a holomorphic (algebraic) map* $f : X \to Y$ *of complex analytic (algebraic) spaces. Denote by*

$$D_{C-?}(X) \quad \text{and} \quad D_{C-?}(Y) \quad (\text{with } ? = \emptyset \text{ or } ? = perf)$$

the corresponding category of **complex analytically (algebraically) constructible** *complexes of sheaves (all of whose stalk complexes are* **perfect** *in the case* $? = perf$, *with* R *a commutative base ring). Then these categories are* **stable** *under the functors*

$$f^*,\ f^!, \quad \text{and for } f \text{ proper also under } Rf_! = Rf_*\,.$$

Suppose R *is commutative. Then* $D_{C-?}(X)$ *is stable under* $\otimes^L$, $D_{C-perf}(X)$ *is stable under* $Rhom(\cdot,\cdot)$, *and* $Rhom(\cdot,\cdot)$ *maps*

$$D^-_C(X) \times D^+_C(X) \quad \text{into} \quad D^+_C(X)\,.$$

Moreover, if we work in the complex **algebraic** *context, then these categories are also* **stable** *under $Rf_!$ and Rf_* (with f not necessarily proper). Similarly, let $f : X \to \mathbb{C}$ be a holomorphic (algebraic) map. Then the* **nearby** *and* **vanishing cycle functor** *ψ_f and ϕ_f map $D_{C-?}(X)$ into $D_{C-?}(\{f=0\})$.*

In the final section we prove especially the following **base change theorem** for constructible sheaves, which is the "cohomological counterpart" of the corresponding "geometric transversality condition":

Proposition 4.0.3. *Let $X_\bullet$ be a smooth* **$\mathbf{d}^0$-regular** *stratification of the closed subset X of the manifold M, and let $g : M \to N$ be a smooth map of manifolds. Fix a point $n \in N$ such that $g : S \to N$ is a* **submersion** *in all points of $S \cap \{g = n\}$, for all strata S of $X_\bullet$. Denote by $j : U \to X$ the inclusion of an open subset, which is a union of strata, and by $i : X' := X \backslash U \to X$ the inclusion of the complement. Consider the cartesian diagram*

$$\begin{array}{ccccc} Y' := X' \cap \{g=n\} & \xrightarrow{i'} & Y := X \cap \{g=n\} & \xleftarrow{j'} & U' := U \cap \{g=n\} \\ \downarrow{\scriptstyle k'} & & \downarrow{\scriptstyle k} & & \downarrow{\scriptstyle k''} \\ X' & \xrightarrow{i} & X & \xleftarrow{j} & U\,. \end{array}$$

Then one has the following **base change isomorphisms***:*

$$k^* Rj_* \mathcal{F} \simeq Rj'_* k''^* \mathcal{F} \quad \text{for} \quad \mathcal{F} \in D(U_\bullet)\,. \tag{4.7}$$

$$k'^* i^! \mathcal{F} \simeq i'^! k^* \mathcal{F} \quad \text{for} \quad \mathcal{F} \in D(X_\bullet)\,. \tag{4.8}$$

These **base change isomorphisms** are very important for our applications of the "(stratified) Morse theory for constructible sheaves", which we develop in the next chapters of this book. We also prove similar **base change isomorphisms** for the **nearby** and **vanishing cycle functor**.

Finally, we explain in section 4.3 also the construction of new stratifications (like "products, transversal pull-back and transversal intersections"). Moreover, we relate this to some "micro-local" results of Kashiwara-Schapira. Especially, we comment on the three operations **proper push-down, non-characteristic pull-back** and **exterior products** from the viewpoint of stratification theory. For different presentations of these three operations (in other situations), we refer to

- [KS, KS2] for the theory of the **micro-support**.
- [KS, SchVi, Gi2, Sab] in the context of **conic Lagrangian cycles**.
- [ScSc] for the theory of **elliptic pairs**.
- [FuMC, Gi2, Sch] for the theory of **characteristic classes** of singular spaces and Lagrangian cycles.

4.1 Stratification theory

Let us fix our notation. All topological spaces are locally compact with countable topology and of finite cohomological dimension (over $\mathbb{Z}$). For a ring R (associative with unit) let $D^*(X) := D^*(X,R)$ for $* = \emptyset, b, +, -$ denote the derived category of (bounded/from below/above) complexes of sheaves of R-modules.

4.1.1 A cohomological version of the first isotopy lemma

First we state the following special case of the **non-characteristic deformation lemma** of Kashiwara ([KS, prop. 2.7.2], [KS2, thm. 1.4.3]):

Lemma 4.1.1. *Let $a, b \in \mathbb{R} \cup \infty$ with $a < b$ and $\mathcal{F} \in ob(D^+([a,b[\,))$ be given. Assume $(R\Gamma_{\{x \geq c\}}(\mathcal{F}))_c = 0$ for all $c \in \,]a,b[$. Then the restriction map*

$$R\Gamma([a,b[\,,\mathcal{F}) \to \mathcal{F}_a$$

is an isomorphism.

Remark 4.1.1. One can give an easy proof of the above isomorphism, if the cohomology sheaves of $\mathcal{F}$ are locally constant on $\,]a,b[\backslash\Sigma$ for a finite (or discrete) subset $\Sigma \subset [a,b[$ (e.g. $\mathcal{F}$ is constructible with respect to an o-minimal structure on $\mathbb{R}$ or a similar "geometric category").

Corollary 4.1.1. *Let $f : X \to [a,b[$ be a* **proper** *map and fix $\mathcal{F} \in ob(D^+(X))$. Assume $(R\Gamma_{\{f \geq f(x)\}}(\mathcal{F}))_x = 0$ for all x with $f(x) > a$. Then the restriction map*

$$R\Gamma(X,\mathcal{F}) \to R\Gamma(\{f = a\}, \mathcal{F})$$

is an isomorphism.

Proof. The above restriction map is isomorphic to the map

$$R\Gamma([a,b[\,, Rf_*\mathcal{F}) \to (Rf_*\mathcal{F})_a \,.$$

Moreover

$$(R\Gamma_{\{x \geq c\}}(Rf_*\mathcal{F}))_c \simeq R\Gamma(\{f = c\}, R\Gamma_{\{f \geq c\}}(\mathcal{F}))\,,$$

since f is proper ([KS, prop. 3.1.9(ii)]). Therefore the claim follows from lemma 4.1.1 (compare also with the proof of [KS2, prop. 4.1.1] and [KS, prop. 5.4.4]). □

Corollary 4.1.1 implies the following important extension of proposition 4.0.1 (which corresponds to the case $N = M$):

Proposition 4.1.1. *Let $k : N \hookrightarrow M$ denote the inclusion of a* **closed** *submanifold N into the manifold M, where we fix a degree $k = 1, \ldots, \infty, \omega$ of smoothness. For $\mathcal{F} \in ob(D^+(N))$ fixed, the following conditions are equivalent:*

1. *All cohomology sheaves of $\mathcal{F}$ are locally constant.*

2. $(R\Gamma_{\{f\geq 0\}}(Rk_*\mathcal{F}))_x = 0$ *for* **each** *germ of a* C^k*-function* $f : (M,x) \to (\mathbb{R},0)$ *such that* $x \in N$ *and* $f|N : (N,x) \to (\mathbb{R},0)$ *is a* **submersion** *in* x.

Remark 4.1.2. If we use the notion of the **micro-support** in the sense of Kashiwara-Schapira as in definition 4.0.4, then the second condition is equivalent to

$$\mu supp(Rk_*\mathcal{F}) \subset T^*_N M \ .$$

Therefore proposition 4.1.1 is a very special case of [KS2, prop. 4.1.1, 4.1.2] and [KS, prop. 5.4.4, 5.4.5].

But we give a direct

Proof of proposition 4.1.1. For the implication $2 \Rightarrow 1$ we can assume

$$M = \mathbb{R}^{n+m} \quad \text{and} \quad i : N = \mathbb{R}^n \times \{0\} \hookrightarrow \mathbb{R}^n \times \mathbb{R}^m = \mathbb{R}^{n+m}$$

is the natural inclusion (by the inverse function theorem). It is enough to show, that the **adjunction morphism**

$$c^*Rc_*\mathcal{F} \to \mathcal{F}$$

is an isomorphism (with c a constant map on N). This can be checked stalkwise, so we have to show that for all $x \in \mathbb{R}^n$ the restriction map $R\Gamma(N,\mathcal{F}) \to \mathcal{F}_{(x,0)}$ is an isomorphism. But this restriction map is isomorphic to

$$R\Gamma(\mathbb{R}^{n+m}, Rk_*\mathcal{F}) \to (Rk_*\mathcal{F})_{(x,0)} \ .$$

Consider the proper polynomial map

$$f : \mathbb{R}^{n+m} \to [0,\infty[\ , \ y \mapsto ||y-(x,0)||^2$$

(with $||\cdot||$ the Euclidean norm). Since $f|(N\backslash\{x\})$ is a submersion, the claim follows from corollary 4.1.1 (with $\{f=0\} = \{(x,0)\}$) and the assumption 2.

The other implication follows from

$$(R\Gamma_{\{f\geq 0\}}(Rk_*\mathcal{F}))_{(x,0)} \simeq (R\Gamma_{\{f|N\geq 0\}}(\mathcal{F}))_{(x,0)}$$

and the following example 4.1.2. □

Remark 4.1.3. Assume that M,N are definable with respect to an o-minimal structure on $(\mathbb{R},+,\cdot)$ or an analytic (Nash) geometric category (compare chapter 2). Then the above proof gives also the implication $2 \Rightarrow 1$, if we consider in the condition 2 only germs of **definable** C^k-functions (for the inverse function theorem in the o-minimal context compare [vDr, ch. 7.2]).

A similar remark applies also to the following improvement of the **cohomological version of the first isotopy lemma**:

Corollary 4.1.2. *Let M be a C^k-manifold ($k = 1, \ldots, \infty, \omega$), X a topological space and $g : X \to M$ a* **proper** *map. Assume X is endowed with a sheaf of real valued C^k-functions such that g is a C^k*-**morphism**, *i.e. for each C^k-function $g : U \to \mathbb{R}$ with U open in M we have: $f \circ g : g^{-1}(U) \to \mathbb{R}$ is a C^k-function on $g^{-1}(U)$. Suppose g is a* **submersion** *in the following sense: One has for $x \in X$ a notion of a submersive C^k-germ $f : (X, x) \to (\mathbb{R}, 0)$ such that $f \circ g : (g^{-1}(U), x) \to (\mathbb{R}, 0)$ is a submersion for* **each** *submersive C^k-germ $f : (U, g(x)) \to (\mathbb{R}, 0)$.*

Let $\mathcal{F} \in ob(D^+(X))$ be given, with $(R\Gamma_{\{f\geq 0\}}(\mathcal{F}))_x = 0$ for **each submersive** *germ of a C^k-function $f : (X, x) \to (\mathbb{R}, 0)$. Then all cohomology sheaves of $Rg_*\mathcal{F}$ are locally constant.*

Proof. This follows from proposition 4.1.1 (with $M = N$) and the isomorphism

$$(R\Gamma_{\{f\geq 0\}}(Rg_*\mathcal{F}))_{g(x)} \simeq R\Gamma(\{g = g(x)\}, R\Gamma_{\{f\circ g\geq 0\}}(\mathcal{F}))\,. \qquad \square$$

Example 4.1.1. *Let X be a closed subset of the smooth manifold N. A smooth map $g : X \to M$ into a smooth manifold is by definition a map, which is* **locally** *at each point $x \in X$ the restriction $g|U \cap X$ of a smooth map $g : U \to M$, with U an open neighborhood of x in N. One defines similarly the sheaf of smooth functions on X. Let in addition $X_\bullet$ be a* **smooth stratification** *of X.*

A smooth germ $g : (X, x) \to (M, g(x))$ is called a submersion (with respect to the stratification $X_\bullet$), if

$$g|S : (S, x) \to (M, g(x))$$

is a submersion, with S the stratum of X containing x. If $X_\bullet$ satisfies the property **LSA**, *then we can apply corollary* 4.1.2 *to any $X_\bullet$-constructible complex $\mathcal{F}$ (after taking a suitable truncation), and we get back our corollary* 4.0.1.

Remark 4.1.4. The same conclusion applies also to a stratification $X_\bullet$ of a topological space X, which is only **locally** of the type above, i.e. each X^i is a smooth manifold with respect to the induced topology and X is **endowed** with a sheaf of smooth functions such that $(X, X_\bullet)$ with this "structure sheaf" is locally isomorphic to a smooth stratification with the sheaf of functions induced by the smooth functions of the ambient manifold. This includes especially the stratified spaces with a **smooth structure** as studied in the monograph [Pflaum] (compare [Pflaum, Def. 1.3.3, p. 27] and also with [Hiro2]. Note that the reference [Hiro2] is missing in [Pflaum]). Then one can easily extend our definition 4.0.3 to these "abstract" stratifications. For example "abstract" Whitney **b-regular** stratifications (compare [Pflaum, p. 36/37]) satisfy this **LSA**-property (since this is a local condition). Important examples of such "abstract" Whitney **b-regular** stratifications come from the real and complex analytic (algebraic) context [Hiro, Hiro2], and from suitable quotient spaces for proper Lie group actions [BaLer, SjLer], [Pflaum, chap. 4].

In order to verify the assumption of corollary 4.1.2, we use the following condition (for S a point):

Lemma 4.1.2. *Let $f : X \to \mathbb{R}$ continuous, $S \subset \{f = 0\}$ compact and $\mathcal{F} \in ob(D(X))$ be given. Assume that there is a fundamental system $K_n \supset K_{n+1}$ of compact neighborhoods of S in X and a sequence of real numbers $\epsilon_n \searrow 0$ $(n \in \mathbb{N})$ such that all direct image sheaves*

$$R^i f_*(\mathcal{F}|K_n \cap \{|f| \le \epsilon_n\})$$

are locally constant on $[-\epsilon_n, \epsilon_n]$. Then $R\Gamma(S, R\Gamma_{\{f\ge 0\}}(\mathcal{F})) = 0$.

Proof. Let $K'_n := K_n \cap \{|f| \le \epsilon_n\}$. The horizontal maps in the commutative diagram

$$\begin{array}{ccc} R\Gamma(K_n \cap \{|f| \le \epsilon_n\}, \mathcal{F}) & \longrightarrow & R\Gamma(K_n \cap \{-\epsilon_n \le f < 0\}, \mathcal{F}) \\ \downarrow \wr & & \downarrow \wr \\ R\Gamma([-\epsilon_n, \epsilon_n], Rf_*(\mathcal{F}|K'_n)) & \longrightarrow & R\Gamma([-\epsilon_n, 0[\, , Rf_*(\mathcal{F}|K'_n)) \end{array}$$

are isomorphisms by the assumption, since the restriction maps

$$R\Gamma(I_n, Rf_*(\mathcal{F}|K'_n)) \to (Rf_*(\mathcal{F}|K'_n))_{-\epsilon_n}$$

are isomorphisms for $I_n = [-\epsilon_n, \epsilon_n], [-\epsilon_n, 0[$. Denote the inclusion

$$j : \{f < 0\} \to X .$$

Since the $K_n \cap \{|f| \le \epsilon_n\}$ build a fundamental system of neighborhoods of S in X, we get for all $i \in \mathbb{Z}$ a commutative diagram

$$\begin{array}{ccc} \varinjlim R^i\Gamma(K_n \cap \{|f| \le \epsilon_n\}, \mathcal{F}) & \xrightarrow{\sim} & \varinjlim R^i\Gamma(K_n \cap \{-\epsilon_n \le f < 0\}, \mathcal{F}) \\ \downarrow \wr & & \downarrow \wr \\ R^i\Gamma(S, \mathcal{F}) & \longrightarrow & R^i\Gamma(S, Rj_*j^*\mathcal{F}). \end{array}$$

Then the claim follows from the distinguished triangle

$$R\Gamma(S, R\Gamma_{\{f\ge 0\}}(\mathcal{F})) \longrightarrow R\Gamma(S, \mathcal{F}) \longrightarrow R\Gamma(S, Rj_*j^*\mathcal{F}) \xrightarrow{[1]} .$$

□

Example 4.1.2. *Consider a C^k-manifold X and a submersive C^k-germ*

$$f : (X, x) \to (\mathbb{R}, 0) .$$

Let $\mathcal{F} \in ob(D(X))$ have only locally constant cohomology sheaves. Then

$$(R\Gamma_{\{f\ge 0\}}(\mathcal{F})_x = 0 .$$

By the inverse function theorem we can assume $(X, x) = (\mathbb{R}^n, 0)$ and f is the projection onto the first coordinate. Then we can apply lemma 4.1.2 to the **trivial fibrations**

$$f = x_1 : K_n := \{(x_1, \dots, x_n) : |x_i| \le \epsilon_n, i = 1, \dots, n\} \to [-\epsilon_n, \epsilon_n]$$

(compare with remark 1.2.1(1) *on page* 42*). Therefore corollary* 4.1.2 *implies for a* **proper submersion** *of manifolds* $g : X \to M$*, that all cohomology sheaves of* $Rg_*\mathcal{F}$ *are locally constant. This is the cohomological version of the* **Ehresmann fibration theorem**.

Remark 4.1.5. The assumption on $\mathcal{F}$ in corollary 4.1.2 is a real counterpart of the notion of a "locally acyclic" morphism (with respect to $\mathcal{F}$) used in [SGA4$\frac{1}{2}$, p. 54, p. 242]. Example 4.1.2 corresponds to the "local acyclicity" of a smooth morphism [SGA4$\frac{1}{2}$, thm. 2.1, p. 58] and [SGA4$\frac{1}{2}$, thm. 3.1, p. 62].

Let us come back to the case of a "proper stratified submersion". There are many different regularity conditions, which imply for a stratification the property **LSA**. Let us begin with a corresponding "geometric" condition:

Definition 4.1.1. *The stratification* $X_\bullet$ *satisfies the* **local stratified triviality** *property (abbreviated* **LST***), if for* **each** *smooth germ of a submersion*

$$g : (X, x) \to (M, g(x))$$

(with M *a smooth manifold), there exists open neighborhoods* U *of* x *in* X*,* V *of* $g(x)$ *in* M *and a filtration* **preserving** *homeomorphism*

$$h : U \to V \times (U \cap \{g = g(x)\}) \,, \quad \textit{with} \quad g = pr_V \circ h \,.$$

Here $pr_V : V \times (U \cap \{g = g(x)\}) \to V$ *denotes the projection. So the homeomorphism* h *satisfies*

$$h(U \cap X_i) = V \times (U \cap X_i \cap \{g = g(x)\}) \quad \textit{for all } i \,.$$

Remark 4.1.6. Definition 4.1.1 is an adaption to the smooth case of a notion introduced in [BMM, def. 4.1.1] and [Tibăr, def. 2.8] in the complex analytic context. They consider a complex analytic stratification in the sense of remark 4.0.1, which is Whitney **a-regular** and satisfies the **frontier condition** (4.1). If $g : (X, x) \to (M, g(x))$ is a holomorphic germ of a submersion, then $U \cap \{g = g(x)\}$ gets for small enough U an induced stratification with the same properties. The frontier condition follows from [Ch].

But in the context of a smooth **a-regular** stratification it is a priori not clear, that the induced stratification of $U \cap \{g = g(x)\}$ satisfies the frontier condition, if this is true for $X_\bullet$. This is indeed the case, as we will see in example 4.3.4. If we do not assume **a-regularity**, then the "pieces" $U \cap \{g = g(x)\} \cap X^i$ of the induced filtration are in general not (sub)manifolds !

Lemma 4.1.3. *A stratification* $X_\bullet$ *with the property* **LST** *satisfies also the condition* **LSA**.

Proof. Fix $\mathcal{F} \in ob(D(X_\bullet))$, a point $x \in X^i$ and consider a germ of a smooth function

$$f : (N, x) \to (\mathbb{R}, 0)$$

such that $f|X^i$ is a submersion in x. Then there exists a filtration preserving homeomorphism

$$h : U \to V \times (U \cap \{g = 0\}) , \quad \text{with} \quad g = pr_V \circ h$$

as in the definition of the condition **LST**. By considering $Rh_*\mathcal{F}$ (instead of $\mathcal{F}$), we can assume

$$f = pr_V : V \times (U \cap \{g = 0\}) \to V \subset \mathbb{R} .$$

Take a compact neighborhood K of x in $U \cap \{g = 0\}$ and a compact interval I in V, containing 0 in its interior. Then $pr_V : I \times K \to I$ is a trivial fibration and $Rh_*\mathcal{F}|I \times K$ is constructible with respect to the filtration $I \times (K \cap X_i)(i = 1, \ldots, n)$. Therefore all cohomology sheaves of

$$Rf_*(\mathcal{F}|h^{-1}(I \times K)) \simeq Rpr_{V*}(Rh_*\mathcal{F}|I \times K)$$

are locally constant on I (compare with remark 1.2.1(1) on page 42), and the claim follows from lemma 4.1.2. □

Remark 4.1.7. For the corresponding result in the complex analytic context compare with (the proof of) [BMM, prop. 4.2.3].

Usually, one deduces the condition **LST** for a given stratification from the proof of the "first isotopy lemma of Thom" by integrating suitable vector fields. This applies for example to Whitney **b-regular** stratifications ([Mather, GWPL, Verona, Shiota, Pflaum]), to the **C-regular** stratifications of [Bekka1, Bekka4], and to **w-regular** stratifications in the sense of Verdier ([Ver, TrWi]). Note, that all these stratifications are Whitney **a-regular**. Most of these proofs work only in the **embedded** case of a stratified subspace of a manifold. Only the approach of Verdier [Ver] applies to "abstract" stratified spaces.

But our approach to the corresponding **cohomological** result applies also to "abstract" stratified spaces, which are only locally isomorphic to embedded stratified spaces as in remark 4.1.4 (e.g. to a sub-(complex-)analytic Whitney **b-regular** stratification of a real (complex) analytic space, or to the quotient space of a smooth manifold with respect to a proper smooth action of a Lie group).

Moreover, we can prove the property **LSA** for our **d^0-regular** stratified spaces directly without any use of the "first isotopy lemma of Thom" (or a suitable theory of vector fields).

Proposition 4.1.2. *A* **d^0-regular** *stratification $X_\bullet$ satisfies the condition* **LSA**.

Proof. The proof is by induction on $depth(X_\bullet)$. The case $depth(X_\bullet) = 0$ correspondes to a closed submanifold X of N and follows from example 4.1.2.

For the induction step, we can assume (by the inverse function theorem) that

$$(X^i, x) = (\mathbb{R}^n \times 0, 0) , \; N = \mathbb{R}^n \times \mathbb{R}^m$$

and f is the (restriction of the) projection $pr_1 := x_1$ onto the first coordinate. Consider the smooth retraction

$$\pi : \mathbb{R}^n \times \mathbb{R}^m \to \mathbb{R}^n \times 0 \ , \ (x_1, \ldots, x_{n+m}) \mapsto (x_1, \ldots, x_n, 0, \ldots, 0) \ .$$

Choose a function $\rho : U' \to \{x \in \mathbb{R} | x \geq 0\}$ as given by the defininition of **$\mathbf{d}^0$-regularity**. Then $U' \cap \{(\pi, \rho) = (0,0)\} = \{0\}$. Therefore, we can choose an open neighborhood V of 0 in X^i and an $\epsilon > 0$ such that

$$(\pi, \rho) : U'' := U' \cap \pi^{-1}(V) \cap \{\rho < \epsilon\} \to V \times [0, \epsilon[$$

is proper, and $(\pi, \rho) : U'' \cap X^i \to V \times \{0\}$ is an isomorphism.

Denote the inclusions $i : V \times \{0\} \to V \times [0, \epsilon[$ and $j : V \times]0, \epsilon[\to V \times [0, \epsilon[$. Then $i^*(R(\pi, \rho)_*(\mathcal{F}|U''))$ has only locally constant cohomology sheaves for $\mathcal{F} \in ob(D(X_\bullet))$. The same is true for $j^*(R(\pi, \rho)_*(\mathcal{F}|U''))$ by the following argument:

$$(\pi, \rho) : U'' \backslash X^i \to V \times]0, \epsilon[$$

is a **proper stratified submersion** on the induced stratification of $(U'' \cap X) \backslash X^i$ as a closed subset of a suitable open subset of the manifold $N \backslash X^i$. But the **depth** of this induced **$\mathbf{d}^0$-regular** stratification is **less** than $depth(X_\bullet)$. So we can apply the induction hypothesis and corollary 4.1.2.

Take a compact neighborhood K of 0 in $V \times [0, \epsilon[$ of the form

$$K = I \times K' \times [0, c] \ ,$$

with $I \subset \mathbb{R}$ a compact interval, K' a compact neighborhood of 0 in $\mathbb{R}^{n-1}$, and $0 < c < \epsilon$. Then

$$pr := x_1 : (K, I \times K' \times \{0\}) \to I$$

is a **trivial fibration of pairs** and $(R(\pi, \rho)_*(\mathcal{F}|U''))|K$ is constructible with respect to the filtration $I \times K' \times \{0\} \subset I \times K' \times [0, c] = K$. Therefore

$$Rf_*(\mathcal{F}|(\pi, \rho)^{-1}(K)) \simeq Rpr_*((R(\pi, \rho)_*(\mathcal{F}|U''))|K)$$

has locally constant cohomology sheaves on I (compare with remark 1.2.1(1) on page 42), and the claim follows from lemma 4.1.2. □

Remark 4.1.8. The proof above shows that a **$\mathbf{d}^0$-regular** stratification is especially a **ccs**-stratification in the sense of definition 4.0.6.

4.1.2 Comparison of different regularity conditions

To see that proposition 4.1.2 is a useful result, we compare the **$\mathbf{d}^0$-regularity** with other well-known regularity conditions:

Definition 4.1.2. *Consider a smooth stratification $X_\bullet$ of $X \subset N$, where we fix a degree $k = 1, \ldots, \infty, \omega$ of smoothness. Let $\rho : U \to \{x \in \mathbb{R} |\, x \geq 0\}$ be a continuous function defined on an open neighborhood U of $x \in X^i$ in N, with $\{\rho = 0\} = U \cap X^i$ and $\rho|(U \backslash X^i)$ C^k-smooth.*

1. *$X_\bullet$ is Whitney* **b-regular** *at $x :\Leftrightarrow$ if $x_n \in X^j$ (for $i < j$) and $y_n \in X^i$ are sequences converging to $x \in X^i$ such that the tangent planes $T_{x_n} X^j$ converge to some limiting plane τ, and the secant lines $l_n = \overline{x_i, y_i}$ (with respect to some local coordinates) converge to some limiting line l, then $l \subset \tau$.*

2. *$X_\bullet$ is* **b^k-regular** *(or* **a^k-regular***) at $x :\Leftrightarrow$ Given any C^k-chart for X^i as a submanifold of N at x:*

$$\phi : (U, U \cap X^i, x) \to (\mathbb{R}^n, \mathbb{R}^i \times \{0\}, 0),$$

there is a neighborhood of x in which (π_ϕ, ρ_ϕ) (or π_ϕ) is a submersion on each $U \cap X^j$ $(i < j)$ for $(\pi_\phi, \rho_\phi) := (\phi^{-1} \circ \pi_i \circ \phi, \rho_i \circ \phi)$, with

$$\pi_i(x_1, \ldots, x_n) := (x_1, \ldots, x_i, 0, \ldots, 0) \quad \text{and} \quad \rho_i(x_1, \ldots, x_n) := \sum_{j=i+1}^{n} x_j^2 .$$

3. *$X_\bullet$ is* **w-regular** *at $x :\Leftrightarrow$ there exists a neighborhood U' of x and a constant $C > 0$ such that (in suitable local coordinates)*

$$d(T_{x'} X^i, T_y X^j) < C \cdot ||y - x'||$$

for all $x' \in U' \cap X^i$ and all $y \in U' \cap X^j$ $(i < j)$. Here $d(\cdot, \cdot)$ is the "vector space distance" of two subvector spaces $V, W \subset \mathbb{R}^N$:

$$d(V, W) := \sup_{v \in V,\, ||v|| = 1} \inf \{\, ||v - w|| \;:\; w \in W \} .$$

4. *$X_\bullet$ is* **locally C^0_k-regular** *(with respect to ρ) at $x :\Leftrightarrow$ there exists an open neighborhood U' of x in U such that*

$$\rho : U' \cap X^j \to \{x \in \mathbb{R} |\, x > 0\}$$

is a submersion for all $j > i$ and the following property holds: if $x_n \in X^j$ $(i < j)$ is a sequence converging to x such that

$$kern(d\rho(x_n)) \cap T_{x_n} X^j$$

converges to some limiting plane τ, then $T_x X^i \subset \tau$ (i.e. X^j is Thom regular over X^i at x relative to ρ [GWPL, p. 23]).

5. *The stratification $X_\bullet$ of $X \subset N$ is by definition* **$*$-regular***, for $*$ one of the conditions as before, if it is $*$-regular at* **all** *$x \in X$.*

Especially, $X_\bullet$ is by definition **locally $\mathbf{C}^0_k$-regular** iff it is locally C^0_k-regular at all $x \in X$ with respect to some function ρ as above **depending** on x (compare with [Bekka2, Bekka3]). This is the difference from the **$\mathbf{C}^0_k$-regularity** of [Bekka4], where one uses for all $x \in X^i$ a fixed "global" function ρ defined on an open neighborhood U of X^i !

If we can choose this ρ as a C^k-smooth function on an open neighborhood U (of x in N) as in [Bekka1], then we call $X_\bullet$ **(locally) $\mathbf{C}_k$-regular** (at x). If the degree k is fixed, we also use the notion of (local) C-regularity (at x).

Moreover, in the case of **w-regularity** one has for an intrinsic definition to assume $k \geq 2$.

Remark 4.1.9. All these regularity conditions have in addition the following **stability** property: Let $X_\bullet$ be a stratification of the closed subset $X \subset N$ with the regularity condition $*$. Suppose Y is a closed subspace of X, which is a union of strata of $X_\bullet$. Then the induced stratification $Y_\bullet$ of $Y \subset N$ is also $*$-regular. This applies especially to $* = a, b, w, a^k, b^k, C^{(0)}, d^{(0)}$, and also to the properties **LSA, LST**.

One has the following relations between all these regularity conditions **(rc)**:

$$\begin{array}{ccccccccc}
\mathbf{b^k} & \longleftarrow & \mathbf{a+b^k} & \longleftarrow & \mathbf{b} & \underset{\text{categ.}}{\overset{\text{in geom.}}{\longleftarrow}} & \mathbf{w} & = \! = \! = & \mu-\mathbf{reg}. \\
\downarrow & & \downarrow & & \downarrow & & \downarrow & & \downarrow \\
\mathbf{d_k^{(0)}} & \longleftarrow & \mathbf{loc.}-\mathbf{C_k^{(0)}} & \longleftarrow & \mathbf{C_k^{(0)}} & \longrightarrow & \mathbf{LST} & \longrightarrow & \mathbf{LSA} \\
\downarrow & & \downarrow & & & & & & \\
\mathbf{a^k+LSA} & & \mathbf{a} & & & & & & \\
\downarrow & & \downarrow & & & & & & \\
\mathbf{a^k} & = \! = \! = & \mathbf{a^k}\,. & & & & & &
\end{array}$$

Let us now explain these implications:

- The implications $\mathbf{a^k} \Leftarrow \mathbf{d_k^{(0)}} \Leftarrow \mathbf{b^k} \Leftarrow \mathbf{a+b^k}$ and $\mathbf{C_k^{(0)}} \Rightarrow \mathbf{loc.}-\mathbf{C_k^{(0)}}$ are trivial.

- The implications $\mathbf{d_k^{(0)}} \Rightarrow \mathbf{LSA} \Leftarrow \mathbf{LST}$ follow from lemma 4.1.3 and proposition 4.1.2.

- Using the definitions, one also has $\mathbf{a} \Leftarrow \mathbf{loc.}-\mathbf{C_k^{(0)}}$:

$$T_xX^i \subset \lim_{n\to\infty} Kern(d\rho(x_n)) \cap T_{x_n}X^j \subset \lim_{n\to\infty} T_{x_n}X^j\,.$$

- By definition one has $\mathbf{w} \Rightarrow \mathbf{a}$, since **a-regularity** can be reformulated as

$$d(T_{x'}X^i, T_yX^j) \to 0 \quad \text{for} \quad x', y \to x\,.$$

In general **w-regularity** does **not** imply **b-regularity** for a smooth stratification (compare [Ver, p. 297]). But this is **always** true if the stratification is defined in a "geometric category" as in definition 4.0.5. Compare with [Ta1] for a proof, and also with [Ver] for the "subanalytic context".

- As already remarked before, the implications $\mathbf{C_k^{(0)}} \Rightarrow \mathbf{LST} \Leftarrow \mathbf{w}$ follow from the proof of the "first isotopy lemma of Thom" by integrating suitable vector fields [Bekka1, Bekka4, Mather, TrWi, Ver, Verona].
- We do not recall the "micro-local" definition of **μ-regularity** introduced in [KS, def. 8.3.19, p. 334], since by [Tr4] it is equivalent to **w-regularity**. Especially, one can get in this way a proof of $\mu -$ **reg.** $\Rightarrow$ **LSA**, using the "first isotopy lemma of Thom" (corresponding to the right square in diagram (rc)). For another purely sheaf-theoretic proof compare with [KS, prop. 8.4.1]. This is based on some functoriality results of Kashiwara-Schapira about their **micro-support** (especially [KS, prop. 6.3.2]).

In the following, we recall the well-known (and easy) proof of the implications (compare for example with [Bekka1, BeKoi, GWPL, Mather, Pflaum]):

$$\mathbf{d_k^{(0)}} \longleftarrow \mathbf{loc.} - \mathbf{C_k^{(0)}} \longleftarrow \mathbf{a} + \mathbf{b^k} \longleftarrow \mathbf{b} \longrightarrow \mathbf{a} \longrightarrow \mathbf{a^k}\,.$$

These are the only ones, that we use in this book ! Moreover, this proof also gives the missing implication $\mathbf{b} \Rightarrow \mathbf{C_k}$, if one uses a corresponding "uniqueness result about tubular neighborhoods" [Mather, prop. 6.1] (compare [Pflaum, 3.1.8(3), p. 95, cor. 3.4.3]), which shows that **any** "tubular neighborhood" is locally of the form (π_ϕ, ρ_ϕ) (as in the definition of $\mathbf{b}^k$) for a **suitable** C^k-chart ϕ.

We use the notation from definition 4.1.2, and show these relations already for the corresponding local conditions at a fixed point $x \in X^i \subset \mathbb{R}^N$ (in some local coordinates), with the standard scalar product and norm on $\mathbb{R}^N$.

$\mathbf{a} \Rightarrow \mathbf{a}^k$ **:** Suppose that the retraction π_ϕ is not a stratified submersion in a neighborhood of x. Then there is a sequence $x_n \in X^j$ (for some $i < j$) converging to $x \in X^i$ such that $T_{x_n}X^j \to \tau$, with

$$rg(d\pi_\phi | T_{x_n}X^j) < i\,.$$

Then one gets the contradiction

$$rg(d\pi_\phi|\tau) < i\,,\; T_xX^i \subset \tau \quad \text{and} \quad d\pi_\phi|T_xX^i = id\,.$$

$\mathbf{b} \Rightarrow \mathbf{a}$ **:** Let $x_n \in X^j$ (for $i < j$) be a sequence converging to $x \in X^i$ such that $T_{x_n}X^j \to \tau$. Fix a tangent vector $\nu = \gamma'(0) \in T_xX^i$, with

$$\gamma : (]-1,1[\,,0) \to (X^i, x)$$

a smooth path. We can assume that $w_n := (x_n - x)/t_n$ converges to some $w \in \mathbb{R}^N$, with $t_n := ||x_n - x||$. Then the sequence of vectors

$$\frac{\gamma(t_n) - x}{t_n} - w_n = \frac{\gamma(t_n) - x_n}{t_n}$$

converges to $\nu - w$. By Whitney **b-regularity** we get $\nu - w, w \in \tau$ so that also $\nu \in \tau$.

b $\Rightarrow$ **b**^{k} **:** Suppose that there is C^k-chart ϕ such that $(\pi, \rho) := (\pi_\phi, \rho_\phi)$ is not a submersion on X^j (for some $i < j$) in a neighborhood of x. Working in this chart, we can assume $\phi = id$ and $x = 0$. Since $\mathbf{b} \Rightarrow \mathbf{a} \Rightarrow \mathbf{a^k}$, there exists a sequence $x_n \in X^j$ converging to $0 \in X^i = \mathbb{R}^i \times \{0\}$ such that

$$K_n := kern(d\pi(x_n)) \cap T_{x_n}X^j \subset kern(d\rho(x_n)) .$$

Especially, the secant line $l_n := \overline{x_n, \pi(x_n)}$ is orthogonal to K_n and to $T_0X^i = \mathbb{R}^i \times \{0\}$. Moreover, we can assume $K_n \to K$ and $l_n \to l$. Then

$$T_{x_n}X^j \to \tau = K \oplus (\mathbb{R}^i \times \{0\})$$

so that l is orthogonal to τ. This is a contradiction to the Whitney **b-condition** $l \subset \tau$.

a+b^k $\Rightarrow$ **loc.-C**$_k$ **:** We show actually a slightly stronger result, since we only use a **fixed** C^k-chart ϕ such that $(\pi, \rho) := (\pi_\phi, \rho_\phi)$ is a stratified submersion outside X_i in a neighborhood of x (as in the condition **m** of [BeKoi, def. 2.3]). Again we assume $\phi = id$, $x = 0$, and we show that $X_\bullet$ is **C**$_k$**-regular** at 0 with respect to this function ρ. Let $x_n \in X^j$ $(i < j)$ be a sequence converging to 0 such that

$$K_n := kern(d\rho(x_n)) \cap T_{x_n}X^j \to \tau .$$

Then the condition **b**^{k} implies, that there is a one-dimensional subspace

$$L_n \subset kern(d\pi(x_n)) \cap T_{x_n}X^j$$

orthogonal to $T_0X^i = \mathbb{R}^i \times \{0\}$ and $kern(d\rho(x_n)) \cap kern(d\pi(x_n)) \cap T_{x_n}X^j$. By

$$kern(d\pi(y))^\perp \subset kern(d\rho(y)) \quad \text{for} \quad y \in \mathbb{R}^n ,$$

L_n is therefore orthogonal to K_n. We can also assume $L_n \to L$. Then

$$T_{x_n}X^j \to \tau' = L \oplus \tau ,$$

and by the Whitney **a-condition** we get $T_xX^i \subset \tau'$. But L is orthogonal to T_xX^i so that finally $T_xX^i \subset \tau$.

loc.-$\mathbf{C}_k^{(0)} \Rightarrow \mathbf{d}_k^{(0)}$: Suppose $X_\bullet$ is $\mathbf{C}_k^{(0)}$**-regular** at x with respect to the function ρ. Let

$$\pi : (U, x) \to (X^i, x)$$

be a germ of a smooth retraction, with U an open neighborhood of x in N. We claim that (π, ρ) is locally near x a submersion on X^j (for all $i < j$). Otherwise there is a sequence $x_n \in X^j$ (for some $i < j$) converging to x such that

$$K_n := kern(d\rho(x_n)) \cap T_{x_n}X^j \to \tau\,, \quad \text{with} \quad rg(d\pi|K_n) < i\,.$$

Then one gets the contradiction

$$rg(d\pi|\tau) < i\,,\; T_xX^i \subset \tau \quad \text{and} \quad d\pi_\phi|T_xX^i = id\,.$$

Let us finish our comparison of these different regularity conditions with the following observations:

- In the proof of **loc.-$\mathbf{C}_k^{(0)} \Rightarrow \mathbf{d}_k^{(0)}$** we showed that (π, ρ) is locally near $x \in X^i$ a submersion on X^j (for all $i < j$) for any germ of a smooth retraction π. One can show (compare with the proof of [Bekka1, prop. 1.2]), that this is in fact **equivalent** to the **local $\mathbf{C}_k^{(0)}$-regularity** of $X_\bullet$ at x with respect to ρ. So the difference between these two regularity conditions is the fact, that for $\mathbf{d}_k^{(0)}$**-regularity** we allow the corresponding function ρ to depend on the choice of the smooth retraction !

- $\mathbf{b}^k$ for $k \geq 2$ does **not** imply **a-regularity** (compare [Tr3, KamTr, Perkal, Hajto]). Especially, $\mathbf{d}_k^{(0)}$ for $k \geq 2$ or the condition **LSA** does **not** imply **a-regularity**, and $\mathbf{d}_k^{(0)}$ does **not** imply **loc.-$\mathbf{C}_k^{(0)}$**. It is a deeper result of Trotman [Tr3] (compare [Perkal, Hajto]), that

$$\mathbf{a} \Leftrightarrow \mathbf{a^1} \quad \text{and} \quad \mathbf{b} \Leftrightarrow \mathbf{b^1}\,.$$

  Especially, $\mathbf{d}_1^{(0)} \Rightarrow \mathbf{a}^1 \Rightarrow \mathbf{a}$, and **b-regularity** is a C^1-invariant condition ! For a simple proof of C^2-invariance compare with [Pflaum, lem. 1.4.4].

- **a** does **not** imply the condition **LSA** (and therefore also not the condition **LST**). A simple example is given by the "Whitney cusp" (compare [TrWi, p. 335] and [Pflaum, p. 20/21]):

$$X := \{(x, y, z) \in \mathbb{R}^3 |\, x^2 + z^2 \cdot (z - y^2) = 0\}\,.$$

  Then the filtration $X_1 := \{(x, y, z) \in \mathbb{R}^3 |\, x = 0 = z\} \subset X_2 := X$ is Whitney **a-regular**, and the function $f(x, y, z) := y$ is a stratified submersion with

$$(R\Gamma_{\{f|X \geq 0\}}(\mathbb{Z}_X))_0 \neq 0\,.$$

  More explicitly, the cohomology of $(R\Gamma_{\{f|X \geq 0\}}(\mathbb{Z}_X))_0$ is isomorphic to the reduced cohomology of a "left Milnor fiber" of $f|X$ in 0 (compare chapter 1 for more on this). In our case this "left Milnor fiber" has the homotopy type of a circle.

4.1.3 Micro-local characterization of constructible sheaves

We come back to the other results stated in the introduction. First we explain the

Proof of lemma 4.0.1. The assumption implies by (4.3) that $supp(\mathcal{G}) \subset X$. The proof is by induction on $d := min\{i \mid supp(\mathcal{G}) \subset X_i\}$. Then $\mathcal{G} \simeq Rk_* k^* \mathcal{G}$, with k the closed inclusion $X_d \to N$. The case $d = -1$ is trivial, since $X_{-1} := \emptyset$ implies $\mathcal{G} = 0$.

For the induction step, we consider the cartesian diagram of inclusions:

$$\begin{array}{ccc} X_d & \xrightarrow{k} & N \\ \uparrow j & & \uparrow j' \\ X^d & \xrightarrow{k'} & N' , \end{array}$$

with $N' := N \backslash X_{d-1}$. Then

$$j'^* \mathcal{G} \simeq Rk'_* j^* k^* \mathcal{G} \,,\; supp(j'^* \mathcal{G}) \subset X^d \quad \text{and} \quad \mu supp(j'^* \mathcal{G}) \subset T^*_{X^d} N' \,.$$

But X^d is a **closed** submanifold of N'. Therefore all cohomology sheaves of $j^* k^* \mathcal{G}$ are locally constant by remark 4.1.2 and proposition 4.1.1. Then $Rj_! j^* k^* \mathcal{G}$ is $X_\bullet$-constructible, and this implies by **a-regularity** and corollary 4.0.2:

$$\mu supp(Rk_* Rj_! j^* k^* \mathcal{G}) \subset \bigcup_i T^*_{X^i} N \,.$$

Denote the closed inclusion $i : X_{d-1} \to X_d$. The distinguished triangle

$$Rk_* Rj_! j^* k^* \mathcal{G} \longrightarrow Rk_* k^* \mathcal{G} \longrightarrow Rk_* Ri_* i^* k^* \mathcal{G} \xrightarrow{[1]}$$

implies by the definition 4.0.4 of the micro-support the estimate

$$\mu supp(Rk_* Ri_* i^* k^* \mathcal{G}) \subset \mu supp(Rk_* k^* \mathcal{G}) \cup \mu supp(Rk_* Rj_! j^* k^* \mathcal{G}) \subset \bigcup_i T^*_{X^i} N \,.$$

But $supp(Rk_* Ri_* i^* k^* \mathcal{G}) \subset X_{d-1}$ so that $(Rk_* Ri_* i^* k^* \mathcal{G})|X$ is $X_\bullet$-constructible by induction hypothesis. Then $\mathcal{G}|X \simeq (Rk_* k^* \mathcal{G})|X$ is also $X_\bullet$-constructible (by the distinguished triangle above). □

Next we work in the context of definition 4.0.5 and give the

Proof of theorem 4.0.1. The proof is similar to the proof of [KS2, thm. 8.2.6].

1. $\Rightarrow$ 2.: Since $\mathcal{F}$ is $\mathcal{S}$-constructible, there exists a finite filtration $X_\bullet$ by definable closed subsets such that $\mathcal{F}$ is $X_\bullet$-constructible and $X_\bullet$ is a Whitney **b-regular** stratification by C^k-submanifolds X^i of M, with $2 \leq k < \infty$ (see [Ta1], [vDrMi, 1.19, 4.8] or [Shiota, (II.1.14)]). By corollary 4.0.2 we get

$$\mu supp(\mathcal{F}) \subset \bigcup_i T^*_{X^i} M =: \Lambda \subset T^* M \,.$$

But Λ is a **closed conic definable isotropic** subset of T^*M. It is **definable** by [vDrMi, 1.9, 2.3], and contains the following open and **dense** (definable) isotropic C^1-submanifold (cf. [vDrMi, 1.7, 1.16, 4.7]):

$$\Lambda' := \bigcup_i \left(T^*_{X^i}M \backslash \bigcup_{i \neq j} cl(T^*_{X^j}M) \right) . \tag{4.9}$$

1. $\Rightarrow$ 2.: Let Λ be a **closed conic definable isotropic** subset of T^*M. By the next proposition 4.1.3, there exists a definable Whitney **b-regular** C^k-stratification $X_\bullet$ as above such that

$$\Lambda \subset \bigcup_i T^*_{X^i}M \, .$$

Then the claim follows from lemma 4.0.1. □

Remark 4.1.10. Let $\mathcal{F} \in ob(D^+(M))$ be $X_\bullet$-constructible with respect to a definable Whitney **a-regular** C^k-stratification $X_\bullet$ as above. Assume that this stratification also has the property **LSA**. By using the **involutivity** of $\mu supp(\mathcal{F})$ ([KS2, thm. 6.4.1] and [KS, thm. 6.5.4]) one can show as in [KS2, prop. 8.2.4], that $\mu supp(\mathcal{F})$ is a union of closures of connected components of Λ' as in (4.9). Especially, $\mu supp(\mathcal{F})$ is a **closed conic definable Lagrangian** subset of T^*M. We will give an **exact** description of $\mu supp(\mathcal{F})$ in the next chapter (compare proposition 5.0.1), by using our main results about "Morse theory for constructible sheaves".

We also use the following lemma (compare [KS2, prop. 8.2.2]):

Lemma 4.1.4. *Let M be a real analytic (Nash) manifold, and fix a 1-form θ on M. Define $\theta|\Lambda \equiv 0$ for a* **locally closed definable** *subset $\Lambda \subset M$, iff there is a dense open $\Lambda^0 \subset \Lambda$ such that Λ^0 is a C^1-submanifold of M with $\theta|\Lambda^0 \equiv 0$. Then*

$$\theta|\Lambda \equiv 0 \quad \Rightarrow \quad \theta|\Lambda' \equiv 0$$

for any locally closed definable subset $\Lambda' \subset \Lambda$. Especially, any **locally closed conic definable** *subset Λ' of a* **closed conic definable isotropic** *set $\Lambda \subset T^*M$ is also* **isotropic** *!*

Proof. There exists a definable Whitney **b-regular** C^1-stratification $\Lambda_\bullet$ of Λ such that Λ' is a union of strata (see [Ta1], [vDrMi, 1.19, 4.8] or [Shiota, (II.1.14)]). It is enough to show that $\theta|S \equiv 0$ for all strata S of $\Lambda_\bullet$, because the union of all strata, which are open in Λ', is **dense** in Λ'. Here we already use the fact that the strata satisfy the **frontier condition** by proposition 4.0.2 and remark 4.1.8. We show $\theta|T_pS = 0$ for $p \in S$ by decreasing induction on $dim(S)$. If S is a neighborhood of p in Λ, especially if S is top-dimensional, this follows from $\theta|\Lambda \equiv 0$. Suppose S is not a neighborhood of p in Λ. Therefore $p \in cl(S')$ for some stratum $S' \neq S$ with $dim(S') > dim(S)$. Then there is a sequence $p_n \in S'$ converging to p such that

$$T_{p_n}S' \to \tau \quad \text{and} \quad \theta|T_{p_n}S' = 0 \, .$$

By **a-regularity** we get $T_pS \subset \tau$ and $\theta|\tau = 0$. □

The final result of this section is the following

Proposition 4.1.3. *Let* Λ *be a* **closed conic definable isotropic** *subset of* T^*M*. Then there exists a definable Whitney b-regular* C^k*-stratification* $X_\bullet$ *of* M $(2 \leq k < \infty)$ *such that*

$$\Lambda \subset \bigcup_i T^*_{X^i}M =: \Delta \, .$$

Proof. The proof is the same as in [KS2, proof of prop. 8.2.2, p. 144]. Consider the projection onto the "sphere-bundle"

$$\pi' : T^*M \backslash T^*_M M \to S(T^*M) := (T^*M \backslash T^*_M M)/\mathbb{R}^+ \, .$$

Then $S(T^*M)$ is a **definable** real analytic manifold such that π' is a **definable submersion**. The proof of this fact is more or less the same as for the projection onto the corresponding "projective bundle":

$$T^*M \backslash T^*_M M \to P(T^*M) := (T^*M \backslash T^*_M M)/\mathbb{R}^* \, .$$

Then $C := \pi'(\Lambda \backslash T^*_M M)$ is a closed definable subset of $S(T^*M)$, since Λ is a closed conic definable set. Moreover, the projection

$$p : S(T^*M) \to M$$

is a **proper definable** map so that it can be **stratified** in the following way (see [vDrMi, 1.19, 4.8] or [Shiota, (II.1.14)]):

There exist definable Whitney b-regular C^k-stratifications $C_\bullet$ and $X_\bullet$ of $S(T^*M)$ and M $(2 \leq k < \infty)$, with C a union of strata of $C_\bullet$ such that p maps each stratum S of $C_\bullet$ into a stratum T of $X_\bullet$ with $p : S \to T$ a C^k-submersion. We claim $\Lambda \subset \Delta$ for such a stratification $X_\bullet$. Since $X_\bullet$ is a Whitney **a-regular** stratification of the whole manifold M, we have $T^*_M M \subset \Delta$. So it is enough to show

$$S' := \pi'^{-1}(S) \subset \Delta$$

for any stratum S of $C_\bullet$ contained in C. But S' is a C^k-submanifold of T^*M such that the restriction $\pi : S' \to T$ (with T as before) is a submersion. Here $\pi : T^*M \to M$ is the canonical map. Moreover, $S' \subset \Lambda$ is a locally closed conic definable subset so that

$$\omega | S' \equiv 0$$

by lemma 4.1.4. Then one gets $S' \subset T^*_T M \subset \Delta$.

This can be checked in local coordinates $(x_1, \ldots, x_n)$, with

$$T = \{x_1 = \cdots = x_k = 0\} \quad \text{for some} \quad 0 < k \leq n \, .$$

If $(x_1, \ldots, x_n, \xi_1, \ldots, \xi_n)$ are the associated coordinates on T^*M, then (compare [KS, p. 481]):

$$\omega | \pi^{-1}(T) = \sum_{i=k+1}^{n} \xi_i dx_i \, .$$

But $dx_{k+1}, \dots, dx_n$ are linearly independent on T and therefore also on S', since $\pi : S' \to T$ is a submersion. This implies $\xi_{k+1} = \dots = \xi_n = 0$ on S', i.e.

$$S' \subset \{x_1 = \dots = x_k = 0,\ \xi_{k+1} = \dots = \xi_n = 0\} = T^*_T M . \qquad \square$$

Remark 4.1.11. The proof above applies also to the complex analytic (algebraic) context, if one uses the projection onto the corresponding "projective bundle":

$$T^*M \backslash T^*_M M \to P(T^*M) := (T^*M \backslash T^*_M M)/\mathbb{C}^* .$$

Moreover, let in this context $C \subset T^*M$ be an **irreducible** conic isotropic subset. By "generic smoothness" of the projection $\pi : C \to \pi(C) =: S$ (i.e. the restriction to the regular parts $\pi : C_{reg} \to S_{reg}$ is "generically" a submersion) one gets as before:

$$C \subset cl(T^*_{S_{reg}} M) =: T^*_S M ,$$

and this inclusion is an equality if C is in addition **Lagrangian** ([ChGi, lem. 1.3.27, p. 39]). This argument works in the algebraic context over any field of characteristic zero ([Ken, lem. 3]), and in this context it can be used as the starting point of the theory of **conic Lagrangian cycles** (compare with chapter 5 and [Ken, Sab]).

4.2 Constructible sheaves on stratified spaces

4.2.1 Cohomologically cone-like stratifications

We now come back to other important properties of stratifications, which are **d^0-regular** or satisfy the property **LST**. Since these results are of **topological** nature, we introduce first the corresponding notions for topological stratifications.

Definition 4.2.1. *Let $X_\bullet$ be a topological stratification. We call $X_\bullet$ a*

1. **locally cone-like** *or* **cs-stratification** *([HS, def. 1.1]), if for all $x \in X^i$ $(0 \leq i \leq n)$ there exist an open neighborhood U of x in $X \backslash X_{i-1}$ and a filtration preserving homeomorphism $U \simeq B \times c(L)$, where B is an open ball of dimension i and*

$$c(L) := L \times [0,1[\,/(y,0) \sim (y',0)$$

is the open cone on a **compact** *topological space L with filtration*

$$\emptyset = L_{-1} \subset L_0 \subset \ldots \subset L_{n-i-1} = L \quad \textit{and vertex} \quad v := [(y,0)] .$$

L is called a **link** *(with $c(\emptyset) := \{v\}$), U has the induced filtration and*

$$(B \times c(L))_j := \begin{cases} B \times c(L_{j-i-1}) & \textit{for} \quad i \leq j \leq n , \\ \emptyset & \textit{for} \quad j < i . \end{cases}$$

2. **locally product-like** *or* **ps-stratification** *(compare also with [Pflaum, 1.4.1, p. 35]), if for all* $x \in X^i$ $(0 \leq i \leq n)$ *there exist an open neighborhood* U *of* x *in* $X \backslash X_{i-1}$ *and a filtration preserving homeomorphism* $U \simeq B \times P$*, where* B *is an open ball of dimension* i *and* P *is a topological space with filtration*

$$\emptyset = P_{-1} \subset P_0 = \{v\} \subset \ldots \subset P_{n-i} = P\,.$$

Here $v \in P$ *is some point, and*

$$(B \times P)_j := \begin{cases} B \times P_{j-i} & \textit{for} \quad i \leq j \leq n\,, \\ \emptyset & \textit{for} \quad j < i\,. \end{cases}$$

A neighborhood U *of* x *as above is called a* **distinguished neighborhood***.*

Compared to the previous notions of regularity conditions for a smooth stratification, one has the following implications, which are sometimes called "local topological triviality" of the stratification.

$$\mathbf{b} \longrightarrow \mathbf{C} \longrightarrow \mathbf{cs} \longrightarrow \mathbf{ps} \longleftarrow \mathbf{LST}\,.$$

- **cs** $\Rightarrow$ **ps** is trivial, and for **ps** $\Leftarrow$ **LST** take a submersive smooth germ $g : (X, x) \to (M, g(x))$, with M of dimension i for $x \in X^i$. So a **w-regular** stratification is for example a **ps-stratification**.
- For a proof of **b** $\Rightarrow$ **cs** $\Leftarrow$ **C** compare with [Mather, cor. 10.6], [GWPL, p. 59, thm. 5.4] and [Bekka4], [Bekka1, proof of lem. 2].

Usually, this "local topological triviality" follows from the "first isotopy lemma of Thom" (by integration of suitable vector fields).

In the following, we explain **cohomological** versions of this "local topological triviality", based on our definition 4.0.6 of a **ccs-stratification**. This includes our smooth $\mathbf{d}^{(0)}$**-regular** stratifications, as well as the geometric **cs-stratifications**:

$$\mathbf{cs} \longrightarrow \mathbf{ccs} \longleftarrow \mathbf{d^{(0)}}\,.$$

In this way we can prove all our results using only the **cohomological version of the first isotopy lemma**, without any use of a theory of suitable vector fields.

For the implication *cs* $\Rightarrow$ *ccs* take a **distinguished neighborhood** U of $x \in X^i$, and define

$$(\pi, \rho) : (U, U \cap X^i) \to (B \times [0, \epsilon[\,, B \times \{0\})$$

as the composition

$$U \xrightarrow{\ \sim\ } B \times c(L) \xrightarrow{(pr_B, r)} B \times [0, 1[\,,$$

with $pr_B : B \times c(L) \to B$ the projection and r the mapping

$$r : c(L) \to [0, 1[\,;\ [(b, t)] \mapsto t\,.$$

Then $\{r = 0\} = \{v\}$ is the vertex of the cone, with

$$r : c(L)\backslash\{v\} \simeq L\times\,]0,1[\, \to\,]0,1[$$

(isomorphic to) the projection so that the condition (2.) of definition 4.0.6 follows from remark 1.2.1(1) on page 42.

Note that the **link** L is then homeomorphic to

$$\{(\pi,\rho) = (b,t)\} \quad \text{for any} \quad (b,t) \in B\times\,]0,1[\ .$$

If $B' \subset B$ is another open ball containing x, then $U' := (\pi,\rho)^{-1}(B' \times [0,r'[\,)$ is for $0 < r' << 1$ also a **distinguished neighborhood** of $x \in X^i$. This is obvious for a **cs-stratification**, but it is also true a **ccs-stratification**. Here one uses:

1. The map (π,ρ) in definition 4.0.6 is **proper**.
2. The **proper base change theorem** [KS, prop. 2.6.7].
3. The fact that condition (2.) of definition 4.0.6 is only asked for all $\mathcal{G} \in ob(D(X\backslash X_i))$, which are constructible with respect to the induced stratification of $X\backslash X_i$ (and not for all $\mathcal{G} \in ob(D(U\backslash X_i))$, which are constructible with respect to the induced stratification of $U\backslash X_i$) !

Especially, any point of a **cs-** or **ccs-stratification** has a countable fundamental system of **distinguished neighborhoods**.

Remark 4.2.1. Our definition of a **ccs-stratification** is **not** a "local definition", because condition (2.) of definition 4.0.6 is only asked for all $\mathcal{G} \in ob(D(X\backslash X_i))$, which are constructible with respect to the induced stratification of $X\backslash X_i$. So it is not clear that the induced stratification of an open subset is again a **ccs-stratification**. To get this property, one should ask in definition 4.0.6 for a corresponding fundamental system of distinguished neighborhoods U such that condition (2.) holds for all $\mathcal{G} \in ob(D(U\backslash X_i))$, which are constructible with respect to the induced stratification of $U\backslash X_i$. Note that a $\mathbf{d^{(0)}}$**-regular** or **cs-stratification** satisfies this stronger property !

Nevertheless, let $X_\bullet$ be a **ccs-stratification** and A a **locally closed** subset of X, which is a union of strata. Then the induced stratification of A is also a **ccs-stratification**. Take a distinguished neighborhood $U \subset X$ of $x \in A$ such that $A\cap U$ is closed in U (i.e. take U small enough). Then $A \cap U$ is also a distinguished neighborhood of x with respect to $A_\bullet$. Just look at the commutative diagram of inclusions

$$\begin{array}{ccc} A\cap U & \xrightarrow{\ i'\ } & U \\ {\scriptstyle j'}\downarrow & & \downarrow{\scriptstyle j} \\ A & \xrightarrow[\ i\ ]{} & X\ , \end{array}$$

with

$$Ri'_{*}j'^{*} \simeq i'_{!}j'^{*} \simeq j^{*}i_{!} \quad \text{and} \quad i_{!} : D(A_\bullet) \to D(X_\bullet)\,.$$

A similar argument applies to the following property of a stratification:

Definition 4.2.2. *Let $X_\bullet$ be a stratification of the topological space X, and denote the open inclusion*

$$j : V' := X\backslash X_i \to X\backslash X_{i-1} =: V \quad \textit{for a fixed } i \ (0 \leq i \leq n)\,.$$

Then $X_\bullet$ satisfies the property $(*_i)$, if $Rj_*\mathcal{G}$ is constructible with respect to the induced stratification of V for any $\mathcal{G} \in ob(D(V'_\bullet))$.

So if $X_\bullet$ satisfies the property $(*_i)$, then the same is true for the induced stratification $A_\bullet$ of a **locally closed** subset A of X, which is a union of strata. Just look at the commutative diagram of inclusions

$$\begin{array}{ccc} A\backslash A_i & \xrightarrow{j'} & A\backslash A_{i-1} \\ {\scriptstyle k}\downarrow & & \downarrow{\scriptstyle l} \\ X\backslash X_i & \xrightarrow{j} & X\backslash X_{i-1}\,, \end{array}$$

with

$$l^*Rj_*k_! \simeq Rj'_* \quad \text{and} \quad k_! : D((A\backslash A_i)_\bullet) \to D((X\backslash X_i)_\bullet)\,.$$

Now we explain some important properties of a **ccs-stratification**, and some of them already hold under weaker assumptions:

Proposition 4.2.1. *Let $X_\bullet$ be a stratification of X, and consider $\mathcal{F} \in ob(D(X_\bullet))$.*

1. *Suppose $X_\bullet$ is a* **ps-** *or* **ccs-stratification**. *Then $X_\bullet$ has for all i the property $(*_i)$ above.*

2. *Suppose $X_\bullet$ has for all i the property $(*_i)$.*

   (a) *The strata of $X_\bullet$ (i.e. the connected components of the X^i) satisfy the* **frontier condition** (4.1).

   (b) *Denote by $k : S \to X$ the inclusion of a stratum S. Then all cohomology sheaves of $k^!\mathcal{F}$ are locally constant.*

3. *Suppose $X_\bullet$ is a* **ccs-stratification**. *Fix a* **distinguished neighborhood** *U of the point $x \in X^i$, and consider the open inclusion j of definition 4.2.2 with $\mathcal{G} \in ob(D(V'_\bullet))$.*

   (a) *The natural morphisms $R\Gamma(U, \mathcal{F}) \to \mathcal{F}_x$ and $R\Gamma_{\{x\}}(\mathcal{F}) \to R\Gamma_c(U, \mathcal{F})$ are isomorphisms.*

(b)
$$R\Gamma_c(U\backslash X_i, \mathcal{G})[i+1] \simeq (Rj_*\mathcal{G})_x \simeq R\Gamma((\pi,\rho)^{-1}(\{(b,t)\}), \mathcal{G})$$
for any $(b,t) \in B\times\,]0,\epsilon[\,.$

(c) *The stratification is* **locally finite** *(i.e. a compact subset of* X *meets only finitely many strata).*

(d) *Suppose we consider sheaves of* R*-modules for a* **commutative** *ring* R*. If all stalk complexes* $\mathcal{F}_y \in ob(D(R))$ *for* $y \in X$ *(or* $\mathcal{G}_y \in ob(D(R))$ *for* $y \in V'$*) are perfect, then* $R\Gamma_{\{x\}}(\mathcal{F})$ *(or* $(Rj_*\mathcal{G})_x$*) is also perfect.*

Proof. **1.** Since the statement is a local result, we can use a **distinguished neighborhood** U of $x \in X^i$. Let us first consider the case of a **ps-stratification**. Then we can assume $j: U' = B\times P' \to B\times P = U$, with $P' := P\backslash\{v\}$ and $x = v$. Consider the cartesian diagram

$$\begin{array}{ccc} U' = B\times P' & \xrightarrow{\;j\;} & B\times P = U \\ {\scriptstyle pr'}\downarrow & & \downarrow{\scriptstyle pr} \\ P' & \xrightarrow{\;j'\;} & P\,, \end{array}$$

with pr, pr' the obvious projections. By assumption, $\mathcal{G} \in ob(D(U'_\bullet))$ has locally constant cohomology sheaves on the fibers of pr'. Therefore

$$\mathcal{G} \simeq pr'^* Rpr'_*\mathcal{G}$$

by the **Vietoris-Begle theorem** [KS, prop. 2.7.8]. Let $s: P' \to B\times P'$ be a continuous section of pr' (i.e. $pr'\circ s = id$). Then

$$s^*\mathcal{G} \simeq s^*pr'^*Rpr'_*\mathcal{G} \simeq Rpr'_*\mathcal{G}\,.$$

But pr is a topological submersion so that the **smooth base change theorem** [BL, thm. 1.8(ii), p. 10] implies:

$$Rj_*\mathcal{G} \simeq Rj_*pr'^*s^*\mathcal{G} \simeq pr^*Rj'_*s^*\mathcal{G}\,.$$

For more general versions of the **smooth base change theorem** compare with [IC, (2) on p. 80] or [BL, p. 56, lem. C1]. Therefore $Rj_*\mathcal{G}$ has locally constant cohomology sheaves on the fibers of pr, especially on

$$pr^{-1}(\{v\}) = U\backslash U' = U\cap X_i\,.$$

Moreover

$$(Rj_*\mathcal{G})_x \simeq (Rj'_*s^*\mathcal{G})_v \quad \text{for} \quad x \in U\cap X_i\,.$$

This implies our claim in this case.

Consider now the case of a **ccs-stratification**. Fix a distinguished neighborhood U of $x \in X^i$. Then we have a cartesian diagram

$$\begin{array}{ccccc} U\backslash X^i & \xrightarrow{j} & U & \xleftarrow{i} & U \cap X_i \\ \downarrow{\scriptstyle (\pi,\rho)'} & & \downarrow{\scriptstyle (\pi,\rho)} & & \wr\downarrow{\scriptstyle \pi'} \\ B\times\,]0,\epsilon[& \xrightarrow{j'} & B \times [0,\epsilon[& \xleftarrow{i'} & B \times \{0\}\,. \end{array}$$

Since $\pi' := (\pi,\rho) : U \cap X^i \to B \times \{0\}$ is a homeomorphism, it is enough to show, that all cohomology sheaves of $R\pi'_* i^* Rj_* \mathcal{G}$ are locally constant on $B \times \{0\}$. Since (π,ρ) is proper, we get

$$R\pi'_* i^* Rj_* \mathcal{G} \simeq i'^* R(\pi,\rho)_* Rj_* \mathcal{G} \simeq i'^* Rj'_* R(\pi,\rho)'_* \mathcal{G}\,.$$

$R(\pi,\rho)'_* \mathcal{G}$ has by assumption locally constant cohomology sheaves on $B\times\,]0,\epsilon[$. But the filtration $B \times \{0\} \subset B \times [0,\epsilon[$ is of course a **ps-stratification**. Then the first claim follows from the case before.

3.(b) It implies moreover the stalk-formula

$$(Rj_* \mathcal{G})_x \simeq (Rk_* s^* R(\pi,\rho)'_* \mathcal{G})_0\,,$$

with $k :]0,\epsilon[\to [0,\epsilon[$ and $s :]0,\epsilon[\to B\times\,]0,\epsilon[;\ t \mapsto (b,t)$ (for $b \in B$ fixed). Since $\mathcal{G}' := s^* R(\pi,\rho)'_* \mathcal{G}$ has locally constant cohomology sheaves, we get:

$$(Rk_* \mathcal{G}')_0 \xleftarrow{\sim} R\Gamma(\,]0,\epsilon[\,,\mathcal{G}') \xrightarrow{\sim} \mathcal{G}'_t \quad \text{for } t \in\,]0,\epsilon[\,.$$

All together this implies already the second statement of 3.(b):

$$(Rj_* \mathcal{G})_x \simeq \mathcal{G}'_t \simeq (R(\pi,\rho)'_* \mathcal{G})_{(b,t)} \simeq R\Gamma((\pi,\rho)^{-1}(\{(b,t)\}),\mathcal{G})\,.$$

Since $(\pi,\rho)'$ is proper, we get also the first statement of 3.(b):

$$R\Gamma_c(U\backslash X_i,\mathcal{G}) \simeq R\Gamma_c(B\times\,]0,\epsilon[\,, R(\pi,\rho)'_* \mathcal{G})\,,$$

and the **Vietoris-Begle theorem** [KS, prop. 2.7.8] implies as before:

$$R(\pi,\rho)'_* \mathcal{G} \simeq c^*((R(\pi,\rho)'_* \mathcal{G})_{(b,t)}\,,$$

with c the constant map $B\times\,]0,\epsilon[\to \{(b,t)\}$. So the claim follows from the **projection formula** (proposition 1.4.1 on page 76):

$$R\Gamma_c(B\times\,]0,\epsilon[\,, c^*(R(\pi,\rho)'_* \mathcal{G})_{(b,t)}) \simeq R\Gamma_c(B\times\,]0,\epsilon[\,,\mathbb{Z}) \otimes^L_Z (R(\pi,\rho)'_* \mathcal{G})_{(b,t)}\,.$$

Of course we use the isomorphism $R\Gamma_c(B\times\,]0,\epsilon[\,,\mathbb{Z}) \simeq \mathbb{Z}[-(i+1)]\,.$

For a later application we remark, that the same argument applies to a smaller open ball $B' \subset B$ and interval $]a, a'[$, with $0 < a < a' < \epsilon$ in such a way that the following diagrams commute (with $W := (\pi, \rho)^{-1}(B' \times]a, a'[)$):

$$\begin{array}{ccc} R\Gamma(U \backslash X_i, \mathcal{G}) & \longrightarrow & R\Gamma(W, \mathcal{G}) \\ \downarrow \wr & & \downarrow \wr \\ R\Gamma(B \times]0, \epsilon[, R(\pi, \rho)'_* \mathcal{G}) & \xrightarrow{\sim} & R\Gamma(B' \times]a, a'[, R(\pi, \rho)'_* \mathcal{G}) \,, \end{array}$$

and

$$\begin{array}{ccc} R\Gamma_c(W, \mathcal{G}) & \longrightarrow & R\Gamma_c(U \backslash X_i, \mathcal{G}) \\ \downarrow \wr & & \downarrow \wr \\ (R(\pi, \rho)'_* \mathcal{G})_{(b,t)}[-(i+1)] & \Longrightarrow & (R(\pi, \rho)'_* \mathcal{G})_{(b,t)}[-(i+1)] \end{array}$$

for $(b, t) \in B' \times]a, b[$. We get especially

(**) There is an open W with compact closure $\overline{W} \subset U \backslash X_i$ such that the natural morphisms

$$R\Gamma(U \backslash X_i, \mathcal{G}) \to R\Gamma(W, \mathcal{G}) \quad \text{and} \quad R\Gamma_c(W, \mathcal{G}) \to R\Gamma_c(U \backslash X_i, \mathcal{G})$$

are isomorphisms.

2.(a) Let S, T be strata of $X_\bullet$. Denote the inclusions $j : S \to X$ and $i : T \to X$. Let $\mathcal{F} \neq 0$ be a constant sheaf on S. Inductive application of the property $(*_i)$ implies that $Rj_*\mathcal{F}$ is $X_\bullet$-constructible. In particular $i^* j_* \mathcal{F}$ is locally constant, and $supp(i^* j_* \mathcal{F})$ is therefore open and closed in T. But $(j_*\mathcal{F})_x \neq 0 \Leftrightarrow x \in \overline{S}$. Therefore

$$T \cap \overline{S} \neq \emptyset \;\Rightarrow\; supp(i^* j_* \mathcal{F}) \neq \emptyset \;\Rightarrow\; supp(i^* j_* \mathcal{F}) = T \;\Rightarrow\; T \subset \overline{S} \,.$$

2.(b) Let S be a connected component of X^i, and denote the inclusions $k' : S \to X_i$, $l : X_i \to X$ and $j : X \backslash X^i \to X$. Then

$$k^! \mathcal{F} \simeq k'^! l^! \mathcal{F} \simeq k'^* l^! \mathcal{F} \,,$$

since S is open in X_i. Inductive application of the property $(*_i)$ implies that $Rj_* j^* \mathcal{F}$ is $X_\bullet$-constructible. Then the distinguished triangle

$$Rl_* l^! \mathcal{F} \longrightarrow \mathcal{F} \longrightarrow Rj_* j^* \mathcal{F} \xrightarrow{[1]}$$

implies that $Rl_* l^! \mathcal{F}$ is also $X_\bullet$-constructible. Therefore $k^! \mathcal{F} \simeq k'^* l^* Rl_* l^! \mathcal{F}$ has only locally constant cohomology sheaves.

3.(a) Take isomorphisms

$$\alpha : (B, \pi(x)) \to (\mathbb{R}^i, 0) \quad , \quad \alpha' : ([0, \epsilon[, 0) \to ([0, \infty[, 0)$$

and consider the proper function

$$g : \mathbb{R}^i \times [0,\infty[\to [0,\infty[\,;\ (x_1,\ldots,x_i,t) \mapsto t^2 + \sum_{l=1}^{i} x_l^2 \,.$$

Note that

$$g : \Big((\mathbb{R}^i \times [0,\infty[)\backslash\{0\}, (\mathbb{R}^i\backslash\{0\}) \times \{0\} \Big) \to]0,\infty[$$

is a **proper trivial fibration of pairs**. Let β be the following composition:

$$U \xrightarrow{(\pi,\rho)} B \times [0,\epsilon[\xrightarrow{(\alpha,\alpha')} \mathbb{R}^i \times [0,\infty[\xrightarrow{\ g\ } [0,\infty[\,.$$

Then β is a proper map with $\beta^{-1}(\{0\}) = \{x\}$. Therefore one has commutative diagrams

$$\begin{array}{ccc} R\Gamma(U,\mathcal{F}) & \longrightarrow & \mathcal{F}_x \\ \downarrow\wr & & \downarrow\wr \\ R\Gamma([0,\infty[\,, R\beta_*\mathcal{F}) & \longrightarrow & (R\beta_*\mathcal{F})_0 \,, \end{array}$$

and

$$\begin{array}{ccc} R\Gamma_{\{x\}}(\mathcal{F}) & \longrightarrow & R\Gamma_c(U,\mathcal{F}) \\ \downarrow\wr & & \downarrow\wr \\ R\Gamma_{\{0\}}(R\beta_*\mathcal{F}) & \longrightarrow & R\Gamma_c([0,\infty[\,, R\beta_*\mathcal{F}). \end{array}$$

But the lower horizontal morphisms are isomorphisms, since $R\beta_*\mathcal{F}$ has only locally constant cohomology sheaves on $]0,\infty[$ (by remark 1.2.1(1) on page 42).

3.(c),(d) We prove these simultaneously by induction on $depth(X_\bullet)$. The case $depth(X_\bullet) = 0$ is obvious, since X^i is a topological manifold of dimension i for exactly one index i (and $X^j = \emptyset$ for $j \neq i$). Especially $R\Gamma_{\{x\}}(\mathcal{F}) \simeq \mathcal{F}_x[-i]$ for $x \in X^i$ (compare the proof of 3.(b), or with remark 2.2.1 on page 100).

For the induction step let us consider a distinguished neighborhood U of $x \in X^i$. Then the induced stratification of $V' := X\backslash X_i$ is a **ccs-stratification** of lower depth.

First we prove the induction step for 3.(c). Fix $j > i$, a stratum S of X^j and consider the inclusion $k : S \to V'$. Let us denote the base-ring by $\mathbb{K}$, and consider $\mathcal{G} := Rk_*(\mathbb{K}_S)$. $\mathcal{G}$ is constructible on V' by (iterated) application of 1. Take W as in condition (**) above so that $R\Gamma(U\backslash X_i, \mathcal{G}) \simeq R\Gamma(W,\mathcal{G})$. This implies especially

$$\Gamma(U\backslash X_i, k_*(\mathbb{K}_S)) \simeq \Gamma(W, k_*(\mathbb{K}_S)) \,.$$

But

$$\Gamma(U\backslash X_i, k_*(\mathbb{K}_S)) \simeq \Gamma(U\cap S,\mathbb{K}) \quad \text{and} \quad \Gamma(W, k_*(\mathbb{K}_S)) \simeq \Gamma(W\cap S,\mathbb{K}) \,.$$

Therefore $W \cap S \neq \emptyset \Leftrightarrow U \cap S \neq \emptyset$. But the closure of W is compact in $U \cap X^j$. By induction hypothesis W and therefore also U meets only finitely many S. This implies 3.(c).

Now we prove the induction step for 3.(d). By 3.(b) we have

$$R\Gamma_c(U \backslash X_i, \mathcal{G})[i+1] \simeq (Rj_*\mathcal{G})_x \,.$$

Moreover, the induced stratification of $U \backslash X_i$ has by the above argument only finitely many strata. Since the stalks of $\mathcal{G}$ are also perfect, this implies that $\mathcal{G}$ is of finite tor-dimension (i.e. isomorphic to a bounded complex of flat sheaves). Moreover, 3.(a) and the induction hypothesis for 3.(d) imply, that $\mathcal{G}$ is **perfect** in the sense of [SHS, exp. 9, p. 43, (4.0)]. But by the property (**) above, we have

$$R\Gamma_c(W, \mathcal{G}) \xrightarrow{\sim} R\Gamma_c(U \backslash X_i, \mathcal{G})$$

with $\overline{W} \subset U \backslash X_i$ compact. Therefore $R\Gamma_c(U \backslash X_i, \mathcal{G})$ is perfect by [SHS, exp. 9, p. 47, thm. 4.3].

This implies the claim for $\mathcal{G}$. If we apply this to $\mathcal{G} := j^*k^*\mathcal{F}$ (with $k : V \to X$ the inclusion), then the claim for $\mathcal{F}$ follows from 3.(a),(b) and the distinguished triangle

$$R\Gamma_c(U \backslash X_i, \mathcal{F}) \longrightarrow R\Gamma_c(U, \mathcal{F}) \longrightarrow R\Gamma_c(U \cap X_i, \mathcal{F}) \xrightarrow{[1]} \,,$$

together with $R\Gamma_c(U \cap X_i, \mathcal{F}) \simeq R\Gamma_{\{x\}}(U \cap X_i, \mathcal{F}) \simeq \mathcal{F}_x[-i]$. □

Remark 4.2.2. 1. Proposition 4.2.1 is a generalization of [IC, p. 76-79, 3.9-3.11] and [HS, prop. A.6]. The only difficult point is the condition 3.(d). Our proof depends on the deep theorem [SHS, exp. 9, p. 47, thm. 4.3]. But assume that the base-ring R is a noetherian ring of finite cohomological dimension (which is enough for most applications). Then a complex of R-modules is **perfect** iff it is cohomologically bounded with finitely generated cohomology modules. Then we can apply the simpler result [SHS, exp. 7, p. 4, thm. 3] or [IC, p. 70, thm. 3.5].

2. Consider $\mathcal{F} \in ob(D(X_\bullet))$ with perfect stalk complexes as in 3.(d). Then 3.(a) together with 3.(d) implies that $\mathcal{F}$ is **cohomologically constructible** in the sense of [KS, Def. 3.4.1] (and **perfect** in the sense of [SHS, exp. 9, p. 43, (4.0)]). Especially, the **biduality result** [KS, prop. 3.4.3(ii)] (or [SHS, Exp. 10, thm. 2.1]) applies. Moreover, if X is compact we get (as in 1.):

$$R\Gamma(X, \mathcal{F}) \quad \text{is } \textbf{perfect}. \tag{4.10}$$

3. For special cases of 2.(a) and 3.(c) in the context of **b-** or **C-regular** stratifications compare with [Mather, cor. 10.5], [GWPL, p. 61, thm. 5.6, cor. 5.7] and [Bekka1, (proof of) lem. 2].

4. Note that 2.(b) implies that the cohomology of a **link** L is independent of the choice of the **distinguished neighborhood**:

$$(Rj_*\mathcal{G})_x \simeq R\Gamma(L,\mathcal{G}), \quad \text{with} \quad L := (\pi,\rho)^{-1}(\{(b,t)\}) \quad ((b,t) \in B\times]0,\epsilon[)\,.$$

4.2.2 Stability results for constructible sheaves

Proposition 4.2.1 implies some stability results for complexes of sheaves, which are constructible with respect to a suitable topological stratification $X_\bullet$.

First we introduce some notation. Let $D^*_?(X_\bullet)$ (with $* = \emptyset, +, -, b$ and $? = \emptyset, perf$) be the triangulated subcategory of $D(X)$, whose objects are the $X_\bullet$-constructible complexes with the appropriate cohomological boundedness condition $*$ in the case $? = \emptyset$, and whose stalk complexes $\mathcal{F}_x \in ob(D(R))$ are in addition **perfect** for all $x \in X$ in the case $? = perf$ (with R commutative).

Let $f : X \to Y$ be a continuous map, where X, Y are endowed with topological stratifications $X_\bullet, Y_\bullet$. Then f is a **stratified** map, if the locally closed subset

$$f^{-1}(S) \subset X$$

is a union of strata for all strata S of $Y_\bullet$ (compare also with [IC, 10.12(i), p. 163]). In "good" cases, a **stratified** map f satisfies also the following properties for the induced map $f_S : f^{-1}(S) \to S$ (with $? = \emptyset$ or $? = perf$):

strat$_!$: $Rf_{S!}$ maps $D_?((f^{-1}(S))_\bullet)$ to $D_?(S_\bullet)$ for all strata S of $Y_\bullet$.

strat$_*$: Rf_{S*} maps $D_?((f^{-1}(S))_\bullet)$ to $D_?(S_\bullet)$ for all strata S of $Y_\bullet$.

Note that these two conditions are the same, if f is a **proper** map. Moreover, $S_\bullet$ has just one stratum so that $ob(D(S_\bullet))$ is given by the complexes, whose cohomology sheaves are locally constant.

The simplest example of a **stratified** map $f : X \to Y$ with the properties **strat$_!$** and **strat$_*$** is the inclusion $f : X \hookrightarrow Y$ of locally closed subsets of a stratified space, which are both a union of strata (as in proposition 4.0.2(3)).

To unify the notation in the following results, let us introduce the following variant of definition 4.2.2: The stratification $X_\bullet$ satisfies the property $(*_{i,perf})$, if Rj_* maps $D_{perf}(V'_\bullet)$ into $D^*_{perf}(V_\bullet)$, with j the open inclusion

$$j : V' := X\backslash X_i \to X\backslash X_{i-1} =: V \quad \text{for a fixed } i \ (0 \le i \le n)\,.$$

For example a **ccs-stratification** (especially a **d$^{(0)}$-regular** stratification) satisfies the property $(*_{i,perf})$ for all i (by proposition 4.2.1).

Corollary 4.2.1. *Consider topological stratifications $X_\bullet$ of X, $Y_\bullet$ of Y, and let $f : X \to Y$ be a continuous* **stratified** *map. Then one has for $? = \emptyset$ or $? = perf$:*

1. *f^* maps $D^*_?(Y_\bullet)$ into $D^*_?(X_\bullet)$.*

2. *Suppose f satisfies the property* **strat$_!$**. *Then $Rf_!$ maps $D_?(X_\bullet)$ into $D_?(Y_\bullet)$.*

3. *If the base ring R is commutative, then $D_?(X_\bullet)$ is stable under $\otimes^L$.*

4. *Suppose $X_\bullet$ and $Y_\bullet$ have for all i the property $(*_i)$ in the case $? = \emptyset$ (or $(*_{i,perf})$ in the case $? = perf$).*

   (a) *$f^!$ maps $D_?(Y_\bullet)$ into $D_?(X_\bullet)$.*

   (b) *Suppose f satisfies the property* **strat$_*$**. *Then Rf_* maps $D_?(X_\bullet)$ into $D^*_?(Y_\bullet)$.*

   (c) *If the base ring R is commutative, then $D_{perf}(X_\bullet)$ is stable under $Rhom(\cdot,\cdot)$. Moreover $Rhom(\cdot,\cdot)$ maps $D^-(X_\bullet)\times D^+(X_\bullet)$ into $D^+(X_\bullet)$.*

Proof. Recall that we assumed all spaces of finite cohomological dimension. Moreover, $\otimes^L$, $Rhom(\cdot,\cdot)$ and $f^!$ are defined even for unbounded complexes by [Sp] (and remark 2.2.9 on page 117).

1. is trivial, and 2. follows from the property **strat$_!$** together with the **base change isomorphism** ([KS, prop. 2.6.7]):

$$i_S^* Rf_! \simeq Rf_{S!} i^* ,$$

with $i_S : S \to Y$ and $i : f^{-1}(S) \to X$ the inclusions. The claim 3. is local on any stratum of $X_\bullet$. So we can assume that this stratum is an open ball and then the proof is the same as for lemma 2.2.4 on page 111. Similarly, the proof of 4.(c) is the same as for corollary 2.2.3 on page 113.

The other results will be proved by induction over $d := dim(Y)$, using the cartesian diagram

$$\begin{array}{ccccc} f^{-1}(Y_{d-1}) & \xrightarrow{i'} & X & \xleftarrow{j'} & f^{-1}(Y^d) \\ \downarrow{\scriptstyle f''} & & \downarrow{\scriptstyle f} & & \downarrow{\scriptstyle f'} \\ Y_{d-1} & \xrightarrow{i} & Y & \xleftarrow{j} & Y^d = Y\backslash Y_{d-1} . \end{array}$$

Note that j, j' are open embeddings. First we prove 4.b. By the property **strat$_*$** we have for $\mathcal{F} \in ob(D_?(X_\bullet))$:

$$j^* Rf_*\mathcal{F} \simeq Rf'_* j'^* \mathcal{F} \in ob(D_?(Y^d_\bullet)) .$$

Since $Y_\bullet$ has for all i the property $(*_i)$ (or $(*_{i,perf})$), we get

$$Rj_*j^*Rf_*\mathcal{F} \simeq Rj_*Rf'_*j'^*\mathcal{F} \simeq Rf_*Rj'_*j'^*\mathcal{F} \in ob(D_?(Y_\bullet)) .$$

Let us now assume that $f : X \to Y$ is the inclusion of a locally closed subset so that $dim(X) \leq d$. Since $X_\bullet$ has for all i the property $(*_i)$ (or $(*_{i,perf})$), we get in the same way by $f^{-1}(Y^d) = X^d = X \backslash X_{d-1}$:

$$Rj'_* j'^* \mathcal{F} \in ob(D_?(X_\bullet)) . \tag{4.11}$$

Using the distinguished triangle

$$Ri'_* i'^! \mathcal{F} \longrightarrow \mathcal{F} \longrightarrow Rj'_* j'^* \mathcal{F} \xrightarrow{[1]} ,$$

we get $Ri'_* i'^! \mathcal{F} \in ob(D_?(X_\bullet))$. Then the claim follows from

$$Rf_* Ri'_* i'^! \simeq Ri_* Rf''_* i'^!$$

and the induction hypothesis applied to f'' and $i'^! \mathcal{F}$. So 4.b is proved in the case of an inclusion! In the general case of a **stratified** map f, we can now apply 4.b to the inclusion j' of the locally closed subset $f^{-1}(Y^d) \subset X$ to get (4.11). Then the claim follows by the same argument as before.

For the proof of 4.a, let us start with $\mathcal{F} \in ob(D_?(Y_\bullet))$. It is enough to show

$$j'^* Rf^! \mathcal{F} \simeq Rf'^! j^* \mathcal{F} \in ob(D_?(f^{-1}(Y^d)_\bullet)) ,$$

because then we can argue as before, using the distinguished triangle

$$Ri_* i^! \mathcal{F} \longrightarrow \mathcal{F} \longrightarrow Rj_* j^* \mathcal{F} \xrightarrow{[1]} ,$$

together with ([KS, prop. 3.1.9(ii)]):

$$Rf^! Rj_* j^* \simeq Rj'_* Rf'^! j^* \quad \text{and} \quad Rf^! Ri_* i^! \simeq Ri'_* Rf''^! i^! .$$

Here we also use the previous result 4.b for the inclusion j' of the locally closed subset $f^{-1}(Y^d) \subset X$. So we can assume $Y = Y^d$. Now we argue by induction on $d' := dim(X)$, using the inclusions

$$X_{d'-1} \xrightarrow{i''} X \xleftarrow{j''} X^{d'} = X \backslash X_{d'-1} .$$

Then we can also assume $X = X^{d'}$, using the distinguished triangle

$$Ri''_* i''^! f^! \mathcal{F} \longrightarrow f^! \mathcal{F} \longrightarrow Rj''_* j''^* f^! \mathcal{F} \xrightarrow{[1]} ,$$

together with

$$j''^* Rf^! \simeq (f \circ j'')^! \quad \text{and} \quad i''^! Rf^! \simeq (f \circ i'')^! .$$

Since the claim is a local result, we can finally assume that X, Y are open balls of dimension d, d'. Then we get by the **Vietoris-Begle theorem** ([KS, prop. 2.7.8]):

$$\mathcal{F} \simeq c^* Rc_* \mathcal{F} \quad \text{for} \quad \mathcal{F} \in ob(D(Y_\bullet)) ,$$

with $c : Y \to \{pt\}$ a constant map. Moreover,

$$c^! \simeq c^*[d] \quad \text{and} \quad (c \circ f)^! \simeq (c \circ f)^*[d']$$

by **Poincaré-Verdier duality** (see [KS, prop. 3.3.2] and the next remark). Together this implies

$$f^! \simeq f^*[d' - d] \; : \; D_?(Y_\bullet^d) \to D_?(X_\bullet^{d'}) \, . \qquad \square$$

Remark 4.2.3. In the above argument we used (a special case of) the following version of **Poincaré-Verdier duality**: let $f : X \to Y$ be a **topological submersion** (i.e. f is locally on X isomorphic to a projection $Y \times \mathbb{R}^N \to Y$). Then the natural morphism

$$f_Z^!(\mathbb{Z}_Y) \otimes_Z^L f^*(\cdot) \to f_R^!(\cdot) \tag{4.12}$$

is an isomorphismn. Here $f_R^!$ (or $f_Z^!$) is the **right adjoint** to

$$Rf_! : D(X, R) \to D(Y, R) \quad (\text{ or to } \quad Rf_! : D(X, \mathbb{Z}) \to D(Y, \mathbb{Z}) \,) \, .$$

Moreover, the morphism (4.12) is by adjunction induced from the morphism

$$Rf_!\Big(f_Z^!(\mathbb{Z}_Y) \otimes_Z^L f^*(\cdot)\Big) \xrightarrow[\sim]{proj.form.} \Big(Rf_! f_Z^!(\mathbb{Z}_Y)\Big) \otimes_Z^L (\cdot) \xrightarrow{ad_f} \mathbb{Z}_Y \otimes_Z^L (\cdot) \simeq (\cdot) \, .$$

Here the first isomorphism is the **projection formula** proposition 1.4.1 on page 76. This **projection formula** implies by adjunction the isomorphism (compare [KS, prop. 3.1.7])

$$forget \circ f_R^! = f_Z^! \circ forget$$

so that it enough to prove the **Poincaré-Verdier duality isomorphism** in the case $R = \mathbb{Z}$. After truncation, it is enough to consider bounded complexes and the statement follows from [KS, prop. 3.3.2]. Note that $c^!\mathbb{Z} \simeq \mathbb{Z}_Y[d]$ for Y on open ball of dimension d and $c : Y \to \{pt\}$ a constant map.

Corollary 4.2.2. *Assume the base ring R is commutative, and let $f : X \to Y$ be a continuous map.*

1. $f^!(D_Y\mathcal{F}) \simeq D_X(f^*\mathcal{F})$ *for* $\mathcal{F} \in ob(D(Y))$.

2. $Rf_*(D_X\mathcal{F}) \simeq D_Y(Rf_!\mathcal{F})$ *for* $\mathcal{F} \in ob(D(X)$.

3. *Let $X_\bullet$ be a stratification of X satisfying the property $(*_{i,perf})$ for all i. Then the* **dualizing complex** *$c^! R_{pt}$ on X belongs to $ob(D_{perf}(X_\bullet))$ (with $c : X \to \{pt\}$ a constant map). Moreover, if $X_\bullet$ is a* **ccs-stratification**, *then we have the* **biduality isomorphism**

$$\mathcal{F} \xrightarrow{\sim} D_X(D_X\mathcal{F}) \quad \textit{for all } \mathcal{F} \in ob(D_{perf}(X_\bullet)) \, , \tag{BD}$$

with $D_X := Rhom(\cdot, c^! R_{pt})$ the **duality functor** *on X.*

Assume $f : X \to Y$ is a **stratified** *map for two stratifications $X_\bullet$ of X and $Y_\bullet$ of Y, which satisfy the property $(*_{i,perf})$ for all i, and also the* **biduality isomorphism (BD)** *(e.g. two* **ccs-stratifications***).*

4. *$f^*(D_Y\mathcal{F}) \simeq D_X(f^!\mathcal{F})$ for $\mathcal{F} \in ob(D_{perf}(Y_\bullet))$.*

5. *Suppose f satisfies also the property* **strat**$_!$ *for $? = perf$. Then one has $Rf_!(D_X\mathcal{F}) \simeq D_Y(Rf_*\mathcal{F})$ for $\mathcal{F} \in ob(D_{perf}(X_\bullet))$.*

Proof. 3. $c^! R_{pt} \in ob(D_{perf}(X_\bullet))$ is a special case of corollary 4.2.1(4.a). The **biduality** statement is a local result, since the natural morphism $\mathcal{F} \to D_X(D_X\mathcal{F})$ is globally defined. So to show that it is an isomorphism, we can assume that $X_\bullet$ has only finitely many strata (by proposition 4.2.1(3.c)). Then $\mathcal{F} \in D_{perf}(X_\bullet)$ is **cohomologically constructible** in the sense of [KS, def. 3.4.1] (or **perfect** in the sense of [SHS, exp. 9, p. 43, (4.0)]). As noticed in remark 4.2.2(2), the claim follows then from [KS, prop. 3.4.3(ii)] (or [SHS, exp. 10, thm. 2.1]).

The proof of the other results of the corollary is exactly the same as the proof of corollary 2.2.7 on page 119. Here one uses in 4. and 5. the **biduality isomorphism (BD)** in the following cases:

4. For $\mathcal{F} \in D_{perf}(Y_\bullet)$ and $f^*(D_Y\mathcal{F}) \in D_{perf}(X_\bullet)$ (by corollary 4.2.1).

5. For $\mathcal{F} \in D_{perf}(X_\bullet)$ and $Rf_!(D_X\mathcal{F}) \in D_{perf}(Y_\bullet)$ (by corollary 4.2.1). □

Let us now explain some examples of **stratified** maps, where the conditions **strat**$_!$ and **strat**$_*$ are satisfied (for $? = \emptyset$ or $? = perf$).

A sufficient **geometric** condition on the **stratified** map f is the assumption, that f is a **locally trivial stratified fibration** in the following sense (compare with [IC, 10.12(ii), p. 163]):

loc.triv.: For all y in a stratum S of Y, there exists an open neighborhood U of y in S, and a filtration preserving homeomorphism

$$h : f^{-1}(U) \to U \times F\,, \quad \text{with} \quad f = pr_U \circ h\,.$$

Here $pr_U : U \times F \to U$ is the projection, and F is a topological space endowed with a filtration. So h satisfies

$$h(f^{-1}(U) \cap X_i) = U \times F_i \quad \text{for all } i.$$

Of course F is then isomorphic to the fiber $\{f = y\} \subset X$ with its induced filtration.

The property **strat**$_!$ for $? = \emptyset$ follows then from the **Vietoris-Begle theorem** ([KS, prop. 2.7.8]) and remark 1.2.1(1) on page 42. Similarly, the property **strat**$_*$ for $? = \emptyset$ follows from the **Vietoris-Begle theorem** and the **smooth base change theorem**

[BL, thm. 1.8(ii), p. 10] (compare with [IC, (2) on p. 80] or [BL, p. 56, lem. C1] for more general versions). Just look at the cartesian diagram

$$\begin{array}{ccc} U \times F & \xrightarrow{p:=pr_U} & U \\ \downarrow q & & \downarrow c \\ F & \xrightarrow{p'} & \{pt\} . \end{array}$$

Then one gets for $\mathcal{F} \in D((U \times F)_\bullet)$ by the **Vietoris-Begle theorem**:

$$\mathcal{F} \simeq q^* Rq_* \mathcal{F} ,$$

and the **smooth base change theorem** for the **topological submersion** c implies:

$$Rp_*\mathcal{F} \simeq Rp_* q^*(Rq_*\mathcal{F}) \simeq c^* Rp'_*(Rq_*\mathcal{F}) .$$

Especially, all cohomology sheaves of $Rp_*\mathcal{F}$ are constant. Moreover, if one uses the section $s : F \to U \times F;\ x \mapsto (y, x)$ for a fixed $y \in U$, then one gets

$$s^*\mathcal{F} \simeq s^* q^* Rq_*\mathcal{F} \simeq (q \circ s)^* Rq_*\mathcal{F} \simeq Rq_*\mathcal{F}$$

so that

$$(Rp_*\mathcal{F})_y \simeq (Rp'_* Rq_*\mathcal{F})_y \simeq R\Gamma(\{p = y\}, \mathcal{F}) . \tag{4.13}$$

So for the property **strat**$_*$ in the case $? = perf$ one has in addition to assume for all $y \in Y$:

$$R\Gamma(\{f = y\}, \mathcal{F}) \quad \text{is } \textbf{perfect} \text{ for all } \mathcal{F} \in ob(D_{perf}(X_\bullet)) .$$

Similarly, for the property **strat**$_!$ in the case $? = perf$ one has in addition to assume for all $y \in Y$:

$$(Rp_!\mathcal{F})_y \simeq R\Gamma_c(\{f = y\}, \mathcal{F}) \quad \text{is } \textbf{perfect} \text{ for all } \mathcal{F} \in ob(D_{perf}(X_\bullet)) .$$

This is for example the case, if one can **compactify** the fibers $\{f = y\}$ for all $y \in Y$ in the following way (compare [IC, 10. 13, p. 163, thm. 10.16(v), p. 167]):

comp.: One has an embedding $j : \{f = y\} \to Z$ of the fiber as a **locally closed** subset of a **compact** space Z, which is endowed with a **ccs-stratification** $Z_\bullet$ such that $\{f = y\} \cap X^i$ is a union of strata of Z for all i (e.g. f is proper and the induced filtration of the compact fiber is a **ccs-stratification**).

Then one gets by corollary 4.2.1 for $\mathcal{F} \in ob(D_{perf}(X_\bullet))$:

$$Rj_!(\mathcal{F}|\{f = y\}) , \ Rj_*(\mathcal{F}|\{f = y\}) \in ob(D_{perf}(Z_\bullet))$$

so that the claim follows from (4.10) in remark 4.2.2(2), since

$$R\Gamma_c(\{f = y\}, \mathcal{F}) \simeq R\Gamma\big(Z, Rj_!(\mathcal{F}|\{f = y\})\big)$$

and

$$R\Gamma(\{f=y\},\mathcal{F}) \simeq R\Gamma\big(Z, Rj_*(\mathcal{F}|\{f=y\})\big)\,.$$

As an important application, we get the constructibility of the **nearby** and **vanishing cycle functor**. Let us first recall the definition of these functors (compare with chapter 1 for details).

Let $f : X \to \mathbb{C}$ be a continuous function and consider the following cartesian diagram of maps induced by the inclusions $\{0\} \subset \mathbb{C} \supset \mathbb{C}^*$ and the universal covering $\widetilde{\mathbb{C}^*} \to \mathbb{C}^*$:

$$\begin{array}{ccccccc}
X' & \xrightarrow{\ i\ } & X & \xleftarrow{\ j\ } & \{f \neq 0\} & \xleftarrow{\ p\ } & \widetilde{X^*} \\
\downarrow & & \downarrow f & & \downarrow & & \downarrow \tilde{f} \\
\{0\} & \longrightarrow & \mathbb{C} & \longleftarrow & \mathbb{C}^* & \longleftarrow & \widetilde{\mathbb{C}^*}\,.
\end{array} \tag{4.14}$$

The **nearby cycle** ψ_f of f is the functor

$$\psi_f := i^* R\tilde{\pi}_* \tilde{\pi}^* : D(X) \to D(X')\,,$$

with $\tilde{\pi} := j \circ p$. The **vanishing cycle** ϕ_f of f is given as "the" cone of the natural morphism

$$sp : i^* \to \psi_f$$

induced from the adjunction map $id \to R\tilde{\pi}_*\tilde{\pi}^*$.

Lemma 4.2.1. *Let $f : X \to \mathbb{C}$ be a continuous function, and $X_\bullet$ a stratification of X such that $X' := \{f = 0\}$ is a union of strata. Suppose $X_\bullet$ has locally for all i the property $(*_i)$ of definition 4.2.2, i.e. each point $x \in X$ has a fundamental system of open neighborhoods U such that the induced stratification of U has for all i the property $(*_i)$. Consider $\mathcal{F} \in ob(D(X_\bullet))$. Then $\psi_f(\mathcal{F})$ and $\phi_f(\mathcal{F})$ are constructible with respect to the induced stratification of X'.*

Proof. Let us first remark, that $X_\bullet$ has also globally for all i the property $(*_i)$. By the definition above, we only have to consider the case of the nearby cycle functor. Then we have

$$\psi_f \simeq i^* Rj_* Rp_* p^* j^*\,.$$

By corollary 4.2.1 (applied to the inclusion j), we only have to show that Rp_*p^* maps $D(U_\bullet)$ into itself, with $U_\bullet$ the induced stratification of $U := \{f \neq 0\}$. By assumption, $U_\bullet$ has also locally for all i the property $(*_i)$. But this is a local condition and $p : \widetilde{X^*} \to U$ is a **covering map**. Therefore the induced stratification of $\widetilde{X^*}$ has the same property. For each stratum S of $U_\bullet$, $p^{-1}(S)$ is a union of strata such that

$$p_S = p : p^{-1}(S) \to S$$

is a covering. More precisely, it is a **locally trivial fibration** with fibers isomorphic to $\mathbb{Z}$. As explained before, it therefore has the property **strat**$_*$ in the case $? = \emptyset$. Then our claim follows from corollary 4.2.1 applied to $p : \widetilde{X^*} \to U$. □

Note, that $p : \widetilde{X^*} \to U$ does **not** satisfy the property $\mathbf{strat}_*$ in the case $? = perf$, because its fiber is isomorphic to $\mathbb{Z}$! Nevertheless, suppose $X_\bullet$ is a Whitney **b-regular** stratification of the closed subset X of the manifold M, with f induced from a smooth map $f : M \to \mathbb{C}$. Assume that this stratification satisfies also for all $x \in X'$ the $\mathbf{a}_f$**-condition** of Thom as in example 1.1.3 on page 36.

Let $\mathcal{F}$ be $X_\bullet$-constructible, and define in local coordinates $(M, x) \simeq (\mathbb{R}^m, 0)$ the distance function r by $r(x) := \sum_{i=1}^m x_i^2$. Then one gets by example 1.1.3 and lemma 1.1.1:

$$(\psi_f(\mathcal{F}))_x \simeq R\Gamma(X \cap \{r \leq \delta, f = w\}, \mathcal{F}) \quad \text{for} \quad 0 < |w| << \delta << 1 .$$

Moreover, the compact set $X \cap \{r \leq \delta, f = w\}$ gets an induced Whitney **b-regular** stratification (as a transversal intersection). So if $\mathcal{F}$ has **perfect** stalks, we get by (4.10) in remark 4.2.2(2) and the above stalk description, that $\psi_f(\mathcal{F})$ and therefore also $\phi_f(\mathcal{F})$ has **perfect** stalks.

By the references of example 1.1.3, this applies especially to the complex analytic (algebraic) context as in theorem 4.0.2. Here we do not use a global embedding into manifolds (since it is a local result)!

Remark 4.2.4. One important difference between ψ_f and ϕ_f is the following property:

- Suppose the stratification $X_\bullet$ of $X \subset M$ is $\mathbf{d}^0$**-regular** or satisfies the property **LST** (but we do not assume that X' is a union of strata), and $f : M \to \mathbb{C}$ is a smooth map. Consider $\mathcal{F} \in ob(D(X_\bullet))$. Then $\left(\phi_{f-f(x)}(\mathcal{F})\right)_x = 0$, if $x \in X^s$ is **not** a critical point of $f|X^s$.

The proof is the same as the one of lemma 4.1.2 (compare also with lemma 1.1.1 on page 27). This is the main point for the introduction of the condition **LST** (in the holomorphic context) in [BMM].

Usually one needs the **first isotopy lemma of Thom** to show, that a **stratified** map is a **locally trivial stratified fibration**. But for the most important applications, we can avoid this **geometric** assumption!

Example 4.2.1. *Suppose X, Y are closed subsets of the smooth manifolds M, N, and $f : M \to N$ is a smooth map with $f(X) \subset Y$ and $f|X$* **proper**. *Assume X and Y are endowed with* **smooth stratifications** *$X_\bullet$ and $Y_\bullet$ such that f is a* **stratified submersion**, *i.e. f maps all strata of $X_\bullet$ submersively into strata of $Y_\bullet$. Especially, $f : X \to Y$ is a* **stratified** *map.*

1. *If $X_\bullet$ satisfies the property* **LSA** *(e.g. it is a* $\mathbf{d}^{(0)}$**-regular** *stratification), then $f : X \to Y$ has the property* $\mathbf{strat}_! = \mathbf{strat}_*$ *for $? = \emptyset$. Fix a stratum S of $Y_\bullet$, and choose on open subset $U \subset M$ such that the locally closed subset $f^{-1}(S)$ of X is a closed subset of U. Then the induced stratification of $f^{-1}(S)$ as a subset of U satisfies also the property* **LSA**. *Moreover, $f : f^{-1}(S) \to S$ is a* **proper stratified submersion** *so that the claim follows corollary 4.0.1.*

2. *If* $X_\bullet$ *is a* $\mathbf{d}^{(0)}$**-regular** *stratification, then the same is true for the induced stratification of all fibers* $\{f = y\} \cap X \subset U$ *for* $y \in S$*, with*

$$\{f = y\}_{i-dim(S)} := \{f = y\} \cap X_i \quad \textit{for } dim(S) \leq i \leq n\,.$$

This will be explained in the next section. So this induced stratification of the fibers is especially a **ccs-stratification**. *Moreover, these fibers are compact, since* $f : X \to Y$ *is proper. Then the property* $\mathbf{strat}_! = \mathbf{strat}_*$ *for* $? = perf$ *follows from* 1. *together with* (4.10) *in remark* 4.2.2(2).

Assume, that the stratifications $X_\bullet$ and $Y_\bullet$ in example 4.2.1(1) are Whitney **a-regular** and satisfy the property **LSA**. Then one can also give a "micro-local" proof of the fact, that $Rf_* = Rf_!$ (for $f : X \to Y$) maps $D^+(X_\bullet)$ into $D^+(Y_\bullet)$.

Introduce the following commutative diagram, whose right square is cartesian (compare [KS, (4.3.2), p. 199] and [KS2, ((1.1.3), p. 14]):

$$\begin{array}{ccccc} T^*M & \xleftarrow{\;{}^t f'\;} & f^*(T^*N) & \xrightarrow{\;f_\pi\;} & T^*N \\ \Big\downarrow{\scriptstyle \pi_M} & & \Big\downarrow{\scriptstyle \pi} & & \Big\downarrow{\scriptstyle \pi_N} \\ M & = & M & \xrightarrow{\;f\;} & N\,. \end{array} \tag{4.15}$$

Here ${}^t f'$ is the dual of the differential of f.

Then the conic subsets

$$\Lambda_Y := \bigcup_j T^*_{Y^j} N \subset T^*N \quad \text{and} \quad \Lambda_X := \bigcup_i T^*_{X^i} M \subset T^*M$$

are **closed** by **a-regularity**. Moreover,

$$\Lambda' := f_\pi(\,{}^t f'^{-1}(\Lambda_X)\,) \subset \Lambda_Y\,, \tag{4.16}$$

since $f : X \to Y$ is a **stratified submersion**. Finally, Λ' is a **closed** subset of T^*N, since $f : X \to Y$ is **proper**. So our claim follows from corollary 4.0.3, lemma 4.0.1 and the following easy proposition of Kashiwara-Schapira ([KS, prop. 5.4.4], [KS2, prop. 4.1.1(i)]):

Proposition 4.2.2. *Let* $f : M \to N$ *be a smooth map of smooth manifolds. Assume* f *is* **proper** *on* $supp(\mathcal{F})$ *for a given* $\mathcal{F} \in ob(D^+(M))$. *Then*

$$\mu supp(Rf_*(\mathcal{F})) \subset f_\pi(\,{}^t f'^{-1}(\mu supp(\mathcal{F}))\,)\,. \tag{4.17}$$

Proof. Let us denote by Λ' the right-hand side of this inclusion. Then Λ' is closed, since f is **proper** on the support of $\mathcal{F}$. Consider a smooth function germ $g : (N, y) \to (\mathbb{R}, 0)$ with $dg_y \notin \Lambda'$. Then

$$d(g \circ f)_x \notin \mu supp(\mathcal{F}) \quad \text{for all } x \in \{f = y\}\,.$$

So we get $(R\Gamma_{\{g\circ f\geq 0\}}(\mathcal{F}))_x = 0$ for all $x \in \{f = y\}$, and this implies

$$(R\Gamma_{\{g\geq 0\}}(Rf_*\mathcal{F}))_y \simeq R\Gamma(\{f = y\}, R\Gamma_{\{g\circ f\geq 0\}}(\mathcal{F})) = 0\,.$$

Compare also with the proof of corollary 4.1.2. □

In particular, these arguments and example 4.2.1 apply to the context of a "geometric category" $\mathcal{S}$ as in definition 4.0.5.

Assume X, Y are **closed definable** subsets of the real analytic (Nash) manifolds M, N, and $f : M \to N$ is a **definable** smooth map, with $f(X) \subset Y$ and $f|X$ **proper**. Then there exists definable Whitney b-regular C^k-stratifications of X and Y ($1 \leq k < \infty$) such that $f : X \to Y$ is a **stratified submersion**. One can even ask these stratifications to be compatible with a given finite family of definable subsets of X and Y (i.e. these subsets are unions of strata)! For more details on this compare with [Ta1], [vDrMi, 1.19, 4.8] or [Shiota, (II.1.14)].

Remark 4.2.5. By (the proof of) proposition 4.1.3 and the estimate (4.16) we get especially the following application:

- If $\Lambda \subset T^*M$ is a **closed conic definable isotropic** subset contained in $T^*M|X$, then $\Lambda' := f_\pi(\,{}^tf'^{-1}(\Lambda)\,) \subset T^*N$ is a **closed conic definable isotropic** subset contained in $T^*N|Y$.

The same argument applies also to the complex analytic (algebraic) context! Moreover, it is closely related to the "proper push-down" in the theory of **conic Lagrangian cycles** (compare with [Sch4, sec. 4.6], [KS, prop. 9.4.2] and [SchVi, Gi1, Gi2, Ken, Sab]).

Using the fact ${}^tf'^*\omega_M = f_\pi^*\omega_N$ (with ω_M, ω_N the canonical 1-forms on M and N), one can also argue as follows (compare [KS, prop. 8.3.11] and [KS2, proof of prop. 8.3.1]):

$$\omega_M|\Lambda \equiv 0 \Rightarrow {}^tf'^*\omega_M|\,{}^tf'^{-1}(\Lambda) \equiv 0 \Rightarrow f_\pi^*\omega_N|\,{}^tf'^{-1}(\Lambda) \equiv 0 \Rightarrow \omega_N|\Lambda' \equiv 0\,.$$

The last implication follows from lemma 4.1.4, if one stratifies the proper map $f_\pi : {}^tf'^{-1}(\Lambda) \to T^*N$.

For the application of example 4.2.1 we have to start with a definable **smooth** map f (of the ambient manifolds). But assume only, that $f : X \to Y$ is a **proper definable** map (i.e. with a definable graph). Then the graph embedding

$$h := (id, f) : X \to g(X) =: X' \subset M \times N$$

is a definable isomorphism onto the closed definable subset $X' \subset M \times N$, with $h^{-1} = pr_M : X' \to X$ and $f = pr_N \circ h^{-1} : X \to Y$. Here pr_M and pr_N are the **smooth definable** projections of $M \times N$ onto M and N, whose restriction to X' are proper. Especially, $pr_N : X' \to Y$ can be stratified as before, and we get the

Theorem 4.2.1. *Fix a "geometric category" $\mathcal{S}$ as in definition* 4.0.5, *and consider a continuous definable map $f : X \to Y$ between closed definable subsets X, Y of the real analytic (Nash) manifolds M, N. Denote by*

$$D_?(X,\mathcal{S}) \quad \textit{and} \quad D_?(Y,\mathcal{S}) \quad (\textit{with } ? = \emptyset \textit{ or } ? = perf)$$

*the corresponding category of $\mathcal{S}$-***constructible** *complexes of sheaves (all of whose stalk complexes are* **perfect** *in the case $? = perf$, with R a commutative base ring). Then these categories are* **stable** *under the functors*

$$f^*,\ f^!, \quad \textit{and for } f \textit{ proper also under } Rf_! = Rf_* \,.$$

Suppose R is commutative. Then $D_?(X,\mathcal{S})$ is stable under $\otimes^L$, $D_{perf}(X,\mathcal{S})$ is stable under $Rhom(\cdot,\cdot)$, and $Rhom(\cdot,\cdot)$ maps

$$D^-(X,\mathcal{S}) \times D^+(X,\mathcal{S}) \quad \textit{into} \quad D^+(X,\mathcal{S})\,.$$

If we work in the context of an **o-minimal structure** *$\mathcal{S}$ on the real field, then it is enough to assume that X, Y are* **locally closed** *definable subsets, and these categories are also* **stable** *under $Rf_!$ and Rf_* (with f not necessarily proper). Finally, also the corresponding "duality results" of corollary* 4.2.2 *apply to this context of $\mathcal{S}$-***constructible** *complexes of sheaves.*

Here the stability under $Rf_!$ and Rf_*, in the context of an **o-minimal structure** $\mathcal{S}$ on the real field, follows from a suitable "compactification" of the map (compare also with the proof of corollary 2.2.5(1) on page 116):

Consider a continuous definable map $f : X \to Y$, with X, Y locally closed definable subsets. Applying corollary 4.2.1 to the inclusion $Y \hookrightarrow N$ (for a suitable Whitney **b-regular** stratification of N adapted to Y), we can assume $Y = N$.

Consider an inclusion $M \subset \bar{M}$ as a locally closed definable subset of a **compact** analytic Nash manifold $\bar{M}$, i.e. in our context M is some affine space $\mathbb{R}^n$, and we can take as $\bar{M} := \mathbb{P}^n(\mathbb{R})$ the projective space. If more generally M is a Nash manifold, then one can either use the **embedding theorem** [vDr, thm. 10.2.15], or one reduces the above result to the affine case by a "Mayer-Vietoris argument" and induction on the number of subsets of a suitable "open affine covering"! Here one uses the distinguished triangle [KS, (2.6.30), p. 114] (or [KS, (2.6.28), p. 114]) for the functor $Rf_!$ (or Rf_*). Then one factorizes f as the composition

$$X \xrightarrow[\sim]{h=(id,f)} h(X) \xrightarrow{\ i\ } cl(h(X)) =: X' \subset \bar{M} \times N \xrightarrow{pr_N} N\,.$$

Then one applies as before corollary 4.2.1 to the inclusion $i : h(X) \hookrightarrow X'$ of the locally closed definable subset $h(X) \subset \bar{M} \times N$, and example 4.2.1 to the **proper smooth definable** projection pr_N.

In this way we can reprove by "stratification theory" some results of chapter 2, which we deduced there (in grater generality) from the **cell decomposition** and **triangulation theorem** for **o-minimal structures** [vDr, thm. 3.2.11, thm. 8.2.9].

If one works with appropriate fixed definable stratifications, then one gets as a "refined information" some control on the strata, where the corresponding cohomology sheaves are **locally constant**!

Moreover, the same argument as before also works for a **proper holomorphic (algebraic)** map $f : X \to Y$ of complex analytic (algebraic) spaces, if $X_\bullet, Y_\bullet$ are complex analytic (algebraic) Whitney b-regular stratifications and f is a stratified submersion as before. Here we do not use any embedding into manifolds (by corollary 4.1.2, example 4.1.1 and remark 4.1.4)!

By a result of Nagata, one can embed any complex algebraic space X (i.e. the analytic space associated to a seperated scheme of finite type over $spec(\mathbb{C})$) as a Zariski open subset into a **compact** algebraic space $\bar{X}$.

By the known existence results about such complex analytic (algebraic) Whitney stratifications (see for example [GM, p. 43, thm. 1.7] and [Hiro, Te, Ver]), one gets in the same way a proof of theorem 4.0.2. Moreover, one can in the algebraic case avoid the embedding theorem of Nagata, if one reduces the above result to the affine case by a "Mayer-Vietoris argument" and induction on the number of subsets of a suitable "open affine covering"! In the affine case one can of course use an embedding into the complex projective space (as in the "o-minimal context").

Finally, also the corresponding "duality results" of corollary 4.2.2 apply to this context of **complex analytically (algebraically) constructible** complexes of sheaves. For other approaches to these results compare with

- [Ver3], where **resolution of singularities** is used.
- [KS, sec. 8.5] for a "micro-local" approach in the embedded context, based on the characterization of the constructibility in terms of the **micro-support** (as in theorem 4.0.1).

  Especially, this "micro-local" approach allows one to prove (under suitable assumptions on $\mathcal{F}$ and f) the constructibility of $Rf_!\mathcal{F}$ and $Rf_*\mathcal{F}$ in the complex analytic context for a **non-proper holomorphic** map f (compare [KS, prop. 8.5.8] and [KS2, prop.8.6.1]). This is based on the following beautiful characterization ([KS, thm. 8.5.5], [KS2, thm. 8.5.2]):

  Consider a subanalytically constructible complex $\mathcal{F} \in ob(D^+(X))$ on the complex manifold X. Then $\mathcal{F}$ is **complex analytically** constructible if and only if $\mu supp(\mathcal{F}) \subset T^*X$ is contained in a closed conic isotropic subanalytic subset of T^*X, which is **stable** under the natural $\mathbb{C}^*$-action, given by multiplication on the fibers on the complex vector bundle $\pi : T^*X \to X$.

4.3 Base change properties

In this section we prove some **base change properties**, which will be important for our later applications of Morse theory for constructible sheaves in the next chapters of this book.

4.3.1 Some constructions for stratifications

First we explain some important examples, how to build new stratifications from given ones. We also discuss, which regularity conditions are "stable" under these "operations". Here we fix for **smooth stratifications** a degree of smoothness $k = 1, \ldots, \infty, \omega$. Let us start with the **product** of two stratifications.

Example 4.3.1. *Let $X_\bullet, X'_\bullet$ be filtrations of X and X'. This induces a product filtration $(X \times X')_\bullet$ of $X \times X'$:*

$$(X \times X')_k := \bigcup_{i+j=k} (X_i \times X'_j) .$$

Then $(X \times X')^k$ is the disjoint union of the $X^i \times X'^j$ (with $i+j = k$), which are also open in $(X \times X')^k$. Especially, if $X_\bullet, X'_\bullet$ are topological (or smooth) stratifications of the (closed sub-) sets X, X' (of the smooth manifolds M, M'), then $(X \times X')_\bullet$ is **also** *a topological (or smooth) stratification of the (closed sub-) set $X \times X'$ (of the smooth manifold $M \times M'$). Moreover, the strata are the products of the strata of $X_\bullet$ and $X'_\bullet$.*

Suppose X, X' are closed subsets of the smooth manifolds M, M', and $X_\bullet, X'_\bullet$ are **smooth stratifications**. Then the following regularity conditions $*$ are "stable" under **products**, i.e. if both are $*$-regular (at x, x'), then the same is true for the product stratification (at (x, x')):

- It is an easy exercise for $* =$ **a, b, w**.
- It is also true for (local) **C-regularity**. Suppose $X_\bullet, X'_\bullet$ are **C-regular** at $x \in X^i$ and $x' \in X^j$ with respect to the **smooth** function(-germs) ρ, ρ'. Then $(X \times X')_\bullet$ is **C-regular** at $(x, x') \in X^i \times X^j$ with respect to the **smooth** function(-germ) $\rho \oplus \rho'$ defined as $\rho \oplus \rho'(x, x') := \rho(x) + \rho'(x')$ (compare with [Bekka1, lem. 1, p. 57, prop. 2.7]).

  Note, that we can in general **not** conclude that $\rho \oplus \rho'$ is smooth outside $X^i \times X^j$, if ρ, ρ' are only smooth outside X^i, X^j ! So this argument does **not** work for (local) $\mathbf{C}^0$**-regularity**, except for the case $X^i = M$ or $X'^i = M'$.

The last exceptional case just means that (local) $\mathbf{C}^0$**-regularity** is "stable" under "smooth pullbacks", which we treat next. Note, that we **cannot** control "stability" of $\mathbf{d}^{(0)}$**-regularity** with respect to products, because a **general** smooth germ of a retraction onto $X^i \times X^j$ needs not to be of "product-type"!

Suppose we have a (derived) tensor product (for $* = \emptyset, b, -, +$):

$$\otimes^L : D^*(X, R_1) \times D^*(X, R_2) \to D^*(X, R_3) ,$$

commuting with pullbacks (as in section 1.4). For example $R_1 = \mathbb{Z}$ and $R_2 = R_3$, or $R = R_1 = R_2 = R_3$ is commutative ring of finite homological dimension. Then one gets for two filtrations $X_\bullet, X'_\bullet$ an induced "exterior product"

$$\boxtimes^L : D^*(X_\bullet, R_1) \times D^*(X'_\bullet, R_2) \to D^*((X \times X')_\bullet, R_3) ,$$

which is also related to the following

Remark 4.3.1. The product of two **isotropic** C^1-submanifolds in T^*M and T^*M' is again an **isotropic** C^1-submanifold in $T^*M \times T^*M' = T^*(M \times M')$. Especially, the product of two **closed conic isotropic** subsets is again **isotropic**, if we work in the context of "geometric categories", or in the complex analytic (algebraic) context.

In this way one gets an exterior product for **conic Lagrangian cycles** (compare [KS, (9.4.1), p. 378]), which is also closely related to the following general estimate for the **micro-support** of an "exterior product" ([KS, prop. 5.4.1], [KS2, prop. 4.2.1]):

$$\mu supp(\mathcal{F} \boxtimes^L \mathcal{F}') \subset \mu supp(\mathcal{F}) \times \mu supp(\mathcal{F}') \,. \tag{4.18}$$

In the next chapter we will get for constructible complexes of sheaves (in "geometric categories") more precise information on this inclusion. In particular, it will be an equality if the base ring R is a field.

We study now the "pull-back" of stratifications.

Let $X_\bullet$ be a filtration of X and consider a continuous map $f : Y \to X$. Then one gets an induced filtration $Y_\bullet$ of Y by $Y_i := f^{-1}(X_i)$ such that f^* maps $D^*(X_\bullet)$ to $D^*(Y_\bullet)$. Of course, if $X_\bullet$ is a **topological stratification**, this need not to be the case for $Y_\bullet$.

Even if the $Y^i = f^{-1}(X^i)$ are topological manifolds, we will have in general $dim(Y^i) \neq i$. But suppose that f is of "relative dimension d" (for a fixed $d \in \mathbb{Z}$) in the sense that all $Y^i = f^{-1}(X^i)$ are topological manifolds of dimension $i+d$, or empty (especially $Y^i = \emptyset$ for $i + d < 0$). In this case we get an **induced topological stratification** $Y_\bullet$ by $Y_{i+d} := f^{-1}(X_i)$.

Example 4.3.2. *Assume* $X_\bullet$ *is a* **smooth stratification** *of the closed subset* X *of the smooth manifold* N. *Consider a* C^k*-map* $f : M \to N$ *of smooth manifolds, which is* **transversal** *to* $X_\bullet$, *i.e. transversal to all* X^i:

$$\forall i : \; df(x)(T_xM) + T_{f(x)}X^i = T_xN \quad \textit{for all} \quad x \in f^{-1}(X^i)\,. \tag{4.19}$$

Then all $Y^i = f^{-1}(X^i)$ *are smooth submanifolds of* M, *whose local dimension in* x *is given by*

$$dim_x(Y^i) = i + dim_x(M) - dim_{f(x)}(N)\,.$$

So if the local relative dimension $d = dim_x(M) - dim_{f(x)}(N)$ *is constant on* Y *(e.g.* N *and* M *are pure-dimensional), we get an induced* **smooth stratification** *defined by* $Y_{i+d} := f^{-1}(X_i)$. *Since this local relative dimension is always locally constant on* M, *we get in general an induced* **smooth stratification** *by* $(Y \cap M')_{i+d(M')} := f^{-1}(X_i)$, *where* M' *runs through the connected components of* M.

The above transversality assumption can in the language of cotangent bundles be reformulated in the following equivalent ways (compare [KS, def. 4.1.5(ii)]), with ${}^tf'$ and f_π as in diagram (4.15) on page 249:

1. The restriction of ${}^t f'$ to all $f_\pi^{-1}(\Lambda_i) = M \times_N \Lambda_i$ is **injective**, with $\Lambda_i := T^*_{X^i} N$.

2. One has for all i the estimate

$$f_\pi^{-1}(A) \cap kern({}^t f') \subset M \times_N T^*_N N \,, \tag{4.20}$$

with $A := \Lambda_i$, $kern({}^t f')$ the corresponding kernel bundle, and $M \times_N T^*_N N$ the zero-section of $f^*(T^*N)$.

If we have the estimate (4.20) for a **closed conic** subset $A \subset T^*N$, then f is called **non-characteristic** with respect to A ([KS, def. 5.4.12]).

So for an **a-regular** stratification $X_\bullet$ the following notions are equivalent:

1. $f : M \to N$ is **transversal** to $X_\bullet$.

2. f is **non-characteristic** with respect to the **closed** conic subset

$$\Lambda_X := \bigcup_i T^*_{X^i} N \subset T^*N \,.$$

Note that a **submersion** $f : M \to N$ of manifolds is always **transversal** to a stratification $X_\bullet$. Similarly, it is **non-characteristic** with respect to any closed conic subset $A \subset T^*N$ (since $kern({}^t f') = 0$) !

Let $f : M \to N$ be C^k-map of smooth manifolds, and consider the induced maps ${}^t f'$ and f_π as in diagram (4.15) on page 249.

Lemma 4.3.1. *Let $A \subset T^*N$ be a* **closed conic** *subset. Then the following conditions are equivalent:*

1. *f is* **non-characteristic** *with respect to A.*

2. *The map ${}^t f' : f_\pi^{-1}(A) \to T^*M$ is* **proper**.

Proof. $2 \Rightarrow 1$. Fix a point $x \in M$. Then

$$f_\pi^{-1}(A) \cap kern({}^t f' : (f^*T^*N)|_{\{x\}} \to (T^*M)|_{\{x\}}) \subset \{0\} \subset (f^*T^*N)|_{\{x\}} \,,$$

since it is a **compact conic** subset of the vector space $(f^*T^*N)|_{\{x\}}$.

$1 \Rightarrow 2$. The claim is local on T^*M so that it is enough to work in local coordinates x, y on M and N. Let $(x, \xi), (y, \eta)$ be the associated coordinates on T^*M and T^*N, with $|| \cdot ||$ a corresponding norm (in these coordinates). Then

$${}^t f'(x, \eta) \neq 0 \quad \text{on} \quad \{(x, \eta) \in f_\pi^{-1}(A) \,:\, ||\eta|| = 1\}$$

implies, that for a given compact subset $K \subset M$, there exists a $c > 0$ with

$$||{}^t f'(x, \eta)|| \geq c \quad \text{for} \quad \{(x, \eta) \in f_\pi^{-1}(A) \,:\, ||\eta|| = 1,\, x \in K\} \,.$$

Then $f_\pi^{-1}(A) \cap {}^t f'^{-1}(T^*M|K)$ is contained in the closed set

$$C := \{(x,\eta) \in f^*(T^*N)|K \;:\; ||{}^t f'(x,\eta)|| \geq c \cdot ||\eta||\}$$

(since $f_\pi^{-1}(A)$ is conic), with ${}^t f' : C \to T^*M$ proper. □

If for example $f : M \to N$ is **transversal** to the Whitney **a-regular** stratification $X_\bullet$ of the closed subset $X \subset N$, then

$${}^t f'(f_\pi^{-1}(\Lambda_X)) = \Lambda_{f^{-1}(X)} ,$$

with Λ_X (or $\Lambda_{f^{-1}(X)}$) the union of conormal spaces to the strata of $X_\bullet$ (of the induced stratification of $f^{-1}(X)$). Especially, $\Lambda_{f^{-1}(X)}$ is a closed subset of T^*M so that the induced stratification of $f^{-1}(X)$ is also Whitney **a-regular**.

Similarly, let us consider a **definable** C^k-map $f : M \to N$ of real analytic (Nash) manifolds, where we work in a fixed "geometric category". Suppose f is **non-characteristic** with respect to the **closed conic isotropic definable** subset $\Lambda \subset T^*N$. Using the fact ${}^t f'^* \omega_M = f_\pi^* \omega_N$ (with ω_M, ω_N the canonical 1-forms on M and N), one gets for $\Lambda' := {}^t f'(f_\pi^{-1}(\Lambda))$ the implication (compare [KS, prop. 8.3.11]):

$$\omega_N|\Lambda \equiv 0 \;\Rightarrow\; f_\pi^*\omega_N|f_\pi^{-1}(\Lambda) \equiv 0 \;\Rightarrow\; {}^t f'^*\omega_M|f_\pi^{-1}(\Lambda) \equiv 0 \;\Rightarrow\; \omega_M|\Lambda' \equiv 0\,.$$

The last implication follows from lemma 4.3.1 and lemma 4.1.4, if one stratifies the proper map ${}^t f' : f_\pi^{-1}(\Lambda) \to T^*M$.

Remark 4.3.2. We get especially the following application:

- Suppose f is **non-characteristic** with respect to the **closed conic definable isotropic** subset $\Lambda \subset T^*N$, then $\Lambda' := {}^t f'(f_\pi^{-1}(\Lambda)) \subset T^*N$ is also a **closed conic definable isotropic** subset.

The same argument applies also to the complex analytic (algebraic) context! Moreover, it is closely related to the "non-characteristic pull-back" in the theory of **conic Lagrangian cycles** (compare with [KS, prop. 9.4.3] and [SchVi, Gi1, Gi2, Sab]). For a corresponding result about the **micro-support**, compare with [KS, cor. 5.4.11, prop. 5.4.13] and [KS2, prop. 5.3.2].

Let us come back to the "stablility" of regularity conditions with respect to a "transversal pull-back".

The regularity conditions $* =$ **a, b, w** and (local) $\mathbf{C^{(0)}}$**-regularity** are "stable" under "pull-back" by a **submersion** $f : M \to N$ of manifolds, i.e. if $X_\bullet$ is $*$-regular (at x), then the same is true for $f^{-1}(X)_\bullet$ (at all points in $\{f = x\}$). In fact, this can be checked locally so that one can assume $f = pr_N : M' \times N \to N$ a projection. Then one is back to the case of a product with a smooth manifold M'.

Remark 4.3.3. Using deeper results about the "micro-support" ([KS, prop. 5.4.5], [KS2, prop. 4.1.2]), one can show that also the regularity condition **a + LSA** is "stable" under "pull-back" by a **submersion**.

Finally, the regularity conditions $* =$ **a, b, w** and (local) **C-regularity** are "stable" under "transversal pull-back". Using a graph embedding, it is enough to consider the case $M \subset N$ is a closed submanifold transversal to the stratification $X_\bullet$ of the closed subset $X \subset N$. This again is a special case of the **tranversal intersection** of two stratifications $X_\bullet, Y_\bullet$ of the closed subsets $X, Y \subset N$, i.e. we assume that all X^i and Y^j intersect transversally. Equivalently, the diagonal embedding $d : N \to N \times N$ is **transversal** to the **product stratification** $(X \times Y)_\bullet$ of $X \times Y$. Especially, we get an induced stratification of $X \cap Y$, whose strata are the connected components of the intersection of the strata of X and Y. The "stability" of **a, b** and **w-regularity** (or (local) **C-regularity**) is for example explained in [OTr2, thm. 2.4] (or [Bekka1, prop. 2.7]).

Note, that the product stratification $(X \times Y)_\bullet$ is just the **transversal intersection** of the two stratifications of $X \times N$ and $N \times Y$ induced by the projection of $N \times N$ onto its factors (these projections are submersions).

Let $f : M \to N$ be a C^k-map of smooth manifolds, and consider two stratifications $X_\bullet, Y_\bullet$ of the closed subsets $X \subset M$ and $Y \subset N$. Assume that $f : X^i \to N$ is transversal to $Y_\bullet$ for all i. Especially, $f : M \to N$ is transversal to $Y_\bullet$ in all points $x \in X$ (i.e. we have the condition (4.19) for all points $x \in X$). Then one would like to get an induced stratification of $X \cap f^{-1}(Y)$.

This is indeed possible, if $f : U \to N$ is transversal to $Y_\bullet$ for an open neighborhood U of X in M. Then we get an induced stratification of $U \cap f^{-1}(Y)$, which is transversal to $X_\bullet$ (in U) so that one gets an induced stratification of $X \cap f^{-1}(Y) \subset U \subset M$. Such an open neighborhood U exists for example, if $Y_\bullet$ is Whitney **a-regular**. Otherwise there would be i, j, k and a sequence $x_n \in M$ converging to $x \in X^i$, with $f(x_n) \in Y^j, f(x) \in Y^k$ and

$$df(T_{x_n}M) + T_{f(x_n)}Y^j \neq T_{f(x_n)}N \,.$$

We can also assume that $T_{f(x_n)}Y^j \to \tau$, with $T_{f(x)}Y^k \subset \tau$ by **a-regularity** (for $j \neq k$. This is also obvious for the case $Y^j = Y^k$). Then one gets the contradiction

$$df(T_xX^i) + T_{f(x)}Y^k \subset df(T_xM) + T_{f(x)}Y^k \subset df(T_xM) + \tau \neq T_{f(x)}N \,.$$

Especially, if $X_\bullet$ and $Y_\bullet$ are both **a, b, w** or (locally) **C-regular**, then we get an induced stratification of $X \cap f^{-1}(Y)$, which is also **a, b, w** or (locally) **C-regular**.

Example 4.3.3. *Let $g : M \to N$ be a C^k-map of smooth manifolds, and consider a stratification $X_\bullet$ of the closed subset $X \subset M$. Fix a point $n \in N$ and assume that $g : X^i \to N$ is a submersion in all points of $X^i \cap \{g = n\}$ (for all i). Then $g : X^i \to N$ is transversal to the Whitney* **a-regular** *stratification $Y_0 := \{n\} \subset Y_d := N$ of N, with $d = dim_n(N)$. Here we work locally near $\{g = n\}$ so that we* *can assume N is pure-dimensional. Therefore we get a new induced stratification of $X = X \cap f^{-1}(N)$ such that $X \cap \{g = n\}$ is a union of strata. Of course, outside $\{g = n\}$ it coincides with the given stratification $X_\bullet$.*

As remarked before, the general "stability results" above do **not** apply to **$\mathbf{d}^{(0)}$-regularity**. Nevertheless, we have the following useful

Lemma 4.3.2. *Consider the context of example* 4.3.3.

1. *Assume $X_\bullet$ is $d^{(0)}$-regular. Then $X \cap \{g = n\}$ with its induced stratification is also $d^{(0)}$-regular (as a subset of M).*

2. *Assume $X_\bullet$ is d-regular. Then the induced new stratification (as above) is also d-regular.*

Proof. 1. Fix a point $x \in X^i \cap \{g = n\}$, and a germ of a smooth retraction

$$\pi : (U, x) \to (X^i \cap \{g = n\}, x) ,$$

with U an open neighborhood of x in M. Then

$$(\pi, g) : (X^i, x) \to (X^i \cap \{g = n\}, x) \times (N, n)$$

is the germ of a diffeomorphism. Define the germ

$$\pi' : (U, x) \to (X^i, x) \quad \text{as} \quad \pi' := ((\pi, g)|X^i)^{-1} \circ (\pi, g) .$$

This is then by construction the germ of a smooth retraction onto X^i. Since $X_\bullet$ is $d^{(0)}$-regular, there exists an open neighborhood U'' of x in $U \cap X$ and a continuous function $\rho' : U'' \to \{y \in \mathbb{R}|\, y \geq 0\} =: \mathbb{R}^{\geq 0}$ with the properties:

a. $\{\rho' = 0\} = U'' \cap X^i$, and $\rho'|(U'' \backslash X^i)$ is smooth (or $\rho' : U'' \to \mathbb{R}$ is smooth in the context of **d**-regularity).

b. $(\pi', \rho') : U'' \cap X^j \to X^i \times \mathbb{R}^{>0}$ is a submersion for all $j > i$.

Consider the continuous function $\rho := \rho'|U' : U' \to \mathbb{R}^{\geq 0}$, defined on the open neighborhood $U' := U'' \cap \{g = n\}$ of x in $U \cap X \cap \{g = n\}$. Then

$$\{\rho = 0\} = U' \cap X^i \cap \{g = n\} ,$$

and $\rho|(U' \backslash (X^i \cap \{g = n\}))$ (or $\rho|U'$) is smooth. Since

$$(((\pi, g)|X^i) \circ \pi', \rho') = (\pi, g, \rho') : U'' \cap X^j \to (X^i \cap \{g = n\}) \times N \times \mathbb{R}^{>0}$$

is a submersion for all $j > i$, we finally also have that

$$(\pi, \rho) : U' \cap (X^j \cap \{g = n\}) \to (X^i \cap \{g = n\}) \times \mathbb{R}^{>0}$$

is a submersion for all $j > i$.

2. First remark, that the d-regularity in a point x of the open subset $\{g \neq n\}$ is obvious. Fix a point $x \in X^i \cap \{g = n\}$, and as in 1. a germ of a smooth retraction

$$\pi : (U, x) \to (X^i \cap \{g = n\}, x) .$$

We follow the proof of 1. and use the same notation π', U'', ρ', but this time with $\rho' : U'' \to \mathbb{R}$ smooth. Choose in addition (in some local coordinates) a smooth function $r : (N, n) \to \mathbb{R}^{\geq 0}$ with

$$\{r = 0\} = \{n\} \quad \text{and} \quad r : N \backslash \{n\} \to \mathbb{R}^{>0} \quad \text{a submersion.}$$

Define $U' := U''$ and $\rho := \rho' + (r \circ g) : U' \to \mathbb{R}^{\geq 0}$. Then ρ is smooth in U', and

$$\{\rho = 0\} = \{\rho' = 0\} \cap \{r \circ g = 0\} = U' \cap X^i \cap \{g = n\} .$$

Since $\rho|(U' \cap \{g = n\}) = \rho'|(U' \cap \{g = n\})$, we have by 1. that

$$(\pi, \rho) : U' \cap (X^j \cap \{g = n\}) \to (X^i \cap \{g = n\}) \times \mathbb{R}^{>0}$$

is a submersion for all $j > i$. Moreover, we also have by 1. that

$$(\pi, g, \rho') : U'' \cap X^j \to (X^i \cap \{g = n\}) \times N \times \mathbb{R}^{>0}$$

is a submersion for all $j > i$. By the definition of ρ we get that

$$(\pi, \rho) : U' \cap (X^j \backslash \{g = n\}) \to (X^i \cap \{g = n\}) \times \mathbb{R}^{>0}$$

is a submersion for all $j > i$. Finally we can take U' small enough so that $U' \cap X^j = \emptyset$ for $j < i$, and

$$(\pi, \rho) \circ ((\pi, g)|X^i)^{-1} = (pr_1, r) : (X^i \cap \{g = n\}) \times N \to (X^i \cap \{g = n\}) \times \mathbb{R}^{\geq 0}$$

is by definition of r a submersion on $(X^i \cap \{g = n\}) \times (N \backslash \{n\})$. This implies that (π, ρ) is a submersion on $X^i \backslash \{g = n\}$. □

Remark 4.3.4. We leave it to the reader to prove the result corresponding to lemma 4.3.2(1) for a stratification with the property **LST**.

4.3.2 Base change isomorphisms

After these general results about stratifications, we switch now to the study of important **base change isomorphisms**.

Lemma 4.3.3. *Let $f : X \to Y$ be a continuous map and $B \subset \mathbb{R}^n$ be an open ball. Let $p : Y' := Y \times B \to Y$ be the projection, and $s : Y \to Y'$ be a continuous section ($p \circ s = id$). Consider the induced cartesian diagram:*

$$\begin{array}{ccccc} X & \xleftarrow{p'} & X' & \xleftarrow{s'} & X \\ \downarrow f & & \downarrow f' & & \downarrow f \\ Y & \xleftarrow{p} & Y' & \xleftarrow{s} & Y . \end{array}$$

Suppose all cohomology sheaves of $\mathcal{F} \in D(X')$ are **locally constant** *on the fibers of the projection p'. Then the natural* **base change morphism**

$$\varphi_s : s^* Rf'_*(\mathcal{F}) \to Rf_* s'^*(\mathcal{F})$$

is an **isomorphism**.

Proof. By the assumption on $\mathcal{F}$ and the **Vietoris-Begle theorem** ([KS2, prop. 2.7.8]) we have $\mathcal{F} \simeq p'^* Rp'_* \mathcal{F}$. Consider a commutative diagram of continuous maps:

$$\begin{array}{ccccc} X & \xleftarrow{g'} & X' & \xleftarrow{h'} & X'' \\ \downarrow f & & \downarrow f' & & \downarrow f'' \\ S & \xleftarrow{g} & S' & \xleftarrow{h} & S'' . \end{array}$$

Then the **base change morphism** $\varphi_g : g^* Rf_* \to Rf'_* g'^*$ is by adjunction induced from $ad_{g'} : Rf_* \to Rf_* Rg'_* g'^* \simeq Rg_* Rf'_* g'^*$ (compare [SGA4$\frac{1}{2}$, p. 84]). Then the composition

$$(g \circ h)^* Rf_* \xrightarrow{\varphi_{g \circ h}} Rf''_*(g' \circ h')^* \xrightarrow{\sim} Rf''_* h'^* g'^*$$

is equal to ([SGA4$\frac{1}{2}$, p. 85, 86, prop. 4.4]):

$$(g \circ h)^* Rf_* \xrightarrow{\sim} h^* g^* Rf_* \xrightarrow{\varphi_g} h^* Rf'_* g'^* \xrightarrow{\varphi_h} Rf''_* h'^* g'^* .$$

This relation is stated in loc.sit. only for a cartesian diagram, but of course it is also true for a commutative diagram. This follows for example from the abstract characterization of the base change morphism given in [SGA4$\frac{1}{2}$, p. 33, prop. 2.1.3] (we need this more general case in our approach to Morse theory for constructible sheaves). Applied to the cartesian diagram of the theorem, we get that the composed morphism

$$Rf_* \simeq s^* p^* Rf_* \xrightarrow{\varphi_p} s^* Rf'_* p'^* \xrightarrow{\varphi_s} Rf_* s'^* p'^* \simeq Rf_*$$

is the identity. But φ_p is an isomorphism by the **smooth base change theorem**. For more general versions of the smooth base change theorem compare with [IC, (2) on p. 80] or [BL, p. 56, lem. C1]. □

This lemma together with lemma 4.3.2 will be used in the proof of the following extension of proposition 4.0.3:

Proposition 4.3.1. *Let $X_\bullet$ be a smooth stratification of the closed subset X of the manifold M, and let $g : M \to N$ be a smooth map of manifolds. Fix a point $n \in N$ such that $g : S \to N$ is a* **submersion** *in all points of $S \cap \{g = n\}$, for all strata S of $X_\bullet$. Denote by $j : U \to X$ the inclusion of an open subset, which is a union of strata, and by $i : X' := X \backslash U \to X$ the inclusion of the complement. Consider the cartesian diagram*

$$\begin{array}{ccccc} Y' := X' \cap \{g = n\} & \xrightarrow{i'} & Y := X \cap \{g = n\} & \xleftarrow{j'} & U' := U \cap \{g = n\} \\ \downarrow{\scriptstyle k'} & & \downarrow{\scriptstyle k} & & \downarrow{\scriptstyle k''} \\ X' & \xrightarrow{i} & X & \xleftarrow{j} & U\,. \end{array}$$

Assume $X_\bullet$ satisfies the property **LST**, *or is* $\mathbf{d}^0$**-regular**. *Then one has the following* **base change isomorphisms***:*

$$k^* Rj_* \mathcal{F} \xrightarrow{\sim} Rj'_* k''^* \mathcal{F} \quad \text{for} \quad \mathcal{F} \in D(U_\bullet)\,. \tag{4.21}$$

$$k'^* i^! \mathcal{F} \xrightarrow{\sim} i'^! k^* \mathcal{F} \quad \text{for} \quad \mathcal{F} \in D(X_\bullet)\,. \tag{4.22}$$

Proof. 1. We show more precisely, that the **base change morphism**

$$\varphi : k^* Rj_* \mathcal{F} \to Rj'_* k''^* \mathcal{F}$$

is an isomorphism. This can be checked on the stalks and is always true for $x \in U'$ (since U is open). Consider a point $x \in X^i \cap Y'$.

First we consider the case, that $X_\bullet$ has the property **LST**. Then there exist open neighborhoods W of x in X, V of n in N and a filtration preserving homeomorphism

$$h : W \to V \times (W \cap \{g = n\})\,,$$

i.e. $h(W \cap X_i) = V \times (W \cap X_i \cap \{g = n\})$, with $g = pr_V \circ h$. Here

$$pr_V : V \times (W \cap \{g = n\}) \to V$$

denotes the projection. We can therefore assume

$$X = V \times (W \cap \{g = n\})$$

with V an open ball around n (in some coordinates), and $g = pr_V$. Since h is filtration preserving and U is an union of strata, we get under this identification

$$U = V \times (U \cap \{g = n\})\,.$$

Moreover, $\mathcal{F}|U$ is constructible with respect to the product filtration. Especially, all cohomology sheaves of $\mathcal{F}|U$ are locally constant on the fibers of the projection

$$U = V \times (U \cap \{g = n\}) \to (U \cap \{g = n\})\,.$$

The claim follows then from lemma 4.3.3 applied to the cartesian diagram

$$\begin{array}{ccc} U = V \times (U \cap \{g = n\}) & \xleftarrow{s'=k''} & \{n\} \times (U \cap \{g = n\}) = U' \\ \big\downarrow j & & \big\downarrow j' \\ X = V \times (W \cap \{g = n\}) & \xleftarrow{s=k} & \{n\} \times (W \cap \{g = n\}) = X' \,. \end{array}$$

Let us now consider the case that $X_\bullet$ is **$\mathbf{d}^0$-regular**. We prove the claim by induction on the **depth** of the stratification (starting with the empty set). By the proof of proposition 4.1.2 and lemma 4.3.2(1), there exists an arbitrarily small open neighborhood V of x in X such that V (or $V \cap \{g = n\}$) is a **distinguished neighborhood** with respect to $X_\bullet$ (or the induced stratification of $X \cap \{g = n\}$). Here we mean distinguished in the sense of a **ccs-stratification**. More precisely, the corresponding map

$$(\pi', \rho') : (V, V \cap X^i) \to (B \times [0, \epsilon[, B \times \{0\})$$

is of the form $\pi' = (\pi, g)$, with B an open ball (in local coordinates) around (x, n) in $(X^i \cap \{g = n\}) \times N$.

Since U is a union of strata and $x \notin U$, we can assume $V \cap U \subset V \backslash X^i$. We therefore get a cartesian diagram of inclusion maps:

$$\begin{array}{ccccc} V' := V \cap \{g = n\} & \xleftarrow{j_2'} & V' \backslash X^i & \xleftarrow{j_1'} & U' := U \cap \{g = n\} \\ \big\downarrow k & & \big\downarrow k' & & \big\downarrow k'' \\ V & \xleftarrow{j_2} & V \backslash X^i & \xleftarrow{j_1} & U \,, \end{array}$$

with $j = j_2 \circ j_1$ and $j' = j_2' \circ j_1'$. The base change morphism φ commutes with restriction to open subsets. By proposition 4.2.1 we get a commutative diagram

$$\begin{array}{ccccc} (Rj'_* k''^* \mathcal{F})_x & \xleftarrow{\sim} & R\Gamma(V', Rj'_{2*} Rj'_{1*} k''^* \mathcal{F}) & \xrightarrow{\sim} & R\Gamma(V' \backslash X^i, Rj'_{1*} k''^* \mathcal{F}) \\ \big\uparrow \varphi & & & & \big\uparrow \alpha \\ (k^* Rj_* \mathcal{F})_x & & & & R\Gamma(V' \backslash X^i, k'^* Rj_{1*} \mathcal{F}) \\ \big\uparrow \wr & & & & \big\uparrow \beta \\ (Rj_* \mathcal{F})_x & \xleftarrow{\sim} & R\Gamma(V, Rj_{2*} Rj_{1*} \mathcal{F}) & \xrightarrow{\sim} & R\Gamma(V \backslash X^i, Rj_{1*} \mathcal{F}) \,. \end{array}$$

But α is induced by the base change morphism $\varphi : k'^* Rj_{1*} \mathcal{F} \to Rj'_{1*} k''^* \mathcal{F}$. It is therefore an isomorphism by the induction hypothesis.

Similarly, β is an isomorphism, since

$$\gamma : R\Gamma(V' \backslash X^i, k'^* Rj_{1*} \mathcal{F}) \to R\Gamma((\pi, g, \rho')^{-1}(\{(x, n, t)\}), Rj_{1*} \mathcal{F}) \,,$$

and also the composition

$$\gamma \circ \beta : R\Gamma(V \backslash X^i, Rj_{1*}\mathcal{F}) \to R\Gamma((\pi, g, \rho')^{-1}(\{(x, n, t)\}), Rj_{1*}\mathcal{F})$$

(with $0 < t << 1$) is an isomorphism by (the proof of) proposition 4.2.1(3.b).

2. The morphism $ad_i : i'_* k'^* i^! \simeq k^* i_* i^! \to k^*$ induces by adjunction the morphism $\varphi : k'^* i^! \to i'^! k^*$, and we show that it is an isomorphism for $\mathcal{F} \in D(X_\bullet)$. By using the distinguished triangle

$$i_* i^! \longrightarrow id \longrightarrow Rj_* j^* \xrightarrow{[1]} ,$$

it is enough to show this for $i_* i^! \mathcal{F}$ and $Rj_* j^* \mathcal{F}$. But φ applied to $i_* i^!$ is isomorphic to the identity transformation

$$k'^* i^! \simeq k'^* i^! i_* i^! \to i'^! k^* i_* i^! \simeq i'^! i'_* k'^* i^! \simeq k'^* i^! .$$

Moreover, $k'^* i^! Rj_* j^*$ is always zero. By 1. this is also true for

$$i'^! k^* Rj_* j^* \mathcal{F} \simeq i'^! Rj'_* k''^* j^* \mathcal{F} . \qquad \square$$

Remark 4.3.5. What we used in the first step of the proof is the fact that, under the assumption of the **LST-property**, the inclusion of the transversal intersection

$$X \cap \{g = n\} \to X$$

is locally a stratum preserving **normally nonsingular inclusion** in the sense of [GM, p. 48, def. 1.11].

Example 4.3.4. *Assume $X_\bullet$ is induced from a stratification in the sense of remark 4.0.1, which is Whitney* **a-regular** *and satisfies the* **frontier condition** (4.1). *But we do* **not** *assume that the "pieces" S are connected. Suppose $X_\bullet$ has the property* **LST** *or is* **d^0-regular**, *and consider a submersive germ*

$$g : (X, x) \to (N, n) ,$$

with N a manifold. By **a-regularity**, *we can locally assume that $g : X \to N$ is a submersion as in proposition 4.3.1. Then the induced stratification of $X \cap \{g = n\}$ (with "pieces" $S \cap \{g = n\}$) satisfies also the* **frontier condition**.

This follows from the **base change property** of proposition 4.3.1 in the following way:

Let S, T be "pieces" of the stratification, with $T \subset cl(S)$. Since the induced stratification of $cl(S)$ has the same properties as $X_\bullet$, we can assume $X = cl(S)$ so that S is open in X. Let $j : U := S \to X$ be the inclusion, and let $\mathcal{F} \neq 0$

be a constant sheaf on S. Apply proposition 4.3.1 to this sheaf and the cartesian diagram

$$\begin{array}{ccc} X \cap \{g = n\} & \xleftarrow{j'} & U \cap \{g = n\} \\ \downarrow{\scriptstyle k} & & \downarrow{\scriptstyle k''} \\ X & \xleftarrow{j} & U = S\,. \end{array}$$

Then $0 \neq (k^* Rj_* \mathcal{F})_x \simeq (Rj'_* k''^* \mathcal{F})_x$ for $x \in T \cap \{g = n\}$, and this implies

$$T \cap \{g = n\} \subset cl(S \cap \{g = n\})\,.$$

Remark 4.3.6. Consider a stratification as in proposition 4.3.1 or example 4.3.4. Instead of the intersection with $\{g = n\}$, consider now the intersection with a submanifold M' of M, which is **transversal** to all strata of the stratification. Then locally M' is of the form $\{g = n\}$ for a suitable g, which is submersive along $\{g = n\}$.

Therefore the results of proposition 4.3.1 and example 4.3.4 remain true for the intersection with M'. For the corresponding conclusion of example 4.3.4 in the context of Whitney **b-regular** stratifications compare with [Mather, cor. 10.4] and [Ch]. But our result applies in particular also to **w-regular** or (locally) $\mathbf{C}^{(0)}$**-regular** stratifications!

Next we discuss some **base change properties** of the **nearby** and **vanishing cycle functor**.

Let $f : X \to \mathbb{C}$ be a continuous function and $k : Y \to X$ a continuous map. Let $f' := f \circ k$. We claim that there is a natural **base change morphism**

$$b : k'^* \psi_f \to \psi_{f'} \circ k^*\,,$$

with k' the induced map of the zero sets. For this we consider the following cartesian diagram:

$$\begin{array}{ccccccc} Y_0 & \xrightarrow{i'} & Y & \xleftarrow{j'} & \{f' \neq 0\} & \xleftarrow{p'} & \widetilde{Y^*} \\ \downarrow{\scriptstyle k'} & & \downarrow{\scriptstyle k} & & \downarrow{\scriptstyle \kappa} & & \downarrow{\scriptstyle \kappa'} \\ X_0 & \xrightarrow{i} & X & \xleftarrow{j} & \{f \neq 0\} & \xleftarrow{p} & \widetilde{X^*} \\ \downarrow & & \downarrow{\scriptstyle f} & & \downarrow & & \downarrow{\scriptstyle \tilde{f}} \\ \{0\} & \longrightarrow & \mathbb{C} & \longleftarrow & \mathbb{C}^* & \longleftarrow & \widetilde{\mathbb{C}^*}\,. \end{array}$$

Then b is defined as the composition of

$$k'^* i^* Rj_* Rp_* p^* j^* \simeq i'^* k'^* Rj_* Rp_* p^* j^* \xrightarrow{b_k} i'^* Rj'_* \kappa^* Rp_* p^* j^* \xrightarrow{b_\kappa}$$

$$i'^* Rj'_* Rp'_* \kappa'^* p^* j^* \simeq i'^* Rj'_* Rp'_* p'^* j'^* k^*\,,$$

where b_k and b_κ are the corresponding base change morphisms (as in the proof of lemma 4.3.3).

Remark 4.3.7. 1. The following diagram **commutes** by this definition of the base change morphism b:

$$\begin{array}{ccc} k'^* i^* & \xrightarrow{sp_f} & k'^* \psi_f \\ \downarrow \wr & & \downarrow b \\ i'^* k^* & \xrightarrow{sp_{f'}} & \psi_{f'} \circ k^* . \end{array}$$

2. If k (and therefore also κ) is a **topological submersion**, then b is an isomorphism by the **smooth base change theorem**. Then one gets by 1. also an isomorphism $k'^* \phi_f \simeq \phi_{f'} \circ k^*$.

3. One defines similarly a **base change morphism**

$$\psi_f \circ Rk_* \to Rk'_* \circ \psi_{f'} ,$$

"compatible" with $i^* Rk_* \to Rk'_* i'^*$ as in 1. Especially, for k **proper** one gets by the **proper base change theorem** ([KS, prop.2.6.7]) the isomorphisms

$$\psi_f \circ Rk_* \simeq Rk'_* \circ \psi_{f'} \quad \text{and} \quad \phi_f \circ Rk_* \simeq Rk'_* \circ \phi_{f'} .$$

Lemma 4.3.4. *Let $X_\bullet$ be a smooth stratification of the closed subset X of the manifold N. Assume X_0 is a union of strata, and let $M \subset N$ be a submanifold which is transversal to all strata of this stratification. Let $k : Y := X \cap M \to X$ be the induced inclusion. Assume $X_\bullet$ has the property* **LST** *or is* $\mathbf{d}^0$**-regular**. *Then the base change morphism*

$$b : k'^* \psi_f(\mathcal{F}) \to \psi_{f'}(k^* \mathcal{F})$$

is an isomorphism for all $\mathcal{F} \in ob(D(X_\bullet))$. Especially,

$$k'^* \phi_f(\mathcal{F}) \simeq \phi_{f'}(k^* \mathcal{F}) \quad \textit{for all} \quad \mathcal{F} \in ob(D(X_\bullet)) .$$

Proof. The last isomorphism for the vanishing cycle functor follows already from remark 4.3.7(1). By the proof of lemma 4.2.1 we know that $Rp_* p^* j^* \mathcal{F}$ is constructible with respect to the induced stratification of $\{f \neq 0\}$. Therefore the base change morphism b_k is an isomorphism by proposition 4.3.1. We therefore only have to show that the base change morphism

$$b_\kappa : \kappa^* Rp_* p^* \mathcal{G} \to Rp'_* \kappa'^* p^* \mathcal{G} \simeq Rp'_* p'^* \kappa^* \mathcal{G}$$

is an isomorphism for $\mathcal{G}$ constructible with respect to the induced stratification of $\{f \neq 0\}$.

Let us first consider the case that the stratification $X_\bullet$ and therefore also the induced stratification of $\{f \neq 0\}$ has the property **LST**. Since the statement is a local result, we can work on an open subset $V \subset \{f \neq 0\}$ such that the covering $p : p^{-1}(V) \to V$ is trivial. Since the stratification has the property **LST**, we can locally assume that the inclusion κ is of the form (compare with the first step of the proof of proposition 4.3.1):

$$V \cap M \simeq \{n\} \times (V \cap M) \to D \times (V \cap M),$$

with D an open ball around n in the stratum containing the point $n \in V \cap M$. Here the stratification of $D \times (V \cap M)$ is the product stratification. We therefore have a cartesian diagram

$$\begin{array}{ccc} D \times (V \cap M) \times \mathbb{Z} & \xleftarrow{\kappa'=s'} & (V \cap M) \times \mathbb{Z} \\ \downarrow{\scriptstyle p} & & \downarrow{\scriptstyle p'} \\ D \times (V \cap M) & \xleftarrow{\kappa=s} & V \cap M, \end{array}$$

and the claim follows from lemma 4.3.3.

Let us now consider the case that the stratification $X_\bullet$ and therefore also the induced stratification of $\{f \neq 0\}$ is **$\mathbf{d}^0$-regular**. Fix a point $x \in \{f \neq 0\}$. As in the second step of the proof of proposition 4.3.1, there exists an arbitrarily small open neighborhood V of x in $\{f \neq 0\}$ such that V (or $V \cap M$) is a **distinguished neighborhood** with respect to $X_\bullet$ (or the induced stratification of $X \cap M$). Here we mean distinguished in the sense of a **ccs-stratification**.

By the proof of lemma 4.2.1 we know that $Rp_*p^*\mathcal{G}$ (or $Rp'_*p'^*\kappa^*\mathcal{G}$) is constructible with respect to the induced stratification of $\{f \neq 0\}$ (or $\{f \neq 0\} \cap M$). We take again V small enough so that the covering p is trivial over V. Then we get by proposition 4.2.1(3.a) a commutative diagram

$$\begin{array}{ccc} (Rp_*p^*\mathcal{G})_x \simeq (\kappa^* Rp_*p^*\mathcal{G})_x & \xrightarrow{b_\kappa} & (Rp'_*p'^*\kappa^*\mathcal{G})_x \\ \uparrow\wr & & \uparrow\wr \\ R\Gamma(V, Rp_*p^*\mathcal{G}) & \longrightarrow & R\Gamma(V \cap M, Rp'_*p'^*\kappa^*\mathcal{G}) \\ \downarrow\wr & & \downarrow\wr \\ \prod_{i\in\mathbb{Z}} R\Gamma(V, \mathcal{G}) & \longrightarrow & \prod_{i\in\mathbb{Z}} R\Gamma(V \cap M, \kappa^*\mathcal{G}) \\ \downarrow\wr & & \downarrow\wr \\ \prod_{i\in\mathbb{Z}} \mathcal{G}_x & \xrightarrow{\sim} & \prod_{i\in\mathbb{Z}} (\kappa^*\mathcal{G})_x\,. \end{array}$$

□

We close this section with the **cohomological version** of the result of Goresky-MacPherson about "moving the wall" ([GM, I. thm. 4.3, thm. 4.4]). This is not really important for our treatment, but it fits nicely with our preceding results. Moreover, it allows the reader who is familiar with the book [GM], to translate the arguments of Goresky-MacPherson into sheaf theoretic terms.

Lemma 4.3.5. *Let $I \subset \mathbb{R}$ be an interval and X a topological space. Consider a locally closed subspace $A \subset X \times I$ and the commutative diagram*

$$\begin{array}{ccccc} A & \xrightarrow{i} & X \times I & \xrightarrow{pr_2} & I \\ \downarrow{\scriptstyle pr} & & \downarrow{\scriptstyle pr_1} & & \\ X & = & X . & & \end{array}$$

Here i is the inclusion, pr_1, pr_2 are the projections and $pr := pr_1|A$. Then one considers A as a family A_t $(t \in I)$ of locally closed subsets

$$A_t := pr(A \cap \{pr_2 = t\}) \subset X .$$

Let $\mathcal{F} \in ob(D(X))$ be given. Assume all cohomology sheaves of $Rpr_{2!}i_!pr^*\mathcal{F}$ are **locally constant**. *Then*

1. $R\Gamma_c(A_t, \mathcal{F}) \simeq R\Gamma_c(A_{t'}, \mathcal{F})$ *for all* $t, t' \in I$.
2. *Suppose $t \leq t'$, $A_t \subset A_s$ for all $s \in]t, t']$, and $A_t \times [t, t'] \subset A$ is closed. Then the natural morphism $R\Gamma_c(A_t, \mathcal{F}) \to R\Gamma_c(A_{t'}, \mathcal{F})$ is an isomorphism. If moreover A is a closed subset and $pr_2 \circ i$ proper, then the restriction map $R\Gamma(A_{t'}, \mathcal{F}) \to R\Gamma(A_t, \mathcal{F})$ is also an isomorphism.*

Proof. 1. Since $Rpr_{2!}i_!$ commutes with restriction, we can assume $t \leq t'$ and $I = [t, t']$. Consider for $s \in I$ the section $i_s : X \to X \times I, x \mapsto (x, s)$ of pr_1. This induces a homeomorphism

$$i_s : A_s \to A \cap \{pr_2 = s\} = A_s \times \{s\} ,$$

and by $s^*pr^*\mathcal{F} \simeq \mathcal{F}$ one gets the isomorphism

$$R\Gamma_c(A_t, \mathcal{F}) \simeq R\Gamma_c(A_s \times \{s\}, pr^*\mathcal{F}) .$$

By the assumption on $Rpr_{2!}i_!pr^*\mathcal{F}$, the morphisms in the second line of the following commutative diagram are isomorphisms:

$$\begin{array}{ccccc} R\Gamma_c(A_t \times \{t\}, pr^*\mathcal{F}) & \longleftarrow & R\Gamma_c(A, pr^*\mathcal{F}) & \longrightarrow & R\Gamma_c(A_{t'} \times \{t'\}, pr^*\mathcal{F}) \\ \downarrow \wr & & \downarrow \wr & & \downarrow \wr \\ (Rpr_{2!}i_!pr^*\mathcal{F})_t & \xleftarrow{\sim} & R\Gamma(I, Rpr_{2!}i_!pr^*\mathcal{F}) & \xrightarrow{\sim} & (Rpr_{2!}i_!pr^*\mathcal{F})_{t'} . \end{array}$$

2. If we take instead of A the subset $A_t \times I$, then one has a cartesian diagram

$$\begin{array}{ccc} A_t \times I & \xrightarrow{pr_2 \circ i} & I \\ \downarrow{\scriptstyle pr} & & \downarrow{\scriptstyle c} \\ A_t & \xrightarrow[c']{} & \{t\} \end{array}$$

so that $Rpr_{2!}i_!pr^*\mathcal{F} \simeq c^*Rc'_!\mathcal{F}$ has always locally constant cohomology sheaves. Together with the argument of 1. we get therefore a commutative diagram

$$\begin{array}{ccccc} R\Gamma_c(A_t \times \{t\}, pr^*\mathcal{F}) & \xleftarrow{\sim} & R\Gamma_c(A, pr^*\mathcal{F}) & \xrightarrow{\sim} & R\Gamma_c(A_{t'} \times \{t'\}, pr^*\mathcal{F}) \\ \| & & \uparrow & & \uparrow \\ R\Gamma_c(A_t \times \{t\}, pr^*\mathcal{F}) & \xleftarrow{\sim} & R\Gamma_c(A_t \times I, pr^*\mathcal{F}) & \xrightarrow{\sim} & R\Gamma_c(A_t \times \{t'\}, pr^*\mathcal{F}) , \end{array}$$

and the right vertical map is isomorphic to $R\Gamma_c(A_t, \mathcal{F}) \to R\Gamma_c(A_{t'}, \mathcal{F})$. One proves the proper case in the same way. □

Example 4.3.5. *Let $X_\bullet$ be Whitney* **a-regular** *stratification of the closed subset X of the manifold M. Let $f : M \to N$ be a smooth map such that*

$$f \times id : X \times \mathbb{R} \to N \times \mathbb{R}$$

is **transversal** *to a Whitney* **a-regular** *stratification of the closed subset $Y \subset N \times \mathbb{R}$ (i.e. for each stratum S of $X_\bullet$, $f \times id : S \times \mathbb{R} \to N \times \mathbb{R}$ is transversal to all strata of the stratification of Y). Then*

$$A := (X \times \mathbb{R}) \cap (f \times id)^{-1}(Y)$$

gets an induced Whitney **a-regular** *stratification. Suppose that $pr_2 \circ i : A \to \mathbb{R}$ is a* **proper stratified submersion** *with respect to this induced stratification. Assume that this induced stratification is also* **$\mathbf{d}^0$-regular** *or satisfies the property* **LST**. *Then the assumptions of lemma 4.3.5 are true, if $\mathcal{F}$ is $X_\bullet$-constructible (because then $pr^*\mathcal{F}$ is constructible with respect to this induced stratification of A).*

This is the situation of [GM, I, thm. 4.3]. Of course they consider the case of Whitney **b-regular** stratifications and formulate the transversality assumptions in terms of a condition on the "characteristic covectors" of the map f. The important point here is, that the **product stratification** of $X \times \mathbb{R}$ and also such a **transversal intersection** of Whitney **b-regular** stratifications is again Whitney **b-regular**, especially **d-regular**.

As explained in the beginning of this section, other interesting classes of regularity conditions with this stability property are **w-regular** stratifications (so that the intersection has the property **LST**), or (locally) **C-regular** stratifications in the sense of Bekka (so that the intersection is **d-regular**).

Chapter 5

Morse theory for constructible sheaves

Introduction

In this chapter we prove the basic results about **stratified Morse theory** in the framework of constructible sheaves. We follow the notions of Goresky-MacPherson so that one can easily compare our results with those of [GM]. Moreover, we explain the relations to the general results of Kashiwara-Schapira [KS, KS2].

We do not use the "Thom-Mather isotopy theorem". Instead of this, we work with a **cohomological version** of this result (cf. corollary 4.0.1), that we proved in chapter 4 based on the **non-characteristic deformation lemma** of Kashiwara. These results are true for stratifications with a weaker regularity condition than Whitney **b-regularity**, e.g. "distance regular" stratifications (called **d-regular** stratifications), or those with the "local stratified acyclicity" property **LSA**.

Here we do not give the definition of these regularity conditions, but refer for the details to chapter 4. We only recall the following implications:

$$\mathbf{a+LSA} \longleftarrow \mathbf{a+d} \longleftarrow \mathbf{loc.-C} \longleftarrow \mathbf{b}\,,$$

with **a** the Whitney **a-regularity** and **loc.-C** the **local C-regularity** of Bekka.

We use the following notations and conventions:

Definition 5.0.1. *Let X be a locally compact topological space.*

1. *A* **filtration** $X_\bullet$ *of X is a filtration $\emptyset = X_{-1} \subset X_0 \subset \cdots \subset X_n = X$ by closed subsets. We call $depth(X_\bullet) := \sharp\{i|\ X^i \neq \emptyset\} - 1$ the* **depth** *of $X_\bullet$, with $X^i := X_i \backslash X_{i-1}$ $(i = 0, \ldots, n)$.*

2. *A complex of sheaves* $\mathcal{F} \in ob(D(X))$ *is called* **constructible** *with respect to* $X_\bullet$ *(or* $X_\bullet$*-constructible), if the restriction of* $\mathcal{F}$ *to all* $X^i := X_i \backslash X_{i-1}$ $(i = 0, \ldots, n)$ *has locally constant cohomology sheaves. Let us denote by* $D(X_\bullet)$ *the corresponding full subcategory of all* $X_\bullet$*-constructible complexes.*

3. *A* **topological stratification** *of* X *is a filtration* $X_\bullet$ *of* X *such that* X^i $(i = 0, \ldots, n)$ *is a topological manifold of pure dimension* i *(or empty). We call* $dim(X) := sup\{i | X^i \neq \emptyset\}$ *the* **dimension** *of* X *(with* $dim(\emptyset) := -\infty$*). The connected components of the* X^i *are called the* **strata** *(of this stratification).*

4. *Let* X *be a closed subset of the smooth manifold* N*, where we fix a degree* $k = 1, \ldots, \infty, \omega$ *of smoothness (with* $\omega =$ *real analytic). A* **smooth stratification** *of* X *is a topological stratification* $X_\bullet$ *of* X *such that* X^i $(i = 0, \ldots, n)$ *is a smooth submanifold of* N*.*

The above regularity conditions are defined for a smooth stratification $X_\bullet$ of the closed subset X of the smooth manifold M (and some of them depend on the degree k of smoothness, that we include in the notation if neccessary). From now on we fix a degree $k = 2, \ldots, \infty, \omega$ of smoothness. (Sub-)manifold and smooth map or function always means a C^k-(sub)manifold and C^k-smooth map or function.

5.0.1 Real stratified Morse theory

We fix a smooth stratification $X_\bullet$ of the closed subset X of the smooth manifold M and formulate the main results of **stratified Morse theory for constructible sheaves**. To simplify the assumptions in the following results, we consider in this introduction only the case of a Whitney **b-regular** stratification (as in [GM]). This is already enough for the most important applications of this theory (e.g. as in chapter 6)!

Nevertheless, in the main body of this chapter we formulate and prove these results under weaker assumptions. Here one has at least to assume that the stratification is **a-regular**, satisfies the property **LSA** and all strata of $X_\bullet$ are **b-regular** over the one-point set $\{x\}$ for all (critical) points $x \in X$. For most results we need for a **self-contained approach** even **d-** or **local C-regularity** (instead of the weaker property **LSA**).

This can be avoided, if one is willing to use deeper results of the "micro-local" sheaf theory of Kashiwara-Schapira [KS, KS2] (i.e. estimates of the "micro-support" under some standard operations of sheaf theory).

Lemma 5.0.1. *Consider an* $X_\bullet$*-constructible complex* $\mathcal{F} \in ob(D(X))$*. Assume that* $f : X \to]c, d[\subset \mathbb{R}$ *is a* **proper** *smooth map, and a* **stratified submersion** *in all points of* X*. Then all cohomology sheaves of* $Rf_*\mathcal{F}$ *are* **locally constant***. Especially*

$$R\Gamma(\{\{a \leq f \leq b\}, \{f = a\}, \mathcal{F}) = 0$$

for $[a, b] \subset]c, d[$ *.*

This is a special case of our **cohomological version of the first isotopy lemma** corollary 4.0.1, and corresponds to the "first fundamental result of stratified Morse theory" [GM, p. 6, thm. SMT Part A, p. 62, prop. 3.2].

Here a function $f : X \to \mathbb{K}$ (with $\mathbb{K} = \mathbb{R}, \mathbb{C}$, or more generally with values in a smooth manifold) is called smooth if it is induced locally at each point $x \in X$ by the restriction of a smooth function germ $f : (M, x) \to \mathbb{K}$. Moreover, such an f is by definition a **stratified submersion** in $x \in X^i \subset X$ (with respect to the stratification $X_\bullet$), if $f : (X^i, x) \to \mathbb{K}$ is a submersion in x. Otherwise x is called a **critical point** of f (with respect to the stratification $X_\bullet$).

The next step is to **localize** the relative cohomology near an **isolated** critical point (compare lemma 5.1.2):

Lemma 5.0.2. *Make the same assumptions as in lemma 5.0.1, but this time we assume that f has at most one critical point x in $\{a \le f \le b\}$, with $a < f(x) < b$. Then*

$$R\Gamma(\{\{a \le f \le b\}, \{f = a\}, \mathcal{F}) \simeq \left(R\Gamma_{\{f \ge f(x)\}}(\mathcal{F})\right)_x .$$

This isomorphism is functorial in $\mathcal{F} \in ob(D(X_\bullet))$.

The stalk complex

$$LMD(\mathcal{F}, f, x) := \left(R\Gamma_{\{f \ge f(x)\}}(\mathcal{F})\right)_x \tag{5.1}$$

is the sheaf theoretic counterpart of the **"local Morse datum"** of [GM, p. 63, def. 3.5.2].

Let us now recall two notions of [GM].

Definition 5.0.2. *Fix a point x in the stratum X^s.*

1. *x is called a* **stratified Morse critical point** *of the smooth function germ $f : (M, x) \to \mathbb{R}$ (with respect to $X_\bullet$), if $f|X^s$ has in x a Morse critical point in the classical sense (i.e. its Hessian in x is nondegenerate), and the covector df_x is* **nondegenerate** *in the sense of [GM, p. 44, def. 1.8], i.e. it does not vanish on any generalized tangent space*

   $$\tau := \lim_{x_n \to x} T_{x_n} X^j \quad \text{at } x,$$

   with x_n a sequence in X^j (for some $j > s$).

2. *A* **normal slice** *N at x is a locally closed submanifold N of M, with $N \cap X^s = \{x\}$ such that N intersects X^s* **transversally** *in x (Our definition is slightly different from [GM, p. 41, def.1.4], who use the intersection of $N \cap X$ with a small ball around x).*

Note that a **normal slice** N intersects all strata transversally near x (by **a-regularity**). Therefore $N \cap X$ gets an induced **b-regular** stratification near x (compare section 4.3). Similarly, a **stratified Morse critical point** of f is an isolated critical point of f.

Theorem 5.0.1. *Let $f : (M, x) \to \mathbb{R}$ be a smooth function germ with a* **stratified Morse critical point** *in $x \in X^s$. Take a* **normal slice** *N in x, and fix $\mathcal{F} \in ob(D(X_\bullet))$.*

1. *Then one has an isomorphism*

$$\left(R\Gamma_{\{f \geq f(x)\}}(\mathcal{F})\right)_x \simeq \left(R\Gamma_{\{f|N\cap X \geq f(x)\}}(\mathcal{F}|N \cap X)\right)_x[-\tau],$$

   with τ the **Morse index** *of $f|X^s$ (i.e. its Hessian in x has τ negative eigenvalues). Moreover, this isomorphism is functorial in $\mathcal{F} \in ob(D(X_\bullet))$.*

2. *The isomorphism class of*

$$\left(R\Gamma_{\{f|N\cap X \geq f(x)\}}(\mathcal{F}|N \cap X)\right)_x$$

   is **independent** *of the choice of the normal slice N. Moreover, this isomorphism class* **depends only** *on the connected component of the set of nondegenerate covectors in the conormal bundle $T^*_{X^s}M$: Let $f_i : (M, x_i) \to \mathbb{R}$ (with $x_i \in X^s$ for $i = 1, 2$) be two smooth function germs with $d(f_i)_{x_i}|T_{x_i}X^s = 0$ and $d(f_i)_{x_i}$ a* **nondegenerate** *covector ($i = 1, 2$). Suppose these two covectors belong to the same* **connected component** *of the set of nondegenerate covectors in the conormal bundle $T^*_{X^s}M$. Then*

$$\left(R\Gamma_{\{f_1|N_1\cap X \geq f_1(x_1)\}}(\mathcal{F}|N_1 \cap X)\right)_{x_1} \simeq \left(R\Gamma_{\{f_2|N_2\cap X \geq f_2(x_2)\}}(\mathcal{F}|N_2 \cap X)\right)_{x_2}.$$

The isomorphism class of

$$NMD(\mathcal{F}, df_x) := NMD(\mathcal{F}, f, x) := \left(R\Gamma_{\{f|N\cap X \geq f(x)\}}(\mathcal{F}|N \cap X)\right)_x \tag{5.2}$$

is the sheaf theoretic counterpart of the **"normal Morse data"** of [GM, p. 65, def. 3.6.1]. Part 1. of the theorem corresponds to [GM, p. 8, thm. SMT Part B, p. 65 Main thm. 3.7], and part 2. to [GM, p. 93, thm. 7.5.1] (compare also with [GM, p. 223, prop. 6.A.1]).

Part 1. of theorem 5.0.1 is a special case of the more general theorem 5.3.3, where we also use weaker assumptions on $f|X^s$. In this more general context one has to **replace** $NMD(\mathcal{F}, f, x)[-\tau]$ by

$$\left(R\Gamma_{\{f|X^s \geq f(x)\}}(\mathbb{Z}_{X^s})\right)_x \otimes^L_Z \left(R\Gamma_{\{f|N\cap X \geq f(x)\}}(\mathcal{F}|N \cap X)\right)_x.$$

Here

$$TMD(\mathcal{F}, f, x) := \left(R\Gamma_{\{f|X^s \geq f(x)\}}(\mathbb{Z}_{X^s})\right)_x \tag{5.3}$$

is the sheaf theoretic counterpart of the **"tangential Morse data"** of [GM, p. 65, def. 3.6.1]. Our proof of this result is completely **different** from the proof of the "Main theorem" given in [GM, I. chapter 8]. By the study of some "local direct image sheaves", we reduce this result to our general **Thom-Sebastiani Theorem for**

constructible sheaves theorem 1.2.2. On the other hand, part 2. of theorem 5.0.1 is a special case of corollary 5.3.3.

The assumptions on $f|X^s$ in theorem 5.3.3 are for example satisfied, if $f|X^s$ is real analytic (in suitable coordinates near $x \in X^s$).

Let $k : X \to M$ be the closed inclusion into the ambient manifold M. Then the more general theorem 5.3.3 implies especially the following explicit description of the **micro-support** $\mu supp(Rk_*\mathcal{F})$ for $\mathcal{F} \in D^+(X_\bullet)$, constructible with respect to a **definable** Whitney **b-regular** stratification $X_\bullet$ in the context of a "geometric category" as in definition 4.0.5 on page 213 (i.e. all X_i are definable).

By corollary 4.0.3 we have the estimate

$$\mu supp(Rk_*\mathcal{F}) \subset \bigcup_i T^*_{X^i}M =: \Lambda \subset T^*M\,. \tag{5.4}$$

By using the **involutivity** of $\mu supp(Rk_*\mathcal{F})$ (compare [KS2, thm. 6.4.1] and [KS, thm. 6.5.4]) one can show as in [KS2, prop. 8.2.4], that $\mu supp(Rk_*\mathcal{F})$ is a union of closures of connected components of the **set of nondegenerate covectors**

$$\Lambda^\circ := \bigcup_i \Lambda_i^\circ\,, \quad \text{with} \quad \Lambda_i^\circ := T^*_{X^i}M \backslash \bigcup_{i \neq j} cl(T^*_{X^j}M). \tag{5.5}$$

So $\mu supp(Rk_*\mathcal{F})$ is a **closed conic definable Lagrangian** subset of T^*M.

Proposition 5.0.1. *Fix a* **connected component** $\Lambda^\circ_{s,i}$ *of the set* Λ°_s *of nondegenerate covectors in* $T^*_{X^s}M$*. Then*

$$cl(\Lambda^\circ_{s,i}) \subset \mu supp(Rk_*\mathcal{F}) \;\Leftrightarrow\; NMD(\mathcal{F}, df_x) \neq 0 \quad \textit{for} \quad df_x \in \Lambda^\circ_{s,i}\,.$$

For the applications of stratified Morse theory for constructible sheaves, it is important to get information about this "normal Morse datum"

$$NMD(\mathcal{F}, f, x) = \big(R\Gamma_{\{f|N\cap X \geq f(x)\}}(\mathcal{F}|N \cap X)\big)_x\,.$$

This is contained in the following result (compare corollary 5.1.2):

Proposition 5.0.2. *Fix a point* $x \in X^s$*,* $\mathcal{F} \in ob(D(X_\bullet))$*, and a* **normal slice** N *in* x*. Let* $f : (M, x) \to (\mathbb{R}, e)$ *be a smooth function germ with* $df_x|T_xX^s = 0$ *and* df_x *a* **nondegenerate** *covector. Choose local coordinates* $(M, x) \simeq (\mathbb{R}^m, 0)$ *and define in these coordinates the distance function* r *by* $r(x) := \sum_{i=1}^m x_i^2$*. If* $k_x : \{x\} \to N \cap X$ *is the inclusion, then there exist two distinguished triangles (functorial in* $\mathcal{F} \in ob(D(X_\bullet))$*):*

$$R\Gamma(l^-_X, \mathcal{F})[-1] \longrightarrow NMD(\mathcal{F}, f, x) \longrightarrow k_x^*(\mathcal{F}|N \cap X) \xrightarrow{[1]}\,,$$

and

$$k_x^!(\mathcal{F}|N \cap X) \longrightarrow NMD(\mathcal{F}, f, x) \longrightarrow R\Gamma(l^+_X, \partial l^+_X, \mathcal{F}) \xrightarrow{[1]}\,.$$

Here we use the notation (with $0 < w << \delta << 1$*)*

$$(l^*_X, \partial l^*_X) := (X \cap N \cap \{r \leq \delta, f - e = *w\}, X \cap N \cap \{r = \delta, f - e = *w\})$$

for the **upper/lower halflink** *($* = +, -$) of X with respect to f ([GM, p. 66, def. 3.9.1]). Moreover, the isomorphism class of*

$$R\Gamma(l^-_X, \mathcal{F}) \quad or \quad R\Gamma(l^+_X, \partial l^+_X, \mathcal{F})$$

depends only on the **connected component** *of the set of nondegenerate covectors in the conormal bundle $T^*_{X^s}M$ (in the same sense as before).*

The last part (about the isomorphism classes) can be reduced to the corresponding statement in theorem 5.0.1 (compare with the isomorphisms (5.49) and (5.50)).

In the applications, we use these distinguished triangles to get information about the "normal Morse datum" $NMD(\mathcal{F}, f, x)$ from the properties of the **stalk** $k^*_x(\mathcal{F}|N \cap X)$ or **costalk** $k^!_x(\mathcal{F}|N \cap X)$. Here one gets the information about

$$R\Gamma(l^-_X, \mathcal{F}) \quad \text{or} \quad R\Gamma(l^+_X, \partial l^+_X, \mathcal{F})$$

by **inductive application** of stratified Morse theory to the function r or $-r$ on the stratified space $X \cap N \cap \{f - e = \pm w\}$. This is the same idea as in **all** inductive proofs in [GM, partII], especially for the complex analytic counterpart that follows soon.

But let us first remark, that all these statements can also be proved with the help of the "micro-local theory of sheaves" developed by Kashiwara-Schapira [KS, KS2] (at least for complexes bounded from below). As we will explain, the "normal Morse datum" $NMD(\mathcal{F}, x)$ is isomorphic to the "local type of $\mathcal{F}$ at the covector df_x" in the notion of Kashiwara-Schapira (compare with the isomorphism (5.52)).

What is missing in [KS, KS2], are the distinguished triangles above to get information (for example by induction) on this "local type of $\mathcal{F}$".

But our approach to these results is **independent** (!) of [GM, KS, KS2]. We use the geometric ideas of Goresky-MacPherson, and (a simpler form of) the sheaf theoretic language of Kashiwara-Schapira. As in [KS, KS2], everything depends on the **non-characteristic deformation lemma** of Kashiwara. But we used it in chapter 4 for the development of stratification theory in terms of constructible sheaves (which is close to the original geometric ideas of stratification theory). Such a "geometric picture" is not possible in the more general context of [KS, KS2].

Of course, each approach has its advantages:

- Using the deep theory of Kashiwara-Schapira about **contact transformation for sheaves** ([KS2, ch.6], [KS, ch.VII]), we show (under some weak restrictions) that the "local type" of $\mathcal{F} \in ob(D^+(M))$ at a non-zero covector

$\omega \in T^*_{X^s}M$ is isomorphic to the stalk

$$\big(\mu_{X^s}(\mathcal{F})\big)_\omega \quad \text{of the } \textbf{micro-localization } \mu_{X^s}(\mathcal{F}).$$

Compare with the isomorphism (5.76) on page 356 (and with [D'AMZ]).

This explains the "micro-local nature" of the "normal Morse data". This result is missing in [KS, KS2], and we only use a "little geometry" on X to deduce this result from the abstract machinery of Kashiwara-Schapira. But we should also mention, that we do **not** use this result in our book !

- The geometric results of Goresky-MacPherson in [GM] apply to the study of the **homotopy type** of the underlying stratified spaces.
- The stratified Morse theory for constructible sheaves developed in this chapter allows one to compare the different "languages" of Goresky-MacPherson and Kashiwara-Schapira.

  Moreover, it unifies most of the (co)homological results of [GM]: e.g. one gets at the same time results about the **cohomology** (with compact support), (Borel-Moore) **homology** or **intersection (co)homology** (with compact support), just by applying this theory to the study of

$$R\Gamma_{(c)}(X,\mathcal{F}) \quad \text{for } \mathcal{F} = \begin{cases} \mathbb{Z}_X & \text{the constant sheaf.} \\ D_X & \text{the dualizing complex.} \\ IC_X & \text{the intersection (co)homology complex.} \end{cases}$$

  One also does not need the generalization of stratified Morse theory to the "relative and/or nonproper case" as in [GM], which one gets back by application of our theory to the complexes $R\pi_*\mathcal{F}$ or $R\pi_!\mathcal{F}$ for $\pi : X \to Z$ a suitable (Whitney) **stratified map** (and $\mathcal{F}$ as before). Note that these direct image complexes are then constructible by our results of section 4.2.

  Similarly, most of the corresponding (co)homological results of [GM, Part II] in the complex analytic (algebraic) context follow from a **few basic vanishing theorems** related to the "middle perversity t-structure", which we prove in chapter 6 (in a "generalized form"), based on the results of this chapter.

Let us now illustrate proposition 5.0.2 in some simple but important special cases.

Example 5.0.1. *For simplicity, we just state the cohomological counterparts of some results of [GM].*

1. *If we take $\mathcal{F} = \mathbb{Z}_X$, then we get by the first distinguished triangle for the cohomology of the "normal Morse datum":*

$$\big(R\Gamma^k_{\{f|N\cap X \geq f(x)\}}(\mathcal{F}|N\cap X)\big)_x \simeq \begin{cases} \mathbb{Z} & \text{for } k = 0, \\ 0 & \text{for } k \neq 0, \end{cases}$$

if $l_X^- = \emptyset$*, e.g.* X^s *is a neighborhood of* x *in* X *(especially for* $dim(X) = s$*). Otherwise we get the shifted "reduced cohomology of the lower halflink":*

$$\left(R\Gamma^k_{\{f|N\cap X\geq f(x)\}}(\mathcal{F}|N\cap X)\right)_x \simeq \tilde{H}^{k-1}(l_X^-,\mathbb{Z})\,.$$

This corresponds to [GM, p. 67, cor. 3.11.2].

2. *Consider now the case* $\mathcal{F} = Rj_*\mathbb{Z}_U$ *for* $j : U \to X$ *the inclusion of an open subspace, which is a union of strata. Assume* $x \notin U$ *so that we consider a "non-proper function"* $f|U$*. Then one gets by the second distinguished triangle for the cohomology of the "normal Morse datum":*

$$\left(R\Gamma^k_{\{f|N\cap X\geq f(x)\}}(\mathcal{F}|N\cap X)\right)_x \simeq H^k(l_X^+,\partial l_X^+,Rj_*\mathbb{Z}_U)\,.$$

Moreover, by proposition 4.0.3 and the property **LSA** *one gets*

$$H^k(l_X^+,\partial l_X^+,Rj_*\mathbb{Z}_U) \simeq H^k(l_U^+,\partial l_U^+,\mathbb{Z})\,,$$

with $(l_U^+,\partial l_U^+) := (l_X^+,\partial l_X^+)\cap U$ *the* **upper halflink of** U *(as in [GM, p. 120, 10.3]). This corresponds to the second case of [GM, p. 122, thm. 10.8].*

3. *Consider finally the following generalization of 2.:* $\mathcal{F} = R\pi_*\mathbb{Z}_U$*, with* $j : U \to Y$ *the inclusion of an open union of strata of a Whitney* **b-regular** *stratified space* Y*,* $\bar{\pi} : Y \to X$ *a* **proper stratified submersion** *and* $\pi := \bar{\pi}\circ j$*. In the case* $\bar{\pi}^{-1}(\{x\}) \subset U$ *one gets by* **proper base change** *a distinguished triangle:*

$$R\Gamma(l^{\pi-},\mathbb{Z})[-1] \longrightarrow NMD(\mathcal{F},f,x) \longrightarrow R\Gamma(\{\pi = x\},\mathbb{Z}) \xrightarrow{[1]}\,,$$

with $l^{\pi-} := \pi^{-1}(l_X^-) = \bar{\pi}^{-1}(l_X^-)$*. In the case* $\bar{\pi}^{-1}(\{x\}) \subset Y\backslash U$ *one gets as in 2., that the "normal Morse datum"* $NMD(\mathcal{F},f,x)$ *is isomorphic to*

$$R\Gamma(l_X^+,\partial l_X^+,R\pi_*\mathbb{Z}_U) \simeq R\Gamma(l_U^{\pi+},\partial l_U^{\pi+},\mathbb{Z})\,,$$

with $(l_U^{\pi+},\partial l_U^{\pi+}) := \pi^{-1}(l_X^+,\partial l_X^+) = U\cap\bar{\pi}^{-1}(l_X^+,\partial l_X^+)$*. These results correspond to [GM, p. 117, prop. 9.7, p. 126, thm. 11.7.2].*

Another simple example is **Morse theory on a manifold with boundary**.

Example 5.0.2. *Consider a smooth function* $g : M \to \mathbb{R}$*, which is a submersion in* $\{g = 0\}$*. Let* X *be the* **manifold with boundary** $\{g \geq 0\}$ *stratified by* $\{g = 0\} =: \partial X =: X_{m-1} \subset X_m := X$*, with* $m := dim(M)$*. Then the covector* $df_x \in T^*_{\partial X}M$ *is* **nondegenerate***, if and only if*

$$df_x = \lambda\cdot dg_x \quad \textit{for some } \lambda\in\mathbb{R}^*\,.$$

In this case, the **lower halflink** l_X^- *of* X *with respect to* f *is given by*

$$l_X^- = \begin{cases}\emptyset & ,\ \textit{for } \lambda > 0.\\ \{pt\} & ,\ \textit{for } \lambda < 0.\end{cases}$$

Consider the constant sheaf $\mathcal{F} = \mathbb{Z}_X$ *on* X *so that* $H^*(X, \mathcal{F}) = H^*(X, \mathbb{Z})$. *This corresponds to the study of* **Morse theory for the manifold with boundary** X. *Then one gets:*

$$NMD(\mathbb{Z}_X, df_x) = \begin{cases} \mathbb{Z} & , \textit{for } \lambda > 0. \\ 0 & , \textit{for } \lambda < 0. \end{cases}$$

Consider the constructible sheaf $\mathcal{F} = j_! \mathbb{Z}_U$, *with* $j : U := X \backslash \partial X \to X$ *the open inclusion of the interior of* X *so that* $H^*(X, \mathcal{F}) = H^*(X, \partial X, \mathbb{Z})$. *This corresponds to the study of* **Morse theory for the manifold modulo the boundary** ∂X. *Then one gets:*

$$NMD(j_! \mathbb{Z}_U, df_x) = \begin{cases} 0 & , \textit{for } \lambda > 0. \\ \mathbb{Z}[-1] & , \textit{for } \lambda < 0. \end{cases}$$

By proposition 5.0.1 *(applied in suitable coordinates), we get especially the following description of the corresponding* **micro-supports** *([KS, prop.* 5.3.3*]):*

$$\begin{aligned} \mu supp(\mathbb{Z}_X) &= \{(x, \lambda \cdot dg_x) \mid \lambda \cdot g(x) = 0, \lambda \geq 0, g(x) \geq 0\} \quad , \textit{ and} \\ \mu supp(j_! \mathbb{Z}_U) &= \{(x, \lambda \cdot dg_x) \mid \lambda \cdot g(x) = 0, \lambda \leq 0, g(x) \geq 0\} \, . \end{aligned}$$

For the study of the "normal Morse data" of a suitable **stratified space with boundary**, we need the following **multiplicativity** of the "normal Morse data" (compare with proposition 5.1.1):

Let $X_\bullet, X'_\bullet$ be Whitney **b-regular stratifications** of the closed subsets X, X' of the smooth manifolds M, M'. This induces a **product stratification** $(X \times X')_\bullet$ of $Y := X \times X' \subset M \times M' =: M''$:

$$(X \times X')_k := \bigcup_{i+j=k} (X_i \times X'_j) \, .$$

Then $(X \times X')^k$ is the disjoint union of the submanifolds $X^i \times X'^j$ (with $i + j = k$), which are also open in $(X \times X')^k$. Especially, the strata are the products of the strata of $X_\bullet$ and $X'_\bullet$. This product stratification is also Whitney **b-regular** (as explained in section 4.3).

Consider $(x, x') \in Y^{i,j} := X^i \times X'^j$. Then

$$T_{(x,x')} Y^{i,j} = T_x X^i \oplus T_{x'} X'^j \subset T_x M \oplus T_{x'} M' = T_{(x,x')} M'' \, ,$$

and a covector

$$(\omega, \omega') \in (T^*_{X^i} M)_x \oplus (T^*_{X'^j} M')_{x'} = (T^*_{Y^{i,j}} M'')_{(x,x')}$$

is **nondegenerate** if and only if ω **and** ω' are **nondegenerate**. In this case we get

$$NMD(\mathcal{F}, \omega) \otimes^L NMD(\mathcal{F}', \omega') \simeq NMD(\mathcal{F} \boxtimes^L \mathcal{F}', (\omega, \omega')) \, , \tag{5.6}$$

for $\mathcal{F} \in ob(D(X_\bullet))$ and $\mathcal{F}' \in ob(D(X'_\bullet))$.

Here $\boxtimes^L$ denotes the (derived) exterior tensor product: if $pr_1 : X \times X' \to X$ and $pr_2 : X \times X' \to X'$ are the projections, then $(\cdot) \boxtimes^L (\cdot) := pr_1^*(\cdot) \otimes^L pr_2^*(\cdot)$. We allow two cases for the (derived) tensor product on a space S:

1. The base ring R is commutative, with $\otimes_R^L : D(S,R) \times D(S,R) \to D(S,R)$.
2. The base ring R is arbitrary, with $\otimes_Z^L : D(S,\mathbb{Z}) \times D(S,R) \to D(S,R)$.

As a first application of (5.6), we get by proposition 5.0.1 the following estimate for the micro-support of

$$Rk''_*(\mathcal{F} \boxtimes^L \mathcal{F}') \simeq (Rk_*\mathcal{F}) \boxtimes^L (Rk'_*\mathcal{F}') ,$$

with $k : X \to M$, $k' : X' \to M'$ and $k'' : Y = X \times X' \to M \times M' = M''$ the closed inclusions:

$$\mu supp(Rk''_*(\mathcal{F} \boxtimes^L \mathcal{F}')) \subset \mu supp(Rk_*\mathcal{F}) \times \mu supp(Rk'_*\mathcal{F}') , \tag{5.7}$$

for $\mathcal{F} \in ob(D^+(X_\bullet))$ and $\mathcal{F}' \in ob(D^+(X'_\bullet))$.

Here we work of course in the **definable** context of a "geometeric category" as in proposition 5.0.1 (i.e. all X_i and X'_j are definable). This inclusion (5.7) is just a special case of the corresponding general result [KS2, prop. 4.2.1] and [KS, prop. 5.4.1]. But if the base ring $R = K$ is a **field**, then our approach gives the following **refinement**:

$$\mu supp(Rk''_*(\mathcal{F} \boxtimes^L \mathcal{F}')) = \mu supp(Rk_*\mathcal{F}) \times \mu supp(Rk'_*\mathcal{F}') , \tag{5.8}$$

for $\mathcal{F} \in ob(D^+(X_\bullet, K))$ and $\mathcal{F}' \in ob(D^+(X'_\bullet, K))$.

A second application of (5.6) deals with an extension of example 5.0.2 to the case of a suitable **stratified space with boundary**.

Let $X_\bullet$ be a Whitney **b-regular stratification** of the closed subset X of the smooth manifold M. Fix a point $x \in X^s$ and a C^k-map $g : (M,x) \to (\mathbb{R},0)$ such that $g|X^s$ is a **submersion** in x. Then $g : U \cap X^j \to \mathbb{R}$ is for all j a submersion in an open neighborhood U of x in M (by **a-regularity**), and we assume $U = M$ in the following (since we work locally).

Then the **manifold with boundary** $B := \{g \geq 0\}$ has a Whitney **b-regular stratification** $\partial B \subset B$ as in example 5.0.2. Moreover, the strata of X intersect the strata of B **transversally** in M so that $\{g \geq 0\} \cap X$ gets an induced **b-regular stratification**. We call the space $\{g \geq 0\} \cap X$ with this induced stratification a **stratified space with boundary**. As explained in section 4.3, this induced stratification is just the "transversal pullback" of the **product stratification** of $B \times X$ under the diagonal embedding $M \to M \times M$. Let $d : B \cap X \to B \times X$ be the induced diagonal map.

Note, that the point x lies in the "boundary stratum" $S := \{g = 0\} \cap X^s$. Consider a smooth function germ $f : (M,x) \to (\mathbb{R},0)$, with $df_x|T_xS = 0$, and

$df_x \in T^*_x M$ **nondegenerate** with respect to the induced stratification of $B \cap X$. This implies that also $d(f|X^s)_x \in T^*_x X^s$ is nondegenerate with respect to the induced stratification of the **manifold with boundary** $\{g \geq 0\} \cap X^s \subset X^s$. Especially

$$d(f|X^s)_x = \lambda \cdot d(g|X^s)_x \quad \text{for some } \lambda \in \mathbb{R}^* \,. \tag{5.9}$$

As in example 5.0.2, we consider the following two cases:

1. $j : U := B \to B$ is the identity, with $j_! \mathbb{Z}_U = \mathbb{Z}_B$. Then $d^*(j_!\mathbb{Z}_U \boxtimes^L \mathcal{F}) \simeq \mathcal{F}|B \cap X$ for $\mathcal{F} \in ob(D(X))$ so that

$$R\Gamma(\, B \cap X, d^*(j_!\mathbb{Z}_U \boxtimes^L \mathcal{F})\,) \simeq R\Gamma(B \cap X, \mathcal{F})\,.$$

2. $j : U := B \backslash \partial B \to B$ is the open inclusion of the interior of B. Then $d^*(j_!\mathbb{Z}_U \boxtimes^L \mathcal{F})$ is the extension by zero of $\mathcal{F}|\{g > 0\} \cap X$ for $\mathcal{F} \in ob(D(X))$ so that

$$R\Gamma(\, B \cap X, d^*(j_!\mathbb{Z}_U \boxtimes^L \mathcal{F})\,) \simeq R\Gamma(B \cap X, \partial B \cap X, \mathcal{F})\,.$$

We are interested in the "normal Morse datum" $NMD(\mathcal{G}, f, x)$ of f in x with respect to this sheaf complex

$$\mathcal{G} := d^*(j_!\mathbb{Z}_U \boxtimes^L \mathcal{F}) \simeq (j_!\mathbb{Z}_U \otimes^L \mathcal{F})|B \cap X \quad \text{for } \mathcal{F} \in ob(D(X_\bullet))\,.$$

And we would like to relate this to the "normal Morse datum" of the complex $\mathcal{F}$. But the **problem** is that $df_x|T_x Z^s \neq 0$, because otherwise df_x would **not** be a nondegenerate covector with respect to the induced stratification of $B \cap X$ (look at a sequence $x_n \in Z^s \cap \{g > 0\}$ with $x_n \to x$). Nevertheless, we can prove the following result (compare with theorem 5.1.1 for a more general result in the context of a **stratified space with corners**):

Theorem 5.0.2. *Assume that the set of* **nondegenerate** *covectors with respect to the stratification $X_\bullet$ is* **dense** *in the fiber $(T^*_{X^s} M)_x$. Then there exists a smooth function germ $h : (M, x) \to (\mathbb{R}, 0)$, with $dh_x|T_x X^s = 0$ and dh_x* **nondegenerate** *with respect to $X_\bullet$ such that*

$$NMD(\mathcal{G}, f, x) \simeq NMD(j'_!\mathbb{Z}_{U'}, f|X^s, x) \otimes^L NMD(\mathcal{F}, h, x)\,, \tag{5.10}$$

with

$$j' := \begin{cases} id_{U'} \quad \text{for } U' := X^s \cap \{g \geq 0\} & \text{in the case 1.} \\ U' := X^s \cap \{g > 0\} \hookrightarrow X^s \cap \{g \geq 0\} & \text{in the case 2.} \end{cases}$$

Moreover, one can choose this h so that $d(\mu \cdot g + h)_x \in (T^*_S M)_x$ is arbitrary close to df_x for a suitable $\mu \in \mathbb{R}^*$, with $\mu/\lambda \in \mathbb{R}^+$.

Using (5.6), the proof goes as follows: Take a covector $dh_x \in (T^*_{X^s}M)_x$, which is **nondegenerate** with respect to $X_\bullet$. Then $(\mu \cdot dg_x, dh_x)$ is for any $\mu \in \mathbb{R}^*$ also a **nondegenerate** covector with respect to the **product stratification** of $B \times X$. Moreover, the corresponding (product-like) "normal Morse datum" doesn't change under the "transversal pull-back" for the diagonal embedding $d : M \to M \times M$ (since d induces an isomorphism of suitable normal slices):

$$NMD(\, j_! \mathbb{Z}_U \boxtimes^L \mathcal{F}, (\mu \cdot dg_x, dh_x)\,) \simeq NMD(\mathcal{G}, d^*(\mu \cdot dg_x, dh_x))\,,$$

with $d^*(\mu \cdot dg_x, dh_x) = d(\mu \cdot g + h)_x$. By assumption, the **nondegenerate** covectors are **dense** in the fiber $(T^*_{X^s}M)_x$ so that one can get $d(\mu \cdot g + h)_x$ arbitrarily close to

$$df_x \in (T^*_S M)_x \simeq d^*\big(\, (T^*_{\{g=0\}}M)_x \oplus (T^*_{X^s}M)_x\,\big)\,.$$

Especially

$$NMD(\mathcal{G}, d^*(\mu \cdot dg_x, dh_x)) \simeq NMD(\mathcal{G}, df_x)\,.$$

Similarly $NMD(j_! \mathbb{Z}_U, \mu \cdot dg_x) \simeq NMD(j'_! \mathbb{Z}_{U'}, \mu \cdot d(g|X^s)_x) \simeq$

$$NMD(j'_! \mathbb{Z}_{U'}, d(\mu \cdot g + h|X^s)_x) \simeq NMD(j'_! \mathbb{Z}_{U'}, d(f|X^s)_x)\,,$$

with $\mu \cdot d(g|X^s)_x = d(\mu \cdot g + h|X^s)_x$ close to $d(f|X^s)_x)$ so that μ has the same sign as λ in (5.9).

This theorem implies especially the following **vanishing** of the "normal Morse data" for λ as in (5.9):

$$NMD(\mathcal{G}, f, x) = 0 \begin{cases} \text{for } \lambda < 0 \text{ in the case 1.} \\ \text{for } \lambda > 0 \text{ in the case 2.} \end{cases}$$

This kind of vanishing results is also related to [KS2, prop. 4.3.2] or [KS, prop. 5.4.8]. But of course our theorem contains **more** information (compare with [Sch2] for a powerful application). In the case $NMD(j'_! \mathbb{Z}_{U'}, f|X^s, x) \neq 0$ we get

$$NMD(\mathcal{G}, f, x) \simeq \begin{cases} NMD(\mathcal{F}, h, x) & \text{for } \lambda > 0 \text{ in the case 1.} \\ NMD(\mathcal{F}, h, x)[-1] & \text{for } \lambda < 0 \text{ in the case 2.} \end{cases}$$

The **problem** with this formula comes from the fact, that we do **not** know to which connected component of the set of nondegenerate covectors dh_x belongs. For applications we must have information about **all** the possible "normal Morse data" of the sheaf complex $\mathcal{F}$ (with respect to the stratification $X_\bullet$).

But if the set of **nondegenerate** covectors in $T^*_S M$ is **connected**, then there exists only one isomorphism class of "normal Morse data" on the stratum S. This is especially true for a **complex analytic b-regular stratification**. Moreover, in this **complex analytic context** the nondegenerate covectors are **dense in all** fibers $(T^*_{X^s}M)_x$ (compare with beginning of the proof of $i) \Rightarrow ii)$ in [Te, p. 455, thm. 1.2]).

Another general result about the "normal Morse data" is related to the **duality** observed in [GM] about their results in stratified Morse theory. In the framework of stratified Morse theory for constructible sheaves, this is a special case of the general **Poincaré-Verdier duality** for constructible sheaves on "nice" stratified spaces.

As before, consider a Whitney **b-regular stratifcation** $X_\bullet$ of the closed subset X of the smooth manifold M. Fix a point $x \in X^s$ and a **nondegenerate** covector $df_x \in T^*_{X^s}M$. Finally assume that the base ring R is commutative so that one has a **duality functor** (as explained in chapter 4):

$$D := D_X : D^-(X_\bullet) \to D^+(X_\bullet)\,.$$

Then we explain later on the following relation between **duality** and the "normal Morse data":

$$D(\,NMD(\mathcal{F}, -df_x)\,) \simeq NMD(D(\mathcal{F}), df_x)[-s]\,. \tag{5.11}$$

5.0.2 Complex stratified Morse theory

Let us now formulate the main results about stratified Morse theory for constructible sheaves in the **complex analytic** context.

We assume that M is a complex analytic manifold, and all X_i are closed complex analytic subsets so that the strata or X^{2i} are locally closed complex analytic manifolds.

Fix a point $x \in X^{2s}$ and a **complex analytic** normal slice N in x. We already know that the isomorphism class of the "normal Morse data" depends only on the covector df_x. Moreover, one can use a **holomorphic** function germ $g : (M, x) \to (\mathbb{C}, 0)$ with $df_x = d(re(g))_x$. Then one has the (compare corollary 5.4.3):

Proposition 5.0.3. *We make the same assumptions as in proposition* 5.0.2*, but this time in the complex analytic context with N a complex analytic normal slice. Choose a holomorphic function germ $g : (M, x) \to (\mathbb{C}, 0)$ with $df_x = d(re(g))_x$. If $k_x : \{x\} \to N \cap X$ is the inclusion, then there exist two distinguished triangles:*

$$R\Gamma(l_X, \mathcal{F})[-1] \longrightarrow NMD(\mathcal{F}, re(g), x) \longrightarrow k_x^*(\mathcal{F}|N \cap X) \xrightarrow{[1]}\,,$$

and

$$k_x^!(\mathcal{F}|N \cap X) \longrightarrow NMD(\mathcal{F}, re(g), x) \longrightarrow R\Gamma(l_X, \partial l_X, \mathcal{F})[-1] \xrightarrow{[1]}\,.$$

Here we use the notation (with $0 < |w| << \delta << 1$)

$$(l_X, \partial l_X) := (X \cap N \cap \{r \le \delta, g = w\}, X \cap N \cap \{r = \delta, g = w\})$$

for the **complex link** *of X with respect to g ([GM, p.* 161*, def.* 2.2*]). Moreover, the isomorphism class of $R\Gamma(l_X, \mathcal{F})$ or $R\Gamma(l_X, \partial l_X, \mathcal{F})$ depends only on the stratum $S \subset X^{2s}$.*

The last part (about the isomorphism class) can be reduced to the corresponding statement in theorem 5.0.1 (compare with the isomorphisms (5.95) and (5.96)).

Since proposition 5.0.3 is a local result, it can also be used in the more general context of a complex analytic Whitney **b-regular stratification** of a complex analytic space X. But also the results of lemma 5.0.1 and lemma 5.0.2 are true in this context, **without** a global embedding of X into a complex manifold (since they follow in the same way from corollary 4.1.2, which applies also to the non-embedded context).

As before, these distinguished triangles can be used to get (inductive) information about the "normal Morse data". The advantage in the complex analytic context comes from the fact, that $X \cap N \cap \{g = w\}$ is a complex analytic space with a **complex analytic stratification**. Moreover, the function r used for the stratified Morse theory on $X \cap N \cap \{g = w\}$ (in the induction step) is now **strongly plurisubharmonic** so that one has nice estimates for the Morse index (of an approximation by a Morse function) of r and $-r$ (compare with chapter 6 for details).

We already explained the "micro-local nature" of the "normal Morse datum" $NMD(\mathcal{F}, dg_x)$. Note that in the complex context the set of nondegenerate covectors in $T^*_S M$ is **connected** for any complex analytic stratum S so that this "normal Morse datum" depends only on the stratum S !

Moreover, the distinguished triangles of proposition 5.0.3 show, that the **complex link** "measures the difference" between the **local (co)stalk** information and the **micro-local** information of the "normal Morse datum". This corresponds to the following remark of Goresky-MacPherson [GM, p.17]: "... "local" in the complex link is called *microlocal* in X ...".

Example 5.0.3. *Again, we just state the cohomological counterparts of some results of [GM].*

1. *If we take $\mathcal{F} = \mathbb{Z}_X$, then we get by the first distinguished triangle for the cohomology of the "normal Morse datum":*

$$\left(R\Gamma^k_{\{re(g)|N\cap X\geq 0\}}(\mathcal{F}|N\cap X)\right)_x \simeq \begin{cases} \mathbb{Z} & \text{for } k=0 \text{ and } l_X=\emptyset, \\ 0 & \text{for } k\neq 0 \text{ and } l_X=\emptyset, \\ \tilde{H}^{k-1}(l_X,\mathbb{Z}) & \text{for } l_X\neq\emptyset. \end{cases}$$

   This corresponds to the "first part of the fundamental theorem of complex stratified Morse theory" [GM, p. 16, thm. CSMT Part A, p. 166 cor. 2.4.1].

2. *Consider now the case $\mathcal{F} = Rj_*\mathbb{Z}_U$ for $j : U \to X$ the inclusion of an open subspace, which is a union of strata. Assume $x \notin U$. Then one gets by the second distinguished triangle for the cohomology of the "normal Morse datum":*

$$\left(R\Gamma^k_{\{re(g)|N\cap X\geq 0\}}(\mathcal{F}|N\cap X)\right)_x \simeq H^{k-1}(l_X, \partial l_X, Rj_*\mathbb{Z}_U)\,.$$

Moreover, by proposition 4.0.3 *and the property* **LSA** *one gets*

$$H^{k-1}(l_X, \partial l_X, Rj_*\mathbb{Z}_U) \simeq H^{k-1}(l_U, \partial l_U, \mathbb{Z}),$$

with $(l_U, \partial l_U) := (l_X, \partial l_X) \cap U$ *the* **complex link of** U*. This corresponds to the "second part of the fundamental theorem of complex stratified Morse theory" [GM, p.* 18*, thm. CSMT Part B, p.* 169*, cor.* 2.6.1*].*

Proposition 5.0.3 is related to the theory of the **vanishing cycle functor** of Deligne [De]:

$$NMD(\mathcal{F}, re(g), x) \simeq \left(\phi_{g|N\cap X}[-1](\mathcal{F}|N\cap X)\right)_x . \tag{5.12}$$

More generally, one has for a holomorphic function germ $g : (X, x) \to (\mathbb{C}, 0)$ and $\mathcal{F} \in ob(D(X_\bullet))$ an isomorphism:

$$LMD(\mathcal{F}, re(g), x) := \left(R\Gamma_{\{re(g)\geq 0\}}(\mathcal{F})\right)_x \simeq \left(\phi_g[-1](\mathcal{F})\right)_x . \tag{5.13}$$

We explain in the last section of this chapter the relation between the **vanishing cycle functor** and our **stratified Morse theory for constructible sheaves**. Here we just mention the beginning of the introduction of [Bry, p.4]: "...de nombreaux mathématiciens se sont rendu compte des relations existant entre "cycles évanescents" (pour des faisceaux constructibles, ou même pervers) et "microlocalisation" ...".

As already pointed out in [GM, p. 9], for the application of this **stratified Morse theory** "... we must have information about *both* the normal Morse data and about the tangential Morse data ...". But there is now an important application, where one does **not** need any a priori information about these "tangential and normal Morse data". This is in the context of the theory of **Lagrangian cycles**, which we now explain in the rest of this introduction (compare with [KS, chap. IX] and [Ka3, GrM, SchVi, Sch2] for the case $G = \mathbb{Z}$).

5.0.3 Introduction to characteristic cycles

Here one only needs the "Euler characteristic version" of our theory. Let us assume that we work in the context of a "geometric category" as in definition 4.0.5 on page 213. Suppose $X_\bullet$ is a **definable** Whitney **b-regular** stratification, i.e. all X_i are definable, and $f : (M, x) \to \mathbb{R}$ is a **definable** smooth function germ with $df_x | T_x X^s = 0$ and df_x **nondegenerate**.

As in section 2.3, we fix a commutative base ring R, a strict full **triangulated** subcategory T of $D(R)$ and an **additive** map μ on T with values in the abelian group G. Denote by $D_T(X_\bullet)$ the category of $X_\bullet$-constructible complexes of sheaves $\mathcal{F}$, with $\mathcal{F}_x \in ob(T)$ for all $x \in X$.

For example $T = perf$ is the subcategory of **perfect** complexes. Recall that a complex in $ob(D(R))$ is called **perfect**, if it is isomorphic to a bounded complex of finitely generated projective R-modules. For R a field this just means that the total cohomology $H^*(\cdot)$ is finite dimensional. For such complexes one can define an **additive Euler characteristic** (compare [SGA5, p. 376]):

$$\chi = Tr(id) : D_{perf}(R) \to R\,,$$

which in the case of a field is just the usual definition, i.e. the alternating sum $\sum_i (-1)^i \cdot dim(H^i(\cdot))$ (multiplied by $1 \in R$). Of course, in the last case we can also consider χ as an additive function with values in $\mathbb{Z}$.

Then one gets by lemma 2.0.2 (or corollary 2.2.2):

$$R\Gamma(l_X^-, \mathcal{F}) \in ob(T) \quad \text{for } \mathcal{F} \in ob(D_T(X_\bullet)).$$

So by the first distinguished triangle of proposition 5.0.2 one gets, that also the "normal Morse datum" $NMD(\mathcal{F}, f, x)$ belongs to $ob(T)$, with

$$\mu(NMD(\mathcal{F}, f, x)) = \mu(\mathcal{F}_x) - \mu(R\Gamma(l_X^-, \mathcal{F}))\,. \tag{5.14}$$

Especially, for $T = perf$ with R a field and $\mu = \chi$ one gets for $\mathcal{F} = R_X$:

$$\chi(NMD(R_X, f, x)) = 1 - \chi(H^*(l_X^-, R))\,. \tag{5.15}$$

Moreover, by the results of chapter 2, the value $\mu(NMD(\mathcal{F}, f, x))$ depends only on the **constructible function** α on X defined by $\alpha(x) := \mu(\mathcal{F}_x)$:

$$\chi(NMD(\alpha, f, x)) := \mu(NMD(\mathcal{F}, f, x)) = \alpha(x) - \int_{l_X^-} \alpha\, d\chi\,. \tag{5.16}$$

This Euler characteristic $\chi(NMD(\alpha, f, x)) =: \chi(NMD(\alpha, df_x))$ depends by theorem 5.0.1(2) only on the **connected component** $\Lambda^\circ_{s,i}$ (containing df_x) of the set Λ°_s of nondegenerate covectors in $T^*_{X^s}M$.

The formula (5.16) can also be reformulated in another way, by introducing the following **topological invariants** of the stratification (with $df_x \in \Lambda^\circ_{s,i}$):

$$c(\Lambda^\circ_{s,i}, S) := -\chi(NMD(1_S, df_x)) = \int_{l_X^-} 1_S\, d\chi = \chi(H^*_c(l_X^- \cap S, \mathbb{Q}))\,, \tag{5.17}$$

for each **connected component** $\Lambda^\circ_{s,i}$ of the set of nondegenerate covectors in $T^*_{X^s}M$, and each **stratum** S with $dim(S) > s$. Then one gets the **multiplicity formula**:

$$m(\Lambda^\circ_{s,i}) := \chi(NMD(\alpha, df_x)) = \alpha(x) - \sum_{S,\, dim(S)>s} c(\Lambda^\circ_{s,i}, S) \cdot \alpha(S)\,. \tag{5.18}$$

Note that there are only finitely many strata S with $c(\Lambda^\circ_{s,i}, S) \neq 0$, since the stratification is **locally finite**.

In the language of **Lagrangian cycles** it gives the **multiplicity** $m(\Lambda^\circ_{s,i})$ of the **characteristic cycle** $CC(\mathcal{F}) = CC(\alpha)$ of $\mathcal{F}$ or α on this connected component $\Lambda^\circ_{s,i}$, as we now explain.

For simplicity we assume, that the ambient manifold M is **oriented**. Let m be the (real) dimension of M. Then one gets an **injection**:

$$L(M, X_\bullet, G) := H^{BM}_m(\Lambda, G) \hookrightarrow H^{BM}_m(\Lambda^\circ, G) = \prod H^{BM}_m(\Lambda^\circ_{s,i}, G)\,, \qquad (5.19)$$

since the set of **degenerate covectors** $\Lambda \backslash \Lambda^\circ$ is a closed **definable** subset of dimension $< m$ so that $H^{BM}_m(\Lambda \backslash \Lambda^\circ, G) = 0$. Here $\Lambda, \Lambda^\circ, \Lambda^\circ_s$ are as in (5.4), (5.5), and the product is over the **connected components** $\Lambda^\circ_{s,i}$ of Λ°. Moreover,

$$H^{BM}_m(Y, G) := H^{-m}(Y, c^! G)$$

(with $c : Y \to \{pt\}$ a constant map) is the **Borel-Moore homology** (or homology with closed support) of a locally compact space Y so that the above claim follows from the corresponding long exact homology sequence.

Note that $\Lambda \subset T^*M$ is **closed** by Whitney **a-regularity**. In the following we assume, that M is pure-dimensional (or one should work on the connected components of M). Next we use the following **orientation conventions** (compare [GrM, SchVi]):

- T^*M is oriented by taking first the orientation of the base, and then the orientation of the fibers, which are oriented as the real duals of the oriented vector spaces T_xM.
- Similarly, $T^*_{X^s}M$ is canonically oriented by taking first a local orientation for X^s, and then the induced dual orientation of the fibers $(T_xM/T_xX^s)^*$. Here (T_xM/T_xX^s) is oriented so that in this order $T_xX^s \oplus (T_xM/T_xX^s) \simeq T_xM$ as oriented vector spaces.

This induces especially also an orientation of all $\Lambda^\circ_{s,i}$. We denote by

$$[\Lambda^\circ_{s,i}]_{GrM} \in H^{BM}_m(\Lambda^\circ_{s,i}, \mathbb{Z})$$

the corresponding **fundamental class** (which is also used in [GrM]). Similarly,

$$[\Lambda^\circ_{s,i}]_{SV} := (-1)^{m-s} \cdot [\Lambda^\circ_{s,i}]_{GrM}$$

is the fundamental class used in [SchVi]. Then $H^{BM}_m(\Lambda^\circ_{s,i}, \mathbb{Z})$ is a free group of rank 1 generated by these fundamental classes.

Note, that f has in $x \in X^s$ a **stratified Morse critical point** if and only if

$$\begin{cases} df_x \in \Lambda^\circ_s\,, \quad \text{and} \\ \text{the graph } df(M) \subset T^*M \text{ of } df \text{ intersects } T^*_{X^s}M \textbf{ transversally} \text{ in } df_x\,. \end{cases}$$

This can easily be checked in local coordinates (compare [KS, p. 311]):

Choose local coordinates $(M, x) \simeq (\mathbb{R}^m, 0)$, with $X^s = \{x_{s+1} = \cdots = x_m = 0\}$. Let $(x, \xi) = (x_1, \ldots, x_n, \xi_1, \ldots, \xi_n)$ be the associated coordinates on T^*M. Then

$$T_{df_x} df(M) = \{(x, \xi) \mid \xi_j = \sum_k \partial^2_{x_j x_k} f(0) \cdot x_k \}$$

and

$$T_{df_x} T^*_{X^s} M = \{(x, \xi) \mid x_{s+1} = \cdots = x_m = 0 = \xi_1 = \cdots = \xi_s \}.$$

So the intersection of these two tangent spaces is $\{0\}$ if and only if the Hesse matrix

$$(\, \partial^2_{x_j x_k} f(0) \,)_{1 \leq j,k \leq s}$$

is **nondegenerate**. Moreover, by our orientation conventions one gets in this case for the corresponding **local intersection number** :

$$\sharp_{df_x} \left([T^*_{X^s} M]_{SV} \cap [df(M)] \right) = (-1)^\lambda , \tag{5.20}$$

with λ the **Morse index** of $f|X^s$ in x. Here $[df(M)] \in H^{BM}_m(df(M), \mathbb{Z})$ is the fundamental class of the manifold $df(M)$, oriented by the induced orientation of M. This is the reason why the fundamental class $[T^*_{X^s} M]_{SV}$ is preferred from the viewpoint of stratified Morse theory.

The local intersection number is defined similarly to the following **global intersection number**

$$\sharp : H^{BM}_m(A, G) \times H^{BM}_m(B, \mathbb{Z}) \to G$$

for two **closed** subsets $A, B \subset T^*M$ with $A \cap B$ **compact**:

$$\begin{array}{ccccc} H^m_A(T^*M, G) \times H^m_B(T^*M, \mathbb{Z}) & \xrightarrow{\cup} & H^{2m}_c(T^*M, G) & \xrightarrow{tr} & G \\ {\scriptstyle PD}\downarrow \wr & & {\scriptstyle PD}\downarrow \wr & & \| \\ H^{BM}_m(A, G) \times H^{BM}_m(B, \mathbb{Z}) & \xrightarrow{\cap} & H_0(T^*M, G) & \longrightarrow & G . \end{array}$$

Here PD is **Poincaré duality** given by the cap-product with the fundamental class $[T^*M]$ of the oriented manifold T^*M.

Let $CF(X_\bullet, G)$ be the group of functions $\alpha : X \to G$, which are constructible with respect to $X_\bullet$, i.e. which are constant on the strata. Using the cap-product

$$\otimes = \cap : H^0(\Lambda^\circ_{s,i}, G) \times H^{BM}_m(\Lambda^\circ_{s,i}, \mathbb{Z}) \to H^{BM}_m(\Lambda^\circ_{s,i}, G) ,$$

we define the **characteristic cycle** map (or **micro-local Euler class**) by

$$\begin{aligned} CC = \mu eu : CF(X_\bullet, G) &\to H^{BM}_m(\Lambda^\circ, G) ; \\ \alpha &\mapsto \prod \chi(NMD(\alpha, df_x)) \otimes [\Lambda^\circ_{s,i}]_{SV} \quad , \text{ with } df_x \in \Lambda^\circ_{s,i} . \end{aligned} \tag{5.21}$$

Consider for example $\alpha = 1_X$ for the case of a (sub)manifold with boundary X as in example 5.0.2. By the corresponding calculation of the "normal Morse data" one gets, that $CC(1_X) \in H_m^{BM}(\Lambda, G) \hookrightarrow H_m^{BM}(\Lambda^\circ, G)$, i.e. $CC(1_X)$ is a **cycle**. This is obvious in the case $X = [a, b] \subset \mathbb{R}$ a compact interval:

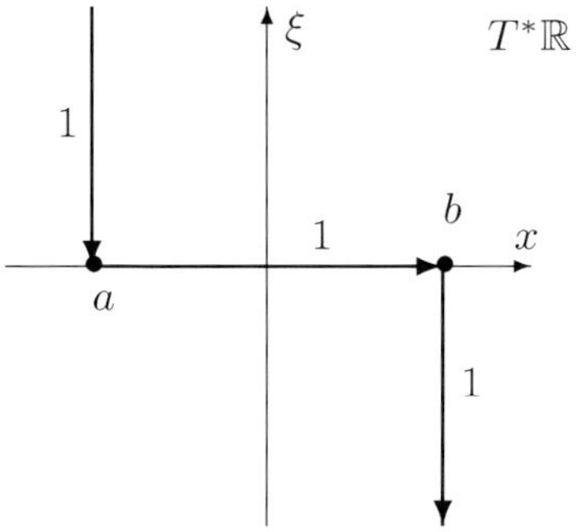

Figure 5.1: Characteristic cycle of $1_{[a,b]}$

It is a result of Kashiwara ([Ka3, thm. 4.1] and [KS, chapter IX]), that $CC(\alpha)$ is **always a cycle**, i.e. the characteristic cycle map CC factorizes as

$$CC = \mu eu : CF(X_\bullet, G) \to H_m^{BM}(\Lambda, G) = L(M, X_\bullet, G)\,. \tag{5.22}$$

Moreover, this map CC is then easily seen to be a linear isomorphism. Just argue by induction on $n = dim(X)$, and use the fact that

$$\chi(NMD(\alpha, df_x)) = \alpha(x)$$

for x in a top-dimensional stratum (since the corresponding "lower halflink" l_X^- is empty. Compare also with the proof of [KS, thm. 9.7.10, p. 405/406]).

Note that Kashiwara considers in loc.cit. the **subanalytic** context, but the methods of [KS, chapter IX] apply also to our context of more general "geometric categories".

Let $c^! \mathbb{Z} =: \omega_X \simeq \mathbb{Z}_X[m]$ (with $c : X \to \{pt\}$ a constant map) be the **dualizing complex** on X. Then one gets by [KS, def. 9.4.1] a map

$$CC_{KS} : K_0(D_{perf}(X_\bullet)) \to H^0_\Lambda(T^*M, \pi^{-1}\omega_X) \simeq H^m_\Lambda(T^*M, \mathbb{Z})\,.$$

Here $\pi : T^*M \to M$ is the projection, and $K_0(D_{perf}(X_\bullet))$ is the Grothendieck group of the triangulated category of $X_\bullet$-constructible complexes over a base field $R = k$ (of characteristic zero), whose stalk complexes are finite dimensional. For the definition of the map CC_{KS} in loc.cit. one needs especially, that $\mathcal{F} \in ob(D_{perf}(X_\bullet))$ is **cohomologically constructible** in the sense of [KS, def. 3.4.1, p. 158] (so that one can apply [KS, prop. 3.4.4]). But this was already explained in chapter 2 (and also in chapter 4 in the context of a **ccs-stratification**). Moreover, CC_{KS} maps to $H^m_\Lambda(T^*M, \mathbb{Z})$ by the estimate (5.4) of the micro-support. Finally, the following diagram commutes:

$$\begin{array}{ccccc} K_0(D_{perf}(X_\bullet)) & \xrightarrow{CC_{KS}} & H^m_\Lambda(T^*M, \mathbb{Z}) & \xrightarrow{(-1)^m \cdot PD} & H^{BM}_m(\Lambda, \mathbb{Z}) \\ \chi_X \downarrow & & & & \| \\ CF(X_\bullet, \mathbb{Z}) & \xrightarrow{CC} & H^{BM}_m(\Lambda^\circ, \mathbb{Z}) & \xleftarrow{\supset} & H^{BM}_m(\Lambda, \mathbb{Z})\,. \end{array}$$

Here χ_X is given by taking stalkwise the Euler characteristic. Then the commutativity of this diagram follows from two observations:

- The "multiplicities" for the map CC_{KS} are given by the Euler characteristic of the corresponding "local type" of $\mathcal{F}$ at a nondegenerate covector df_x (compare [KS, (9.4.10), p. 382]). But as already remarked before, this is just $\chi(NMD(\mathcal{F}, df_x))$.
- The local cohomology class $[T^*_{X^s}M]$ used in loc.sit. corresponds by **Poincaré duality** to $(-1)^m \cdot [T^*_{X^s}M]_{SV}$. This follows from [KS, (9.5.18), p. 388], where (the dual of) the intersection number (5.20) is also calculated as $(-1)^\lambda$, but for the opposite order of the two intersecting cycles !

Remark 5.0.1. The composed transformation $CC \circ \chi_X$ corresponds also to the characteristic cycle map CC_{Ka} used in [Ka3]. Kashiwara uses the same "multiplicities" [Ka3, def. 3.4], and also the corresponding local intersection numbers [Ka3, (5.1.4), p. 201]. But be aware that he uses the **symplectic orientation** of T^*M, which differs by the factor $(-1)^{m(m+1)/2}$ from our orientation of T^*M !

So in this way one can deduce from the "micro-local" theory of Kashiwara-Schapira the characteristic cycle map (5.22) for the case $G = \mathbb{Z}$, i.e. for integer valued constructible functions. But then $CC(\alpha)$ is a **cycle** for any $\alpha \in CF(X_\bullet, G)$. Since this is a local result, it is enough to show

$$CC(g \cdot 1_S) \in H^{BM}_m(\Lambda, G)$$

for all strata S and $g \in G$. But then one has by the **multiplicity formula** (5.18):

$$CC(g \cdot 1_S) = g \otimes CC(1_S)\,, \quad \text{with } \otimes = \cap : H^0(\Lambda, G) \times H^{BM}_m(\Lambda, \mathbb{Z}) \to H^{BM}_m(\Lambda, G)\,.$$

Let us finally explain some other conventions used in the literature for the characteristic cycle map:

- The composed transformation (as above)

$$K_0(D_{perf}(X_\bullet)) \xrightarrow{\chi_X} CF(X_\bullet, \mathbb{Z}) \xrightarrow{CC} H^{BM}_m(\Lambda, \mathbb{Z})$$

  is the characteristic cycle map CC_{SV} used in [SchVi]. They use the same orientation conventions, and their "multiplicity" is given by $\chi(LMD(\mathcal{F}, f, x))$

for $x \in S$ a **stratified Morse critical point** with Morse index $\lambda = 0$ for $f|S$ (compare [SchVi, (2.4),(2.5), p. 454]). But

$$\chi(LMD(\mathcal{F}, f, x)) = (-1)^{\lambda} \cdot \chi(NMD(\mathcal{F}, f, x))$$

by theorem 5.0.1(1). Especially, $(-1)^m \cdot CC_{SV}$ corresponds by **Poincaré duality** to the characteristic cycle map of [KS].

- Let $a : T^*M \to T^*M$ be the **antipodal map** (i.e. multiplication by -1 on the fibers of the cotangent bundle). Then

$$a_* \circ CC : CF(X_\bullet, \mathbb{Z}) \to H_m^{BM}(\Lambda, \mathbb{Z})$$

is the characteristic cycle map CC_{GrM} used in [GrM]. This follows from $a_*[T^*_{X^s}M]_{SV} = [T^*_{X^s}M]_{GrM}$, and the fact that their "multiplicity" is given by $\chi(NMD(\mathcal{F}, -f, x))$ as in (5.16) (compare [GrM, 2.5]). They use the "upper half link" l_X^+ of X with respect to f, which is the same as the "lower half link" of $-f$.

- It is also the composition $a_* \circ CC$, or more precisely the induced map

$$N^* : CF(X_\bullet, \mathbb{Z}) \xrightarrow{a_* \circ CC} H_m^{BM}(\Lambda \backslash T^*_M M, \mathbb{Z}) \xrightarrow[\sim]{(\pi'^*)^{-1}} H_{m-1}^{BM}(C, \mathbb{Z}) ,$$

which is used in the context of **curvature measures** for definable sets (compare [Fu, FuMC, BrBe, BrKu]). Here π' is the projection onto the "sphere-bundle"

$$T^*M \backslash T^*_M M \to S(T^*M) := (T^*M \backslash T^*_M M)/\mathbb{R}^+ ,$$

with $C := \pi'(\Lambda \backslash T^*_M M)$.

Especially, the approach of [Fu] gives **another way** of proving the fact, that the characteristic cycle is indeed a **cycle**. He uses the language and results of "geometric measure theory", representing the "normal cycle" $N^*(1_X)$ for X a closed (or compact) definable subset of M by a suitable "integral current". We explain later the following direct "geometric approach", based on the ideas of Fu, but written down in the language of **homology**:

Theorem 5.0.3. *Let $f : M \to \mathbb{R}^{\geq 0}$ be a non-negative* **definable** *C^2-function. Assume $X_0 := \{f = 0\}$ is compact and (everywhere) of smaller dimension than M. By the* **curve selection lemma**, *there exists $0 < \epsilon << 1$ such that f is a submersion in $\{0 < f < \epsilon\}$, if we work on a suitable open neighborhood of X_0. Then*

$$CC(1_{X_0}) = \lim_{t \to 0^+} CC(1_{\{f \leq t\}}) \tag{5.23}$$

as cycles in the real analytic (Nash) manifold T^*M. Here "$\lim_{t \to 0^+}$" is the **specialization** *of a family of homology classes in the sense of [SchVi, sec. 3]. Note that $CC(1_{\{f \leq t\}})$ is a* **cycle** *for $0 < t < \epsilon$, since $\{f \leq t\}$ is a* **manifold with boundary**.

From this theorem we then also deduce, that the characteristic cycle is indeed a **cycle**. In this way we can avoid the use of the sophisticated "micro-local machinery" of Kashiwara-Schapira, especially the use of the functor μhom.

By our definition, the characteristic cycle map CC just codifies the "Euler characteristics" $\chi(NMD(\alpha, df_x))$ of the "normal Morse data". Especially, the **multiplicativity of the "normal Morse data"** (5.6) implies for $G = R$ a commutative ring the **multiplicativity of the characteristic cycles** (compare [KS, (9.4.1), p. 378]):

$$CC(\alpha \times \beta) = CC(\alpha) \times CC(\beta) \tag{5.24}$$

for $\alpha \in CF(X_\bullet, R)$ and $\beta \in CF(X'_\bullet, R)$, with $\alpha \times \beta \in CF((X \times X')_\bullet, R)$ defined as $\alpha \times \beta(x, x') := \alpha(x) \cdot \beta(x')$.

Similarly, the relation (5.11) between **duality** and the "normal Morse data" implies the following transformation rule for the characteristic cycle under the **duality transformation D_X for constructible functions** as in section 2.3 (compare with [FuMC, thm. 3.12], and also with [Ka3, prop. 7.2]):

$$CC(\,D_X(\alpha)\,) = (-1)^m \cdot a_*(\,CC(\alpha)\,)\,, \tag{5.25}$$

with $m = dim(M)$ and $a : T^*M \to T^*M$ the **antipodal map**. Note that D_X maps $CF(X_\bullet)$ to itself by the corresponding result for constructible sheaves. Especially, $D_X = id$ in the complex analytic (algebraic) context. The transformation

$$(-1)^m \cdot a_* : H_m^{BM}(\Lambda, G) \to H_m^{BM}(\Lambda, G)$$

corresponds under **Poincaré duality** to

$$a^* : H^m_\Lambda(T^*M, G) \to H^m_\Lambda(T^*M, G)\,.$$

This follows from the **projection formula** and $a_*([T^*M]) = (-1)^m \cdot [T^*M]$. So (5.25) is also in agreement with [KS, prop. 9.4.4, p. 380].

The additional information, that $CC(\alpha)$ is a **cycle** allows us to use techniques from homology theory, to study these cycles. For example intersection theory is used in the following beautiful application:

Theorem 5.0.4. *Let $f : M \to [a, d[\,\subset \mathbb{R}$ be a smooth function ($d \in \mathbb{R} \cup \{\infty\}$), with $f|X$* **proper**. *Suppose that $\Lambda \cap df(M)$ is* **compact**, *with Λ the union of conormal spaces to the strata as in (5.4). Then we have for all $\mathcal{F} \in ob(D_T(X_\bullet))$:*

$$R\Gamma(X, \mathcal{F}) \in ob(T), \;\; \textit{with} \;\; \mu(R\Gamma(X, \mathcal{F})) = \sharp(\,CC(\chi_X(\mathcal{F})) \cap [df(M)]\,)\,. \tag{5.26}$$

$$R\Gamma_c(X, \mathcal{F}) \in ob(T), \;\; \textit{with} \;\; \mu(R\Gamma_c(X, \mathcal{F})) = \sharp(\,CC(\chi_X(\mathcal{F})) \cap [-df(M)]\,)\,. \tag{5.27}$$

Remark 5.0.2. Consider the classical case: The base-ring is a field, the objects of T are the (bounded) complexes with finite dimensional cohomology and $\chi : T \to \mathbb{Z}$ is the usual Euler characteristic. Then the above intersection theorem goes back in the **complex analytic** context to Dubson ([Du2]), Sabbah ([Sab]) and Ginsburg ([Gi1], in the context of holonomic D-modules). The corresponding intersection theorem in the **subanalytic** context is due to Kashiwara ([Ka3, thm. 4.2, 4.3], [KS, thm. 9.5.3, cor. 9.5.4]).

Assume that $f : M \to [a,d[\subset \mathbb{R}$ is a **definable** smooth function, with $f|X$ **proper** and $\pi(\Lambda \cap df(M)) \subset [a,b]$ for some $b \in [a,d[$ (with π the projection $T^*M \to M$). Especially, $\Lambda \cap df(M)$ is **compact**.

Then one also has the following counterpart of theorem 5.0.4 for a constructible function $\alpha \in CF(X_\bullet, G)$:

$$\int 1_{\{f \leq r\}} \cdot \alpha \, d\chi = \sharp(\, CC(\alpha) \cap [df(M)]\,) \qquad \text{for all } r \in]b,d[\,. \tag{5.28}$$

$$\int 1_{\{f < r\}} \cdot \alpha \, d\chi = \sharp(\, CC(\alpha) \cap [-df(M)]\,) \quad \text{for all } r \in]b,d[\,. \tag{5.29}$$

Note, that $1_{\{f \leq r\}} \cdot \alpha$ and $1_{\{f<r\}} \cdot \alpha$ are constructible functions with **compact** support (for $r < d$) so that the left sides of these equations are well defined by section 2.3.

Let us give some examples, where the conditions on f are satisfied:

1. X is compact. Then we can take for f a constant function so that $df(M)$ is the zero-section. Then we get

$$\int \alpha \, d\chi = \sharp(\, CC(\alpha) \cap [T^*_M M]\,)\,. \tag{5.30}$$

   This formula generalizes the classical **Poincaré-Hopf theorem**, which corresponds to $\alpha = 1_M$ for M a compact oriented manifold.

2. We work in the context of an **o-minimal structure** on the real field (e.g. in the **semialgebraic context**), with $M = \mathbb{R}^m$ (or a real analytic Nash manifold). If $f : M \to [a,d[$ is a **definable** smooth map, with $f|X$ proper, then

$$f \circ \pi(\Lambda \cap df(M)) \subset [a,d[$$

   is definable. Moreover, by the **curve selection lemma** it is **finite** so that there is a $b \in [a,d[$ with $\pi(\Lambda \cap df(M)) \subset [a,b]$. Then one gets for r big enough:

$$\int \alpha \, d\chi' = \int 1_{\{f \leq r\}} \cdot \alpha \, d\chi = \sharp(\, CC(\alpha) \cap [df(M)]\,)\,, \tag{5.31}$$

   and

$$\int \alpha \, d\chi = \int 1_{\{f < r\}} \cdot \alpha \, d\chi = \sharp(\, CC(\alpha) \cap [-df(M)]\,)\,. \tag{5.32}$$

   Here χ (or χ') is the **Euler characteristic with compact (or closed) support** as in section 2.3.

3. We work in the context of an **analytic (Nash) geometric category** (e.g. in the **subanalytic (or locally semialgebraic) context**), with M a real analytic (Nash) manifold. If $f : M \to [0, d[$ is a **definable** smooth map, with $f|X$ proper, then
$$f \circ \pi(\, \Lambda \cap df(M)\,) \subset [0, d[$$
is definable. Moreover, by the **curve selection lemma** it is **discrete**. Especially,
$$\pi(\, \Lambda \cap df(U)\,) \subset X \cap \{f = 0\}\,,$$
if we **restrict** to $f : U := \{f < \epsilon\} \to [0, \epsilon[$, with $0 < \epsilon << 1$. Then one gets for $0 < r$ small enough:
$$\int_{X\cap\{f=0\}} \alpha \, d\chi = \int 1_{\{f\leq r\}} \cdot \alpha \, d\chi = \sharp(\, CC(\alpha) \cap [df(U)]\,)\,. \tag{5.33}$$
This is of course also true in the **o-minimal** context.

The most important special case of 3. is the local situation $(M, x) \simeq (\mathbb{R}^m, 0)$ with $f(x) := r(x) := \sum_{i=1}^m x_i^2$. In this case we get for $0 < \epsilon << 1$:
$$\alpha(x) = \sharp_{dr_x}(\, CC(\alpha) \cap [dr(\{r < \epsilon\})]\,)\,. \tag{5.34}$$

Especially, this local intersection number is independent of the choice of the function r with $\{r = 0\} = \{x\}$. Moreover, it defines the inverse of the characteristic cycle map CC (compare with [Ka3, thm. 8.3], [KS, thm. 9.7.1] for the subanalytic, and also with [Gi1, thm. 11.7, p. 393] for the holomorphic context):

Corollary 5.0.1. *The inverse of the characteristic cycle map*
$$CC = \mu eu : CF(X_\bullet, G) \to H_m^{BM}(\Lambda, G) = L(M, X_\bullet, G)$$
is given by
$$Eu^\vee : L(M, X_\bullet, G) = H_m^{BM}(\Lambda, G) \to CF(X_\bullet, G)\,;$$
$$[C] \mapsto \alpha\,, \quad \textit{with } \alpha(x) := \sharp_{dr_x}(\, [C] \cap [dr(\{r < \epsilon\})]\,)\,,$$
for r, ϵ as above.

Note, that this corollary is also a **refinement** of the corresponding results in loc.cit., since we work with a **fixed stratification** ! Suppose for example, that $Y \subset M$ is a definable **locally closed** submanifold such that the fundamental class $[T_Y^* M]$ (for a suitable orientation) belongs to
$$H_m^{BM}(cl(T_Y^* M)) \hookrightarrow H_m^{BM}(T_Y^* M)\,.$$
Let us denote this fundamental class of the closure by $C := [cl(T_Y^* M)]$. Then $C \in H_m^{BM}(\Lambda, \mathbb{Z})$ for **any** definable Whitney **b-regular stratification** $X_\bullet$ of $cl(Y)$ so that $Eu^\vee(C)$ is $X_\bullet$-constructible with respect to **any** such stratification! And this

argument can easily be generalized to any Whitney **a-regular stratification** with the property **LSA**.

This applies for example to the case of a pure-dimensional **closed complex analytic** subset X of the complex manifold M, with $Y := X_{reg}$ the regular part of $X = cl(Y)$. Then one gets (as explained later on):

$$Eu^{\vee}(\, [cl(T^*_{X_{reg}}M)]\,) = Eu^{\vee}_X := (-1)^{dim_C(X)} \cdot Eu_X \,, \tag{5.35}$$

with $Eu_X : X \to \mathbb{Z}$ the famous **Euler obstruction** of MacPherson [MP] (compare [Gi1, thm. 11.7, p. 393]). Here $Eu^{\vee}_X$ is the so called **dual Euler obstruction**, which justifies our notion for the inverse of the characteristic cycle map.

Especially, we get without any calculation that the **Euler obstruction** Eu_X is constructible with respect to **any** complex analytic Whitney **b-regular stratification** of X. Since this is a local result, it is then also true for any pure-dimensional reduced complex space X (without any embedding into a complex manifold). This result is due to Dubson ([Du1, prop. 1, thm. 3]) and Brasselet-Schwartz ([BSchwa, p. 125, cor. 10.2]).

Remark 5.0.3. We use in (5.35) the **complex orientation** of $T^*_Y M$ for the definition of the fundamental class. If we choose for the complex manifold M also the complex orientation, then this **complex orientation** of $T^*_Y M$ agrees only up to the factor $(-1)^{dim_C(Y)}$ with our **real orientation conventions**!

The reason for this is just the following simple observation:

Let $V^* = hom_C(V, \mathbb{C})$ be the complex dual space of a (finite dimensional) complex vector space V. If V^R denotes the underlying real vector space, with its induced orientation, then $(V^*)^R$ is canonically isomorphic to $(V^R)^* := hom_R(V^R, \mathbb{R})$. But

$$\text{real dual orientation of } (V^R)^* = (-1)^{dim_C(V)} \cdot \text{ complex orientation of } (V^*)^R \,.$$

So we get a factor $(-1)^{dim_C(M)}$ for the orientations of T^*M, and also a factor $(-1)^{dim_C(M) - dim_C(Y)}$ for the orientations of $T^*_Y M$.

Let us now translate some of our formulae to the **complex analytic context**. We assume that M is a complex analytic manifold, and all X_i are closed complex analytic subsets so that the strata or X^{2i} are locally closed complex analytic manifolds.

If we use proposition 5.0.3, then the **multiplicity formula** (5.18) for the characteristic cycle can also be formulated in terms of the **complex links** of the complex analytic stratification $X_\bullet$. In this case one has to use the following **topological invariants** of the stratification (with dg_x a nondegenerate holomorphic covector in $T^*_{S'}M$):

$$c(S', S) := -\chi(NMD(1_S, dg_x)) = \int_{l_X} 1_S \, d\chi = \chi(H^*_c(l_X \cap S, \mathbb{Q})) \,, \tag{5.36}$$

for all **strata** S', S with $dim_C(S) > dim_C(S')$. Here l_X is the complex link of X at $x \in S'$ (with respect to g). Note also, that

$$\chi(H_c^*(l_X \cap S, \mathbb{Q})) = \chi(H_c^*((l_X \backslash \partial l_X) \cap S, \mathbb{Q})), \tag{5.37}$$

since ∂l_X is a compact Whitney stratified set with **odd-dimensional definable** strata so that $\chi(H_c^*(\partial l_X \cap S, \mathbb{Q})) = 0$ (compare with lemma 5.0.3 and [Su]).

If we use the **complex orientations** for the fundamental classes of the conormal spaces, then one gets the **multiplicity formula** (with $m := dim_C(M)$):

$$CC(\alpha) = \prod m(S') \cdot [cl(T^*_{S'}M)] \in \prod H_{2m}^{BM}(cl(T^*_{S'}M), G) \simeq H_{2m}^{BM}(\Lambda, G)\,;$$

$$\begin{aligned} \text{with } (-1)^{dim_C(S')} \cdot m(S') &= \chi(NMD(\alpha, dg_x)) = \alpha(x) - \int_{l_X} \alpha\, d\chi \\ &= \alpha(x) - \sum_{S,\, dim_C(S) > dim_C(S')} c(S', S) \cdot \alpha(S)\,. \end{aligned} \tag{5.38}$$

Here again l_X is the complex link of X at $x \in S'$ (with respect to g).

For $\mathcal{F} \in ob(D_T(X_\bullet))$ and $\mu : T \to G$ an additive function, the multiplicities of $CC(\chi_X(\mathcal{F}))$ can also be described in terms of **vanishing cycles**:

$$m(S') = (-1)^{dim_C(S')} \cdot \mu\big((\phi_{g|N\cap X}[-1](\mathcal{F}|N \cap X))_x \big) = \mu\big((\phi_g[-1](\mathcal{F}))_x \big), \tag{5.39}$$

with $g : (M, x) \to (\mathbb{C}, 0)$ a holomorphic function germ such that dg_x is **non-degenerate**, and x is a **complex Morse critical point** of $g|S'$ (i.e. $dg(M)$ intersects $T^*_{S'}M$ transversally in dg_x). Then $re(g)|S'$ has in x a classical Morse critical point of **Morse index** $\tau = dim_C(S')$ so that (5.39) follows from theorem 5.0.1(1) and (5.12), (5.13).

Then one gets by (5.34) and (5.35) also the following **local index formula** for a point x of the stratum S':

$$\alpha(x) = (-1)^{dim_C(S')} \cdot m(S') + \sum m(S) \cdot Eu^{\vee}_{cl(S)}(x)\,, \tag{5.40}$$

where the (locally finite) sum is over all strata S with $dim_C(S) > dim_C(S')$.

Remark 5.0.4. This **local index formula** (5.40) goes back to a corresponding **index formula** of Kashiwara [Ka1] (compare [Ka2, thm. 6.3.1]) for the **solution complex** $Rhom_{D_X}(\mathcal{M}, \mathcal{O}_X)$ of a **holonomic** D_X-module on the complex manifold X. Note that this **solution complex** is a **complex analytically constructible** complex of sheaves of $\mathbb{C}$-vector spaces, with finite dimensional stalks ([KS, thm. 11.3.7]). His formula corresponds to (5.40) for the constructible function

$$\alpha := \chi_X(\, Rhom_{D_X}(\mathcal{M}, \mathcal{O}_X)[dim_C(X)]\,)\,,$$

i.e. his **multiplicities** $m(S)$ differ by a factor $(-1)^{dim_{\mathbb{C}}(X)}$ from our conventions. Moreover, he works directly with the **characteristic cycle of a holonomic** D_X-**module** (defined with the help of a "good filtration"). Similarly, he introduced for his **index formula** some topological invariants, which are nothing else than the **Euler obstructions** of the $cl(S)$. But this was only observed later on by Dubson (compare with [BDK], and also with [Ka2, introduction, p. xiii]).

Other references for these formulae are [Du2], [BMM, p. 545] (using the sign-convention of [Ka1]), and [Gi1, p. 331, thm. 8.2, p. 381, thm. 11.7, p. 393] (using our sign-convention, but with a sign-mistake in [Gi1, cor. 6.19(b), thm. 8.2]).

Note, that these references use the Euler characteristic

$$\chi(H^*_c((l_X \backslash \partial l_X) \cap S, \mathbb{Q})) = \chi(H^*((l_X \backslash \partial l_X) \cap S, \mathbb{Q}))\,.$$

This equality follows from **Poincaré duality** for the even-dimensional manifold $(l_X \backslash \partial l_X) \cap S$. Let N be a (pure-dimensional) **relatively compact definable manifold**, with $n := dim(N)$ and $c : N \to \{pt\}$ a constant map. Then

$$\chi(H^*(N, \mathbb{Q})) = \chi(Rc_*\mathbb{Q}_N) \overset{PD}{=} \chi(Rc_! D_N) \overset{(*)}{=}$$

$$\int (-1)^n \cdot 1_N \, d\chi \overset{(*)}{=} (-1)^n \cdot \chi(Rc_!\mathbb{Q}_N) = (-1)^n \cdot \chi(H^*_c(N, \mathbb{Q}))\,.$$

Here we use for $(*)$ the fact, that this Euler characteristic depends only on the corresponding **constructible function**, with

$$\chi((D_N)_x) = \chi((\mathbb{Q}_N[n])_x) = (-1)^n \cdot 1_N\,.$$

Similarly, the equality (5.37) follows from the

Lemma 5.0.3. *Let X be a* **compact** *set, with a* **definable b-regular** *stratification such that all strata are* **odd-dimensional**. *Let N be a finite union of the n-dimensional strata. Then:*

$$\chi(H^*(N, \mathbb{Q})) = \chi(H^*_c(N, \mathbb{Q})) = 0\,.$$

The proof of this lemma is just by induction on the **depth** of the stratification. Note, that the first equality implies the second by **Poincaré duality** as before. In the case depth $= 0$, N is a **compact odd-dimensional** manifold so that the first equality is trivial. In general, let $j : N \to X$ be the locally closed inclusion so that

$$\chi(H^*_c(N, \mathbb{Q})) = \int_X 1_N \, d\chi \quad \text{and} \quad \chi(H^*(N, \mathbb{Q})) = \int_X \chi_X(Rj_*\mathbb{Q}_X) \, d\chi\,.$$

Then we claim:

$$\chi_X(Rj_*\mathbb{Q}_X)(x) := \chi((Rj_*\mathbb{Q}_X)_x) = 1_N(x) = \begin{cases} 1, & \text{for } x \in N. \\ 0, & \text{for } x \notin N. \end{cases}$$

But $H^*(Rj_*\mathbb{Q}_X)_x)$ is for $x \in X^s \backslash N$ just the cohomology of $L \cap N$, with L the **link** of X at x (by proposition 4.0.3 and proposition 4.2.1(3.b)). And in our case of a **b-regular** stratification, the **link** L can be chosen as

$$L = \{\pi = 0\,,\, \rho = r\} \quad (0 < r << 1)\,,$$

with

$$\pi(x_1, \ldots, x_n) := (x_1, \ldots, x_s, 0, \ldots, 0) \quad \text{and} \quad \rho(x_1, \ldots, x_n) := \sum_{j=s+1}^{m} x_j^2$$

in definable local coordinates

$$(M, X^s, x) \simeq (\mathbb{R}^m, \mathbb{R}^s \times \{0\}, 0)\,.$$

By transversality, this is again a **compact definable b-regular** stratified set, with all strata **odd-dimensional** such that $L \cap N$ is a finite union of the $n-s-1$-dimensional strata. So our claim follows by induction.

5.1 Stratified Morse theory, part I

All topological spaces are locally compact with countable topology and of finite cohomological dimension (over $\mathbb{Z}$). For a ring R (associative with unit) let $D^*(X) := D^*(X, R)$ for $* = \emptyset, b, +, -$ denote the derived category of (bounded/from below/above) complexes of sheaves of R-modules.

5.1.1 Local Morse data

Let us begin with the cohomological counterpart of [GM, thm. SMT Part A, p. 6, prop. 3.2, p. 62].

Lemma 5.1.1. *Let $f : X \to \mathbb{R}$ be a continuous function such that*

$$f : \{a \leq f < b\} \to [a, b[$$

is **proper** *for some $a \in \mathbb{R}$, $b \in \mathbb{R} \cup \{\infty\}$, with $a < b$. Let $i : \{f \leq a\} \to X$ be the inclusion, and consider $\mathcal{F} \in ob(D(X))$. Assume that all cohomology sheaves of $(Rf_*(\mathcal{F}))|\,]a, b[$ are* **locally constant**. *Then*

1. *$R\Gamma(\{f < b\}, \mathcal{F}) \to R\Gamma(\{f \leq a\}, \mathcal{F})$ is an isomorphism.*

2. *$R\Gamma_c(\{f \leq a\}, i^!\mathcal{F}) \to R\Gamma_c(\{f < b\}, \mathcal{F})$ is an isomorphism.*

Proof. We can assume $X = \{f < b\}$. Let $j : \{a < f < b\} \to X$ be the inclusion. 1. Then one has a distinguished triangle

$$R\Gamma(X, Rj_!j^*\mathcal{F}) \longrightarrow R\Gamma(X, \mathcal{F}) \longrightarrow R\Gamma(\{f \leq a\}, \mathcal{F}) \xrightarrow{[1]}\,.$$

Since $f : \{a \leq f < b\} \to [a, b[$ is proper, one has

$$R\Gamma(X, Rj_! j^* \mathcal{F}) \simeq R\Gamma([a, b[\,, Rj'_! \mathcal{G}) ,$$

with $j' :]a, b[\to [a, b[$ the inclusion and $\mathcal{G} := (Rf_* \mathcal{F}) | \,]a, b[$. But the assumption on $Rf_* \mathcal{F}$ implies

$$R\Gamma([a, b[\,, Rj'_! \mathcal{G}) \simeq (Rj'_! \mathcal{G})_a \simeq 0 .$$

2. This time we start with the distinguished triangle

$$R\Gamma_c(\{f \leq a\}, i^! \mathcal{F}) \longrightarrow R\Gamma_c(X, \mathcal{F}) \longrightarrow R\Gamma_c(X, Rj_* j^* \mathcal{F}) \xrightarrow{[1]} .$$

Then one gets with the same notation as before

$$R\Gamma_c(X, Rj_* j^* \mathcal{F}) \simeq R\Gamma_c([a, b[\,, Rj'_* \mathcal{G}) \simeq i_a^!(Rj'_* \mathcal{G}) \simeq 0 .$$

Here i_a is the inclusion $\{a\} \to [a, b[$. □

Remark 5.1.1. Assume $\mathcal{F} \in ob(D^+(X))$. Then the **non-characteristic deformation lemma** of Kashiwara [KS, prop. 2.7.2] implies 1. or 2. under the **weaker assumption**

$$\mu supp((Rf_* \mathcal{F}) | \,]a, b[\,) \subset \begin{cases} \{(x, \xi) \in T^* \,]a, b[\mid \xi \leq 0\} & \text{for the case 1.} \\ \{(x, \xi) \in T^* \,]a, b[\mid \xi \geq 0\} & \text{for the case 2.} \end{cases}$$

(compare with the "micro-local Morse lemma" [KS, cor. 5.4.19]). But if we want to get both conclusions at the same time, then these two assumptions together imply our assumption, that all cohomology sheaves of $(Rf_* \mathcal{F}) | \,]a, b[$ are locally constant (by proposition 4.0.1).

Assume $b < \infty$ in lemma 5.1.1, with $f : \{a \leq f \leq b\} \to [a, b]$ proper. If one wants to follow the notation of [GM], it is more natural, to consider in 1. the morphism

$$R\Gamma(\{f \leq b\}, \mathcal{F}) \to R\Gamma(\{f \leq a\}, \mathcal{F}) .$$

By the distinguished triangle

$$R\Gamma(\{f = b\}, R\Gamma_{\{f \geq b\}}(\mathcal{F})) \to R\Gamma(\{f \leq b\}, \mathcal{F}) \to R\Gamma(\{f < b\}, \mathcal{F}) \xrightarrow{[1]} , \qquad (5.41)$$

this is an isomorphism, if we in addition assume

$$0 \simeq (R\Gamma_{\{x \geq b\}}(Rf_* \mathcal{F}))_b \simeq R\Gamma(\{f = b\}, R\Gamma_{\{f \geq b\}}(\mathcal{F}))$$

(e.g. $(Rf_* \mathcal{F}) | \,]a, b]$ has locally constant cohomology sheaves). If we make this additional assumption for a instead of b, then we get also the isomorphism

$$R\Gamma(\{f < b\}, \mathcal{F}) \simeq R\Gamma(\{f < a\}, \mathcal{F}) .$$

Similarly

$$R\Gamma_c(\{f < a\}, \mathcal{F}) \simeq R\Gamma_c(\{f < b\}, \mathcal{F}) ,$$

if the last term in the following distinguished triangle is vanishing:

$$R\Gamma_c(\{f < a\}, \mathcal{F}) \to R\Gamma_c(\{f \leq a\}, i^! \mathcal{F}) \to R\Gamma(\{f = a\}, R\Gamma_{\{f \leq a\}}(\mathcal{F})) \overset{[1]}{\to} . \quad (5.42)$$

By definition 4.0.3 and corollary 4.0.1 we have the

Example 5.1.1. *Let $X_\bullet$ be a stratification of the closed subset X of the manifold M, and consider an $X_\bullet$-constructible complex $\mathcal{F} \in ob(D(X))$. Assume that $f : X \to]c, d[\subset \mathbb{R}$ is a* **proper smooth** *map and that $X_\bullet$ satisfies the property* **LSA**.

1. *If f is a* **stratified submersion** *in all points of $\{a < f < b\}$ (with $]a, b[\subset]c, d[$), then all cohomology sheaves of $(Rf_*\mathcal{F})|]a, b[$ are locally constant.*
2. *If f is a submersion in all points of $\{f = a\}$ ($a \in]c, d[$), then*

$$R\Gamma(\{f = a\}, R\Gamma_{\{f \geq a\}}(\mathcal{F})) \simeq 0 \simeq R\Gamma(\{f = a\}, R\Gamma_{\{f \leq a\}}(\mathcal{F})) ,$$

since

$$R\Gamma_{\{f \geq a\}}(\mathcal{F})_x \simeq 0 \simeq R\Gamma_{\{f \leq a\}}(\mathcal{F})_x \quad \text{for all } x \in \{f = a\}.$$

We see especially, that in this context the difference between working with subsets of the form $\{f < a\}$ or $\{f \leq a\}$ (with a a regular value of f on the strata of $X_\bullet$) does not matter.

This example together with lemma 5.1.1 implies the

Corollary 5.1.1. *Make the same assumptions as in the example above. If f is a* **submersion** *in all points of $\{a \leq f \leq b\} \cap \{f \neq e\}$ (with $[a, b] \subset]c, d[$ and $e \in]a, b[$), then one has distinguished triangles (functorial in $\mathcal{F} \in ob(D(X_\bullet))$):*

$$R\Gamma(\{f = e\}, R\Gamma_{\{f \geq e\}}(\mathcal{F})) \to R\Gamma(\{f \leq b\}, \mathcal{F}) \to R\Gamma(\{f \leq a\}, \mathcal{F}) \overset{[1]}{\to} , \quad (5.43)$$

and

$$R\Gamma_c(\{f < a\}, \mathcal{F}) \to R\Gamma_c(\{f < b\}, \mathcal{F}) \to R\Gamma(\{f = e\}, R\Gamma_{\{f \leq e\}}(\mathcal{F})) \overset{[1]}{\to} . \quad (5.44)$$

Proof. The first distinguished triangle comes from the triangle (5.41) and the commutative diagram

$$\begin{array}{ccc} R\Gamma(\{f \leq b\}, \mathcal{F}) & \longrightarrow & R\Gamma(\{f \leq a\}, \mathcal{F}) \\ \downarrow \wr & & \uparrow \wr \\ R\Gamma(\{f \leq e\}, \mathcal{F}) & \longrightarrow & R\Gamma(\{f < e\}, \mathcal{F}) . \end{array}$$

The second distinguished triangle comes from the triangle (5.42) and the commutative diagram

$$\begin{array}{ccc} R\Gamma_c(\{f < a\}, \mathcal{F}) & \longrightarrow & R\Gamma_c(\{f < b\}, \mathcal{F}) \\ \downarrow \wr & & \uparrow \wr \\ R\Gamma_c(\{f < e\}, \mathcal{F}) & \longrightarrow & R\Gamma_c(\{f \leq e\}, R\Gamma_{\{f \leq e\}}(\mathcal{F})) . \end{array}$$

□

Remark 5.1.2. 1. One should compare the distinguished triangles of the corollary with the proof of [KS, prop. 5.4.20] (especially [KS, (5.4.22, 23)]).

2. $R\Gamma(\{f = e\}, R\Gamma_{\{f \geq e\}}(\mathcal{F}))$ is the cohomological counterpart of the **"coarse Morse datum"** as defined in [GM, p. 62, def. 3.4].

Especially,

$$R\Gamma(\{f = e\}, R\Gamma_{\{f \leq e\}}(\mathcal{F})) \simeq R\Gamma(\{-f = -e\}, R\Gamma_{\{-f \geq -e\}}(\mathcal{F}))$$

is the "coarse Morse datum" of $-f$. So that this "coarse Morse datum" of $-f$ corresponds to the "relative cohomology with compact support" in the triangle (5.44). But the cohomology with compact support is in some sense "dual" to the cohomology with closed support. So this reflects the "duality" observed in [GM, p. 27, 2.7]. In the context of stratified Morse theory for constructible sheaves, this is a special form of **Poincaré-Verdier duality**.

Let us consider the situation of corollary 5.1.1. We make in addition the following assumptions:

- The base ring R is commutative.
- $\mathcal{F} \in ob(D^-(X_\bullet))$ (at least locally).
- The stratification $X_\bullet$ is **d-regular**.

Then we know by proposition 4.2.1 and corollary 4.2.1, that the **duality functor** $D := D_X := Rhom(\cdot, c^! R_{pt})$ (with $c : X \to \{pt\}$ a constant map) induces a map

$$D_X : D^-(X_\bullet) \to D^+(X_\bullet) .$$

Recall also for a continuous map g the general isomorphisms (compare with the proof of corollary 2.2.7(1) on page 119):

$$D(Rg_!(\cdot)) \simeq Rg_*(D(\cdot)) \quad \text{and} \quad g^!(D(\cdot)) \simeq D(g^*(\cdot)) .$$

Application of the duality functor D to the first morphism of the distinguished triangle (5.44) gives then a commutative diagram

$$\begin{array}{ccc} D(R\Gamma_c(\{f<a\},\mathcal{F})) & \longleftarrow & D(R\Gamma_c(\{f<b\},\mathcal{F})) \\ \downarrow\wr & & \downarrow\wr \\ R\Gamma(\{f<a\},D(\mathcal{F})) & \longleftarrow & R\Gamma(\{f<b\},D(\mathcal{F})) \\ \uparrow\wr & & \uparrow\wr \\ R\Gamma(\{f\leq a\},D(\mathcal{F})) & \longleftarrow & R\Gamma(\{f\leq b\},D(\mathcal{F}))\,. \end{array}$$

We get especially

$$D\big(R\Gamma(\{-f=-e\},R\Gamma_{\{-f\geq -e\}}(\mathcal{F}))\big)\simeq R\Gamma\big(\{f=e\},R\Gamma_{\{f\geq e\}}(D(\mathcal{F}))\big)\,. \tag{5.45}$$

Let us come back to the general case of corollary 5.1.1. The next step is to consider a smooth function f with **isolated** critical points in $\{f=e\}$, and to localize the "coarse Morse datum" at these critical points. In terms of sheaf theory, this is quite easy:

Lemma 5.1.2. *Consider the same situation as in corollary* 5.1.1. *Assume in addition that there are finitely many $x_i\in\{f=e\}$ such that $f:(X,x)\to\mathbb{R}$ is a stratified submersion (with respect to $X_\bullet$) for all $x\in\{f=e\}$ with $x\neq x_i$. Then*

$$R\Gamma(\{f=e\},R\Gamma_{\{f\geq e\}}(\mathcal{F}))\simeq\oplus_i\,(R\Gamma_{\{f\geq e\}}(\mathcal{F}))_{x_i}\,. \tag{5.46}$$

This isomorphism is functorial in $\mathcal{F}\in ob(D(X_\bullet))$.

Proof. Since the stratification has the property **LSA**, we have by definition $0\simeq(R\Gamma_{\{f\geq e\}}(\mathcal{F}))_x$, if $x\in\{f=e\}$ is a regular point in the stratified sense. Therefore

$$supp\big((R\Gamma_{\{f\geq e\}}(\mathcal{F}))|\{f=e\}\big)\subset\cup_i\,\{x_i\}\,.$$ □

Remark 5.1.3. $LMD(\mathcal{F},f,x_i):=(R\Gamma_{\{f\geq e\}}(\mathcal{F}))_{x_i}$ is the cohomological counterpart of the **"local Morse datum"** with respect to the isolated "critical" point x_i, as defined in [GM, p. 63, def. 3.5.2] .

To make the relation between the sheaf theory and the notions of [GM] more explicit, we recall the following definition of chapter 1 (for later applications we consider also complex valued functions):

Definition 5.1.1. *Let $f:X\to\mathbb{K}$ continuous (with $\mathbb{K}=\mathbb{R},\mathbb{C}$), $S\subset\{f=0\}$ compact and $\mathcal{F}\in ob(D(X))$ be given. The triple $(f,S,\mathcal{F})$ satisfies a* **cohomological version of a Milnor fibration**, *if there is a fundamental system $K_n\supset K_{n+1}$ of compact neighborhoods of S in X and a sequence of real numbers $\epsilon_n\searrow 0$ $(n\in\mathbb{N})$, with*

1. *The higher direct images sheaves* $R^i f_*(\mathcal{F}|K_n^*)$ *for the map*

$$f : K_n^* := K_n \cap \{|f| \leq \epsilon_n, f \neq 0\} \to \{w \in \mathbb{K}|\ 0 < |w| \leq \epsilon_n\}$$

are locally constant for all $i \in \mathbb{Z}$.

2. $\forall n\ \exists \epsilon$ *with* $0 < \epsilon \leq \epsilon_{n+1}$ *such that*

$$R\Gamma(K_n \cap \{f = w\}, \mathcal{F}) \to R\Gamma(K_{n+1} \cap \{f = w\}, \mathcal{F})$$

is an isomorphism for $|w| \leq \epsilon$.

The relation between this assumption and Morse theory for constructible sheaves follows from lemma 1.1.1, which implies in the case of a real function the

Lemma 5.1.3. *Assume that the triple* $(f, S, \mathcal{F})$ *satisfies a* **cohomological version of a Milnor fibration** *(with* $f : X \to \mathbb{R}$, *and* K_n, ϵ_n *as in definition* 5.1.1*). Let* $i : \{f = 0\} \to X$ *and* $j : \{f < 0\} \to X$ *be the inclusions. Then there exists for* $-\epsilon_n \leq w < 0$ *a commutative diagram*

$$\begin{array}{ccc} R\Gamma(K_n \cap \{|f| \leq \epsilon_n\}, \mathcal{F}) & \longrightarrow & R\Gamma(K_n \cap \{f = w\}, \mathcal{F}) \\ \downarrow \wr & & \downarrow \wr \\ R\Gamma(S, \mathcal{F}) & \xrightarrow{sp} & R\Gamma(S, i^* Rj_* j^* \mathcal{F}) \end{array}$$

(the upper horizontal map is the restriction map, and sp is the map induced from the adjunction morphism $id \to Rj_*j^*$*). By embedding this diagram into an isomorphism of distinguished triangles, one gets*

$$R\Gamma(K_n \cap \{|f| \leq \epsilon_n\}, K_n \cap \{f = w\}, \mathcal{F}) \simeq R\Gamma(S, R\Gamma_{\{f \geq 0\}}(\mathcal{F}))\,.$$

Especially, $R\Gamma(K_n \cap \{|f| \leq \epsilon_n\}, K_n \cap \{f = w\}, \mathcal{F})$ *is independent (up to isomorphism) of the choice of* K_n, ϵ_n.

Example 5.1.2. *Consider the situation of corollary* 5.1.1, *and take a sequence of real numbers* $\epsilon_n \searrow 0$ $(n \in \mathbb{N})$, *with* $[e - \epsilon_n, e + \epsilon_n] \subset]a, b[$. *By example* 5.1.1(1), *the assumptions of definition* 5.1.1 *are true for the function* $f - e$, *with* $S := \{f = e\}$ *and* $K_n := \{|f - e| \leq \epsilon_n\}$. *Therefore*

$$R\Gamma(\{|f - e| \leq \epsilon_n\}, \{f = w\}, \mathcal{F}) \simeq R\Gamma(\{f = e\}, R\Gamma_{\{f \geq e\}}(\mathcal{F}))\,,$$

for $w \in [e - \epsilon_n, e[$. *This is the precise sheaf theoretic conterpart of the* **"coarse Morse datum"** *[GM, p. 62, def. 3.4].*

We get a similar interpretation of the **"local Morse datum"**

$$LMD(\mathcal{F}, f, x) := (R\Gamma_{\{f \geq e\}}(\mathcal{F}))_x$$

for an isolated critical point $x \in \{f = e\}$, if the triple $(f - e, \{x\}, \mathcal{F})$ satisfies a **cohomological version of a Milnor fibration**. This is for example the case, if

$$f - e : (X, \{f = e\}) \to (\mathbb{R}, 0)$$

has a suitable stratification satisfying the a_f-condition of Thom (see example 1.1.3 on page 36). For our applications it is enough to consider a simpler situation. First we recall a definition of [GM, p. 44, def. 1.8, p. 128, def. 12.1]:

Definition 5.1.2. *Let $X_\bullet$ be a smooth stratification of the closed subset X of the smooth manifold M. Consider for $x \in X^s$ a covector*

$$\omega \in \begin{cases} T_x^*M & := T^*M|\{x\} \, . \\ (T_x^*M)^C & := Hom_R(T_xM, \mathbb{C}) \, . \end{cases}$$

The covector ω is called **nondegenerate**, *if it satisfies for all $j > s$ the following property: If $x_n \in X^j$ is a sequence converging to x and the sequence of tangent planes $T_{x_n}X^j$ converges to some limiting plane $\tau \subset T_xM$, then*

$$\omega(\tau) = \begin{cases} \mathbb{R}, & i.e.\ \omega(\tau) \neq \{0\}. \\ \mathbb{C}. & \end{cases}$$

In terms of conormal spaces, the first condition is equivalent to $\omega \notin cl(T^*_{X_j}M)$. Therefore the set of nondegenerate covectors is **open** *in $T^*M|X^i$ (for a fixed i).*

Remark 5.1.4. Assume M is a(n) (almost) **complex manifold** and X^i is a(n) (almost) **complex submanifold** for all i (i.e. T_xX^i is a complex subvector space of T_xM for all $x \in X^i$). Then one has a canonical isomorphism

$$T_x^*M \simeq Hom_C(T_xM, \mathbb{C}) \subset (T_x^*M)^C \, .$$

Under this identification, both definitions for $\omega \in Hom_C(T_xM, \mathbb{C})$ to be **nondegenerate** are equivalent (since τ is a complex subvector space of T_xM and ω is $\mathbb{C}$-linear. Compare also with [GM, p. 44]).

We consider from now on the following situation: $X_\bullet$ is a smooth stratification of the closed subset X of the smooth manifold M.

$$f : (M, x) \to (\mathbb{K}, f(x)) \quad \text{with } x \in X^s \text{ and } \mathbb{K} = \mathbb{R}, \mathbb{C}$$

is a **smooth** function germ such that the covector df_x is **nondegenerate** (with respect to $X_\bullet$). This implies that there exists an open neighborhood U of x in M such that $f : U \cap X^j \to \mathbb{K}$ is a **submersion** for all $j > s$ (otherwise there would be a limiting plane $\tau \subset T_xM$ with $df_x(\tau) \neq \mathbb{K}$). We choose U moreover so small that $U \cap X^j = \emptyset$ for $j < s$. Then

$$(U \cap X \cap \{f = f(x)\}) \backslash X^s$$

gets an induced smooth stratification (as a closed subset of $U \backslash X^s$).

Lemma 5.1.4. *Let $S \subset U$ be a locally closed submanifold with*

$$x \in S \subset U \cap X^s \cap \{f = f(x)\}\ ,$$

e.g. $S = \{x\}$. Assume in addition that X^j is Whitney **b-regular** *over S in x (for some $j > s$, e.g. X^j is Whitney* **b-regular** *over X^s in x). Then $U \cap X^j \cap \{f = f(x)\}$ is also Whitney* **b-regular** *over S in x.*

Proof. The proof is the same as in [King2, lem. 7]. Let

$$y_n \in U \cap X^j \cap \{f = f(x)\} =: Y^j \quad \text{and } x_n \in S$$

be sequences converging to x such that the tangent planes $T_{y_n}Y^j$ converge to some limiting plane τ, and the secant lines $l_n = \overline{y_n, x_n}$ (with respect to some local coordinates) converge to some limiting line l. We can also assume (by taking a subsequence), that $T_{y_n}X^j$ converges to some limiting plane τ'. Then

$$T_{y_n}Y^j = T_{y_n}X^j \cap ker(df_{y_n}) \Longrightarrow \tau \subset (\tau' \cap ker(df_x))\ .$$

Moreover

$$dim(\tau) = dim(\tau' \cap ker(df_x))\ ,$$

since df_x is nondegenerate. Therefore $\tau = \tau' \cap ker(df_x)$. Since $f(y_n) = f(x_n)$ for all n, we also have $l \subset ker(df_x)$. Finally $l \subset \tau'$, since X^j is by assumption Whitney **b-regular** over S in x. Altogether we get $l \subset (\tau' \cap ker(df_x)) = \tau$. □

By this geometric implication we get, under the assumption df_x **nondegenerate**, control of the strata $U \cap X^j \cap \{f = f(x)\}$ for $j > s$. For the control of $U \cap X^s \cap \{f = f(x)\}$ one needs extra information about $f|X^s$.

Example 5.1.3. *Let $X_\bullet$ be a* **b-regular** *stratification of X. Consider a smooth function $f : M \to \mathbb{R}$ such that*

$$S := crit(f|X^s) := \{x \in X^s \mid d(f|X^s)_x = 0\}$$

is a **smooth submanifold** *of X^s, with $S \subset \{f = e\}$ for some $e \in \mathbb{R}$. Assume*

1. *df_x is* **nondegenerate** *for all $x \in S$.*

2. *S is "reasonable" in the sense that $(X^s \cap \{f = e\})\backslash S$ is* **b-regular** *over S (in all points of S).*

Then there exists an open neighborhood U of S in $M\backslash X_{s-1}$ such that $f : U \cap X^j \to \mathbb{R}$ is a **submersion** *for all $j > s$.*

Moreover, the strata of the induced **b-regular stratification** *of $(U \cap \{f \leq e\})\backslash S$ are by lemma 5.1.4 also* **b-regular** *over S (in all points of S) so that we get an induced* **b-regular stratification** *of $U \cap \{f \leq e\}$. Especially, all cohomology sheaves of*

$$(Rj_*j^*\mathcal{F})|S \quad \text{and} \quad (R\Gamma_{\{f \geq e\}}(\mathcal{F}))|S\ ,$$

with $j : \{f < e\} \to \{f \leq e\}$ *the inclusion, are locally constant (by corollary 4.2.1). So the* **local Morse datum** $LMD(\mathcal{F}, f, x)$ *is "locally constant" on* S. *For an application of the special case of a "stratified Morse-Bott critical manifold"* S *compare with [BG, sec. 8].*

For our applications, we only need the case $S = \{x\}$, with weaker assumptions on the stratifcation:

Example 5.1.4. *Let* $X_\bullet$ *be an* **a+d-regular** *stratification, with*

$$f : (M, x) \to (\mathbb{K}, f(x)) \quad \text{for } x \in X^s \text{ and } \mathbb{K} = \mathbb{R}, \mathbb{C}$$

a smooth function germ such that the covector df_x *is* **nondegenerate**. *Assume*

1. X^j *is Whitney* **b-regular** *over* $\{x\}$ *for all* $j > s$ *(e.g. we work with a definable stratification in the context of "geometric categories" [Ta1]).*
2. x *is an isolated critical point of* $f|X^s$*, which is also "reasonable" in the sense that* $(X^s \cap \{f = f(x)\}) \backslash \{x\}$ *is* **b-regular** *over* $\{x\}$ *(e.g.* $\mathbb{K} = \mathbb{R}$ *and* x *is a "classical Morse critical point" of* $f|X^s$*).*

Let $\mathcal{F}$ *be* $X_\bullet$*-constructible, and define in local coordinates* $(M, x) \simeq (\mathbb{R}^m, 0)$ *the distance function* r *by* $r(x) := \sum_{i=1}^m x_i^2$. *If we take sequences* $\epsilon_n, \delta_n \searrow 0$ *with* $0 < \epsilon_n << \delta_n << 1$*, and define*

$$K_n := X \cap \{|f - f(x)| \leq \epsilon_n, r \leq \delta_n\},$$

then the assumptions of definition 5.1.3 *are true for* $(f - f(x), \{x\}, \mathcal{F})$ *and we get a* **cohomological version of a local Milnor fibration**. *Especially*

$$R\Gamma(X \cap \{|f - f(x)| \leq \epsilon_n, r \leq \delta_n\}, X \cap \{r \leq \delta_n, f = w\}, \mathcal{F}) \simeq (R\Gamma_{\{f \geq f(x)\}}(\mathcal{F}))_x ,$$

for $w \in [f(x) - \epsilon_n, f(x)[$ *and* $\mathbb{K} = \mathbb{R}$. *This is the precise sheaf theoretic conterpart of the "local Morse datum" as defined in [GM, p.* 63*, def.* 3.5.2*].*

That $(f - f(x), \{x\}, \mathcal{F})$ satisfies a **cohomological version of a local Milnor fibration** (with K_n, ϵ_n as above) follows from lemma 5.1.4 and example 1.1.2. The argument on page 36 applies by lemma 4.3.2 also to an **a+d-regular** stratification, i.e. the induced stratifications of $X \cap \{r \leq \delta_n\}$ and $X \cap \{f = const.\}$ are also **a+d-regular**. Then one only has to use the following

Lemma 5.1.5. *Let* $X_\bullet$ *be an* **a-regular** *stratification of* X*, which also has the property* **LSA**. *Consider a* **proper smooth** *map* $f : X \to N$ *into the manifold* N*, and an* $X_\bullet$*-constructible complex* $\mathcal{F}$. *Let* Z *be a* **locally closed** *subset of* N *such that the germ* $f : (X, x) \to (N, f(x))$ *is a stratified submersion (with respect to* $X_\bullet$*) for all* $x \in f^{-1}(Z) =: Z'$. *Consider the cartesian diagram*

$$\begin{array}{ccc} Z' & \xrightarrow{i'} & X \\ \downarrow{\scriptstyle f'} & & \downarrow{\scriptstyle f} \\ Z & \xrightarrow{i} & N . \end{array}$$

Then all cohomology sheaves of $Rf'_* i'^* \mathcal{F}$ are locally constant.

Proof. The set C of all points $x \in X$ such that the germ $f : (X, x) \to (N, f(x))$ is not a stratified submersion is closed, since $X_\bullet$ is **a-regular**. Therefore $f(C)$ is closed (since f is proper), and we can choose an open neighborhood U of Z in N with $U \cap f(C) = \emptyset$. We can assume $N = U$ so that f is a proper stratified submersion. Then $Rf_* \mathcal{F}$ has locally constant cohomology sheaves, since $X_\bullet$ has the property **LSA** (by corollary 4.0.1). But

$$Rf'_* i'^* \mathcal{F} \simeq i^* Rf_* \mathcal{F}$$

by the the **proper base change theorem** [KS, prop. 2.6.7]. □

For later applications, we recall the main point of the argument above:

Remark 5.1.5. Consider the situation of example 5.1.4, with $\mathbb{K} = \mathbb{R}$.

1. Choose $\delta > 0$ so small that r has no stratified critical points in $\{0 < r \leq \delta\}$ with respect to the induced stratification of $X \cap \{f = f(x)\}$.
2. After that, choose $\epsilon > 0$ so small that f has no stratified critical points in $\{0 < |f - f(x)| \leq \epsilon\}$ with respect to the induced stratification of $X \cap \{r \leq \delta\}$.

Then the **local Morse datum** $LMD(\mathcal{F}, f, x)$ is given by

$$R\Gamma(X \cap \{|f - f(x)| \leq \epsilon, r \leq \delta\}, X \cap \{r \leq \delta, f = w\}, \mathcal{F}) \simeq (R\Gamma_{\{f \geq f(x)\}}(\mathcal{F}))_x \,,$$

with $w \in [f(x) - \epsilon, f(x)[$.

Corollary 5.1.2. *Consider the situation of example* 5.1.4, *with*

$$f : (M, x) \to (\mathbb{R}, f(x)) \,.$$

Choose $0 < \epsilon << \delta << 1$ *as in the remark before, and define*

$$(K, \partial K) := X \cap (\{r \leq \delta\}, \{r = \delta\}) \,.$$

If $k_x : \{x\} \to X$ is the inclusion, then there exist two distinguished triangles (functorial in $\mathcal{F} \in ob(D(X_\bullet))$):

$$R\Gamma(K \cap \{f = w\}, \mathcal{F})[-1] \to (R\Gamma_{\{f \geq f(x)\}}(\mathcal{F}))_x \to k_x^* \mathcal{F} \xrightarrow{[1]} , \tag{5.47}$$

for $w \in [f(x) - \epsilon, f(x)[$.

$$k_x^! \mathcal{F} \to (R\Gamma_{\{f \geq f(x)\}}(\mathcal{F}))_x \to R\Gamma(K \cap \{f = w\}, \partial K \cap \{f = w\}, \mathcal{F}) \xrightarrow{[1]} , \tag{5.48}$$

for $w \in]f(x), f(x) + \epsilon]$.

Proof. The first distinguished triangle is a special case of lemma 5.1.3. Let

$$\{x\} \xrightarrow{i_x} Y := \{f = f(x)\} \xrightarrow{i} \{f \geq f(x)\} \xleftarrow{j} \{f > f(x)\}\,,$$

and $k : \{f \geq f(x)\} \to X$ be the inclusions of subsets of X. Since $X_\bullet$ is **d-regular** and x is an isolated critical point, we have $supp\big((R\Gamma_{\{f\geq f(x)\}}(\mathcal{F}))|Y\big) \subset \{x\}$. Then

$$(R\Gamma_{\{f\geq f(x)\}}(\mathcal{F}))_x \simeq i_x^*(i^*k^!\mathcal{F}) \xleftarrow{\sim} i_x^!(i^*k^!\mathcal{F})\,,$$

and the last term is part of the distinguished triangle

$$i_x^!(i^!k^!\mathcal{F}) \to i_x^!(i^*k^!\mathcal{F})) \to i_x^!(i^*Rj_*j^*k^!\mathcal{F}) \overset{[1]}{\to}\,.$$

Since $k_x^!\mathcal{F} \simeq i_x^!(i^!k^!\mathcal{F})$, we only have to show

$$i_x^!(i^*Rj_*j^*k^!\mathcal{F}) \simeq R\Gamma(K \cap \{f = w\}, \partial K \cap \{f = w\}, \mathcal{F})\,.$$

$i^*Rj_*j^*k^!\mathcal{F}$ is by corollary 4.2.1 constructible with respect to the induced **d-regular** stratification $Y_\bullet$ of Y (which one gets by adding the point stratum $\{x\}$). Choose $1 >> \rho > 0$ so that

1. $r : Y \to \mathbb{R}$ is a proper stratified submersion over $]0, \delta+\rho]$ with respect to the **d-regular** stratification $Y_\bullet$.
2. (r, f) has no stratified critical points in $\{|r-\delta| \leq \rho,\, 0 < |f| \leq \epsilon\}$ with respect to $X_\bullet$.

Then one gets by the first condition on ρ (compare lemma 5.1.1(2)):

$$i_x^!(i^*Rj_*j^*k^!\mathcal{F}) \simeq R\Gamma(Y \cap \{r \leq \delta\}, Y \cap \{r = \delta\}, i^*Rj_*j^*k^!\mathcal{F})\,.$$

If we take ρ, ϵ with the above properties arbitrarily small, we see that

$$(f - f(x), Y \cap \{r = \delta\}, \mathcal{F}) \quad \text{and} \quad (f - f(x), Y \cap \{r \leq \delta\}, \mathcal{F})$$

satisfy the assumption of definition 5.1.3. Let $i' : Y \to X$ and $j' : \{f > f(x)\} \to X$ be the inclusions. Then $i' = k \circ i$, $j' = k \circ j$, and

$$i^*Rj_*j^*k^!\mathcal{F} \simeq i^*k^*Rk_*Rj_*j^*k^!\mathcal{F} \simeq i'^*Rj'_*j'^*\mathcal{F}\,.$$

By lemma 1.1.1 and remark 1.1.3(3) we get for $w \in]f(x), f(x)+\epsilon]$ a commutative diagram:

$$\begin{array}{ccc}
R\Gamma(Y \cap \{r \leq \delta\}, i'^*Rj'_*j'^*\mathcal{F}) & \longrightarrow & R\Gamma(Y \cap \{r = \delta\}, i'^*Rj'_*j'^*\mathcal{F}) \\
\uparrow\wr & & \uparrow\wr \\
R\Gamma(X \cap \{r \leq \delta+\rho, f = w\}, \mathcal{F}) & \longrightarrow & R\Gamma(X \cap \{|r-\delta| \leq \rho, f = w\}, \mathcal{F}) \\
\downarrow\wr & & \downarrow\wr \\
R\Gamma(X \cap \{r \leq \delta, f = w\}, \mathcal{F}) & \longrightarrow & R\Gamma(X \cap \{r = \delta, f = w\}, \mathcal{F})\,.
\end{array}$$

The last vertical isomorphisms follow by lemma 5.1.1(1) from the fact, that r has no stratified critical points in $\{|r - \delta| \leq \rho\}$ with respect to the induced **d-regular** stratification of $X \cap \{f = w\}$. This implies the claim for $\mathcal{F}$. The functoriality in the $X_\bullet$-constructible complex $\mathcal{F}$ follows from the proof and remark 1.1.3(2). $\square$

Remark 5.1.6. For the proof of the second distinguished triangle (5.48) we used:

1. $\{x\}$ is an **isolated** point in $supp\big((R\Gamma_{\{f \geq f(x)\}}(\mathcal{F}))|Y\big)$ so that
$$i_x^*(i^* k^! \mathcal{F}) \simeq i_x^!(i^* k^! \mathcal{F}) .$$
2. $f : (X, \{f = f(x)\}) \to (\mathbb{R}, f(x))$ has a suitable stratification so that one gets a nice description of the **costalk** $i_x^! i'^* R j'_* j'^* \mathcal{F}$.

If for example all X_i and f are definable with respect to an o-minimal structure on $(\mathbb{R}, +, \cdot)$, then one can deduce the first distinguished triangle (5.47) and also 2. from a suitable stratification satisfying the a_f-condition of Thom (see example 1.1.3). This argument applies also to example 5.1.3, where the a_f-condition follows as in the proof of lemma 5.1.4. In the case of a complex function, one gets by the same argument as in 2. a **geometric** description for the **costalk** $i_x^! \psi_{f-f(x)}(\mathcal{F})$ of the **nearby cycle functor** $\psi_{f-f(x)}$ (compare lemma 5.4.2).

5.1.2 Normal Morse data

Example 5.1.4 and corollary 5.1.2 imply the corresponding results of proposition 5.0.2 about the **"normal Morse datum"** $NMD(\mathcal{F}, f, x)$ under weaker assumptions on the stratification. Let $X_\bullet$ be a smooth stratification, with $x \in X^s$ such that

1. $f : (M, x) \to (\mathbb{K}, f(x))$ for $\mathbb{K} = \mathbb{R}, \mathbb{C}$ is a smooth function germ, with $0 = df_x|T_x X^s$ and df_x **nondegenerate**.
2. $X_\bullet$ is **a+d-regular** in $X \backslash X_s$.
3. X^j is **b-regular** over $\{x\}$, and **a-regular** over X^s in x for all $j > s$.

Take a **normal slice** N to X^s in x, i.e. a locally closed submanifold N of M, with $N \cap X^s = \{x\}$, which is transversal to X^s. Then N is transversal to all strata of $X_\bullet$ in some neighborhood U of x in M (by **a-regularity** over X^s in x). Then
$$Y := U \cap X \cap N \subset U$$
with its induced stratification is also **a+d-regular** in all of U, since $Y^j := U \cap X^j \cap N$ is **b-regular** over $\{x\}$ for all $j > s$:

By the inverse function theorem, there are local coordinates
$$(M, X^s, x) \simeq (\mathbb{R}^m, \mathbb{R}^s \times \{0\}, 0) ,$$

with

$$N = \{\pi = 0\} \quad \text{for } \pi(x_1, \ldots, x_m) := (x_1, \ldots, x_s, 0, \ldots, 0)\,.$$

Take a sequence $y_n \in Y^j$ converging to $x = 0$ such that

$$T_{y_n} Y^j \to \tau\,, \quad l_n = \overline{y_n, x} \to l\,, \quad \text{and} \quad T_{y_n} X^j \to \tau'\,.$$

Then

$$\tau \subset (\tau' \cap kern(d\pi_x))\,, \quad l \in \tau'\,, \quad \text{and} \quad l \perp T_x X^s\,.$$

This implies the claim $l \in \tau$, since

$$T_x X^s \subset \tau' \quad \text{and} \quad d\pi_x | T_x X^s = id \ \Rightarrow\ \tau' = \tau \oplus T_x X^s\,.$$

Moreover, df_x is also **nondegenerate** with respect to this induced stratification of Y. Finally, x is an **isolated "reasonable" critical point**, since $N \cap X^s = \{x\}$. So in the case $\mathbb{K} = \mathbb{R}$, we can apply corollary 5.1.2 to the induced stratification of $Y \subset U$, with

$$NMD(\mathcal{F}, f, x) := (R\Gamma_{\{f|N \cap X \geq f(x)\}}(\mathcal{F}|N \cap X))_x$$

the "normal Morse datum" of f. Moreover,

$$(l_X^*, \partial l_X^*) := (K, \partial K) \cap \{f - f(x) = *w\}$$

$$= X \cap N \cap \{f - f(x) = *w\} \cap (\{r \leq \delta\}, \{r = \delta\})$$

is for $0 < w << \delta << 1$ the **upper/lower halflink** $(* = +, -)$ of X with respect to f ([GM, p. 66, def. 3.9.1]).

The (relative) cohomology of the upper/lower halflink (modulo its "boundary") with respect to $\mathcal{F} \in ob(D(X_\bullet))$ is also a **special case** of a "normal Morse datum", since we started with an **a+d-regular** stratification:

Let $j : U := X \backslash X^s \to X$ be the inclusion of the locally closed complement. Then $Rj_! j^* \mathcal{F}$ and $Rj_* j^* \mathcal{F}$ are also $X_\bullet$-constructible (corollary 4.2.1), and we want to calculate their stalks in $x \in X^s$. Since this is a local result, we can assume that X^s is **closed** such that the normal slice N to X^s in x is transversal to all strata. Consider the cartesian diagram

$$\begin{array}{ccc} X \cap N & \xleftarrow{\ j'\ } & U' := U \cap N \\ \downarrow{\scriptstyle k} & & \downarrow{\scriptstyle k''} \\ X & \xleftarrow{\ j\ } & U\,. \end{array}$$

Then we have

$$k^* j_! j^* \mathcal{F} \simeq j'_! k''^* j^* \mathcal{F}\,,$$

and by proposition 4.3.1 and remark 4.3.6 also

$$k^* Rj_* j^* \mathcal{F} \simeq Rj'_* k''^* j^* \mathcal{F} .$$

Moreover,

$$X \cap N \cap \{f = w\} = U \cap N \cap \{f = w\} \quad \text{for } w \neq f(x),$$

since $X^s \cap N = \{x\}$. By the distinguished triangle (5.47) of corollary 5.1.2 applied to $f|N$, we get

$$R\Gamma(l_X^-, \mathcal{F})[-1] \simeq R\Gamma(l_X^-, Rj_! j^* \mathcal{F})[-1] \simeq NMD(Rj_! j^* \mathcal{F}, f, x) . \tag{5.49}$$

By the distinguished triangle (5.48) of corollary 5.1.2 applied to $f|N$, we get

$$R\Gamma(l_X^+, \partial l_X^+, \mathcal{F}) \simeq R\Gamma(l_X^+, \partial l_X^+, Rj_* j^* \mathcal{F}) \simeq NMD(Rj_* j^* \mathcal{F}, f, x) . \tag{5.50}$$

In the same way one also gets the base-change isomorphisms used in the examples 5.0.1 and 5.0.3.

Let $Y_\bullet$ be an **a+d-regular** stratification of the closed subset Y of the manifold M'. Consider a smooth map $\bar{\pi} : (M', Y) \to (M, X)$ such that

$$\bar{\pi} : Y \to X \quad \text{is a proper stratified submersion.}$$

Finally, let $j : U \to Y$ be the inclusion of an open union of strata. Then we know by section 4.2, that

$$\pi := \bar{\pi} \circ j : U \to X$$

induces $R\pi_* : D(U_\bullet) \to D(X_\bullet)$. Consider in addition

1. a smooth submanifold L of M transversal to $X_\bullet$, and
2. a smooth function $r : M \to \mathbb{R}$ such that δ is a regular value of $r|X \cap L$ with respect to the induced **a+d-regular** stratification of $X \cap L$.

In the examples, $X \cap L$ is the **upper/lower halflink** l_X^+, l_X^- or the **complex link** l_X, with r the distance function to the point x.

Note that $L' := \pi^{-1}(L)$ (or $L'' := \bar{\pi}^{-1}(L)$) is a smooth submanifold of M' transversal to $U_\bullet$ (or $Y_\bullet$) such that δ is a regular value of $r \circ \pi | U \cap L'$ with respect to the induced **a+d-regular** stratification of $U \cap L'$. Consider the cartesian diagram

$$\begin{array}{ccccc} X \cap L \cap \{r \leq \delta\} & \xrightarrow{k} & X \cap L & \xrightarrow{i} & X \\ \uparrow{\scriptstyle \pi''} & & \uparrow{\scriptstyle \pi'} & & \uparrow{\scriptstyle \pi} \\ U \cap L' \cap \{r \circ \pi \leq \delta\} & \xrightarrow{k'} & U \cap L' & \xrightarrow{i'} & U . \end{array}$$

Since a **d-regular** stratification has the property **LSA**, we get

$$(k^! i^* R\pi_* \mathcal{F})|X \cap L \cap \{r = \delta\} \simeq 0 \simeq (k'^! i'^* \mathcal{F})|U \cap L' \cap \{r \circ \pi = \delta\}$$

for $\mathcal{F} \in ob(D(U_\bullet))$. This implies

$$R\Gamma(X \cap L \cap \{r \le \delta\}, k^! i^* R\pi_* \mathcal{F}) \simeq R\Gamma(X \cap L \cap \{r \le \delta\}, X \cap L \cap \{r = \delta\}, R\pi_* \mathcal{F}),$$

and

$$R\Gamma(X \cap L \cap \{r \le \delta\}, R\pi''_* k'^! i'^* \mathcal{F}) \simeq R\Gamma(U \cap L' \cap \{r \circ \pi \le \delta\}, U \cap L' \cap \{r \circ \pi = \delta\}, \mathcal{F}).$$

Finally, these two "relative cohomology complexes" are then **isomorphic** by

$$k^! i^* R\pi_* \mathcal{F} \simeq k^! R\pi'_* i'^* \mathcal{F} \simeq R\pi''_* k'^! i'^* \mathcal{F}.$$

Here the second isomorphism follows from the general isomorphism $k^! R\pi'_* \simeq R\pi''_* k'^!$ [KS, prop. 3.1.9(ii)]. Similarly, the first isomorphism follows from the **proper base change theorem** for the map $\overline{\pi} : Y \to X$, together with proposition 4.3.1 and remark 4.3.6 applied to the cartesian diagram

$$\begin{array}{ccccc} X \cap L & \xleftarrow{\overline{\pi}'} & Y \cap L'' & \xleftarrow{j'} & U \cap L' \\ \downarrow{\scriptstyle i} & & \downarrow{\scriptstyle i''} & & \downarrow{\scriptstyle i'} \\ X & \xleftarrow{\overline{\pi}} & Y & \xleftarrow{j} & U : \end{array}$$

$$i^* R\pi_* \simeq i^* R\overline{\pi}_* Rj_* \simeq R\overline{\pi}'_* i''^* Rj_* \simeq R\overline{\pi}'_* Rj'_* i'^* \simeq R\pi'_* i'^*.$$

Note, that we suppress the dependence on the **normal slice** N in the notation $NMD(\cdot, f, x)$. Of course, the isomorphism class of $NMD(\cdot, f, x)$ doesn't depend on N by theorem 5.0.1(2), which we prove only later on in section 5.3 for **locally C-regular** stratifications. If one is willing to use deep results from the "microlocal sheaf theory" of Kashiwara-Schapira, then one can give a quick proof of theorem 5.0.1 for $\mathcal{F} \in ob(D^+(X_\bullet))$, with $X_\bullet$ a Whitney **a-regular** stratification satisfying the property **LSA**.

Let $k : X \to M$ be the closed inclusion. Then we have by corollary 4.0.3 the estimate (5.4):

$$\mu supp(Rk_* \mathcal{F}) \subset \bigcup_i T^*_{X^i} M \subset T^* M.$$

Choose a **nondegenerate** covector $\omega \in T^*_Y M$, with $Y := X^s$. Since the set of nondegenerate covectors is **open**, there is an open neighborhood of ω in $T^*_Y M$, which does not meet $cl(T^*_{X^j} M)$ $(j > s)$. Then we have

$$\mu supp((Rk_* \mathcal{F}) \subset T^*_Y M \quad \text{in a neighborhood of } \omega.$$

By using local coordinates, we can assume that M and $Y = X^s$ are C^∞-manifolds. By [KS2, prop. 6.2.2] or [KS, prop. 6.6.1(ii)] there exists a complex of R-modules $M^\bullet \in ob(D^+(R))$ such that

$$Rk_* \mathcal{F} \simeq M^\bullet_Y \quad \text{in } D^+(M, \omega). \tag{5.51}$$

Here $M_Y^\bullet$ is the trivial extension by zero of the constant complex on Y with stalk $M^\bullet$, and compare with loc.sit. for the definition of this **localization** $D^+(M, \omega)$ of $D^+(M)$. $M^\bullet$ is also called the **"local type of $Rk_*\mathcal{F}$ at ω"**, and by the definition of $D^+(M, \omega)$, its isomorphism class is "locally constant" on the set of nondegenerate covectors in T_Y^*M.

Let $x := \pi(\omega)$, with $\pi : T_Y^*M \to Y = X^s$ the projection. Take a **transversal slice** $i : N \hookrightarrow M$ with $N \cap X^s = \{x\}$. By **a-regularity** we can assume that N is transversal to all strata in an open neighborhood of x. Since we only work locally, we can assume that this neighborhood is M. Then N is **non-characteristic** with respect to $Rk_*\mathcal{F}$ in the language of Kashiwara-Schapira (compare section 4.3). Moreover, the natural map

$$^t i' : (T_{X^j}^* M)|N \to T^*N$$

is **injective** for all $j \geq s$ (as explained on page 255). We therefore can apply [KS, prop. 6.1.9.(ii)], which implies that the "local type of $i^*Rk_*\mathcal{F}$ at $^t i'(\omega)$" is **also isomorphic** to $M^\bullet$ (this argument is also used in the proof of [KS2, lem. 10.3.9]).If $\omega = df_x$, then $^t i'(\omega) = d(f \circ i)_x$. So this covector is **nondegenerate** with respect to the induced **a-regular** stratification of $X \cap N$, and we are reduced to a point stratum. By the definition of the "local type" one gets

$$M^\bullet \simeq \left(R\Gamma_{\{f|N\cap X \geq f(x)\}}(\mathcal{F}|N \cap X)\right)_x =: NMD(\mathcal{F}, df_x)\,. \tag{5.52}$$

This implies already the second part of theorem 5.0.1. Especially, the "normal Morse datum" $NMD(\mathcal{F}, f, x)$ is independent (up to isomorphism) of the choice of the normal slice N. Moreover, for the proof of the first part of theorem 5.0.1, we can assume (by the same argument as before):

$$Rk_*\mathcal{F} = M_Y^\bullet\,,$$

with $M^\bullet$ the "local type of $Rk_*\mathcal{F}$ at df_x". So it is enough to work with a **constant** sheaf complex on $Y = X^s$. This shows the "power" of the "micro-local theory"! But in this case, theorem 5.0.1(1) reduces to

$$\left(R\Gamma_{\{f\geq f(x)\}}(M_Y^\bullet)\right)_x \simeq \left(R\Gamma_{\{f|Y\geq f(x)\}}(M_Y^\bullet)\right)_x \simeq M^\bullet[-\tau]\,,$$

which is now an easy exercise based on the **Morse lemma** (compare [KS, p.311] and [KS2, p.124/125]).

Let us come back to the study of the properties of this "normal Morse datum".

Example 5.1.4 allows us to apply our Thom-Sebastiani-Theorem 1.2.2(1). In this way get the following generalization of the **multiplicativity** (5.6) of the "normal Morse data":

Proposition 5.1.1. *Let $X_\bullet$ and $X'_\bullet$ be* **a+d-regular** *stratifications. Assume that*

$$f : (M, x) \to (\mathbb{R}, f(x)) \quad \text{and} \quad f' : (M', x') \to (\mathbb{R}, f'(x'))$$

are two smooth function germs such that

$$df_x \in T^*_{X^i} M \quad \text{and} \quad df'_{x'} \in T^*_{X'^j} M'$$

are **nondegenerate**. *Then*

$$d(f \oplus f')_{(x,x')} = (df_x, df'_{x'}) \in (T^*_{X^i} M)_x \oplus (T^*_{X'^j} M')_{x'} = (T^*_{Y^{i,j}} M'')_{(x,x')}$$

is also **nondegenerate** *with respect to the* **product stratification** $(X \times X')_\bullet$, *with* $Y^{i,j} := X_i \times X'_j$, $M'' := M \times M'$ *and*

$$f \oplus f' : M \times M' \to \mathbb{R}\,;\; f \oplus f'(m, m') := f(m) + f'(m')\,.$$

Suppose X^k and X'^l are Whitney **b-regular** *over $\{x\}, \{x'\}$ for $k > i$ and $l > j$. Let N, N' be* **normal slices** *at x, x'. Then $N \times N'$ is a normal slice at (x, x'), with*

$$NMD(\mathcal{F}, f, x) \otimes^L NMD(\mathcal{F}', f', x') \simeq NMD(\mathcal{F} \boxtimes^L \mathcal{F}', f \oplus f', (x, x'))\,) \tag{5.53}$$

for $\mathcal{F} \in ob(D(X_\bullet))$ and $\mathcal{F}' \in ob(D(X'_\bullet))$.

Remark 5.1.7.

1. Here we use for $\boxtimes^L$ the same conventions as on page 277, and these "normal Morse data" are calculated with respect to N, N' and $N \times N'$.
2. As explained in section 4.3, the product stratification is also **a-regular**, but we do not know if it is also **d-regular**. Nevertheless, for the proof of the Thom-Sebastiani-Theorem 1.2.2 we only need that

   $$(f - f(x), \{x\}, \mathcal{F}) \quad \text{and} \quad (f' - f'(x'), \{x'\}, \mathcal{F}')$$

   satisfy a **cohomological version of a Milnor fibration**. This applies for example in good cases also to the "local Morse datum". If $X_\bullet$ and $X'_\bullet$ are **locally C-regular**, then also $(X \times X')_\bullet$ is **locally C-regular**, and these "normal Morse data" depend only on the corresponding nondegenerate covectors, but **not** on the choice of the normal slices.
3. The corresponding "multiplicativity" of the **"local type"** is explained in [KS, prop. 7.5.10(i)].

The following discussion explains the relation between **duality** and the "normal Morse data".

Let us relate the observed "duality" between f and $-f$ to the two distinguished triangles of corollary 5.1.2 and **Poincaré-Verdier duality**. Make the same assumptions as in corollary 5.1.2, and let the base ring R be commutative.

We use the same notation for the different inclusions of subsets used in the proof of corollary 5.1.2. Then the first distinguished triangle for $\mathcal{F} \in ob(D(X_\bullet))$ and the function $-f$ is by lemma 5.1.3 and lemma 1.1.1 isomorphic to the distinguished triangle

$$k_x^* i'^* Rj'_* j'^* \mathcal{F}[-1] \xrightarrow{[-1]} k_x^* i'^* R\Gamma_{\{-f \geq -f(x)\}}(\mathcal{F}) \longrightarrow k_x^* i'^* \mathcal{F} \xrightarrow{sp} .$$

Here we use $\{-f < -f(x)\} = \{f > f(x)\}$. Moreover, we have the following isomorphism of distinguished triangles:

$$\begin{array}{ccccccc} k_x^* i'^* Rj'_* j'^*[-1] & \xrightarrow{[-1]} & k_x^* i'^* R\Gamma_{\{-f \geq -f(x)\}}(\cdot) & \longrightarrow & k_x^* i'^* & \xrightarrow{sp} & \\ \downarrow \wr & & \downarrow & & \downarrow \wr & & \\ i_x^* i^* Rj_* j^* k^*[-1] & \longrightarrow & i_x^* i^* k^! & \longrightarrow & i_x^* i^* k^* & \longrightarrow & , \end{array}$$

where the lower triangle is induced from the **adjunction map** $id \to Rj_* j^*$. If we **dualize** this adjunction map, we get a commutative diagram

$$\begin{array}{ccc} D(Rj_* j^*) & \longrightarrow & D \\ \downarrow \wr & & \| \\ j_! j^* D & \longrightarrow & D\,. \end{array}$$

Here we use the general isomorphisms

$$DRj_* \simeq j_! D \quad \text{and} \quad Dj^* \simeq j^! D \simeq j^* D\,,$$

with $j^! \simeq j^*$, since j is an open inclusion. The map in the lower row is the **adjunction map** $j_! j^* \to id$, since its pullback under j^* is the identity morphism of $j^* D$.

Using the functorial isomorphism $Di^* \simeq i^! D$, we get a commutative diagram

$$\begin{array}{ccc} D(i^* Rj_* j^*[-1]) & \xrightarrow{D([-1])} & D(i^*[-1]) \\ \downarrow \wr & & \downarrow \wr \\ i^! j_! j^*[1] D & \xrightarrow{[1]} & i^![1] D\,, \end{array}$$

where the maps in the rows are induced from the distinguished triangles corresponding to the adjunction maps for j. But one has a **functorial isomorphism**

$$i^! j_! j^*[1] \simeq i^* Rj_* j^*$$

such that the induced map

$$[1] : i^* Rj_* j^* \to i^![1]$$

is the same as the one induced from the distinguished triangle corresponding to the adjunction map $id \to Rj_*j^*$ (see for example [BBD, p.46,47]). By the functorial isomorphisms

$$Di_x^* \simeq i_x^! D \quad \text{and} \quad Dk^* \simeq k^! D \,,$$

we finally get an **isomorphism** of distinguished triangles:

$$\begin{array}{ccccccc} D(k_x^* i^* k^*) & \longrightarrow & D(k_x^* i^* k^!) & \longrightarrow & D(k_x^* i^* Rj_* j^* k^*[-1]) & \xrightarrow{[1]} & \\ \downarrow\wr & & \downarrow & & \downarrow\wr & & \\ i_x^! i^! k^! D & \longrightarrow & i_x^! i^* k^! D & \longrightarrow & i_x^! i^* Rj_* j^* k^! D & \xrightarrow{[1]} & . \end{array}$$

Assume now $\mathcal{F} \in ob(D^-(X_\bullet))$ (at least locally) so that $D(\mathcal{F}) \in ob(D(X_\bullet))$ (as explained in section 4.2). By the proof of corollary 5.1.2, the last triangle is then isomorphic to the second distinguished triangle of corollary 5.1.2 for $D(\mathcal{F})$ and the function f. Especially

$$D(\,LMD(\mathcal{F}, -f, x)\,) \simeq LMD(D(\mathcal{F}), f, x) \quad \text{for } \mathcal{F} \in ob(D^-(X_\bullet)). \tag{5.54}$$

If one wants to conclude the similar relation (5.54) for the "normal Morse data" of $\mathcal{F}$, then one has to use that in this situation the duality D **commutes only up to a shift** with the restriction to a **normal slice** N.

Lemma 5.1.6. *Let $X_\bullet$ be an* **a-regular** *stratification satisfying the property* **LSA**. *Let $k : X \to M$ be the closed inclusion, and $i : N \to M$ the inclusion of a* **normal slice** *N at x. By* **a-regularity** *N is transversal to all strata of $X_\bullet$ in some open neighborhood U of x in M. We assume $U = M$, and that i is a closed inclusion (since we work locally near x). Then the natural morphism*

$$i^!(\mathbb{Z}_M) \otimes^L_Z i^*(Rk_*\mathcal{F}) \to i^!(Rk_*\mathcal{F}) \tag{5.55}$$

is an isomorphism for $\mathcal{F} \in ob(D(X_\bullet))$. Moreover, $i^!(\mathbb{Z}_M) \simeq \mathbb{Z}_N[-s]$ locally near x.

Proof. Here we consider

$$\otimes^L_Z : D(N, \mathbb{Z}) \times D(N, R) \to D(N, R) \,,$$

and the morphism (5.55) is defined in remark 4.2.3. Moreover, by its definition the morphism (5.55) commutes also with the forgetful functor $D(N, R) \to D(N, \mathbb{Z})$. So we can assume $R = \mathbb{Z}$. Since all functors used in (5.55) are of finite cohomological dimension, we are reduced to the case $\mathcal{F} \in ob(D^b(X_\bullet))$. Then N is **non-characteristic** with respect to $\mathcal{F}$ (this was already explained in the discussion of the "local type" on page 310). Then the first claim follows from [KS, cor.5.4.11(ii)]. The isomorphism $i^!(\mathbb{Z}_M) \simeq \mathbb{Z}_N[-s]$ follows from **Poincaré-Verdier duality** as in remark 4.2.3. □

Remark 5.1.8. The used result [KS, cor. 5.4.11(ii)] is a **non-trivial** application of the "micro-local sheaf theory" of Kashiwara-Schapira. We used it, since we can give in this way a purely sheaf theoretic proof of the above lemma. And we will use it in this chapter only for the following "duality result" of the "normal Morse data".

If one uses the more **geometric** assumption, that the stratification satisfies the "local stratified triviality property" **LST** (e.g. **b-, w- or C-regularity** by the "first isotopy lemma of Thom"), then one gets an isomorphism

$$i^*(Rk_*\mathcal{F})[-s] \simeq i^!(Rk_*\mathcal{F})$$

also from the **Vietoris-Begle theorem** [KS, prop. 2.7.8] and **Poincaré-Verdier duality** as in remark 4.2.3.

Corollary 5.1.3. *Make the same assumption as in lemma* 5.1.6, *and consider (locally at x) the closed inclusion $i' : X \cap N \to X$. Then one has for $\mathcal{F} \in ob(D^-(X_\bullet))$ an isomorphism*

$$D(i'^*\mathcal{F}) \simeq i'^*(D(\mathcal{F}))[-s]\,. \tag{5.56}$$

Proof. Consider the cartesian diagram

$$\begin{array}{ccc} X \cap N & \xrightarrow{k'} & N \\ \downarrow{\scriptstyle i'} & & \downarrow{\scriptstyle i} \\ X & \xrightarrow{k} & M\,. \end{array}$$

We always have the functorial isomorphisms $i^*Rk_* \simeq Rk'_*i'^*$, $Di^* \simeq i^!D$ and $DR\kappa_* \simeq R\kappa_*D$ for $\kappa = k, k'$ (since $R\kappa_* = R\kappa_!$ for a closed embedding). This implies

$$Rk'_*(Di'^*) \simeq D(Rk'_*i'^*) \simeq D(i^*Rk_*) \simeq i^!(DRk_*) \simeq i^!(Rk_*D)\,,$$

and

$$(i^*Rk_*D)[-s] \simeq (Rk'_*i'^*D)[-s]\,.$$

But $D(\mathcal{F})$ is $X_\bullet$-constructible for $\mathcal{F} \in ob(D^-(X_\bullet))$. By lemma 5.1.6 we get

$$Rk'_*(D(i'^*\mathcal{F})) \simeq Rk'_*i'^*(D(\mathcal{F}))[-s]\,.$$

Then the claim follows from $k'^*Rk'_* \simeq id$. □

Together with equation (5.54), this implies the following **duality for the "normal Morse data"**:

Corollary 5.1.4. *Let $X_\bullet$ be an* **a+d-regular** *stratification, with*

$$f : (M, x) \to (\mathbb{R}, f(x)) \quad \textit{for } x \in X^s$$

a smooth function germ such that the covector df_x is **nondegenerate**, *with* $0 = df_x|T_xX^s$. *Assume* X^j *is Whitney* **b-regular** *over* $\{x\}$ *for all* $j > s$. *Then one has for* $\mathcal{F} \in ob(D^-(X_\bullet))$ *an isomorphism*

$$D(\,NMD(\mathcal{F}, -f, x)\,) \simeq NMD(D(\mathcal{F}), f, x)[-s]\,. \tag{5.57}$$

5.1.3 Morse theory for a stratified space with corners

We finish this section with the counterparts of example 5.0.2 and theorem 5.0.2 for a **manifold or stratified space with corners**.

Let us start with the classical case of a **manifold with corners**. Since we work locally, we look at the following situation:

$$B := \{(x_1, \dots, x_n) \in \mathbb{R}^n \mid x_j \geq 0, j = 1, \dots, n\} \subset \mathbb{R}^n\,,$$

with the usual Whitney **b-regular** stratification so that B_{n-i} is the set of points in B, which have at least i of its coordinates equal to 0 ($0 \leq i \leq n$).

We are interested in the "normal Morse datum" at the **corner** $x = 0$. Since this is a point stratum $\{0\} = B^0$, this is also the same as the "local Morse datum". If we use the basis $\{dx_j|\, j = 1, \dots, n\}$ of $T^*_{\{0\}}\mathbb{R}^n$, then the set of **nondegenerate** covectors in $T^*_{\{0\}}\mathbb{R}^n$ is given by

$$\{\omega = \sum_{j=1}^{n} \lambda_j \cdot dx_j|\, \lambda_j \neq 0, \forall j = 1, \dots, n\}\,.$$

To include the case of **"Morse theory modulo part of the boundary"**, we make the following choice for the constructible sheaf $\mathcal{F}$ that we use:

For a fixed k with $0 \leq k \leq n$, let Y be the following subspace of X:

$$Y := B \cap \{x_j = 0, j = 1, \dots, k\}\,, \quad \text{with } Y := \emptyset \text{ for } k = 0.$$

Note that Y is a union of strata. Let $j : U := B\backslash Y \to X$ be the inclusion of the open complement. Then we take $\mathcal{F} := j_!\mathbb{Z}_U$, i.e.

$$R\Gamma(B, j_!\mathbb{Z}_U) \simeq R\Gamma(B, Y, \mathbb{Z})$$

calculates the relative cohomology of B modulo Y.

Let $f : (\mathbb{R}^n, 0) \to (\mathbb{R}, 0)$ be a smooth function germ with

$$\sum_{j=1}^{n} \lambda_j \cdot dx_j := df_0 =: \omega$$

nondegenerate. Then we would like to calculate

$$NMD(j_!\mathbb{Z}_U, \omega) \simeq \left(R\Gamma_{\{f \geq 0\}}(j_!\mathbb{Z}_U)\right)_0\,.$$

From the viewpoint of stratified Morse theory for constructible sheaves, this can be done in the following way. First we know, that this depends only on $\omega := df_0$. We can therefore assume $f = \omega$ is a **linear** function. Then we use the distinguished triangle (5.47) of corollary 5.1.2,

$$\mathcal{G} := R\Gamma(B \cap \{r \leq \delta, \omega = w\}, \mathcal{F})[-1] \to (R\Gamma_{\{\omega \geq 0\}}(\mathcal{F}))_0 \to \mathcal{F}_0 \overset{[1]}{\to},$$

with $w \in [-\epsilon, 0[$ for $0 < \epsilon << \delta << 1$. What makes the calculation so easy, is the fact that the spaces

$$B \cap \{r \leq \delta, \omega = w\} \quad \text{and} \quad Y \cap \{r \leq \delta, \omega = w\}$$

are **convex**. Therefore we get

$$\mathcal{G} \simeq \begin{cases} \mathbb{Z}[-1], & \text{for } B \cap \{\omega < 0\} \neq \emptyset \text{ and } Y \cap \{\omega < 0\} = \emptyset, \\ 0, & \text{otherwise.} \end{cases}$$

Similarly

$$\mathcal{F}_0 = \begin{cases} \mathbb{Z}, & \text{if } Y = \emptyset, \text{ i.e. } k = 0, \\ 0, & \text{if } Y \neq \emptyset, \text{ i.e. } k > 0. \end{cases}$$

Moreover, the map sp:

$$\mathcal{F}_0 \overset{\sim}{\longleftarrow} R\Gamma(B \cap \{r \leq \delta, |\omega| \leq \epsilon\}, \mathcal{F}) \longrightarrow R\Gamma(B \cap \{r \leq \delta, \omega = w\}, \mathcal{F})$$

is an **isomorphism** in the case $Y = \emptyset$ and $B \cap \{\omega < 0\} \neq \emptyset$. We therefore get the following result (compare also with [KS2, ex. 3.1.3(3)] or [KS, prop. 5.3.1]):

- We consider Morse theory on the **manifold with corners** B, i.e. $k = 0$ and $Y = \emptyset$. Then

$$NMD(j_!\mathbb{Z}_U, \omega) \simeq \begin{cases} \mathbb{Z}, & \forall i : \lambda_i > 0\,, \\ 0, & \text{otherwise.} \end{cases} \tag{5.58}$$

- We consider Morse theory on the **manifold with corners modulo the part** Y **of the boundary**, i.e. $k > 0$ and $Y \neq \emptyset$. Then

$$NMD(j_!\mathbb{Z}_U, \omega) \simeq \begin{cases} \mathbb{Z}[-1], & \exists j : \lambda_j < 0 \wedge \forall i > k : \lambda_i > 0\,, \\ 0, & \text{otherwise.} \end{cases} \tag{5.59}$$

Next we consider the following slightly more general case: Let

$$g = (g_1, \ldots, g_n) : M \to \mathbb{R}^n$$

be a **submersion** of manifolds. Then

$$B' := g^{-1}(B) = \{\forall i : g_i \geq 0\}$$

gets an induced **b-regular** stratification such that

$$Y' := g^{-1}(Y) = B' \cap \{g_i = 0 : \forall i \leq k\}$$

is a union of strata. Let $j' : U' := B'\backslash Y' \to B'$ be the inclusion of the open complement. Take a point x in the **corner stratum**

$$S := g^{-1}(\{0\}) = \{\forall i : g_i = 0\} .$$

If N is a **normal slice** at x, then

$$g|N : (N, x) \to (\mathbb{R}^n, 0)$$

is locally a **diffeomorphism**. Let $f' : (M, x) \to (\mathbb{R}, 0)$ be a smooth function germ. Then

$$\left(R\Gamma_{\{f'|N\cap X'\geq 0\}}((j'_!\mathbb{Z}_{U'})|N\cap X')\right)_x \simeq \left(R\Gamma_{\{f\geq 0\}}((j_!\mathbb{Z}_U))\right)_0 ,$$

if we define $f := f' \circ (g|N)^{-1}$ with the help of this diffeomorphism. Assume now $df'_x|T_xS = 0$ so that

$$df'_x \in [d(g_1)_x, \ldots, d(g_n)_x] \subset T^*_xM ,$$

with $[d(g_1)_x, \ldots, d(g_n)_x]$ the subvector space with the basis $d(g_1)_x, \ldots, d(g_n)_x$. The natural map $T^*_xM \to T^*_xN$ induces an isomorphism

$$[d(g_1)_x, \ldots, d(g_n)_x] \xrightarrow{\sim} T^*_xN \xleftarrow[\sim]{{}^t d(g|N)} T^*_{\{0\}}\mathbb{R}^n$$

so that

$$df'_x = \sum_{j=1}^{n} \lambda_j \cdot d(g_j)_x \quad \text{corresponds to} \quad \omega = \sum_{j=1}^{n} \lambda_j \cdot dx_j .$$

Since $g : M \to \mathbb{R}^n$ is a **submersion**, one also has the equivalence of:

1. $df'_x \in (T^*_SM)_x$ is nondegenerate.
2. $d(f'|N)_x \in T^*_xN$ is nondegenerate with respect to the induced stratification of $B' \cap N$.
3. $df'_x = \sum_{j=1}^n \lambda_j \cdot d(g_j)_x$, with $\lambda_j \neq 0$ for all j.

So our calculation of before gives us in this case the **"normal Morse datum"**:

$$NMD(j'_!\mathbb{Z}_{U'}, df'_x) \simeq NMD(j_!\mathbb{Z}_U, \omega) .$$

Of course, these results are well known and we only wanted to indicate, how to get them from the application of our abstract theory to this very easy example. But nevertheless, even this simple example will be useful in our application to "stratified spaces with corners", and in the proof of theorem 5.0.3.

Let now $X_\bullet$ be a **locally C-regular stratification** of the closed subset X of the smooth manifold M. Fix a point $x \in X^s$ and a C^k-map $g : (M, x) \to (\mathbb{R}^n, 0)$ such that $g|X^s$ is a **submersion** in x. Then $g : U \cap X^j \to \mathbb{R}^n$ is for all j a submersion in an open neighborhood U of x in M (by **a-regularity**), and we assume $U = M$ in the following (since we work locally). Then the **manifold with corners**

$$B := \{\forall i : g_i \geq 0\}$$

has a Whitney **b-regular stratification** as before. Moreover, the strata of X intersect the strata of B **transversally** in M so that

$$B \cap X = \{\forall i : g_i \geq 0\} \cap X$$

gets an induced **locally C-regular stratification**. We call the space $B \cap X$ with this induced stratification a **stratified space with corners**. As explained in section 4.3, this induced stratification is just the "transversal pullback" of the **product stratification** of $B \times X$ under the diagonal embedding $M \to M \times M$. Let $d : B \cap X \to B \times X$ be the induced diagonal map.

Note, that the point x lies in the "corner stratum" $S := \{g = 0\} \cap X^s$. Consider a smooth function germ $f : (M, x) \to (\mathbb{R}, 0)$, with $df_x | T_x S = 0$, and $df_x \in T^*_x M$ **nondegenerate** with respect to the induced stratification of $B \cap X$. This implies that also $d(f|X^s)_x \in T^*_x X^s$ is nondegenerate with respect to the induced stratification of the **manifold with corners** $B' := B \cap X^s \subset X^s$. Especially

$$d(f|X^s)_x = \sum_{i=1}^{n} \lambda_i \cdot d(g_i|X^s)_x \quad \text{for some } \lambda_i \in \mathbb{R}^* . \tag{5.60}$$

As before, fix a k with $0 \leq k \leq n$ and let $j : U := B \backslash Y \to B$ be the inclusion of the open complement of

$$Y := B \cap \{g_i = 0 : \forall i \leq k\}, \quad \text{with } Y := \emptyset \text{ for } k = 0.$$

Then one gets for $\mathcal{F} \in ob(D(X))$:

$$R\Gamma(\, B \cap X, d^*(j_! \mathbb{Z}_U \boxtimes^L \mathcal{F})\,) \simeq R\Gamma(B \cap X, Y \cap X, \mathcal{F}) .$$

We are interested in the "normal Morse datum" $NMD(\mathcal{G}, f, x)$ of f in x with respect to this sheaf complex

$$\mathcal{G} := d^*(j_! \mathbb{Z}_U \boxtimes^L \mathcal{F}) \simeq (j_! \mathbb{Z}_U \otimes^L \mathcal{F})|B \cap X \quad \text{for } \mathcal{F} \in ob(D(X_\bullet)) .$$

By proposition 5.1.1, the proof of theorem 5.0.2 applies without modification to this more general context, and we get the

Theorem 5.1.1. *Make in addition the following assumptions:*

1. *The set of* **nondegenerate** *covectors with respect to the stratification* $X_\bullet$ *is* **dense** *in the fiber* $(T^*_{X^s} M)_x$.

2. X^j *is* **b-regular** *over* $\{x\}$ *for all* $j > s$.

Then there exists a smooth function germ $h : (M, x) \to (\mathbb{R}, 0)$, with $dh_x | T_x X^s = 0$ and dh_x **nondegenerate** *with respect to $X_\bullet$ such that*

$$NMD(\mathcal{G}, f, x) \simeq NMD(j'_! \mathbb{Z}_{U'}, f|X^s, x) \otimes^L NMD(\mathcal{F}, h, x)\,, \qquad (5.61)$$

with $j' : U' := (B \backslash Y) \cap X^s \to B \cap X^s = B'$ the open inclusion. Moreover, one can choose this h so that

$$\sum_{i=1}^{n} \mu_i \cdot d(g_i)_x + dh_x \in (T^*_S M)_x$$

is arbitrarily close to df_x for suitable $\mu_i \in \mathbb{R}^*$, with $\mu_i / \lambda_i \in \mathbb{R}^+$ for all $i = 1, \ldots, n$. Here the λ_i are given by (5.60).

Remark 5.1.9. If we would like to use this theorem, we have to check the first assumption about the **nondegenerate** covectors (compare [GM, p. 45]).

1. If our stratification is definable with respect to a nice "geometric category", then this assumption is true for **generic** points in X^s (i.e. for points in an open dense definable subset of X^s). This follows from

$$dim(\partial(T^*_{X^j} M)) < dim(T^*_{X^j} M) = dim(M)\,.$$

   But it is **not** known, if this is true for all points in X^s, even if the stratification is Whitney **b-regular**. But nevertheless, we can choose a Whitney **b-regular refinement** of a given definable stratification such that the condition 1. is true for all points with respect to this new stratification (this follows in the standard way from the dimension property above. Compare [OTr]).

2. Suppose that the X^j are **b-regular** over X^s in x. Then we can apply [O1, cor. 3.1.2]: The set of **nondegenerate** covectors in $(T^*_{X^s} M)_x$ is **dense** in this fiber if and only if the stratification is b_{cod1}**-regular** in x.

3. If $X_\bullet$ is **subanalytic and w-regular**, then condition 1. is true for **all** points of X^s ([NaTr, thm. 3.14]). Since **w-regularity** is equivalent to μ**-regularity** ([Tr4]), this follows also from [KS2, cor. 8.2.24].

4. If $X_\bullet$ is a **complex analytic Whitney b-regular** stratification, then condition 1. is true for **all** points of X^s (compare with the beginning of the proof of $i) \Rightarrow ii)$ in [Te, p. 455, thm. 1.2]).

5.2 Characteristic cycles and index formulae

In this section we explain the missing details about the results on "characteristic cycles" stated in the introduction.

5.2.1 Index formulae and Euler obstruction

Proof of theorem 5.0.4. For the following argument compare also with the proof of [KS, prop. 5.4.20] and [Ka3, sec. 5].

1. By **approximation**, we can assume that f has only **stratified Morse critical points** with respect to $X_\bullet$ (since the corresponding intersection number doesn't change by a homotopy argument). The existence of such an approximation follows from the fact, that we work with a **definable** (Whitney b-regular) stratification so that the set of **degenerate** covectors in T^*M is a **closed definable** subset of dimension $< dim(M)$. Especially, it can be stratified with strata of dimension $< dim(M)$. Then the claim follows from [GM, thm. 2.2.3, p. 53], or from [O2, thm. 1, thm. 3] (after taking a proper real analytic embedding $M \hookrightarrow \mathbb{R}^N$. Compare also with [KS, prop. 8.3.27]).

2. By assumption, $\Lambda \cap df(M)$ is **compact** so that f has only **finitely** many stratified Morse critical points. Choose $b \in [a, d[$ with all these critical points contained in $\{a \leq f < b\}$. Then one gets by lemma 5.1.1 and example 5.1.1:
$$R\Gamma(X \cap \{f \leq b\}, \mathcal{F}) \simeq R\Gamma(X, \mathcal{F}) \,,$$
and
$$R\Gamma_c(X \cap \{f < b\}, \mathcal{F}) \simeq R\Gamma_c(X, \mathcal{F}) \,.$$

3. Since T is stable by extensions and μ is additive, we can by induction also assume, that these critical points are all contained in $\{f = e\}$ for some $e \in]a, b[$. By corollary 5.1.1 and lemma 5.1.2, it is enough to show that
$$LMD(\mathcal{F}, \pm f, x) := (R\Gamma_{\{\pm f \geq \pm e\}}(\mathcal{F}))_x \in ob(T)$$
for a stratified Morse critical point $x \in X^s$, with
$$\mu(LMD(\mathcal{F}, \pm f, x)) = \sharp_{\pm\, df_x}\,(\, CC(\chi_X(\mathcal{F})) \cap [\pm\, df(M)]\,)\,.$$
But this follows from theorem 5.0.1:
$$LMD(\mathcal{F}, \pm f, x) \simeq NMD(\mathcal{F}, \pm f, x)[-\tau] \,,$$
with τ the **Morse index** of $\pm f|X^s$ in x. Moreover, $NMD(\mathcal{F}, \pm f, x)$ depends by theorem 5.0.1(2) only on the covector $\pm\, df_x$. If we represent df_x by a **definable smooth** function germ, then we get as in the introduction
$$NMD(\mathcal{F}, \pm f, x) \in ob(T) \,.$$
Finally, by the definition of $CC(\chi_X(\mathcal{F}))$ we have
$$\begin{aligned}\mu(LMD(\mathcal{F}, \pm f, x)) &= (-1)^\tau \cdot \mu(NMD(\mathcal{F}, \pm f, x)) \\ &= \sharp_{\pm\, df_x}\,(\, CC(\chi_X(\mathcal{F})) \cap [\pm\, df(M)]\,)\,.\end{aligned}$$
□

The proof of the corresponding intersection formulae (5.28), (5.29) for a constructible function $\alpha \in CF(X_\bullet, G)$ is similar, or one can reduce it to the results for constructible sheaves, at least for $G = \mathbb{Z}$ (just take $\mu = \chi$ the usual Euler characteristic for $T := perf$ the perfect complexes over a base field $R = k$). Then the general case follows by linearity from

$$\sharp(\, CC(g \cdot 1_S) \cap [\pm\, df(M)]\,) = g \cdot \sharp(\, CC(1_S) \cap [\pm\, df(M)]\,)\,.$$

Remark 5.2.1. 1. If we work with a **fixed** $\mathcal{F} \in ob(D(X_\bullet))$, then the claim of theorem 5.0.4 is true under the weaker assumption

$$m - supp(\mathcal{F}) \cap \pm\, df(M) \quad \text{is compact.}$$

Here

$$m - supp(\mathcal{F}) := \bigcup_{NMD(\mathcal{F},\cdot)\neq 0} cl(\Lambda^\circ_{s,i})$$

is the union over the closures of all connected components $\Lambda^\circ_{s,i}$ of the set of **nondegenerate covectors**, whose corresponding "normal Morse datum" is non-trivial:

$$NMD(\mathcal{F}, df_x) \neq 0 \quad \text{for} \quad df_x \in \Lambda^\circ_{s,i}\,.$$

In this case, a suitable approximation maybe has **countably** many stratified Morse critical points, but only finitely many of them contribute a non-trivial "local Morse datum". If $\mathcal{F} \in ob(D^+(X_\bullet))$, then

$$m - supp(\mathcal{F}) = \mu supp(\mathcal{F})$$

by proposition 5.0.1, and we get the assumptions used in [Ka3, thm. 4.2, thm. 4.3] and in [KS, thm. 9.5.3, cor. 9.5.4].

2. The corresponding intersection formulae (5.28), (5.29) for a constructible function $\alpha \in CF(X_\bullet, G)$ are true under the assumptions

$$supp(CC(\alpha)) \cap \pm\, df(M) \quad \text{is compact.}$$

Here $supp(CC(\alpha))$ is the **support of the characteristic cycle** $CC(\alpha)$:

$$supp(CC(\alpha)) = \bigcup_{\chi(NMD(\alpha,\cdot))\neq 0} cl(\Lambda^\circ_{s,i})\,,$$

which in our case is the union over the closures of all connected components $\Lambda^\circ_{s,i}$ of the set of **nondegenerate covectors**, whose corresponding "normal Morse datum" has a **non-trivial Euler characteristic**:

$$\chi(NMD(\alpha, df_x)) \neq 0 \quad \text{for} \quad df_x \in \Lambda^\circ_{s,i}\,.$$

Especially,

$$supp(\, CC(\chi_X(\mathcal{F}))\,) \subset m - supp(\mathcal{F})\,,$$

and in general this is really a proper inclusion.

As an application of this intersection formula (5.28), let us explain the description (5.35) for the **dual Euler obstruction**.

Let X be a pure-dimensional closed complex analytic subset of the complex manifold M, with $Y := X_{reg}$ the regular part of X and $d := dim_C(X) = dim_C(Y)$. Consider the following commutative diagram:

$$\begin{array}{ccccc} V & \xleftarrow{q} & W := G \times_M T^*M & \xrightarrow{\tilde{\pi}} & T^*M \\ {\scriptstyle p''}\downarrow & & {\scriptstyle p'}\downarrow & & {\scriptstyle p}\downarrow \\ G & = & G & \xrightarrow{\pi} & M\,, \end{array}$$

with $G := G_d(TM)$ the Grassmann bundle of complex d-planes in TM. Then the projection map π is **proper**. Since the right square is cartesian, the same is true for $\tilde{\pi}$. Finally V is the total space of the **universal quotient bundle** of π^*T^*M.

Fix the point $x \in X$, and choose a local distance function r as in corollary 5.0.1. The section dr of the projection p induces by base change also a section

$$\widetilde{dr} \text{ of } p', \quad \text{and} \quad \overline{dr} := q \circ \widetilde{dr} \text{ of } p''\,.$$

Let finally s be the zero-section of V, and $N := N(X)$ be the closure in G of the image of Y under the **Gauss map** $z \mapsto (z, T_zY)$. This is by definition the **Nash blowup** of X. Then we also get a commutative subdiagram by the following subspaces:

$$\begin{array}{ccccc} s(N) & \xleftarrow{q} & N' := q^{-1}(s(N)) & \xrightarrow{\tilde{\pi}} & \Lambda(X) := cl(T^*_Y M) \\ {\scriptstyle p''}\downarrow & & {\scriptstyle p'}\downarrow & & {\scriptstyle p}\downarrow \\ N & = & N & \xrightarrow{\pi} & X\,. \end{array}$$

Note that $\tilde{\pi} : N' \to \Lambda(X)$ is surjective and generically one to one. Since we can choose a Whitney **b-regular** stratification of X containing the point stratum $\{x\}$, there exists an open neighborhood U of X in M with

$$\Lambda(X) \cap dr(U) = \{dr_x\}\,.$$

Since we work locally near x, we can assume $M = U$. Then also

$$N' \cap \widetilde{dr}(G) \quad \text{and} \quad s(N) \cap \overline{dr}(G)$$

are **compact**, with

$$s(N) \cap \overline{dr}(G) \subset s(N_x)\,, \quad N_x := N \cap \{\pi = x\}\,.$$

By the **projection formula**, we get the following equalities of intersection numbers (with $m = dim_C(M), g := dim_C(G)$):

$$\begin{array}{lll} & \sharp(\, [\Lambda(X)] \cup [dr(M)]\,) & ;\sharp : H^{2m}_{\Lambda(X)}(T^*M,\mathbb{Z}) \times H^{2m}_{dr(M)}(T^*M,\mathbb{Z}) \to \mathbb{Z} \\ = & \sharp(\, [N'] \cup [\widetilde{dr}(G)]\,) & ;\sharp : \quad H^{2g}_{N'}(W,\mathbb{Z}) \times H^{2m}_{\widetilde{dr}(G)}(W,\mathbb{Z}) \quad \to \mathbb{Z} \\ = & \sharp(\, [s(N)] \cup [\overline{dr}(G)]\,) & ;\sharp : \quad H^{2g}_{s(N)}(V,\mathbb{Z}) \times H^{2d}_{\overline{dr}(G)}(V,\mathbb{Z}) \quad \to \mathbb{Z} \\ = & \sharp(\, [N] \cap s'^*([\overline{dr}(G)])\,) & ;\sharp : \quad H^{BM}_{2d}(N,\mathbb{Z}) \times H^{2d}_{N_x}(N,\mathbb{Z}) \quad \to \mathbb{Z}\,, \end{array}$$

with $s' : (N, N_x) \to (V, \overline{dr}(G))$ the restriction of the zero-section. Note that $s'(N_x) \subset \overline{dr}(G)$, since $r|X$ has a stratified critical point in x. For the definition of the last intersection number we use the **cap-product**

$$H^{BM}_{2d}(N,\mathbb{Z}) \times H^{2d}_{N_x}(N,\mathbb{Z}) \xrightarrow{\ \cap\ } H_0(N_x,\mathbb{Z})\,.$$

We always use the **complex orientations** and **Poincaré duality** (between local cohomology and Borel-Moore homology) to define the above fundamental classes:

1. $[N] \in H^{BM}_{2d}(N,\mathbb{Z})$ is the fundamental class of the pure d-dimensional complex analytic set N.

2. $[s(N)] \in H^{2g}_{s(N)}(V,\mathbb{Z}) \simeq H^{BM}_{2d}(s(N),\mathbb{Z})$ corresponds under **Poincaré duality** to the proper push-forward $s_*([N])$.

3. $[N'] := q^*([s(N)]) \in H^{2g}_{N'}(W,\mathbb{Z})$ is just the cohomology pull-back. Under **Poincaré duality**, it corresponds to the fundamental class $[N'] \in H_{2m}(N',\mathbb{Z})$ of the pure m-dimensional complex analytic set N', since q is a holomorphic submersion.

4. Finally, $[\Lambda(X)] = \tilde{\pi}_*([N']) \in H_{2m}(\Lambda(X),\mathbb{Z})$, since $\tilde{\pi} : N' \to \Lambda(X)$ is surjective and generically one to one.

More or less by definition, the last intersection number is equal to

$$Eu^{\vee}_X(x) := (-1)^{dim_C(X)} \cdot Eu_X(x)$$

(compare for example with [CEP, p. 12, 13]). The sign $(-1)^{dim_C(X)}$ comes from the fact, that we use the complex orientation of the complex vector bundle V, whereas MacPherson uses in [MP] a corresponding real orientation (compare with remark 5.0.3).

Remark 5.2.2. The same argument as before also works in the real context of "geometric categories" for (co)homology with $\mathbb{Z}_2$-coefficients (in "half of the degrees" as used above), if we assume that $Y \subset M$ is a pure d-dimensional locally closed definable subset, with $X := cl(Y)$ such that the **Nash blowup** $N = N(X)$ of X has a **fundamental class**

$$[N] \in H^{BM}_d(N,\mathbb{Z}_2)\,.$$

By [OP, thm. 2] this is for example the case for X an **arc-symmetric semialgebraic** subset.

If one wants to use (co)homology with $\mathbb{Z}$-coefficients, with M oriented (as we always assumed), then one should use for $G := G_+$ the Grassmann bundle of **oriented** d-planes in TM, together with $N = N(X)$ the **oriented Nash blowup** of X as in [OP]. Then the submersion q has also a relative orientation, but this time $\tilde{\pi} : N' \to \Lambda(X)$ is surjective and generically two to one so that

$$2 \cdot [\Lambda(X)] := \tilde{\pi}_*([N']) \in H_m(\Lambda(X), \mathbb{Z}) .$$

Especially,

$$Eu^\vee(\tilde{\pi}_*([N'])) = 2 \cdot Eu^\vee([\Lambda(X)]) \in CF(X_\bullet, 2\mathbb{Z}) ,$$

if the **oriented Nash blowup** $N = N(X)$ of X has a **fundamental class**

$$[N] \in H_d^{BM}(N, \mathbb{Z}) .$$

In particular, the "local indices" introduced in [OP] by "obstruction-methods" are **constructible functions**.

By corollary 5.0.1, the **dual Euler obstruction** (or any one of the "local indices" of [OP]) is constructible with respect to **any** complex analytic (or definable) Whitney **b-regular** stratification $X_\bullet$ of X.

In fact, this is also true for **any** such stratification $X'_\bullet$ of $X \subset M$, which is **locally C-regular**.

Just look at a corresponding Whitney **b-regular** stratification $X_\bullet$ of X, which is a **refinement** of $X'_\bullet$, i.e. each stratum T of $X_\bullet$ is contained in a stratum S of $X'_\bullet$. Especially, one has an inclusion

$$i : CF(X'_\bullet, G) \hookrightarrow CF(X_\bullet, G) .$$

Then one also gets a closed inclusion

$$\Delta := \bigcup T^*_{X'^s} M \xrightarrow[k]{\subset} \bigcup T^*_{X^t} M =: \Lambda \subset T^*M$$

of the corresponding unions of conormal spaces to the strata of $X'_\bullet$ and $X_\bullet$ (both subsets are closed by **a-regularity**). Let Δ° and Λ° be the open subsets of **nondegenerate** covectors in Δ and Λ.

Consider a covector df_x in the connected component $\Lambda^\circ_{T,j}$ of the set of **nondegenerate** covectors in $T^*_T M$. Suppose that the stratum T of $X_\bullet$ is contained in the stratum S of $X'_\bullet$.

- If T is open in S (i.e. $dim(T) = dim(S)$), then df_x is also **nondegenerate** with respect to $X'_\bullet$. Especially

  $$\Lambda^\circ_{T,j} \subset \Delta^\circ_{S,i} \quad \text{for a suitable connected component}$$

  of the set of **nondegenerate** covectors in $T^*_S M$.

- Suppose T is not open in S. Then x is **not** a critical point of $f|S$.

So one gets the inclusions

$$\Delta^\circ \xleftarrow[l]{\supset} \Lambda^\circ \cap \Delta \xrightarrow[\kappa]{\subset} \Lambda^\circ \,,$$

with l open and κ closed. Moreover, we have a commutative diagram

$$\begin{array}{ccc} L(M, X_\bullet, G) := H_m^{BM}(\Lambda, G) & \xrightarrow{\subset} & H_m^{BM}(\Lambda^\circ, G) = \prod H_m^{BM}(\Lambda^\circ_{T,j}, G) \\ {\scriptstyle k_*}\uparrow & & \uparrow{\scriptstyle \kappa_* l^*} \\ L(M, X'_\bullet, G) := H_m^{BM}(\Delta, G) & \xrightarrow{\subset} & H_m^{BM}(\Delta^\circ, G) = \prod H_m^{BM}(\Delta^\circ_{S,i}, G) \,, \end{array} \tag{5.62}$$

with $\kappa_* l^*$ the linear injection induced by

$$[\Delta^\circ_{S,i}] \mapsto \begin{cases} [\Lambda^\circ_{T,j}], & \text{if } \Lambda^\circ_{T,j} \subset \Delta^\circ_{S,i} \,, \\ 0 & \text{otherwise.} \end{cases}$$

Remark 5.2.3. In particular, k_* is also an injection. Moreover, $[C] \in H_m^{BM}(\Lambda^\circ, G)$ belongs to the subgroup $H_m^{BM}(\Delta, G)$ if and only if it satisfies the following conditions:

1. $[C] \in H_m^{BM}(\Lambda, G)$.
2. The multiplicity of $[C]$ in $H_m^{BM}(\Lambda^\circ_{T,j}, G)$ is 0, if T is not open in a stratum S of $X'_\bullet$.
3. The multiplicities of $[C]$ in $H_m^{BM}(\Lambda^\circ_{T,j}, G)$ and $H_m^{BM}(\Lambda^\circ_{T',j'}, G)$ are **equal**,

$$\text{if } \Lambda^\circ_{T,j}, \Lambda^\circ_{T',j'} \quad \text{belong to the } \textbf{same} \text{ connected component } \Delta^\circ_{S,i} \,,$$

   with T and T' open in the stratum S of $X'_\bullet$.

Since X'^j is for all $j > s$ **b-regular over a generic point** of X'^s, we can choose $X_\bullet$ so that T open in S implies X'^j is **b-regular** over X'^s for all $x \in T$ and $j > s$ (compare [vDrMi, Ver, Te]).

Under this assumption we get

$$CC(\alpha) \in H_m^{BM}(\Delta, G) \quad \text{for} \quad \alpha \in CF(X'_\bullet, G) \,,$$

with CC the **characteristic cycle map** for the **b-regular** stratification $X_\bullet$. Let T be a stratum of $X_\bullet$, which is contained in the stratum S of $X'_\bullet$:

- If T is open in S, then the corresponding "normal Morse data" and their Euler characteristics are the same for both stratifications. This implies the condition 3. above by theorem 5.0.1, which we prove in the next section for a **locally C-regular** stratification over points in such an open subset T (compare corollary 5.3.3).

- Suppose T is not open in S. Consider a covector df_x (with $x \in T$), which is **nondegenerate** with respect to $X_\bullet$. Then x is **not** a critical point of $f|S$. This implies the condition 2. above by the property **LSA** of $X'_\bullet$, because then the corresponding "local Morse data" vanish. Since we can assume that x is a classical Morse critical point of $f|T$, we also get that the "normal Morse data" and their Euler characteristics vanish (by theorem 5.0.1).

So we can apply corollary 5.0.1 also to the stratification $X'_\bullet$ by the following commutative diagram:

$$\begin{array}{ccc} CF(X'_\bullet, G) & \xrightarrow[i]{\subset} & CF(X_\bullet, G) \\ {\scriptstyle CC}\downarrow{\scriptstyle \wr} & & {\scriptstyle CC}\downarrow{\scriptstyle \wr} \\ L(M, X'_\bullet, G) & \xrightarrow[k_*]{\subset} & L(M, X_\bullet, G)\,. \end{array} \tag{5.63}$$

Remark 5.2.4. The condition 3. of remark 5.2.3 is automatically true for **complex analytic** stratifications. So we get in the complex analytic context such a commutative diagram also for a **complex analytic a-regular** stratification $X'_\bullet$ satisfying the property **LSA**.

If one uses the "micro-local theory" of Kashiwara-Schapira [KS], then one gets the same conclusion also for **definable** stratifications in the context of "geometric categories".

The above calculation of the **dual Euler obstruction** (or of the "local indices" of [OP]) is one of the few places, where one has an explicit relation between the "micro-local approach" and methods coming from "obstruction theory". Moreover, it applies also in the corresponding global situation.

Here we consider the same complex analytic context as before, but this time $r : M \to \mathbb{R}$ is a globally defined **subanalytic smooth** function, with $f|X$ **proper** and r_0 a **regular value** of r with respect to the **complex analytic b-regular stratification** $X_\bullet$ of X. Then

$$\Lambda(X) \cap dr(\{r < r_0\}) \subset T^*M|_{\{r<r_0\}}$$

is also **compact** and one gets in the same way

$$\begin{aligned} \sharp(\,[\Lambda(X)] \cup [dr(\{r < r_0\})]\,) &= \sharp(\,[N^\circ] \cap s'^*([\overline{dr}(G^\circ)])\,) \\ &=: (-1)^{dim_{\mathbb{C}}(X)} \cdot Obs(X, \{r < r_0\})\,, \end{aligned} \tag{5.64}$$

with $Obs(X, \{r < r_0\})$ the "global obstruction" introduced in [Du2, sec.2B],

$$N^\circ := N \cap \pi^{-1}(X \cap \{r < r_0\}) \quad \text{and} \quad G^\circ := G \cap \pi^{-1}(\{r < r_0\})\,.$$

For the definition of the second intersection number we use the **cap-product**

$$H_{2d}^{BM}(N^\circ, \mathbb{Z}) \times H_c^{2d}(N^\circ, \mathbb{Z}) \xrightarrow{\;\cap\;} H_0(N^\circ, \mathbb{Z})\,.$$

But as we have seen before

$$CC(Eu_X^\vee) = [\Lambda(X)] \,.$$

So the **intersection formula** (5.28) implies (compare [Du2, thm.1(i)]):

$$\begin{aligned} Obs(X, \{r < r_0\}) &= \int 1_{\{r \leq r_0\}} \cdot Eu_X \, d\chi \\ = \int 1_{\{r < r_0\}} \cdot Eu_X \, d\chi &= \sum_S \chi(S \cap \{r < r_0\}) \cdot Eu_X(S) \,. \end{aligned} \tag{5.65}$$

Here the sum is over all strata S of $X_\bullet$, with

$$\chi(S \cap \{r < r_0\}) := \chi(H_c^*(S \cap \{r < r_0\}, \mathbb{Q})) = \chi(H^*(S \cap \{r < r_0\}, \mathbb{Q}))$$

by **Poincaré duality**. The equality of the two Euler characteristic integrals follows from lemma 5.0.3 (applied to $X \cap \{r = r_0\}$).

5.2.2 A specialization argument

The final application of the **intersection formula** (5.28) is the proof of theorem 5.0.3.

Fix a "geometric category" and a corresponding real analytic (Nash) manifold M, which we assume to be oriented of pure dimension m. Let $f : M \to \mathbb{R}^{\geq 0}$ be a nonnegative **definable** C^2-function. Assume $X_0 := \{f = 0\}$ is compact and (everywhere) of smaller dimension than M. By the **curve selection lemma**, there exists $0 < \epsilon << 1$ such that f is a submersion in $\{0 < f < \epsilon\}$, if we work on a suitable open neighborhood U of X_0. Especially, $X_t := \{f \leq t\} \subset U =: M$ is for $t \in I :=]0, \epsilon[$ a **manifold with boundary** $\partial X_t := \{f = t\}$. In particular,

$$CC(1_{\{f \leq t\}}) \quad \text{is a cycle.}$$

Since it is a local question, this follows by the multiplicativity (5.24) of the characteristic cycle from the special and obvious case $CC(1_{[a,b]})$ for $[a, b] \subset \mathbb{R}$.

From this we would like to deduce by a **specialization argument**, that also $CC(1_{X_0})$ is a cycle. By remark 5.2.3, it is enough to show this for one fixed **definable b-regular** stratification $X_\bullet$ of X_0 (compare with the "proof" of (5.63)). Moreover, by [Ta2] we can then also assume, that $V := \{0 < f < \epsilon\}$ satisfies the $\mathbf{a}_f$**-condition** with respect to $X_\bullet$:

- Let $x_i \in V$ be a sequence converging to x in the stratum X^s in X_0 such that

  $$kern(df(x_i)|T_{x_i}V) = T_{x_i}(V \cap \{f = f(x_i)\})$$

  converges to some limiting plane τ, then $T_x X^s \subset \tau$.

Then we use the following cartesian diagram, with $J := [0, \epsilon[$:

$$\begin{array}{ccc} J \times \mathbb{R}^* \times T^*M & \overset{\supset}{\longleftarrow} & I \times \mathbb{R}^* \times T^*M = T^*(I \times M) \\ \downarrow \pi & & \downarrow \pi' \\ J \times T^*M & \overset{\supset}{\longleftarrow} & I \times T^*M\,, \end{array} \tag{5.66}$$

with π, π' the canonical projections. Moreover, the projection $p : I \times M \to I$ is a stratified submersion with respect to the **manifold with boundary**

$$(B; \partial B) := (\{(t,x) \,|\, f(x) \leq t\}, \{(t,x) \,|\, f(x) = t\}) \subset I \times M\,.$$

Especially, the intersection of $I \times \mathbb{R}^* \times T^*_M M$ and

$$\Lambda_B := T^*_{B \backslash \partial B}(I \times M) \cup T^*_{\partial B}(I \times M)$$

is contained in the zero-section of $T^*(I \times M)$. Therefore

$$\pi' : \Lambda_B \to I \times T^*M$$

is **proper** (compare with the proof of lemma 4.3.1 on page 255). Then

$$\Delta' := \pi'(\Lambda_B) \subset I \times T^*M$$

is a closed definable subset. Let also

$$\Delta := cl(\Delta') \subset J \times T^*M\,.$$

If we use the embedding

$$s_t : T^*M \simeq \{t\} \times T^*M \hookrightarrow J \times T^*M \qquad (\, t \in J\,)\,,$$

then $\Delta_t := s_t^{-1}(\Delta)$ is for $t \in I$ just the union of conormal spaces to the strata of the manifold with boundary $(\{f \leq t\}, \{f = t\}) \subset M$. For $t = 0$ one gets by the a_f**-condition** at least the estimate

$$\Delta_0 \subset \Lambda := \bigcup T^*_{X^s} M \subset T^*M\,.$$

Now we are ready to translate these operations into the language of **homology**:

1. Let $[C'] := \pi'_*(\,CC(1_B)\,) \in H^{BM}_{m+1}(\Delta', \mathbb{Z})$.
2. Let $[C'_t] := s_t^*([C']) \in H^{BM}_m(\Delta_t, \mathbb{Z})$ for $t \in I$, with s_t^* defined by **Poincaré duality**. Here we take for $I \times T^*M$ the product-orientation (in this order), with $I \subset \mathbb{R}$ oriented in the usual way. Then one gets for $t \in I$:

$$[C'_t] = CC(1_{\{f \leq t\}})\,.$$

This follows from the diagram (4.15) applied to $s := s_t$ $(t \in I)$, and the cartesian diagram:

$$\begin{array}{ccc} T^*(I \times M) & \xleftarrow{\;s_\pi\;} & s^*T^*(I \times M) \\ \downarrow{\scriptstyle \pi'} & & \downarrow{\scriptstyle {}^t s'} \\ I \times T^*M & \xleftarrow{\;s\;} & T^*M\,, \end{array}$$

with ${}^t s'$ the dual of the differential of s. Since s is **transversal** to the **b-regular** stratification $\partial B \subset B$ (which is also b_{cod1}-regular), one gets

$$\begin{aligned} s^*\pi'_*(CC(1_B)) &\simeq ({}^t s')_* s_\pi^*(CC(1_B)) \\ &\simeq CC(s^*(1_B)) = CC(1_{\{f \le t\}})\,. \end{aligned}$$

Of course, in our simple case this can also be shown directly.

3. Next we use **Poincaré duality** and the isomorphism

$$H^m_\Delta(J \times T^*M, \mathbb{Z}) \simeq H^m_{\Delta'}(I \times T^*M, \mathbb{Z})\,,$$

to define $[C] \in H^m_\Delta(J \times T^*M, \mathbb{Z})$ as the image of $[C']$ under this isomorphism. This isomorphism comes from the cartesian diagram of inclusions:

$$\begin{array}{ccc} \Delta' & \xrightarrow{\;j'\;} & \Delta \\ {\scriptstyle i'}\downarrow & & \downarrow{\scriptstyle i} \\ I \times T^*M & \xrightarrow[\;j\;]{} & J \times T^*M\,, \end{array}$$

together with the general **base change isomorphism**

$$Rj'_* i'^! \mathbb{Z}_{I \times T^*M} \simeq i^! Rj_* \mathbb{Z}_{I \times T^*M}$$

and $Rj_*\mathbb{Z}_{I \times T^*M} \simeq \mathbb{Z}_{J \times T^*M}$ (because $J \times T^*M$ is a manifold with boundary).

4. Finally, the specialization at 0 is defined as

$$[C_0] := s_0^*([C]) \in H^m_{\Delta_0}(T^*M, \mathbb{Z}) \simeq H^{BM}_m(\Delta_0, \mathbb{Z})\,.$$

We also use the same symbol $[C_0]$ for its image in $H^{BM}_m(\Lambda, \mathbb{Z})$.

The above construction can be summarized as the **specialization**

$$[C_0] := \lim_{t \to 0^+} CC(1_{\{f \le t\}})$$

in the sense of [SchVi, sec.3]. And we claim

$$CC(1_{X_0}) = [C_0] := \lim_{t \to 0^+} CC(1_{\{f \le t\}})\,. \tag{5.67}$$

It is enough to show, that $[C_0]$ has the right multiplicity at any **nondegenerate** covector $dl_x \in T^*_{X^s}M$, with $l : (M,x) \to (\mathbb{R},0)$ smooth function germ such that $l|X^s$ has in x a **classical Morse critical point of Morse index** τ. By theorem 5.0.1 we get

$$\sharp_{dl_x}(\,CC(1_{X_0}) \cup [dl(M)]\,) = (-1)^\tau \cdot \chi(NMD(1_{X_0}, dl_x)) = \chi(LMD(1_{X_0}, l, x))\,.$$

Let r be the distance function to x in some local definable coordinates.

Then one gets for $0 < e << \delta << 1$ and $U := T^*M|\{|l| < e, r < \delta\}$:

$$\begin{aligned}
\sharp_{dl_x}(\,[C_0] \cup [dl(M)]\,) &\overset{(1)}{=} \sharp_U(\,[C_0] \cup [dl(M)]\,)\\
&\overset{(2)}{=} \lim_{t\to 0^+} \sharp_U(\,[C'_t] \cup [dl(M)]\,)\\
&\overset{(3)}{=} \lim_{t\to 0^+} \sharp(\,CC(1_{\{-e<l\leq e,\, r\leq\delta,\, f\leq t\}}) \cup [dl(M)]\,)\\
&\overset{(4)}{=} \lim_{t\to 0^+} \int 1_{\{-e<l\leq e,\, r\leq\delta,\, f\leq t\}}\, d\chi\\
&\overset{(5)}{=} \int 1_{\{-e<l\leq e,\, r\leq\delta,\, f=0\}}\, d\chi\\
&\overset{(6)}{=} \chi(LMD(1_{X_0}, l, x))\,.
\end{aligned}$$

Let us now explain these equalities:

1. By example 5.1.4 and remark 5.1.5 we can choose e, δ, with $0 < e << \delta << 1$ such that

   (a) r has with respect to X_0 no stratified critical point in $\{r = \delta,\ |l| \leq e\}$.

   (b) x is the only stratified critical point of l in $\{|l| \leq e\}$ with respect to the induced **b-regular** stratification of $X_0 \cap \{r \leq \delta\}$.

   Especially, dl_x is the only intersection point in U of the supports of the used cycles. Moreover

   (c) l has with respect to X_0 no stratified critical point in $\{|l| = e,\ r \leq \delta\}$, and similarly for (r, l) in $\{r = \delta,\ |l| \leq e\}$.

2. By the a_f**-condition** one gets the property 1.(c) also for X_t instead of X_0 for all $0 < t << 1$ small enough. By taking ϵ smaller, we can assume that this is the case for all $t \in I$. Especially,

   $$\Delta_t \cap dl(\{|l| < e\}) \cap U \quad \text{is } \textbf{compact} \text{ for all } t \in J.$$

   One also gets a map

   $$H^m_{J\times dl(\{|l|<e\})}(J \times U, \mathbb{Z}) \times H^m_{\Delta\cap J\times U}(J \times U, \mathbb{Z}) \xrightarrow{Rq_!\circ\cup} H^0(J, \mathbb{Z})\,,$$

whose pullback under s_t gives the corresponding intersection number of the "fibers" of q. Here $q : J \times U \to J$ is the projection, and $Rq_!$ is induced by the adjunction map

$$Rq_!(\mathbb{Z}_{J\times U}[2m]) \simeq Rq_! q^! \mathbb{Z}_J \to \mathbb{Z}_J \, .$$

3. Of course we have for $t \in I$:

$$CC(1_{\{-e<l\leq e,\, r\leq\delta,\, f\leq t\}})\,|U = CC(1_{\{f\leq t\}})\,|U = [C'_t]\,|U \, .$$

Moreover,

$$\{|l| \leq e,\, r \leq \delta,\, f \leq t\}) \subset M$$

is by 1. and 2. a **compact manifold with corners**. Especially, the corresponding characteristic cycle is indeed a cycle. This follows by linearity from the same argument as before applied to

$$1_{\{|l|\leq e,\, r\leq\delta,\, f\leq t\}} \quad \text{and} \quad 1_{\{-e=l,\, r\leq\delta,\, f\leq t\}} \, .$$

So it is enough to show that $dl(M)$ does **not intersect the support** of

$$CC(1_{\{-e<l\leq e,\, r\leq\delta,\, f\leq t\}}) \quad \text{over} \quad \{|l| \leq e,\, r = \delta\} \cup \{|l| = e,\, r \leq \delta\} \, .$$

Here we use the description of remark 5.2.1(2) for this **support**, i.e. we show that $dl(M)$ does not intersect the closure of any component of the set of nondegenerate covectors in $T^*_S M$, whose "normal index" is non-trivial

$$\chi(\,NMD(1_{\{-e<l\leq e,\, r\leq\delta,\, f\leq t\}}, \cdot)\,) \neq 0 \, .$$

- Over $\{|l| \leq e,\, r = \delta\}$ this follows from 1.(c) and 2., i.e. l has no stratified critical points on the "boundary" $X_t \cap \{r = \delta\}$.
- Over $\{-e = l,\, r \leq \delta\}$ this follows for a stratum S in this "boundary" from (5.59), i.e. we are there in the case $k = 1$ with x_1 corresponding to $g + e$ so that $\lambda_1 = 1 > 0$. For a stratum S in $\{|l| < e,\, r \leq \delta\}$ this follows from 1.(c) and 2.
- Over $\{e = l,\, r \leq \delta\}$ this follows for a stratum S in this "boundary" from (5.58), i.e. we are there in the case $k = 0$ with x_1 corresponding to $e - g$ so that $\lambda_1 = -1 < 0$. For a stratum S in $\{|l| < e,\, r \leq \delta\}$ this follows from 1.(c) and 2.

4. This is just a special case of the **intersection formula** (5.28).

5. $f : \{|l| \leq e,\, r \leq \delta\} \to \mathbb{R}^{\geq 0}$ is a **proper definable** function. Therefore

$$f_*(1_{\{|l|\leq e,\, r\leq\delta\}}) \quad \text{and} \quad f_*(1_{\{-e=l,\, r\leq\delta\}})$$

are **constructible functions** on $[0, \infty[$. The "cell-decomposition" $\{0,t\}$, $]0,t[$ of $[0,t]$, implies then for $0 < t << 1$ small enough:

$$\int f_*(1_{\{|l|\leq e,\, r\leq\delta\}}) \cdot 1_{[0,t]}\, d\chi = \int f_*(1_{\{|l|\leq e,\, r\leq\delta\}}) \cdot 1_{\{0\}}\, d\chi\,,$$

and

$$\int f_*(1_{-e=l,\, r\leq\delta\}}) \cdot 1_{[0,t]}\, d\chi = \int f_*(1_{\{-e=l,\, r\leq\delta\}}) \cdot 1_{\{0\}}\, d\chi\,.$$

This implies the claim by linearity, the projection formula and functoriality, i.e. $\int \circ f_* = \int$ (compare with section 2.3 for details).

6. The last equality follows from example 5.1.4 and (5.47) in corollary 5.1.2. In terms of constructible functions, this is the counterpart of (5.16), but this time for the Euler characteristic $\chi(LMD(1_{X_0}, l, x))$ of the "local Morse datum".

Remark 5.2.5. 1. The above argument also works for a **continuous** definable function f, if there exists an $\epsilon > 0$ such that $f : M \to \mathbb{R}$ is a **smooth submersion** in $\{0 < f < \epsilon\}$.

2. A similar specialization argument can be used to show, that the characteristic cycle map CC "commutes" with the **specialization of constructible functions**. Here one has to calculate for a "specialization cycle" the corresponding **local intersection number** as in corollary 5.0.1. Again, this is the translation into the "language of homology" of some arguments of Fu as in [FuMC, thm. 3.7] for the real context, and in [Fu2] for the complex context (compare also with [Gi1, thm. 5.5], [Sab, thm. 4.3] and [Ken2]).

3. The proof of the equality (3) above is similarly to the proof of [KS, thm. 9.5.3] and [SchVi, lem. 4.19] (but the proof given in [SchVi] is not correct).

4. Theorem 5.0.3 implies in the context of "geometric categories" the important result (5.22) of Kashiwara and Fu, that the "characteristic cycle" is always a **cycle**. As in the introduction, it is enough to consider the case $G = \mathbb{Z}$ of integer valued constructible functions. Since it is a local result, it is enough to show this for $CC(1_Z)$, with $Z \subset M$ a **compact definable** subset (by linearity). We can also assume that Z is everywhere of smaller dimension than M. Otherwise we take a product with $(\mathbb{R}, \{0\})$ and use the multiplicativity (5.24) of the characteristic cycle, i.e. the isomorphism

$$CC(1_Z \times 1_{\{0\}}) = CC(1_Z) \times [\{0\} \times \mathbb{R}^*]\,.$$

Finally we use the result of [vDrMi, 1.20, D.19, thm. C.11], that Z can be represented as the zero-set $Z = \{f = 0\}$ of a non-negative **definable** C^2-function (By 1. it would also be enough to use the corresponding weaker result of [BrKu, prop. 7.1]).

5.3 Stratified Morse theory, part II

In this section we prove the deepest results of our **Morse theory for constructible sheaves**, i.e. we give the proof of theorem 5.0.1 in a more general context.

5.3.1 Normal Morse data are independent of choices

Let $X_\bullet$ be a smooth stratification of the closed subset X of the manifold M, and fix a point $z \in X^s$. We choose local coordinates

$$(M, X^s, z) \simeq (\mathbb{R}^m, \mathbb{R}^s \times \{0\}, 0) ,$$

and consider (locally) X as a **closed** subset of $U := \{r < r_0\} \subset \mathbb{R}^m$ for $0 < r_0 << 1$, with $X^s = U \cap \mathbb{R}^s \times \{0\}$ and r the distance function:

$$r : \mathbb{R}^m \to \mathbb{R}^{\geq 0}, \quad r(x_1, \dots, x_m) := \sum_{j=1}^{m} x_j^2 .$$

We first prove a basic lemma (compare [GM, p. 82, lem. 6.2, p. 84, lem. 6.5]). For other applications, we prove directly a parametrized version.

Lemma 5.3.1. *Let P be a smooth manifold (the "parameter space"), with $P_c \subset P$ a* **compact** *subset. Consider a smooth map*

$$(\pi, F) : P \times U \to X^s \times \mathbb{K} , \quad \textit{with } \mathbb{K} = \mathbb{R}, \mathbb{C}.$$

Let us denote its components by (π_t, f_t) $(t \in P)$:

$$\pi_t : U \to X^s ;\ x \mapsto \pi(t, x) \quad \textit{and} \quad f_t : U \to \mathbb{K} ;\ x \mapsto F(t, x) .$$

We make the following assumptions for $t \in P_c$:

(cp) $d(f_t)_z | T_z X^s = 0$, i.e. z is a **critical point** *of $f_t | X^s$.*

(nd) $d(f_t)_z$ is **nondegenerate**.

(rt) $\pi_t | X^s = id$, i.e. π_t is a **retraction** *onto X^s.*

Assume X^j is Whitney **a-regular** *over X^s in z for all $j > s$.*

1. *There exists an $0 < r_1 \leq r_0$ such that for* **all** *$t \in P_c$:*

$$(\pi_t, f_t - f_t \circ \pi_t) : (U, x) \to X^s \times \mathbb{K} \quad \textit{is in} \quad \{\pi_t = z, r < r_1\} \backslash X^s$$

a **stratified submersion** *(with respect to $X_\bullet$).*

2. *Assume* X^j *is also Whitney* **b-regular** *over* $\{z\}$ *for all* $j > s$. *Then there exists an* $0 < r_2 \leq r_0$ *such that for* **all** $t \in P_c$:

$$(2.a) \qquad (\pi_t, r) : (U, x) \to X^s \times \mathbb{R} \quad \textit{is in} \quad \{\pi_t = z, r < r_2\} \backslash X^s$$

a **stratified submersion**, *and*

$$(2.b) \qquad h_t := (\pi_t, r, f_t - f_t \circ \pi_t) : (U, x) \to X^s \times \mathbb{R} \times \mathbb{K}$$

is a **stratified submersion** *in* $\{\pi_t = z, r < r_2, f_t - f_t \circ \pi_t = 0\} \backslash X^s$.

3. *Assume* $X_\bullet$ *is also* **a-regular** *in* $X \backslash X_s$, *and* X^j *is* **b-regular** *over* $\{z\}$ *for all* $j > s$. *Fix a* δ *with* $0 < \delta < min(r_1, r_2)$. *Then there exists an* $\epsilon' > 0$ *such that for* **all** $t \in P_c$:

$$h_t := (\pi_t, r, f_t - f_t \circ \pi_t) : (U, x) \to X^s \times \mathbb{R} \times \mathbb{K}$$

is a **stratified submersion** *in* $\{\pi_t = z, r = \delta, |f_t - f_t \circ \pi_t| < \epsilon'\}$.

Proof. We prove the results indirectly.

1. Otherwise there exists a $j > s$ and a sequence $(x_n, t_n) \in X^j \times P_c$ such that

$$rg\big((d(\pi_{t_n})_{x_n}, d(f_{t_n})_{x_n})|T_{x_n}X^j\big) < s + dim_R(\mathbb{K}) ,$$

with

$$x_n \to z, \quad t_n \to t \in P_c, \quad \text{and} \quad T_{x_n}X^j \to \tau .$$

Then we get

$$rg\big((d(\pi_t)_z, d(f_t)_z)|\tau\big) < s + dim_R(\mathbb{K}) .$$

But X^j is **a-regular** over X^s in z. So we get

$$T_zX^s \subset \tau, \quad d(\pi_t)_z|T_zX^s = id \quad \text{and} \quad d(f_t)_z|T_zX^s = 0$$

by the assumption (rt) and (cp). This implies

$$rg\big(d(f_t)_z|\tau\big) < dim_R(\mathbb{K}) ,$$

and gives a contradiction to the assumption (nd).

2. Note that $\pi_t|X^j$ $(t \in P_c)$ is for $j > s$ a submersion near z, by **a-regularity**. Suppose the claim (2.a) is false. Then there exists a $j > s$ and a sequence $(x_n, t_n) \in X^j \times P_c$ such that

$$K_n := kern(\, d(\pi_{t_n})_{x_n}|T_{x_n}X^j\,) \subset kern(\, dr_{x_n}\,) ,$$

with

$$\pi_{t_n}(x_n) = z, \quad x_n \to z, \quad t_n \to t \in P_c, \quad l_n = \overline{x_n, z} \to l, \quad \text{and} \quad T_{x_n}X^j \to \tau .$$

Especially, $l_n \perp K_n$, and we can also assume $K_n \to K$ so that

$$l \perp K, \quad K \subset kern(d(\pi_t)_z), \quad d(\pi_t)_z | T_z X^s = id \quad \text{and} \quad \tau = K \oplus T_z X^s \,.$$

We claim $l \cap \tau = \{0\}$, which gives a contradiction to the **b-regularity** over $\{z\}$. Suppose

$$v = v_1 \oplus v_2 \in l \cap (K \oplus T_z X^s) \,.$$

Then $\pi_{t_n}(x_n) = z = \pi_{t_n}(z)$ for all n implies $d(\pi_t)_z(l) = 0$. So we get

$$d(\pi_t)_z(v) = 0 = d(\pi_t)_z(v_1) \;\Rightarrow\; d(\pi_t)_z(v_2) = 0 \;\Rightarrow v_2 = 0 \,,$$

since $d(\pi_t)_z | T_z X^s = id$. But then $l \perp K \Rightarrow v_1 = 0$.

Suppose the claim (2.b) is false. Then there exists a sequence (x_n, t_n) as before such that

$$rg\big((d(\pi_{t_n})_{x_n}, dr_{x_n}, d(g_{t_n})_{x_n}) | T_{x_n} X^j \big) < s + 1 + dim_R(\mathbb{K}) \,,$$

with $g_t := f_t - f_t \circ \pi_t$. By the result (2.a) we can assume for all n:

$$rg\big((d(\pi_{t_n})_{x_n}, dr_{x_n}) | T_{x_n} X^j \big) = s + 1 \,.$$

This implies

$$rg\big(d(f_{t_n})_{x_n} | (T_{x_n} X^j \cap kern(d(\pi_{t_n})_{x_n}, dr_{x_n})) \big) < dim_R(\mathbb{K}) \,.$$

We finally can assume

$$T_{x_n} X^j \cap kern(\, d(\pi_{t_n})_{x_n}, dr_{x_n} \,) \to \tau' \,.$$

Then we get as before:

$$l \perp \tau', \quad l \cap (\tau' \oplus T_z X^s) = \{0\} \quad \text{and} \quad rg\big(d(f_t)_z | \tau'\big) < dim_R(\mathbb{K}) \,.$$

But by **a-regularity** we have $T_z X^s \subset \tau$, and by **b-regularity** $l \subset \tau$ so that

$$\tau = l \oplus \tau' \oplus T_z X^s \,.$$

By assumption (cp) we have $d(f_t)_x | T_z X^s = 0$, and by $f_{t_n}(x_n) = f_{t_n}(\pi_{t_n}(x_n))$ for all n we get $d(f_t)_z | l = 0$. This implies

$$rg\big(d(f_t)_z | \tau \big) < dim_R(\mathbb{K}) \,,$$

in contradiction to the assumption (nd).

3. h_t is by (2.b) for all $t \in P_c$ a stratified submersion in

$$\{\pi_t = z, r = \delta, f_t - f_t \circ \pi_t = 0\} \,.$$

Then the claim follows easily from a-regularity and the compactness of P_c. □

We consider now the situation of lemma 5.3.1(3), i.e. $X_\bullet$ is **a-regular** in $X\backslash X_s$, X^j is **b-regular** over z and **a-regular** over X^s in z for all $j > s$. We choose $r_1, r_2, \delta, \epsilon'$ as in lemma 5.3.1, together with a (locally closed) submanifold T of P such that $T \subset P_c$. After taking U smaller, we can assume $r_0 = r_1 = r_2$.

We are interested in the subspace

$$Z := (T \times X) \cap \{\pi = z,\, r \leq \delta\} \subset T \times U$$

(with r induced by $r : U \to \mathbb{R}^{\geq 0}$), and in the smooth map

$$h : T \times U \to T \times \mathbb{K}\ ;\ (t, x) \mapsto \big(t, f_t(x) - f_t(\pi_t(x))\big)\,.$$

Corollary 5.3.1. *Consider the closed subspaces*

$$T \times \{z\} \subset Z_0 := (T \times X) \cap \{\pi = z,\, r \leq \delta,\, F - F \circ \pi = 0\} \subset Z\,.$$

1. $Z\backslash Z_0$ *gets an induced* **a-regular** *stratification such that*

$$h : Z \cap h^{-1}(W) \to W := T \times \{w \in \mathbb{K} \,|\, 0 < |w| < \epsilon'\}$$

is a **proper stratified submersion**.

2. $Z_0' := \big(T \times (X\backslash X^s)\big) \cap \{\pi = z,\, F - F \circ \pi = 0\}$ *gets an induced* **a-regular** *stratification such that*

$$id_T \times r : Z_0' \to T\times\,]0, r_0[\ ;\ (t, x) \mapsto (t, r(x))$$

is a **proper stratified submersion**, *with*

$$Z_0\backslash(T \times \{z\}) = Z_0' \cap (T \times \{0 < r \leq \delta\})\,.$$

Proof. 1. $h : Z \cap h^{-1}(W) \to W$ is proper, since already the projection to the first coordinate t is proper on Z. Note, that $Z_0 = Z \cap h^{-1}(T \times \{0\})$. Then $Z\backslash Z_0$ gets an induced a-regular stratification, since the intersection of the stratified spaces

$$T \times (X\backslash X^s) \quad \text{and} \quad \{\pi = z,\, r \leq \delta\}\backslash(T \times X^s)$$

is transversal by lemma 5.3.1(2.a). Similarly, h is a stratified submersion on strata in $T \times \{r < \delta\}$ by lemma 5.3.1(1), and on strata in $T \times \{r = \delta\}$ by lemma 5.3.1(3).

2. The used map is proper, since $r : U \to [0, r_0[$ is proper, with

$$(T \times X^s) \cap \{\pi = z,\, F - F \circ \pi = 0\} = T \times \{z\}\,.$$

Z_0' gets an induced a-regular stratification, since the intersection of the stratified spaces

$$T \times (X\backslash X^s) \quad \text{and} \quad \{\pi = z,\, F - F \circ \pi = 0\}\backslash(T \times X^s)$$

is transversal by lemma 5.3.1(1). Similarly, the used map is a stratified submersion by lemma 5.3.1(2.b). □

Let us now explain the **geometry** behind this construction. Fix a point $t \in T$. Then

$$h^{-1}(\{t\} \times \{|w| < \epsilon'\}) \cap Z = \big(X \cap \{\pi_t = z,\, r \leq \delta,\, |f_t - f_t \circ \pi_t| < \epsilon'\}\big) \times \{t\}$$
$$\simeq X \cap \{\pi_t = z,\, r \leq \delta,\, |f_t - f_t \circ \pi_t| < \epsilon'\}\,.$$

The space $N_t := \{\pi_t = z\}$ is a **normal slice** at $z \in X^s$ (condition (rt) of lemma 5.3.1). If we choose $0 < \epsilon < \epsilon'$, then

$$X \cap \{\pi_t = z,\, r \leq \delta,\, |f_t - f_t \circ \pi_t| \leq \epsilon\}$$

is just the space used in the description of the **normal Morse datum** in z with respect to the function f_t and the normal slice N_t (at least in the case $\mathbb{K} = \mathbb{R}$. We explain the other case in the next section). Moreover, the conditions on (δ, ϵ) used in remark 5.1.5 for the description of this "normal Morse datum" follow from

$$\begin{cases} \text{lemma 5.3.1(2.a) for the assumptions on } \delta, \\ \text{lemma 5.3.1(3) for the assumptions on } \epsilon. \end{cases}$$

So we can use remark 5.1.5, if we in addition assume that $X_\bullet$ is also **d-regular** in $X \backslash X_s$ (up to now we have not assumed the last condition).

With these remarks, it is now easy to prove the first important theorem of our **Morse theory for constructible sheaves**. But first we have to introduce some notation. We use the following commutative diagram:

$$\begin{array}{ccccc} & & T \times \mathbb{K} & \xleftarrow{\;h\;} & Z & \xrightarrow{\;p\;} X \\ & & \uparrow k & & \uparrow k' & \\ \{|w| < \epsilon'\} & \xrightarrow[s = s_t]{} & W' := T \times \{|w| < \epsilon'\} & \xleftarrow[h']{} & h^{-1}(W')\,. & \end{array}$$

Here k is the open inclusion, p is the restriction of the projection $T \times X \to X$, h', k' are the induced maps (in the corresponding cartesian square), and finally $s = s_t$ is for a fixed $t \in T$ the section $w \mapsto (t, w)$ of the projection

$$q : T \times \{|w| < \epsilon'\} \to \{|w| < \epsilon'\}\,.$$

Theorem 5.3.1. *Assume $X_\bullet$ is* **locally C-regular** *in $X \backslash X_s$, X^j is* **b-regular** *over $\{z\}$ and* **a-regular** *over X^s in z for all $j > s$. Fix $\mathcal{F} \in ob(D(X_\bullet))$. Then one has*

1. *$Rh'_* k'^* p^* \mathcal{F}$ is* **constructible** *with respect to the filtration $(T \times \{0\}, W')$. Especially,*
$$\mathcal{G}_t := s_t^* Rh'_* k'^* p^* \mathcal{F} \in ob(D(\{|w| < \epsilon'\}))$$
is **constructible** *with respect to the filtration $(\{0\}, \{|w| < \epsilon'\})$.*

2. *The isomorphism class of* $\mathcal{G}_t \in ob(D(\{|w| < \epsilon'\}))$ *is* **locally independent** *of* $t \in T$*, i.e.* t *has an open neighborhood in* T *such that this isomorphism class is constant on this neighborhood.*

3. *Choose* ϵ, w' *with* $0 < \epsilon < \epsilon'$ *and* $w' \in [-\epsilon, 0[\subset \mathbb{K}$. *The isomorphism class of*

$$R\Gamma(X \cap \{\pi_t = z,\, r \leq \delta,\, |g_t| \leq \epsilon\}, X \cap \{\pi_t = z,\, r \leq \delta,\, g_t = w'\}, \mathcal{F})$$

(with $g_t := f_t - f_t \circ \pi_t$*) is locally independent of* $t \in T$*. Especially, for* $\mathbb{K} = \mathbb{R}$ *is the isomorphism class of the "normal Morse datum"*

$$LMD(\mathcal{F}|X \cap \{\pi_t = z\}, f_t, z) := \left(R\Gamma_{\{f_t \geq f_t(z)\}}(\mathcal{F}|X \cap \{\pi_t = z\})\right)_z$$

locally independent *of* $t \in T$.

Proof. The assumption $X_\bullet$ is **locally C-regular** in $X \backslash X_s$ implies, that the induced stratifications of the spaces used in corollary 5.3.1(1),(2) are also **locally C-regular** (compare section 4.3).

1. Note that $h : h^{-1}(W') \to W'$ is proper. Then the restriction of

$$Rh'_* k'^* p^* \mathcal{F} \simeq k^* Rh_* p^* \mathcal{F}$$

to the open subset $W = W' \backslash (T \times \{0\})$ has locally constant cohomology sheaves by corollary 5.3.1(1) and lemma 5.1.5.

So it is enough to show that

$$(\, Rh'_* k'^* p^* \mathcal{F}\,)|T \times \{0\} \simeq Rh_*(\,(p^*\mathcal{F})|Z_0\,)$$

has locally constant cohomology sheaves. In the last term we consider the restriction

$$h = (id_T, 0) :\; Z_0 \to T \times \{0\}\,.$$

Since $h = (id_T, 0) : T \times \{z\} \to T \times \{0\}$ is a homeomorphism, we only have to show, that

$$Rh_!\big(\,(p^*\mathcal{F})|Z_0 \backslash (T \times \{z\})\,\big)$$

has locally constant cohomology sheaves. Here

$$h : Z_0 \backslash (T \times \{z\}) \to T \times \{0\} \quad \text{is not proper !}$$

Then the claim follows from the commutative diagram

$$\begin{array}{ccc} Z_0 \backslash (T \times \{z\}) & = & Z'_0 \cap (T \times \{0 < r \leq \delta\}) \\ \downarrow{\scriptstyle h} & & \downarrow{\scriptstyle (id_T, r)} \\ T \times \{0\} & \xleftarrow{pr} & T \times]0, \delta]\,, \end{array}$$

with $pr : T \times]0, \delta] \to T \times \{0\}$ the projection map $(t, r) \mapsto (t, 0)$. Then

$$\mathcal{G} := R(id_T, r)_* \big((p^* \mathcal{F}) | Z'_0 \cap (T \times \{0 < r \leq \delta\}) \big)$$

has locally constant cohomology sheaves by corollary 5.3.1(2) and lemma 5.1.5. Finally, $Rpr_! \mathcal{G}$ has locally constant cohomology sheaves by remark 1.2.1(1) on page 42, i.e. these cohomology sheaves are constant after restriction to a compact ball $K = B$ in T.

2. The local independence of the isomorphism class of $\mathcal{G}_t$ follows from a similar argument as in the proof of remark 1.2.1(1). Take an open (or closed) ball $B \subset T$ (in local coordinates of the manifolds T), and consider the projection

$$q : B \times \{|w| < \epsilon'\} \to \{|w| < \epsilon'\} \,.$$

Since

$$\mathcal{G} := (Rh'_* k'^* p^* \mathcal{F}) | B \times \{|w| < \epsilon'\}$$

is constructible with respect to the filtration

$$(B \times \{0\}, B \times \{|w| < \epsilon'\}) \,,$$

we get $\mathcal{G} \simeq q^* Rq_* \mathcal{G}$ by the **Vietoris-Begle theorem** [KS, cor. 2.7.7(iv), prop. 2.7.8]. Then

$$\mathcal{G}_t := s_t^* Rh'_* k'^* p^* \mathcal{F} \simeq s_t^* q^* Rq_* \mathcal{G} \simeq Rq_* \mathcal{G} \qquad (t \in B) \,,$$

since s_t is a section of q, i.e. $q \circ s_t = id$. So this isomorphism class is independent of $t \in B$.

3. Choose ϵ, w' with $0 < \epsilon < \epsilon'$ and $w' \in [-\epsilon, 0[\subset \mathbb{K}$. Fix $t \in T$ and let

$$N' := N'(t, \epsilon) := X \cap \{\pi_t = z, \, r \leq \delta, \, |f_t - f_t(z)| \leq \epsilon\} \,.$$

Then the following diagram is cartesian:

$$\begin{array}{ccc}
W' := T \times \{|w| < \epsilon'\} & \xleftarrow{\;h'\;} & h^{-1}(W') \\
\uparrow{\scriptstyle s = s_t} & & \uparrow{\scriptstyle s'} \\
\{|w| < \epsilon'\} & \xleftarrow{\;f_t - f_t \circ \pi_t\;} & N'' \\
\uparrow{\scriptstyle \kappa} & & \uparrow{\scriptstyle \kappa'} \\
\{|w| \leq \epsilon\} & \xleftarrow{\;f_t - f_t \circ \pi_t\;} & N' \,,
\end{array}$$

with κ, κ' the inclusions and s' the composition

$$X \cap \{\pi_t = z, \, r \leq \delta, \, |f_t - f_t(z)| < \epsilon'\} =: N'' \simeq N'' \times \{t\} \hookrightarrow h^{-1}(W') \,.$$

This implies

$$\kappa^* s_t^* Rh'_* k'^* p^* \mathcal{F} \simeq R(f_t - f_t \circ \pi_t)_* \kappa'^* s'^* k'^* p^* \mathcal{F} \,,$$

with $\kappa'^* s'^* k'^* p^* \mathcal{F} \simeq \mathcal{F}|N'$. But N' is compact and therefore

$$g_t := f_t - f_t \circ \pi_t : N' \to \{|w| \le \epsilon\}$$

is proper. Then the isomorphism class of

$$R\Gamma(N', N' \cap \{g_t = w'\}, \mathcal{F}) \simeq R\Gamma\big(\{|w| \le \epsilon\}, \{w'\}, R(g_t)_*(\mathcal{F}|N')\big)$$

is by 2. locally independent of $t \in T$. Moreover, $X_\bullet$ is by assumption **locally C-regular** in $X \backslash X_s$. We already explained before the theorem, that this implies for $t \in T$ in the case $\mathbb{K} = \mathbb{R}$ (by the choices of δ and ϵ):

$$R\Gamma(N', N' \cap \{g_t = w'\}, \mathcal{F}) \simeq \big(R\Gamma_{\{f_t \ge f_t(z)\}}(\mathcal{F}|X \cap \{\pi_t = z\})\big)_z . \qquad \square$$

Let us now explain the most important special case of theorem 5.3.1 (corresponding to [GM, p. 93, thm. 7.5.1]).

Example 5.3.1. *Consider $z \in X^s$ for a stratification $X_\bullet$ as in theorem* 5.3.1, *and fix $\mathcal{F} \in ob(D(X_\bullet))$. Take a* **nondegenerate** *covector $\omega \in (T^*_{X^s}M)_z$ and consider two smooth function germs $(i = 0, 1)$:*

$$f_i : (M, z) \to (\mathbb{R}, f_i(z)), \qquad \textit{with } d(f_i)_z = \omega .$$

Similarly, take two **normal slices** *N_i at $z \in X^s$.*

By the inverse function theorem, these are of the form $N_i = \{\pi_i = z\}$ for suitable germs of a smooth retraction onto X^s $(i = 0, 1)$. By taking $U = \{r < r_0\}$ small enough, we can assume that all these germs are defined in U. Then we can take $P := \mathbb{R}$ as a "parameter space", with (π, F) defined as

$$F(t, x) := (1 - t) \cdot f_0(x) + t \cdot f_1(x) ; \quad \pi(t, x) := (1 - t) \cdot \pi_0(x) + t \cdot \pi_1(x) .$$

Note that $d(f_t)_z = \omega$, and $\pi_t : U' \to X^s = U \cap \mathbb{R}^s \times \{0\}$ is for $U' \subset U$ a small neighborhood of z and t close to $[0, 1]$ a well-defined retraction.

If we choose a small compact neighborhood P_c of $[0, 1] \subset P$, then we can apply the theorem and get that the "normal Morse data" of f_i with respect to N_i are isomorphic $(i = 0, 1)$:

$$\big(R\Gamma_{\{f_0 \ge f_0(z)\}}(\mathcal{F}|X \cap N_0)\big)_z \simeq \big(R\Gamma_{\{f_1 \ge f_1(z)\}}(\mathcal{F}|X \cap N_1)\big)_z .$$

Especially, the isomorphism class of the "normal Morse datum"

$$NMD(\mathcal{F}, f, z) = NMD(\mathcal{F}, df_z)$$

depends only on the nondegenerate covector $\omega = df_z$, and **not** *on the choice of the normal slice, or on the function f.*

A similar argument works for two **nondegenerate** *covectors $d(f_i)_z = \omega_i \in (T^*_{X^s}M)_z$ $(i = 0, 1)$, which are close to each other, since the set of nondegenerate* *covectors in* $(T^*_{X^s}M)_z \simeq (\mathbb{R}^{m-s})^*$ *is open (i.e. contains* **convex** *neighborhoods around its points). Especially, the isomorphism class of the "normal Morse datum"*

$$NMD(\mathcal{F}, df_z)$$

depends only on the connected component of the set of nondegenerate covectors in $(T^*_{X^s}M)_z$*, which contains* df_z *(compare [GM, p. 95, cor. 7.5.3, cor. 7.5.4]).*

A similar reasoning can also be used to get control for the dependence on the "base point" $z \in X^s$. This time we work with a **fixed** function f and retraction π. We choose local coordinates

$$(M, X^s, z) \simeq (\mathbb{R}^m, \mathbb{R}^s \times \{0\}, 0) ,$$

and consider (locally) X as a **closed** subset of $U \times \{\rho < r_0\} \subset \mathbb{R}^m$ for $0 < r_0 << 1$, with $X^s = U \times \{0\}$ for an open neighborhood U of $0 \in \mathbb{R}^s$, and ρ the distance function:

$$\rho : \mathbb{R}^{m-s} \to \mathbb{R}^{\geq 0}, \quad \rho(x_{s+1}, \ldots, x_m) := \sum_{j=s+1}^{m} x_j^2 .$$

We use the same symbol for the induced function

$$\rho : U \times \mathbb{R}^{m-s} \to \mathbb{R}^{\geq 0} ,$$

which measures the distance to the stratum $X^s = U \times \{0\}$. Finally, we use as a retraction onto X^s the linear projection

$$\pi : U \times \mathbb{R}^{m-s} \to U \times \{0\} ; \; \pi(x_1, \ldots, x_m) := (x_1, \ldots, x_s, 0, \ldots, 0) .$$

Note, that $(\pi, \rho) : U \times \{\rho < r_0\} \to X^s \times [0, r_0[$ is **proper**.

Lemma 5.3.2. *Fix a* **compact** *subset* S *of* $X^s = U \times \{0\}$*, and consider a smooth map*

$$f : U \times \{\rho < r_0\} \to \mathbb{K} , \quad \textit{with } \mathbb{K} = \mathbb{R}, \mathbb{C}.$$

We make the following assumptions for $z \in S$*:*

(cp) $df_z | T_z X^s = 0$*, i.e.* z *is a* **critical point** *of* $f|X^s$*.*

(nd) df_z *is* **nondegenerate***.*

Assume X^j *is for all* $j > s$ *Whitney* **b-regular** *over* X^s *in all points of* S*.*

1. *There exists an open neigborhood* V_1 *of* S *in* $X^s = U \times \{0\} \simeq U$*, and an* $0 < r_1 \leq r_0$ *such that the following maps are* **stratified submersions***:*

   (a) $(\pi, f - f \circ \pi)$ *and* (π, ρ) *in* $V_1 \times \{0 < \rho < r_1\}$*.*

   (b) $(\pi, \rho, f - f \circ \pi)$ *in* $(V_1 \times \{0 < \rho < r_1\}) \cap \{f - f \circ \pi = 0\}$*.*

2. *Assume $X_\bullet$ is also* **a-regular** *in $X\backslash X_s$, and fix a δ with $0 < \delta < r_1$. Then there exists an open neigborhood V_2 of S in $X^s = U \times \{0\} \simeq U$, and an $\epsilon' > 0$ such that $(\pi, \rho, f - f \circ \pi)$ is a* **stratified submersion** *in*

$$(V_2 \times \{\rho = \delta\}) \cap \{|f - f \circ \pi| < \epsilon'\}.$$

Note that (1.a) for the map (π, ρ) follows from the implication **b-regular** $\Rightarrow$ **b^k-regular** proved in section 4.1. Similarly, **b-regular** over X^s in S implies also **a-regular** over X^s in S. Then the proof of the lemma is the same as for lemma 5.3.1, with

$$X^j \ni x_n \to z \in S \quad \text{and} \quad l_n = \overline{x_n, \pi(x_n)} \to l.$$

Consider the situation of lemma 5.3.2(2), i.e. $X_\bullet$ is **a-regular** in $X\backslash X_s$, and X^j is **b-regular** over X^s in S for all $j > s$, with $r_1, \delta, \epsilon', V_1, V_2$ as in lemma 5.3.2. After taking U smaller, we can assume $r_0 = r_1$ and $X^s = V_1 = V_2$.

We are interested in the subspace $Z := X \cap \{\rho \leq \delta\}$, and in the smooth map

$$h := (\pi, f - f \circ \pi) : U \times \{\rho < r_0\} \to X^s \times \mathbb{K}.$$

Then one gets as before the

Corollary 5.3.2. *Consider the closed subspaces*

$$X^s = U \times \{0\} \subset Z_0 := X \cap \{\rho \leq \delta,\ f - f \circ \pi = 0\} \subset Z.$$

1. *$Z\backslash Z_0$ gets an induced* **a-regular** *stratification such that*

$$h : Z \cap h^{-1}(W) \to W := X^s \times \{w \in \mathbb{K} \,|\, 0 < |w| < \epsilon'\}$$

is a **proper stratified submersion**.

2. *$Z_0' := (X\backslash X^s) \cap \{f - f \circ \pi = 0\}$ gets an induced* **a-regular** *stratification such that*

$$(\pi, \rho) : Z_0' \to X^s \times\,]0, r_0[$$

is a **proper stratified submersion**, *with $Z_0\backslash X^s = Z_0' \cap \{0 < \rho \leq \delta\}$.*

We also need the following commutative diagram:

$$\begin{array}{ccccc} & & X^s \times \mathbb{K} & \xleftarrow{h} & Z & \xrightarrow{p} & X \\ & & \uparrow k & & \uparrow k' & & \\ \{|w| < \epsilon'\} & \xrightarrow[s=s_t]{} & W' := X^s \times \{|w| < \epsilon'\} & \xleftarrow[h']{} & h^{-1}(W'). & & \end{array} \tag{5.68}$$

Here k is the open and p the closed inclusion, h', k' are the induced maps (in the corresponding cartesian square), and finally $s = s_t$ is for a fixed $t \in X^s$ the section $w \mapsto (t, w)$ of the projection

$$q : X^s \times \{|w| < \epsilon'\} \to \{|w| < \epsilon'\}.$$

Theorem 5.3.2. *Assume $X_\bullet$ is* **a+d-regular** *in $X\backslash X_s$, and X^j is* **b-regular** *over X^s in S for all $j > s$. Fix $\mathcal{F} \in ob(D(X_\bullet))$. Then one has*

1. *$Rh'_* k'^* p^* \mathcal{F}$ is* **constructible** *with respect to the filtration $(X^s \times \{0\}, W')$. Especially,*
$$\mathcal{G}_t := s_t^* Rh'_* k'^* p^* \mathcal{F} \in ob(D(\{|w| < \epsilon'\}))$$
is **constructible** *with respect to the filtration $(\{0\}, \{|w| < \epsilon'\})$.*

2. *The isomorphism class of $\mathcal{G}_t \in ob(D(\{|w| < \epsilon'\}))$ is* **locally independent** *of $t \in X^s$.*

3. *Choose ϵ, w' with $0 < \epsilon < \epsilon'$ and $w' \in [-\epsilon, 0[\subset \mathbb{K}$. The isomorphism class of*
$$R\Gamma(X \cap \{\pi = t,\, \rho \leq \delta,\, |f - f(t)| \leq \epsilon\}, X \cap \{\pi = t,\, \rho \leq \delta,\, f - f(t) = w'\}, \mathcal{F})$$
is locally independent of $t \in X^s$. Especially, for $\mathbb{K} = \mathbb{R}$ is the isomorphism class of the "normal Morse datum"
$$LMD(\mathcal{F}|X \cap \{\pi = z\}, f, z) := \left(R\Gamma_{\{f \geq f(z)\}}(\mathcal{F}|X \cap \{\pi = z\})\right)_z$$
locally independent *of $z \in S \subset X^s$.*

The proof is the same as for theorem 5.3.1. This time the induced stratifications of $Z\backslash Z_0$ and Z'_0 are **d-regular** by lemma 4.3.2. Similarly, only for $z = t \in S$ we can use remark 5.1.5 for the identification of the corresponding "normal Morse datum" in theorem 5.3.2(3). Here we use of course, that X^j b-regular over X^s in z implies the weaker conditions:

$$\begin{cases} X^j \text{ is a-regular over } X^s \text{ in } z, \text{ and} \\ X^j \text{ is b-regular over } \{z\}. \end{cases}$$

Example 5.3.2. *Consider $z \in S \subset X^s = U \times \{0\}$ for a stratification $X_\bullet$ as in theorem 5.3.2, and fix $\mathcal{F} \in ob(D(X_\bullet))$. Take a* **nondegenerate** *covector $\omega \in (T^*_{X^s} M)_z$ In our* **linearized situation** *this implies $\omega | T_x X^s = 0$ for all $x \in X^s$. Here we use the trivialization*
$$T^*_{X^s}\mathbb{R}^m \simeq X^s \times (\mathbb{R}^{m-s})^*,$$
with $(\mathbb{R}^{m-s})^*$ the dual vector space. Since the set of nondegenerate covectors is open in $T^*_{X^s} M$, there exists an open neighborhood V of $z \in X^s$ such that $V \times \{\omega\}$ is contained in this open set of nondegenerate covectors. So we can apply theorem 5.3.2 to the linear function $f = \omega$. Especially, the isomorphism class of the "normal Morse data"
$$\left(R\Gamma_{\{\omega \geq \omega(z)\}}(\mathcal{F}|X \cap \{\pi = z\})\right)_z$$
is locally independent of $z \in S \cap V$.

Together with example 5.3.1, this implies the

Corollary 5.3.3. *Let $X_\bullet$ be a smooth stratification, which is* **locally C-regular** *in $X \backslash X_s$. Let U be an open subset of X^s such that X^j is Whitney* **b-regular** *over X^s in U for all $j > s$. Fix $\mathcal{F} \in ob(D(X_\bullet))$. Then the isomorphism class of the "normal Morse datum"*

$$NMD(\mathcal{F}, df_z)$$

depends only on the connected component of the set of nondegenerate covectors in $(T^*_{X^s}M)|U$, which contains df_z. Especially, if $X_\bullet$ is Whitney **b-regular**, *we can take $U = X^s$.*

In particular, we get the second part of theorem 5.0.1 (compare also with the proof of [GM, lem.2.3.2, p.163]).

Suppose we consider a complex analytic (algebraic) stratification, or a definable stratification with respect to a "geometric category". Then there exists an **open dense constructible (definable)** subset U of X^s such that X^j is Whitney **b-regular** over X^s in U for all $j > s$ (compare [Te, Ver, vDrMi]). So we can apply this corollary to such a "generic open subset" U of X^s, if $X_\bullet$ is **locally C-regular** (in $X \backslash X_s$). This was already used in section 5.2 for the commutative diagram (5.63).

5.3.2 Splitting of the local Morse data

Now we come to the second main result of our **Morse theory for constructible sheaves**, which generalizes the first part of theorem 5.0.1.

This result describes (under appropriate assumptions) the "local Morse datum" as the (derived tensor) product of the "normal Morse datum" and the "tangential Morse datum" (compare [GM, p. 223, prop. 6.A.1], and [GM, thm. 3.7] for the topological counterpart).

There is a strong resemblance between this result and our **Thom-Sebastiani Theorem** 1.2.2. If one works for example with a Whitney **b-regular** stratification, then X is locally near a point $z \in X^s$ homeomorphic to the product of X^s and the intersection $X \cap N$ of X with a "normal slice" N at z. But the germ

$$f : (X, z) \to (\mathbb{R}, 0)$$

of a smooth function with a **stratified Morse critical point** need **not** correspond to $(f|X^i) \oplus (f|X \cap N)$ under this homeomorphism !

That one can find a homeomorphism as above such that f **also splits** in the right way was shown by King ([King, thm. 5],[King2, thm. 8]) in the context of a Whitney **b-regular** stratification. He uses this to **reduce** [GM, thm. 3.7] to the topological result [King, thm. 4], [King2, thm. 2], which is a **topological version** of our Thom-Sebastiani Theorem for constructible sheaves. It is **not** clear whether his methods also apply to the context of constructible sheaves.

Our approach is similar to the idea of King. We reduce the result to the "sum-construction" (i.e. to the Thom-Sebastiani Theorem for constructible sheaves). But the details are **completely different**. We do **not** prove any product structure like the above. We only show that some local direct image sheaves, corresponding to a **cohomological version of a local Milnor fibration**, are isomorphic to the corresponding local direct image sheaves of the product situation. This is enough for the applications to constructible sheaves.

Moreover, in this approach we do **not** need that $f|X^s$ has an isolated singularity. Similarly, we use weaker assumptions on the stratification.

We work in the same context as in theorem 5.3.2, but this time with $S := \{z\}$ the fixed point $z = 0 \in X^s = U \times \{0\}$. To simplify the notation, we also assume $f(z) = 0$. Consider the map

$$h = (\pi, f - f \circ \pi) : Z \to X^s \times \mathbb{K}$$

as in the commutative diagram (5.68). Choose $0 < \epsilon < \epsilon'$ and a **compact ball** B in X^s, which is a neighborhood of $z = 0 \in X^s$. Consider the induced proper map

$$h = (\pi, f - f \circ \pi) : K \to B \times \{|w| \leq \epsilon\} ,$$

with

$$K := K(B, \delta, \epsilon) := X \cap \pi^{-1}(B) \cap \{\rho \leq \delta,\, |f - f \circ \pi| \leq \epsilon\} .$$

Then $Rh_*(\mathcal{F}|K)$ is by theorem 5.3.2(1) **constructible** with respect to the filtration

$$(B \times \{0\}, B \times \{|w| \leq \epsilon\}) .$$

By the **Vietoris-Begle theorem** [KS, cor. 2.7.7(iv), prop. 2.7.8] one gets an isomorphism, functorial in $\mathcal{F} \in ob(D(X_\bullet))$:

$$Rh_*(\mathcal{F}|K) \xleftarrow{\sim} q^* s^* Rh_*(\mathcal{F}|K) , \quad \text{with}$$

$q : B \times \{|w| \leq \epsilon\} \to \{|w| \leq \epsilon\}$ the projection and $s : \{|w| \leq \epsilon\} \to B \times \{|w| \leq \epsilon\}$ the zero section. Since this h is proper, we get

$$s^* Rh_*(\mathcal{F}|K) \simeq Rf_*(\mathcal{F}|K \cap \{\pi = z\}) , \quad \text{with}$$

$$f : N := K \cap \{\pi = z\}) = X \cap \{\pi = z,\, \rho \leq \delta,\, |f - f \circ \pi| \leq \epsilon\} \to \{|w| \leq \epsilon\} .$$

This N is exactly the space used in remark 5.1.5 for the identification of the corresponding "normal Morse datum" with respect to the **normal slice** $\{\pi = z\}$.

Now we can indicate the **main point** in our approach to the "splitting of the local Morse datum". The composition

$$K \xrightarrow{(\pi, f - f\circ\pi)} B \times \{|w| \leq \epsilon\} \xrightarrow{(f, id)} \mathbb{K} \times \{|w| \leq \epsilon\}$$

is just the map $(f \circ \pi, f - f \circ \pi)$, with $f = (f \circ \pi) + (f - f \circ \pi)$. Moreover

$$R(\pi, f - f \circ \pi)_*(\mathcal{F}|K) \simeq q^* Rf_*(\mathcal{F}|N) \simeq \mathbb{Z}_B \boxtimes^L_Z Rf_*(\mathcal{F}|N)\,.$$

Since we work with proper maps, we get by the **Künneth isomorphism** (compare section 1.4):

$$R(f \circ \pi, f - f \circ \pi)_*(\mathcal{F}|K) \simeq (Rf_*\mathbb{Z}_B) \boxtimes^L_Z Rf_*(\mathcal{F}|N)\,. \tag{5.69}$$

As promised before, these local direct image complexes are isomorphic to those of the corresponding product situation:

$$R(f \times f)_*\big(\mathbb{Z}_B \boxtimes^L_Z (\mathcal{F}|N)\big) \simeq (Rf_*\mathbb{Z}_B) \boxtimes^L_Z Rf_*(\mathcal{F}|N)\,.$$

If $(Rf_*\mathbb{Z}_B)|\{|z| \leq \epsilon\}$ is also **constructible** with respect to the filtration $(\{0\}, \{|z| \leq \epsilon\})$, then we can apply the **Thom-Sebastiani Theorem** to

$$Rf_*\mathbb{Z}_B,\ Rf_*(\mathcal{F}|N) \quad \text{and the function } a : \mathbb{K}^2 \to \mathbb{K},\ (z, w) \mapsto z + w\,.$$

As explained before, we **cannot** apply this theorem directly to $f \circ \pi$ and $f - f \circ \pi$. The only point that we then have to explain, is the relation to a **cohomological version of a Milnor fibration** for $(f, \{z\}, \mathcal{F})$ so that we can calculte the "local Morse datum" with the help of these direct image complexes. Here we would like to use lemma 1.2.2 on page 43. And for this we need the following compability result:

Lemma 5.3.3. *Suppose $B_i, \delta_i, \epsilon_i$ are chosen as before ($i = 1, 2$), with $B_2 \subset B_1$, $\delta_2 \leq \delta_1$ and $\epsilon_2 \leq \epsilon_2$. Let*

$$h_i := (\pi, f - f \circ \pi) : K_i := K(B_i, \delta_i, \epsilon_i) \to B_i \times \{|w| \leq \epsilon_i\}$$

be the corresponding maps ($i = 1, 2$), with

$$j : B_2 \times \{|w| \leq \epsilon_2\} \to B_1 \times \{|w| \leq \epsilon_1\}$$

the inclusion. Then the following diagram commutes for $\mathcal{F} \in ob(D(X_\bullet))$:

$$\begin{array}{ccc} j^* Rh_{1*}(\mathcal{F}|K_1) & \xleftarrow{\ \sim\ } & j^* q_1^* s_1^* Rh_{1*}(\mathcal{F}|K_1) \simeq q_2^* s_2^* j^* Rh_{1*}(\mathcal{F}|K_1) \\ \downarrow{\scriptstyle \phi} & & \downarrow{\scriptstyle q_2^* s_2^*(\phi)} \\ Rh_{2*}(\mathcal{F}|K_2) & \xleftarrow{\ \sim\ } & q_2^* s_2^* Rh_{2*}(\mathcal{F}|K_2)\,, \end{array}$$

with ϕ the natural base change morphism (as in the proof of lemma 4.3.3 on page 260). Here q_1, q_2 and s_1, s_2 are the corresponding projections and zero sections.

Proof. Let $\mathcal{G} \in ob(D(B \times \{|w| \leq \epsilon\}))$ be constructible with respect to the filtration

$$(B \times \{0\}, B \times \{|w| \leq \epsilon\}), \quad \text{with}$$

$q : B \times \{|w| \leq \epsilon\} \to \{|w| \leq \epsilon\}$ the projection and $s : \{|w| \leq \epsilon\} \to B \times \{|w| \leq \epsilon\}$ the zero section. Then the adjunction morphism

$$ad_q : q^* Rq_* \mathcal{G} \to \mathcal{G}$$

is an isomorphism by the **Vietores-Begle theorem**. If we apply s^* to this isomorphism, we get the isomorphism

$$Rq_*\mathcal{G} \simeq s^*q^*Rq_*\mathcal{G} \xrightarrow[\sim]{s^*(ad_q)} s^*\mathcal{G}.$$

Altogether we get the isomorphism $\alpha : \mathcal{G} \xleftarrow{\sim} q^*s^*\mathcal{G}$. This is of course functorial in $\mathcal{G}$. Consider now the commutative diagram

$$\begin{array}{ccc} B_1 \times \{|w| \leq \epsilon_1\} & \xrightarrow{q_1} & \{|w| \leq \epsilon_1\} \\ \uparrow j & & \uparrow j' \\ B_2 \times \{|w| \leq \epsilon_2\} & \xrightarrow{q_2} & \{|w| \leq \epsilon_2\}\,, \end{array}$$

with j, j' the closed inclusions. Take $\mathcal{G} \in ob(D(B_1 \times \{|w| \leq \epsilon_1\}))$ as above. Then $j^*\mathcal{G}$ is constructible with respect to the filtration

$$(B_2 \times \{0\}, B_2 \times \{|w| \leq \epsilon_2\})\,.$$

The adjunction morphism for q_1 induces the morphism

$$q_2^* j'^* Rq_{1*} \simeq j^* q_1^* Rq_{1*} \to j^*\,.$$

By adjunction, this induces $\phi' : j'^* Rq_{1*} \to Rq_{2*} j^*$ such that the diagram

$$\begin{array}{ccc} j^* q_1^* Rq_{1*} & \xrightarrow{ad_{q_1}} & j^* \\ \uparrow \wr & & \uparrow ad_{q_2} \\ q_2^* j'^* Rq_{1*} & \xrightarrow{\phi'} & q_2^* Rq_{2*} j^* \end{array}$$

commutes. By the isomorphisms $\psi : q_2^* s_2^* j^* \simeq j^* q_1^* s_1^*$ and $\beta : q_2^* j'^* \simeq j^* q_1^*$, we get the commutative diagram:

$$\begin{array}{ccccccc} j^*\mathcal{G} & \xleftarrow[\sim]{ad_{q_1}} & j^*q_1^*Rq_{1*}\mathcal{G} & \xleftarrow{\sim} & j^*q_1^*s_1^*q_1^*Rq_{1*}\mathcal{G} & \xrightarrow[\sim]{ad_{q_1}} & j^*q_1^*s_1^*\mathcal{G} \\ \| & & \beta \uparrow \wr & & \psi\circ\beta \uparrow \wr & & \psi \uparrow \wr \\ j^*\mathcal{G} & & q_2^*j'^*Rq_{1*}\mathcal{G} & \xleftarrow{\sim} & q_2^*s_2^*q_2^*j'^*Rq_{1*}\mathcal{G} & & q_2^*s_2^*j^*\mathcal{G} \\ \| & & \downarrow \phi' & & \downarrow \phi' & & \| \\ j^*\mathcal{G} & \xleftarrow[\sim]{ad_{q_2}} & q_2^*Rq_{2*}j^*\mathcal{G} & \xleftarrow{\sim} & q_2^*s_2^*q_2^*Rq_{2*}j^*\mathcal{G} & \xrightarrow[\sim]{ad_{q_2}} & q_2^*s_2^*j^*\mathcal{G}\,. \end{array}$$

This implies the commutativity of

$$\begin{array}{ccc} j^*\mathcal{G} & \xleftarrow{j^*(\alpha_1)} & j^*q_1^*s_1^*\mathcal{G} \\ \| & & \psi \uparrow \wr \\ j^*\mathcal{G} & \xleftarrow{\alpha_2} & q_2^*s_2^*j^*\mathcal{G}\,. \end{array}$$

If we apply this to $\mathcal{G} = Rh_{1*}(\mathcal{F}|K_1)$, we get the claim of the lemma by the functoriality of the isomorphism α_2, applied to the base change morphism

$$\phi : j^*Rh_{1*}(\mathcal{F}|K_1) \to Rh_{2*}(\mathcal{F}|K_2)\,. \qquad \square$$

Let us now explain the **geometry** behind this lemma, especially of the morphism $s_2^*(\phi)$. Consider the commutative diagram

$$\begin{array}{ccccc} N_2 & \xrightarrow{\subset} & K_2 & \xrightarrow{\subset} & K_1 \\ \downarrow f & & \downarrow h_2 & & \downarrow h_1 \\ \{|w| \leq \epsilon_2\} & \xrightarrow{s_2} & B_2 \times \{|w| \leq \epsilon_2\} & \xrightarrow{j} & B_1 \times \{|w| \leq \epsilon_1\}\,, \end{array}$$

with $N_2 := X \cap \{\pi = z, \rho \leq \delta_2, |f| \leq \epsilon_2\}$. Since K_2 is compact and the left square is cartesian, we get

$$s_2^*Rg_{2*}(\mathcal{F}|K_2) \xrightarrow{\sim} Rf_*(\mathcal{F}|N_2)\,.$$

By the choices of δ_2, ϵ_2 we know that we can use N_2 for the calculation of the "normal Morse datum" (let us say in the case $\mathbb{K} = \mathbb{R}$). Recall that we assumed $f(z) = 0$. Moreover, $Rf_*(\mathcal{F}|N_2)$ is constructible with respect to the filtration $(\{0\}, \{|w| \leq \epsilon_2\})$. This corresponds to the first assumption about a **cohomological version of a Milnor fibration** as in definition 5.1.1.

By the composition rule for the base change morphism (compare with the proof of lemma 4.3.3 on page 261, or with [SGA4, p. 85, prop. 4.4]), the morphism

$$s'^*Rh_{1*}(\mathcal{F}|K_1) \simeq s_2^*j^*Rh_{1*}(\mathcal{F}|K_1) \xrightarrow{s_2^*(\phi)} s_2^*Rh_{2*}(\mathcal{F}|K_2) \xrightarrow{\sim} Rf_*(\mathcal{F}|N_2)$$

is the base change morphism of the outer commutative square for the map $s' := j \circ s_2$. Consider now the commutative diagram

$$\begin{array}{ccccc} N_2 & \xrightarrow{\subset} & K_1' & \xrightarrow{\subset} & K_1 \\ \downarrow f & & \downarrow h_1' & & \downarrow h_1 \\ \{|w| \leq \epsilon_2\} & \xrightarrow{s} & B_1 \times \{|w| \leq \epsilon_2\} & \xrightarrow[k]{\subset} & B_1 \times \{|w| \leq \epsilon_1\}, \end{array}$$

with $K_1' := K_1 \cap f^{-1}(\{|w| \leq \epsilon_2\})$ and s the corresponding zero section (h_1' is the induced map of the right cartesian square). Then $k \circ s = s' = j \circ s_2$, and the outer square is the same as before. By the same argument, the base change morphism for the map $k \circ s = s'$ is given as the composition

$$s'^* Rh_{1*}(\mathcal{F}|K_1) \simeq s^* k^* Rh_{1*}(\mathcal{F}|K_1) \xrightarrow{\sim} s^* Rh'_{1*}(\mathcal{F}|K'_1) \xrightarrow{\phi'} Rf_*(\mathcal{F}|N_2)\,.$$

Here ϕ' is the base change morphism for s, and the base change morphism for k is an isomorphism since K_1 is compact.

Let $k' : \{|w| \leq \epsilon\} \to \{|w| \leq \epsilon_2\}$ be the inclusion for $0 < \epsilon \leq \epsilon_2$. Then we know by the choice of δ_1, that $k'^*(\phi')$ is an **isomorphism** for $0 < \epsilon << \epsilon_2$.

This corresponds to the second assumption about a **cohomological version of a Milnor fibration**. We therefore get the following conclusion:

$$k'^* s_2^*(\phi) \text{ is an isomorphism for } 0 < \epsilon << \epsilon_2. \tag{5.70}$$

This will be used in the proof of the following

Theorem 5.3.3. *Assume $X_\bullet$ is* **a+d-regular** *in $X \backslash X_s$, and X^j is* **b-regular** *over X^s in the point $z \in X^s$ for all $j > s$. Consider a smooth function germ*

$$f : (M, z) \to (\mathbb{R}, 0), \quad \text{with } df_z|T_zX^s = 0 \text{ and } df_z \textbf{ nondegenerate}.$$

Fix $\mathcal{F} \in ob(D(X_\bullet))$, and choose local coordinates

$$(M, X^s, z) \simeq (\mathbb{R}^m, \mathbb{R}^s \times \{0\}, 0)\,, \quad \text{with} \quad \pi : (\mathbb{R}^m, 0) \to (\mathbb{R}^s \times \{0\}, 0)$$

the linear projection. Suppose that the triple $(f|X^s, \{z\}, \mathbb{Z}_{X^s})$ satisfies the assumptions of definition 5.1.1 *about a* **cohomological version of a Milnor fibration**, *with $K_n = B_n$ a sequence of* **compact balls** *in $X^s \simeq \mathbb{R}^s \times \{0\}$.*

Then there exists an isomorphism, functorial in $\mathcal{F} \in ob(D(X_\bullet))$:

$$LMD(\mathcal{F}, f, z) := \left(R\Gamma_{\{f \geq 0\}}(\mathcal{F})\right)_z \simeq TMD(\mathcal{F}, f, z) \otimes^L_Z NMD(\mathcal{F}, f, z)\,, \tag{5.71}$$

with

$$TMD(\mathcal{F}, f, z) := \left(R\Gamma_{\{f|X^s \geq 0\}}(\mathbb{Z}_{X^s})\right)_z$$

the "tangential Morse datum", and

$$NMD(\mathcal{F}, f, z) := \left(R\Gamma_{\{f|N \cap X \geq 0\}}(\mathcal{F}|N \cap X)\right)_z$$

the "normal Morse datum" with respect to the normal slice $N := \{\pi = z\}$.

Proof. For applications in the next section, we include at the beginning of the proof also the case of a complex function.

Let $f : (M, z) \to (\mathbb{K}, 0)$ (with $\mathbb{K} = \mathbb{R}, \mathbb{C}$) a smooth function germ such that df_z is nondegenerate, with $df_z|T_zX^s = 0$. Suppose that the assumptions of definition 5.1.1 are true for the triple $(f|X^s, \{z\}, \mathbb{Z}_{X^s})$, with $K_n = B_n$ a sequence of compact balls in $X^s \simeq \mathbb{R}^s \times \{0\}$. Let ϵ_n be the corresponding sequence of real numbers as given by this definition, with $\epsilon_n \searrow 0$.

Note, that the corresponding assumptions are then also true, if we take a subsequence of smaller balls and values ϵ_n. Therefore we can assume, that there is a sequence $(U_n, \delta_n, \epsilon'_n)$, with U_n an open neighborhood of $\{0\}$ in $\mathbb{R}^s$, and $\delta_n, \epsilon'_n \searrow 0$ such that we can apply theorem 5.3.2(1) to each triple $(U_n, \delta_n, \epsilon'_n)$ $(n \in \mathbb{N})$, with $B_n \subset U_n$ and $\epsilon_n < \epsilon'_n$.

As in the discussion before, let

$$K_n := X \cap \pi^{-1}(B_n) \cap \{\rho \leq \delta_n,\ |f - f \circ \pi| \leq \epsilon_n\}$$

and

$$N_n := X \cap \{\pi = z,\ \rho \leq \delta_n,\ |f| \leq \epsilon_n\},$$

with q_n, s_n the corresponding projection/section. To simplify the notation, let

$$h_n := (\pi, f - f \circ \pi) : K_n \to B_n \times \{|w_2| \leq \epsilon_n\}$$

and

$$g_n := (f \circ \pi, f - f \circ \pi) : K_n \to \mathbb{K} \times \{|w_2| \leq \epsilon_n\}.$$

Similarly, if we consider $f|N_n$, then we use the map $f : N_n \to \{|w_2| \leq \epsilon_n\}$.

Then the K_n build a decreasing fundamental system of compact neighborhoods of $\{z\}$ in X with the following properties:

1. $Rh_{n*}(\mathcal{F}|K_n) \simeq q_n^* s_n^* Rh_{n*}(\mathcal{F}|K_n) \simeq \mathbb{Z}_{B_n} \boxtimes_Z^L s_n^* Rh_{n*}(\mathcal{F}|K_n)$.

2. $s_n^* Rh_{n*}(\mathcal{F}|K_n) \simeq Rf_*(\mathcal{F}|N_n)$ is constructible with respect to the filtration $(\{0\}, \{|w_2| \leq \epsilon_n\})$, and $Rg_{n*}(\mathcal{F}|K_n) \simeq (Rf_*\mathbb{Z}_{B_n}) \boxtimes_Z^L Rf_*(\mathcal{F}|N_n)$.

3. The restriction of $Rf_*\mathbb{Z}_{B_n}$ to $\{|w_1| \leq \epsilon_n\}$ is constructible with respect to the filtration $(\{0\}, \{|w_1| \leq \epsilon_n\})$.

Especially, the restriction of $Rg_{n*}(\mathcal{F}|K_n)$ to $\{|w_i| \leq \epsilon_n, i = 1, 2\}$ is constructible with respect to the filtration

$$\{0\} \subset \{w_1 \cdot w_2 = 0, |w_i| \leq \epsilon_n, i = 1, 2\} \subset \{|w_i| \leq \epsilon_n, i = 1, 2\}.$$

We state now the following technical result, which we will be explained in the end of the proof:

$$\text{For } n \in \mathbb{N} \text{ fixed, there exists } 0 < \epsilon \leq \epsilon_{n+1} \text{ so that } \kappa^*(\phi) \text{ is an isomorphism.} \tag{5.72}$$

Here $\kappa : \{|w_i| \leq \epsilon, i = 1, 2\} \to \mathbb{K} \times \{|w_2| \leq \epsilon_{n+1}\}$ is the inclusion, and

$$\phi : k^* Rg_{n*}(\mathcal{F}|K_n) \to Rg_{n+1*}(\mathcal{F}|K_{n+1})$$

is the **base change morphism**, with $k : \mathbb{K} \times \{|w_2| \leq \epsilon_{n+1}\} \hookrightarrow \mathbb{K} \times \{|w_2| \leq \epsilon_n\}$.

Then $g = (f \circ \pi, f - f \circ \pi) : X \to \mathbb{K}^2$, $S := \{z\}$ and $\mathcal{F} \in ob(D(X_\bullet))$ satisfy the assumptions of lemma 1.2.2 on page 43 for the sequences K_n and $1/2 \cdot \epsilon_n$. So we get by this lemma, that

$$f = a \circ (f \circ \pi, f - f \circ \pi) : X \to \mathbb{K},\ S := \{z\} \quad \text{and} \quad \mathcal{F}$$

satisfy the assumptions of definition 5.1.1 about a **cohomological version of a Milnor fibration**, with respect to K_n and $\gamma_n \searrow 0$ any sequence such that $\gamma_n < 1/2 \cdot \epsilon_n$. Here a is the addition map

$$a : \mathbb{K}^2 \to \mathbb{K};\ (w_1, w_2) \mapsto w_1 + w_2 .$$

Assume now $\mathbb{K} = \mathbb{R}$, and choose the sequence γ_n as above. Then we get by lemma 5.1.3 for $w \in [-\gamma_n, 0[$:

$$\begin{aligned}(R\Gamma_{\{f\geq 0\}}(\mathcal{F}))_z &\simeq R\Gamma(K_n \cap \{|f| \leq \gamma_n\}, K_n \cap \{f = w\}, \mathcal{F}) \\ &\simeq R\Gamma\big(K(\gamma_n), K(\gamma_n) \cap \{a = w\}, Rg_{n*}(\mathcal{F}|K_n)\big) ,\end{aligned}$$

with

$$K(\gamma_n) := \{|w_1 + w_2| \leq \gamma_n,\ |w_i| \leq \epsilon_n,\ i = 1, 2\} \subset \mathbb{K}^2 .$$

By the same argument, this is isomorphic to (compare lemma 1.2.1 on page 38):

$$\big(R\Gamma_{\{a\geq 0\}}(Rg_{n*}(\mathcal{F}|K_n)) \big)_0 .$$

This allows us to calculate the "local Morse datum" $LMD(\mathcal{F}, f, z)$ in terms of the direct image complex

$$Rg_{n*}(\mathcal{F}|K_n) \simeq (Rf_*\mathbb{Z}_{B_n}) \boxtimes^L_Z Rf_*(\mathcal{F}|N_n) .$$

Now we can apply the **Thom-Sebastiani Theorem** 1.2.2 to the map a and get the isomorphism

$$(R\Gamma_{\{f\geq 0\}}(\mathcal{F}))_0 \simeq (R\Gamma_{\{w_1\geq 0\}}(Rf_*\mathbb{Z}_{B_n}))_0 \otimes^L_Z \big(R\Gamma_{\{w_2\geq 0\}}(Rf_*(\mathcal{F}|N_n)) \big)_0 .$$

Finally, by the same argument (or by remark 1.1.3(1)) we get

$$(R\Gamma_{\{w_1\geq 0\}}(Rf_*\mathbb{Z}_{B_n}))_0 \simeq (R\Gamma_{\{f|X^s\geq 0\}}(\mathbb{Z}_{X^s}))_z$$

and

$$\big(R\Gamma_{\{w_2\geq 0\}}(Rf_*(\mathcal{F}|N_n)) \big)_0 \simeq (R\Gamma_{\{f|N\cap X\geq 0\}}(\mathcal{F}|N \cap X))_z .$$

This implies the theorem, since the functoriality in $\mathcal{F} \in ob(D(X_\bullet))$ follows from the proof.

Let us now explain the technical result (5.72). Consider the following commutative diagram:

$$\begin{array}{ccccc} K_n & \xrightarrow{h_n} & B_n \times \{|w_2| \leq \epsilon_n\} & \xrightarrow{(f,id)} & \mathbb{K} \times \{|w_1| \leq \epsilon_n\} \\ \uparrow j' & & \uparrow j=j_1\times j_2 & & \uparrow k=k_1\times k_2 \\ K_{n+1} & \xrightarrow{h_{n+1}} & B_{n+1} \times \{|w_2| \leq \epsilon_{n+1}\} & \xrightarrow{(f,id)} & \mathbb{K} \times \{|w_1| \leq \epsilon_{n+1}\} , \end{array}$$

with j_i, k_i the corresponding inclusion of the factors ($i = 1, 2$, especially $k_1 = id$). The base change morphism ϕ is by

$$Rg_{l*} \simeq R(f, id)_* Rh_{l*} \qquad (l = n, n+1)$$

isomorphic to the composition (compare [SGA4, p. 86, prop. 4.4(ii)]):

$$k^* R(f, id)_* Rh_{n*} \xrightarrow{\phi_k} R(f, id)_* j^* Rh_{n*} \xrightarrow{\phi_j} R(f, id)_* Rh_{n+1*} j'^* .$$

Here ϕ_j (or ϕ_k) is the base change map for j (or for k and the right square). But the morphism ϕ_j corresponds by the isomorphisms

$$Rh_{l*}(\mathcal{F}|K_l) \simeq q_l^* s_l^* Rh_{l*}(\mathcal{F}|K_l) \qquad (l = n, n+1) ,$$

to the morphism $q_{n+1}^* s_{n+1}^*(\phi_j)$ (as shown in lemma 5.3.3). Similarly, if we use the isomorphism

$$Rh_{n*}(\mathcal{F}|K_n) \simeq \mathbb{Z}_{B_n} \boxtimes_Z^L s_n^* Rh_{n*}(\mathcal{F}|K_n)$$

and the **Künneth isomorphism** (together with $k^* \simeq k_1^* \boxtimes^L k_2^*$ and $j^* \simeq j_1^* \boxtimes^L j_2^*$), then we get a commutative diagram

$$\begin{array}{ccc} k^* R(f, id)_* \big(\mathbb{Z}_{B_n} \boxtimes_Z^L s_n^* Rh_{n*}(\mathcal{F}|K_n)\big) & \xleftarrow{\sim} & k_1^*(Rf_* \mathbb{Z}_{B_n}) \boxtimes_Z^L k_2^* s_n^* Rh_{n*}(\mathcal{F}|K_n) \\ \downarrow \phi_k & & \phi_{k_1} \boxtimes^L \downarrow \phi_{k_2} \\ R(f, id)_* j^* \big(\mathbb{Z}_{B_n} \boxtimes_Z^L s_n^* Rh_{n*}(\mathcal{F}|K_n)\big) & \underset{\sim}{\xleftarrow{K\ddot{u}}} & Rf_* j_1^*(\mathbb{Z}_{B_n}) \boxtimes_Z^L j_2^* s_n^* Rh_{n*}(\mathcal{F}|K_n) . \end{array}$$

This can for example be deduced from the definition of the base change morphism by the adjunction map $R(f, id)_* \to R(f, id)_* Rk_* k^*$ and a similar splitting result for this map (as in lemma 1.4.2 on page 79). Recall that we work with proper maps and closed inclusions so that $Rg_* = Rg_!$ for these maps.

In our case we have $k_1 = id$, $k_2 = j_2$ and $\phi_{k_2} = id$. If we use $j_2^* s_n^* \simeq s_{n+1}^* j^*$ and the **Künneth isomorphism**, then we get for the composition a commutative diagram

$$\begin{array}{ccc} k^* Rg_{n*}(\mathcal{F}|K_n) & \xleftarrow{\sim} & (Rf_* \mathbb{Z}_{B_n}) \boxtimes_Z^L s_{n+1}^* j^* Rh_{n*}(\mathcal{F}|K_n) \\ \downarrow \phi & & \phi_1 \boxtimes^L \downarrow s_{n+1}^*(\phi_2) \\ Rg_{n+1*}(\mathcal{F}|K_{n+1}) & \underset{\sim}{\xleftarrow{K\ddot{u}}} & Rf_* j_1^*(\mathbb{Z}_{B_n}) \boxtimes_Z^L s_{n+1}^* Rh_{n+1*}(\mathcal{F}|K_{n+1}) , \end{array}$$

with ϕ_1, ϕ_2 the corresponding base change morphisms. If we use the decomposition

$$\kappa_1 \times \kappa_2 := \kappa : \{|w_i| \leq \epsilon, i = 1, 2\} \to \mathbb{K} \times \{|w_2| \leq \epsilon_{n+1}\} ,$$

then we get

$$\kappa^*(\phi) \simeq \kappa_1^*(\phi_1) \boxtimes^L \kappa_2^* s_{n+1}^*(\phi_2) .$$

But $\kappa_2^* s_{n+1}^*(\phi_2)$ is for small ϵ an isomorphism by (5.70). Similarly, $\kappa_1^*(\phi_1)$ is for small ϵ an isomorphism by the second assumption of definition 5.1.1 about a **cohomological version of a Milnor fibration** for $f|X^s$. □

Let us now explain some important examples for a function f such that the assumptions of definition 5.1.1 are true for the triple $(f|X^s, \{0\}, \mathbb{Z}_{X^s})$, with $K_n = B_n$ a sequence of **compact balls** in $X^s \simeq \mathbb{R}^s \times \{0\}$.

Example 5.3.3. 1. *The trivial case that* $f|X^s$ *is the* **zero function**.

2. $\{z\}$ *is an* **isolated critical** *point of* $f|X^s$, *which is also* **"reasonable"** *in the sense that*

$$(X^s \cap \{f = 0\}) \backslash \{z\} \quad \textit{is Whitney } \textbf{b-regular} \textit{ over } \{z\}.$$

For example, the isolated critical point z *of* $f|X^s$ *is* **nondepraved** *in the sense of [GM, p. 54, def. 2.3] (compare [GM, p. 57, lem. 2.5.1]). Especially, if* z *is a* **"classical Morse critical point"** *of* $f|X^s$.

3. *The map* $f : (X^s, X^s \cap \{f = 0\}) \to (\mathbb{R}, 0)$ *has a suitable stratification satisfying the* a_f**-condition** *of Thom. This is especially the case, if* $f|X^s$ *is in some coordinates of* (X^s, z) **real analytic** *(or more generally, definable in some o-minimal structure [Ta2]). Moreover, in this case one can verify the assumptions of definition* 5.1.1 *directly (without the use of the* a_f*-condition, see lemma* 2.2.2 *on page* 106*).*

In two special cases, we can easily **calculate** the "tangential Morse datum"

$$TMD(\mathcal{F}, f, z) := \left(R\Gamma_{\{f|X^s \geq 0\}}(\mathbb{Z}_{X^s})\right)_z .$$

For the trivial case that $f|X^s$ is the **zero function**, we get

$$TMD(\mathcal{F}, f, z) \simeq \mathbb{Z} \quad \text{(viewed as a complex in degree 0)} .$$

The other case deals with a **"classical Morse critical point"** of $f|X^s$. Assume $(f|X^s, z)$ is in some coordinates of X^s isomorphic to

$$g \oplus h : (\mathbb{R}^s, 0) = (\mathbb{R}^m, 0) \times (\mathbb{R}^l, 0) \to (\mathbb{R}, 0) ,$$

with $g(x) := \sum_{i=1}^m x_i^2$ and $h(x) := -\sum_{j=1}^l y_i^2$. Using lemma 5.1.3 with K_n usual balls around 0, one gets

$$\left(R\Gamma_{\{g\geq 0\}}(\mathbb{Z}_{R^m})\right)_0 \simeq \mathbb{Z} \quad \text{and} \quad \left(R\Gamma_{\{h\geq 0\}}(\mathbb{Z}_{R^l})\right)_0 \simeq \mathbb{Z}[-l] .$$

Then one gets by the **Thom-Sebastiani Theorem**:

$$TMD(\mathcal{F}, f, z) := \left(R\Gamma_{\{f|X^s \geq 0\}}(\mathbb{Z}_{X^s})\right)_z \simeq \mathbb{Z}[-l]\,. \tag{5.73}$$

Here l is the **Morse index** of $g \oplus h$ in 0. Since our fixed degree of smoothness k is ≥ 2, this is of course also the **Morse index** of $f|X^s$ in z. Moreover, by the **Morse lemma** one can always find such coordinates as before (for a nice proof of the Morse lemma, which works also in the case $k = 2$, see [LdM]. Note that the coordinates of the Morse lemma are only C^{k-1}). Especially, theorem 5.3.3 implies by this example the first part of theorem 5.0.1.

Using theorem 5.3.3 and example 5.3.3, we can now give the

Proof of proposition 5.0.1. Fix $x \in X^s$. Since this is a local result, we can assume

$$(M, X^s, x) \simeq (\mathbb{R}^m, \mathbb{R}^s \times \{0\}, 0)\,.$$

Let $f : (\mathbb{R}^m, 0) \to (\mathbb{R}, 0)$ be a smooth function germ with $df_0 \in \Lambda^\circ_{s,i}$. For the implication $\Leftarrow$ we have to show:

$$NMD(\mathcal{F}, df_0) = 0 \quad \Rightarrow \quad LMD(\mathcal{F}, f, 0) := \left(R\Gamma_{\{f \geq 0\}}(\mathcal{F})\right)_0 = 0\,.$$

By [KS2, thm. 3.1.1] or [KS, prop. 5.1.1], we only have to show this for a **real analytic** function germ f. But then the claim follows from theorem 5.3.3 and example 5.3.3(3). The implication $\Rightarrow$ follows similarly. Assume $NMD(\mathcal{F}, df_0) \neq 0$ and choose a smooth function germ f with $df_0 \in \Lambda^\circ_{s,i}$ and $f|X^s \equiv 0$. Then we get in the same way by theorem 5.3.3 and example 5.3.3(1):

$$LMD(\mathcal{F}, f, 0) := \left(R\Gamma_{\{f \geq 0\}}(\mathcal{F})\right)_0 \neq 0\,.$$

And this implies $df_0 \in \mu supp(Rk_*\mathcal{F})$ by the definition of the **micro-support** as in definition 4.0.4 on page 211. □

5.3.3 Normal Morse data and micro-localization

We finish this section with some applications of the "micro-local theory" of Kashiwara-Schapira (these results are **not** used somewhere else in this book).

If we assume that $X_\bullet$ is Whitney **a-regular** and satisfies the property **LSA**, then one can deduce theorem 5.3.3 for $\mathcal{F} \in ob(D^+(X_\bullet))$ (and therefore proposition 5.0.1) also from the "micro-local theory" of Kashiwara-Schapira.

Consider a smooth function germ $f : (M, z) \to \mathbb{R}, 0)$, with $df_z|T_zX^s = 0$ for a point $z \in X^s =: Y$ such that $\omega := df_z$ is **nondegenerate**. As we explained in (5.51) on page 310, there exists a complex of R-modules $M^\bullet \in ob(D^+(R))$, with

$$Rk_*\mathcal{F} \simeq M^\bullet_Y \quad \text{in } D^+(M, \omega)$$

in the corresponding **localized** category. We already identified $M^\bullet$ in (5.52) with the "normal Morse datum"

$$NMD(\mathcal{F}, df_z) := \left(R\Gamma_{\{f|N\cap X\geq 0\}}(\mathcal{F}|N\cap X)\right)_z .$$

Suppose that the assumptions of definition 5.1.1 are true for the triple

$$(f|X^s, \{z\}, \mathbb{Z}_{X^s}) ,$$

with $K_n = B_n$ a sequence of **compact balls** in $X^s \simeq \mathbb{R}^s \times \{0\}$. Then this is **also true** for the triple $(f|X^s, \{z\}, M^\bullet_{X^s})$, and one has:

$$\left(R\Gamma_{\{f|X^s\geq 0\}}(M^\bullet_{X^s})\right)_z \simeq \left(R\Gamma_{\{f|X^s\geq 0\}}(\mathbb{Z}_{X^s})\right)_z \otimes^L_Z M^\bullet . \tag{5.74}$$

This follows from the **projection formula**

$$Rf_*(M^\bullet_{K_n}) \simeq Rf_*(\mathbb{Z}_{K_n}) \otimes^L_Z M^\bullet$$

and the fact, that the isomorphism of the projection formula commutes with restriction (as explained in section 1.4). But by the definition of the category $D^+(M, df_z)$, one also has:

$$\left(R\Gamma_{\{f|X^s\geq 0\}}(M^\bullet_{X^s})\right)_z \simeq \left(R\Gamma_{\{f\geq 0\}}(M^\bullet_{X^s})\right)_z \simeq \left(R\Gamma_{\{f\geq 0\}}(\mathcal{F})\right)_z . \tag{5.75}$$

Using the deep theory of Kashiwara-Schapira about **contact transformation for sheaves** ([KS2, ch. 6], [KS, ch. VII]), we can use these two formulae to identify (under some weak restrictions) the "local type" of $\mathcal{F} \in ob(D^+(M))$ at a **non-zero** covector $\omega \in T^*_X M$ with the stalk

$$\left(\mu_X(\mathcal{F})\right)_\omega \quad \text{of the **micro-localization** } \mu_X(\mathcal{F}).$$

Let us assume for the rest of this section, that the base ring $R = A$ is **commutative of bounded weak global dimension** (as in [KS2, ch. 6], [KS, ch. VII]). Since we work locally, we can assume that M, X are C^∞-(sub)manifolds (in local coordinates). Suppose

$$\mu supp(\mathcal{F}) \subset T^*_X M$$

near a non-zero covector $\omega \in T^*_X M$ so that $\mathcal{F}$ is isomorphic to $M^\bullet_X$ in $D^+(M, \omega)$, with $M^\bullet \in ob(D^+(R))$ the "local type of $\mathcal{F}$ at ω". We claim

$$M^\bullet \simeq \left(\mu_X(\mathcal{F})\right)_\omega . \tag{5.76}$$

Also note, that (5.76) is trivial for $X = M$ (locally, with $\omega = 0$). Let us first consider the case that X is a **real hypersurface**. Suppose $\omega \in T^*_x M$, and choose locally a smooth real function f with

$$X = \{f = 0\} \quad \text{and} \quad df_x = \omega \neq 0 .$$

Especially, $f : (M, x) \to (\mathbb{R}, 0)$ is locally a **submersion**. By (5.74) and (5.75) we have

$$M^\bullet \simeq \left(R\Gamma_{\{f\geq 0\}}(\mathcal{F})\right)_x \simeq \left(\mu_X(\mathcal{F})\right)_\omega .$$

The last isomorphism is well known and follows by **Fourier-Sato transformation** from the isomorphism of distinguished triangles [KS, p. 193, proof of thm. 4.2.3.(v)] (for the following argument compare also with [KS, p. 354, 355, proof of prop. 8.6.3]):

Choose a local trivialization $T^*_X M \simeq X \times \mathbb{R}$ given by the section $s' : x \mapsto (x, df_x)$, and let π be the projection onto X. Since $\mu_X(\mathcal{F})$ is $\mathbb{R}^+$-conic, we have

$$s'^{-1}\mu_X(\mathcal{F}) \simeq R\pi_* Rhom\left(A_{X\times U}, \mu_X(\mathcal{F})\right) ,$$

with $U := \{x \in \mathbb{R} \,|\, x > 0\}$. By [KS, p. 172, thm. 3.7.9] we get

$$s'^{-1}\mu_X(\mathcal{F}) \simeq R\tau_* Rhom\left(A^\vee_{X\times U}, \nu_X(\mathcal{F})\right) \simeq R\tau_* Rhom\left(A_{X\times Z}, \nu_X(\mathcal{F})\right) ,$$

with τ the projection of $T_X M \simeq X \times \mathbb{R}$ onto X, and $Z := \{y \in \mathbb{R} \,|\, y \geq 0\}$ (compare [KS, p. 173, lem. 3.7.10.(i)]). From the isomorphism of distinguished triangles [KS, p. 193, proof of thm. 4.2.3.(v)] one finally gets

$$R\tau_* Rhom\left(A_{X\times Z}, \nu_X(\mathcal{F})\right) \simeq \left(R\Gamma_{\{f\geq 0\}}(\mathcal{F})\right)|X .$$

Let us now treat the **general case**. Choose locally a smooth function

$$\phi : (M, x) \to (\mathbb{R}, 0) , \quad \text{with } d\phi_x = \omega$$

such that $\phi|X$ has in x a **classical Morse critical point of index** 0. Especially,

$$\Lambda_\phi := \{(y, d\phi_y) \,|\, y \in M\} \quad \text{is } \textbf{transversal} \text{ to } T^*_X M \text{ at } \omega.$$

By (5.74) and (5.75) we also have

$$M^\bullet \simeq \left(R\Gamma_{\{\phi\geq 0\}}(\mathcal{F})\right)_x .$$

By [KS, p. 486, cor. A.2.7] one can find in a neighborhood of ω a **contact transformation**

$$\chi : T^*M \simeq T^*M' ,$$

whose graph is the conormal bundle to a hypersurface S such that

$$\chi(T^*_X M) = T^*_{X'} M' \quad \text{and} \quad \chi(T^*_{\{\phi=0\}} M) = T^*_{\{\psi=0\}} M' ,$$

with X' a **smooth hypersurface** and ψ a smooth function (compare [KS, p. 311]). We normalize ψ in such a way that

$$d\psi_{x'} = \chi(\omega) , \quad \text{with } \psi(x') = 0 .$$

Since a contact transformation as in loc.sit. is by definition **homogeneous**, we get by [KS2, p. 310, (7.5.6)] that

$$\Lambda_\psi := \{(y, d\psi_y) \mid y \in M'\} \quad \text{is transversal to } T^*_{X'}M' \text{ at } \chi(\omega).$$

Therefore $\psi|X'$ has in x' a **classical Morse critical point**.

By [KS2, thm. 6.3.5, thm. 6.3.9] or [KS, thm. 7.2.1, cor. 7.2.2] one can extend χ locally to an equivalence of categories

$$\Phi_K : D^+(M, \Omega_M) \simeq D^+(M', \Omega_{M'}) \,,$$

with $\Omega_M, \Omega_{M'}$ suitable open neighborhoods of ω and $\chi(\omega)$ such that

$$\chi_*\big(\mu hom(\mathcal{G}, \mathcal{F})\big) \simeq \mu hom\big(\Phi_K(\mathcal{G}), \Phi_K(\mathcal{F})\big) \tag{5.77}$$

for $\mathcal{G} \in ob(D^b(M))$ and $\mathcal{F} \in ob(D^+(M))$. Moreover $K \simeq A_S$ in the above construction. By [KS2, cor. 7.3.4] or [KS, prop. 7.4.6] one gets

$$\Phi_K(A_X) \simeq A_{X'}[\delta] \quad \text{and} \quad \Phi_K(A_{\{\phi=0\}}) \simeq A_{\{\psi=0\}}[\delta']$$

for suitable $\delta, \delta' \in \mathbb{Z}$. By (5.77) and the discription of the stalk of μ_ϕ as above, we get (since $\phi|X$ has in x a classical Morse critical point of index 0):

$$A \simeq \big(\mu_{\{\phi=0\}}(A_X)\big)_\omega \simeq \big(\mu hom(A_{\{\phi=0\}}, A_X)\big)_\omega \simeq$$

$$\big(\mu hom(A_{\{\psi=0\}}, A_{X'})\big)_{\chi(\omega)}[\delta - \delta'] \simeq \big(\mu_{\{\psi=0\}}(A_{X'})\big)_{\chi(\omega)}[\delta - \delta'] \,.$$

Since $\psi|X'$ has in x' a **classical Morse critical point**, we get from this isomorphism that the **Morse index** of $\psi|X'$ is $\delta - \delta'$. By (5.74), (5.75) and the isomorphism (5.76) for the hypersurface X' we get

$$\big(\mu_{\{\psi=0\}}(\Phi_K(\mathcal{F}))\big)_{\chi(\omega)} \simeq \big(\mu_{X'}(\Phi_K(\mathcal{F}))\big)_{\chi(\omega)}[\delta' - \delta] \,.$$

Note that

$$\mu supp(\Phi_K(\mathcal{F})) = \chi(\mu supp(\mathcal{F})) \subset T^*_{X'}M'$$

locally near $\chi(\omega)$. Then (5.77) implies

$$\big(\mu_{\{\phi=0\}}(\mathcal{F})\big)_\omega \simeq \big(\mu hom(A_{\{\phi=0\}}, \mathcal{F})\big)_\omega \simeq \big(\mu hom(A_{\{\psi=0\}}, \Phi_K(\mathcal{F}))\big)_{\chi(\omega)}[-\delta']$$

$$\simeq \big(\mu hom(A_{X'}, \Phi_K(\mathcal{F}))\big)_{\chi(\omega)}[-\delta] \simeq \big(\mu hom(A_X, \mathcal{F})\big)_\omega \simeq \big(\mu_X(\mathcal{F})\big)_\omega \,.$$

But we know that the first term is isomorphic to the "local type of $\mathcal{F}$ at ω", since $\phi|X$ has in x a **classical Morse critical point of index** 0.

Remark 5.3.1. One should compare the above argument with [KS2, cor. 7.4.2] and [KS, lem. 7.5.2, prop. 7.5.3]. But the explicit identification of the "local type" with the stalk of the micro-localization functor as in (5.76) is **not** contained in [KS, KS2]. Besides the results of Kashiwara-Schapira, we only used our isomorphisms (5.74), (5.75) and the two special cases of example 5.3.3(1),(2) for the calculation of the "tangential Morse datum". For another proof of the isomorphism (5.76) compare with [D'AMZ, thm. 2.3]).

Let $U \subset T^*M$ be an open subset, with

$$\mu supp(\mathcal{F}) \cap U \subset T^*_X M \cap U =: V$$

for a fixed $\mathcal{F} \in ob(D^+(M))$. By [KS, thm. 6.4.1] and [KS2, thm. 5.2.1(i,ii)] one gets the estimate:

$$\mu supp(\, \mu_X(\mathcal{F})\,) \cap T^*V \subset T^*_V V \,, \tag{5.78}$$

i.e. the **micro-support** of $\mu_X(\mathcal{F})$ restricted to V is contained in the **zero-section** of T^*V. Therefore all cohomology sheaves of

$$\mu_X(\mathcal{F})|V \quad \text{are locally constant !}$$

This was already explained in proposition 4.0.1 and 4.1.1 (cf. remark 4.1.2). By (5.76), this is a far-reaching generalization of the fact, that the "local type of $\mathcal{F}$" is locally constant on V. Consider especially $\mathcal{F} \in ob(D^+(X_\bullet))$ for a Whitney **a-regular** stratification $X_\bullet$ satisfying the property **LSA**. Let $k : X \to M$ be the closed inclusion. Then all cohomology sheaves of

$$\mu_{X^i}(Rk_*\mathcal{F})|\, \Lambda_i^\circ \quad \text{are locally constant,}$$

with Λ_i° the set of **nondegenerate** covectors in $T^*_{X^i}M$. Especially, the isomorphism class of the "normal Morse datum"

$$NMD(\mathcal{F},\omega) \simeq \big(\mu_{X^i}(Rk_*\mathcal{F})\big)_\omega$$

is locally independent of $\omega \in \Lambda_i^\circ$ (by the Vietoris-Begle theorem).

A very important example comes from the theory of **holonomic** D**-modules**:

Example 5.3.4. *Let M be a complex manifold and*

$$\mathcal{F} = Sol(\mathcal{M}) := Rhom_{D_M}(\mathcal{M}, \mathcal{O}_M)$$

be the **solution complex** *of a* **holonomic** *D-modul $\mathcal{M}$. Then $\mathcal{F}$ is complex analytically constructible ([KS, thm. 11.3.7]). Supppose $X_\bullet$ is a complex analytic Whitney* **a-regular** *stratification of M such that*

$$\mu supp(\mathcal{F}) \subset \bigcup_i T^*_{X^{2i}}M \subset T^*M \,.$$

If $\omega \in T^*_{X^{2s}}M$ *is a* **nondegenerate** *covector, then the stalk of* $\mu_{X^{2s}}(\mathcal{F})$ *at* ω *corresponds to the "micro-solution of* $\mathcal{M}$ *in* ω*" (cf. [MF, cor. 2.4, rem. 4]):*

$$\mu_{X^{2s}}(\mathcal{F}) \simeq Rhom_{\pi^{-1}D_M}(\pi^{-1}\mathcal{M}, \mu_{X^{2s}}(\mathcal{O}_M))\,.$$

Here $\pi : T^*M \to M$ *is the projection. Moreover, the complex* $\mu_{X^{2s}}(\mathcal{O}_M)$ *is concentrated in degree* $d := codim_{\mathbb{C}}(X^{2s}) = dim_{\mathbb{C}}(M) - s$*, with*

$$\mathcal{C}^R_{X^{2s}|M} := H^d(\mu_{X^{2s}}(\mathcal{O}_M))$$

the sheaf of **holomorphic microfunctions** *along* X^{2s} *(compare [KS, prop. 11.4.1, def. 11.4.2]).*

If $X_\bullet$ *is also* **b-regular***, then we know that the "micro-local type of* $\mathcal{F}$ *at* ω*" corresponds to the "normal Morse datum". Therefore* (5.76) *implies the relation between the "normal Morse datum" and the "micro-solution" stated in [GM, p. 224, 6.A.4.(b)]. Moreover, by [Ka2, p. 93, thm. 3.2.42] one has*

$$\mathcal{E}xt^i_{\pi^{-1}D_M}(\pi^{-1}\mathcal{M}, \mathcal{C}^R_{X^{2s}|M})_\omega \simeq$$

$$\mathcal{E}xt^i_{\mathcal{E}_M}(\mathcal{E}_M \otimes_{\pi^{-1}D_M} \pi^{-1}\mathcal{M}, \mathcal{C}^R_{X^{2s}|M})_\omega \simeq \begin{cases} \mathbb{C}^m & i = 0, \\ 0 & i \neq 0. \end{cases}$$

Here m *is the* **multiplicity** *of* $\mathcal{M}$ *at* ω*, with* $\mathcal{E}_M$ *the ring of* **finite-order microdifferential operators** *(which is flat over* $\pi^{-1}D_M$*). This implies directly the relations between the different notions of characteristic cycles as stated in remark* 5.0.4*, i.e.*

$$\chi(NMD(\mathcal{F}, \omega)) = (-1)^d \cdot m\,.$$

We finish this section with a "micro-local" generalization of the relation (5.57) between **duality and the "normal Morse data"**.

Let again $U \subset T^*M$ be an open subset, with

$$\mu supp(\mathcal{F}) \cap U \subset T^*_X M \cap U =: V$$

for a fixed $\mathcal{F} \in ob(D^b(M))$ so that all cohomology sheaves of $\mu_X(\mathcal{F})|V$ are locally constant. Suppose we are in the context of a "geometric category", with X **definable** and $\mathcal{F}$ **constructible** with all stalks **perfect**. Then one gets by [KS, prop.8.4.13(ii)] the following isomorphisms for $i_\omega : \{\omega\} \to V$ the inclusion:

$$\begin{aligned} i^*_\omega(\mu_X(D_M(\mathcal{F}))) &\simeq i^*_\omega(D_{T^*_X M}(\mu_X(\mathcal{F}))^a)\,[dim(X) - dim(M)] \\ &\simeq i^!_\omega(D_{T^*_X M}(\mu_X(\mathcal{F}))^a)\,[dim(X)] \\ &\simeq D(i^*_\omega(\mu_X(\mathcal{F}))^a)\,[dim(X)]\,. \end{aligned}$$

Here $(\cdot)^a$ denotes the direct image by the **antipodal map** a on $T^*_X M$. Moreover, the second isomorphism follows from the fact that (locally on V) all cohomology sheaves of

$$D_{T^*_X M}(\mu_X(\mathcal{F})) \simeq Rhom(\mu_X(\mathcal{F}), R_V[dim(V)])$$

are locally constant (compare [KS, prop. 3.2.3] and remark 2.2.1(2) on page 100). So one gets the following "micro-local" generalization of (5.57):

$$\big(\mu_X(D_M(\mathcal{F}))\big)_{\omega}[-dim(X)] \simeq D\big((\mu_X(\mathcal{F}))_{-\omega}\big)\,. \tag{5.79}$$

5.4 Vanishing cycles

In this section we explain the relation between the **vanishing cycle functor** of Deligne [De] and our **stratified Morse theory for constructible sheaves**. The most important case is for a complex analytic stratification, but the results are true in a more general context.

Let $f : X \to \mathbb{C}$ be a continuous function and consider the following cartesian diagram of maps induced by the inclusions $\{0\} \subset \mathbb{C} \supset \mathbb{C}^*$ and the universal covering $\widetilde{\mathbb{C}^*} \to \mathbb{C}^*$:

$$\begin{array}{ccccccc} X_0 & \xrightarrow{i} & X & \xleftarrow{j} & \{f \neq 0\} & \xleftarrow{p} & \widetilde{X^*} \\ \downarrow & & \downarrow f & & \downarrow & & \downarrow \tilde{f} \\ \{0\} & \longrightarrow & \mathbb{C} & \longleftarrow & \mathbb{C}^* & \longleftarrow & \widetilde{\mathbb{C}^*}\,. \end{array} \tag{5.80}$$

The **nearby cycle** ψ_f of f is the functor

$$\psi_f := i^* R\tilde{\pi}_* \tilde{\pi}^* : D(X) \to D(X_0)\,,$$

with $\tilde{\pi} := j \circ p$. The **vanishing cycle** ϕ_f of f is given as "the" cone of the natural morphism

$$sp : i^* \to \psi_f$$

induced from the adjunction map $id \to R\tilde{\pi}_* \tilde{\pi}^*$.

Let g be a generator of $\pi_1(\mathbb{C}^*) \simeq \mathbb{Z}$ (given by an orientation of $\mathbb{C}^*$) which we interpret as an automorphism of $\widetilde{\mathbb{C}^*}$. Then it acts on $R\tilde{\pi}_*\tilde{\pi}^*$ and we get the automorphism m_f of the distinguished triangle defining ϕ_f :

$$\begin{array}{ccccccc} i^*(\mathcal{F}) & \xrightarrow{sp} & \psi_f(\mathcal{F}) & \xrightarrow{can} & \phi_f(\mathcal{F}) & \xrightarrow{[1]} & \\ \downarrow id & & \downarrow m_f & & \downarrow m_f & & \\ i^*(\mathcal{F}) & \xrightarrow{sp} & \psi_f(\mathcal{F}) & \xrightarrow{can} & \phi_f(\mathcal{F}) & \xrightarrow{[1]} & . \end{array} \tag{5.81}$$

Remark 5.4.1.

1. We use the original definition of [De] for the vanishing cycle functor, in contrast to [KS2, section 8.6], who use our $\phi_f[-1]$ as the vanishing cycle of f.
2. See section 1.1 for a more precise definition for the automorphism m_f of the vanishing cycle functor ϕ_f (which we do not need here).

The **geometric** interpretation of these functors is related to our definition 5.1.1 of a **cohomological version of a Milnor fibration** for a function $f : X \to \mathbb{C}$. By lemma 1.1.1, remark 1.1.3 and corollary 1.1.1 we have the

Lemma 5.4.1. *Assume that the triple $(f, S, \mathcal{F})$ satisfies a* **cohomological version of a Milnor fibration** *(with K_n, ϵ_n as in definition 5.1.1). Then one has:*

1. *There exists for $0 < |w| \le \epsilon_n$ a commutative diagram*

$$\begin{array}{ccc} R\Gamma(K_n \cap \{|f| \le \epsilon_n\}, \mathcal{F}) & \longrightarrow & R\Gamma(K_n \cap \{f = w\}, \mathcal{F}) \\ \downarrow \wr & & \downarrow \wr \\ R\Gamma(S, \mathcal{F}) & \xrightarrow{sp} & R\Gamma(S, \psi_f(\mathcal{F})) \,. \end{array}$$

By embedding this diagram into an isomorphism of distinguished triangles, one gets

$$R\Gamma(K_n \cap \{|f| \le \epsilon_n\}, K_n \cap \{f = w\}, \mathcal{F}) \simeq R\Gamma(S, \phi_f[-1](\mathcal{F})) \,. \tag{5.82}$$

Moreover

$$R\Gamma(S, R\Gamma_{\{re(f) \ge 0\}}(\mathcal{F})) \simeq R\Gamma(S, \phi_f[-1](\mathcal{F})) \,. \tag{5.83}$$

2. *There exists for $0 < |w| \le \epsilon_n$ a commutative diagram*

$$\begin{array}{ccccc} R\Gamma(S, \psi_f(\mathcal{F})) & \xleftarrow{\sim} & R\Gamma(K_n \cap \{f = w\}, \mathcal{F}) & \xrightarrow{\sim} & Rf_*(\mathcal{F}|K_n)_w \\ \downarrow m_f & & \downarrow & & \downarrow mon \\ R\Gamma(S, \psi_f(\mathcal{F})) & \xleftarrow{\sim} & R\Gamma(K_n \cap \{f = w\}, \mathcal{F}) & \xrightarrow{\sim} & Rf_*(\mathcal{F}|K_n)_w \,. \end{array}$$

Here mon is the induced **monodromy** *action (the restriction of $Rf_*(\mathcal{F}|K_n)$ to $\{0 < |z| \le \epsilon_n\}$ has locally constant cohomology sheaves). Moreover one has an isomorphism*

$$R\Gamma(S, \phi_f(\mathcal{F})) \simeq \phi_{id}(Rf_*(\mathcal{F}|K_n))_0 \,, \tag{5.84}$$

which commutes with the (monodromy) automorphisms m_f, m_{id}.

We use this lemma only in the very special case $S = \{x\}$ a point.

Example 5.4.1. *Let us consider again the example 5.1.4, but this time with*

$$f : (M, x) \to (\mathbb{C}, 0)$$

a complex smooth function germ such that the covector df_x is **nondegenerate**. *Then we get by lemma 5.4.1(1) for $\mathcal{F} \in ob(D(X_\bullet))$ an isomorphism:*

$$LMD(\mathcal{F}, re(f), x) := \left(R\Gamma_{\{re(f) \ge 0\}}(\mathcal{F}) \right)_x \simeq \left(\phi_f[-1](\mathcal{F}) \right)_x \,.$$

This is the cohomological version of [GM, p. 134, prop. 12.5]. Consider especially the situation after taking a normal slice N to X^s at x. Then

$$\begin{aligned} NMD(\mathcal{F}, re(f), x) &\simeq \left(\phi_{f|N\cap X}[-1](\mathcal{F}|N\cap X)\right)_x \\ &\simeq R\Gamma(X\cap N\cap\{r\leq\delta_n\}, l'_X, \mathcal{F}) \,. \end{aligned} \tag{5.85}$$

Here we use the notation (with $0<|w|<<\delta<<1$)

$$(l'_X, \partial l'_X) := (X\cap N\cap\{r\leq\delta, f=w\}, X\cap N\cap\{r=\delta, f=w\})$$

for the **quarterlink** *of X with respect to f ([GM, p. 131, def. 12.4.1]). Especially for a* **complex analytic Whitney b-regular** *stratification, we get the relation* (5.12) *between the "normal Morse datum" and the vanishing cycle functor as stated in [GM, p. 223, 6.A.2].*

Remark 5.4.2. Consider a given smooth function germ $f' : (M, x) \to (\mathbb{R}, 0)$, with $df'_x|T_xX^s = 0$ and df'_x **nondegenerate**. If one wants to calculate $NMD(\mathcal{F}, f', x)$ by example 5.4.1 in terms of **vanishing cycles**, then one has to find a smooth function germ $f : (M, x) \to (\mathbb{C}, 0)$ with df_x **nondegenerate** (and $df_x|T_xX^s = 0$) such that $re(f) = f'$. This is possible in the following cases:

1. Assume that M is a(n) (almost) complex manifold, and all X^i are (almost) complex submanifolds of M. Let

   $$\omega \in Hom_C(T_xM, \mathbb{C}) \subset (T^*_xM)^C$$

   correspond to df'_x under the canonical isomorphism $T^*_xM \simeq Hom_C(T_xM, \mathbb{C})$. Choose a smooth function germ $g : (M, x) \to (\mathbb{R}, 0)$ with dg_x the imaginary part of ω. Then we can use $f := f' + i\cdot g$ (compare remark 5.1.4).

   Consider especially a **complex analytic** stratification in a complex analytic manifold M, with the **normal slice** N a **complex** submanifold of M. Suppose that the "normal Morse datum"

   $$NMD(\mathcal{F}, f', x) = NMD(\mathcal{F}, df'_x)$$

   depends only on the **nondegenerate** covector df'_x (e.g. the stratification is locally C-regular). Then we can work in suitable complex analytic coordinates with a **linear** function

   $$f' \in Hom_R(T_xM, \mathbb{R}) \,.$$

   In this case we can take $f = \omega \in Hom_C(T_xM, \mathbb{C})$ (with ω as before). Especially, in this case one can find a **holomorphic** function f with

   $$d(Re(f))_x = df'_x \,.$$

   Then the **quarterlink** is just the **complex link** as in proposition 5.0.3, and this is the intersection of the **complex analytic set** $X\cap N\cap\{f - f(x) = w\}$ with the ball $(B, \partial B) = (\{r\leq\delta_n, r=\delta_n\})$.

This is the construction of [GM, p. 161, lem. 2.1.4] for the "normal projection data" (compare [GM, p. 160, def. 2.1.3]. In part (b) of the definition of loc.sit. one has **not** to assume f proper).

2. Let the Hausdorff dimension of the **Nash fiber** over x, i.e. the **set of limit tangent planes**

$$\tau(X^j, x) := \{T \in G_j^{dim(M)}(T_x M) \mid \exists\, x_n \in X^j :\ x_n \to x,\ T_{x_n} X^j \to T\}$$

be $\leq j - i - 2$ for all $j > i$. Then the set of **nondegenerate** covectors in

$$\{\omega \in (T_x^* M)^C \mid 0 = \omega | T_x X^i\}$$

is dense in this set (cf. [NaTr, cor. 4.3]). Then we can at least **approximate** df_x' by $d(Re(f))_x$, with f as above. Under suitable assumptions, this is enough for a description of the "normal Morse datum" in terms of **vanishing cycles**.

A trivial example is given by the canonical Whitney stratification of $X := \mathbb{R}^m$, induced from a **linear subspace arrangement** $\{A_1, \ldots, A_n\}$ (compare [GM, p. 245]), if one has the estimate $dim(|v'|) \geq dim(|v|) + 2$ for all **distinct flats**

$$|v| = A_{i_1} \cap \ldots \cap A_{i_r} \subset A_{j_1} \cap \ldots \cap A_{j_s} = |v'|\,.$$

If one takes a **linear** normal slice N and

$$\omega \in (\mathbb{R}^{m-i})^* \simeq (T_{X^i}^* \mathbb{R}^m)_x$$

a **nondegenerate** covector, then the **quarterlink** is the intersection of the ball

$$(B, \partial B) = (\{r \leq \delta_n, r = \delta_n\}) \quad \text{with} \quad N \cap \{f - f(x) = w\}\,.$$

The stratification of $N \cap \{f - f(x) = w\}$ is for $f = \omega$ also induced from a **linear subspace arrangement** in N.

By the isomorphism (5.84) of lemma 5.4.1, we can now prove the main results of section 5.3 also in the case of complex functions.

Corollary 5.4.1 (of theorem 5.3.1 and 5.3.2). *Consider the context of theorem 5.3.1 (or of theorem 5.3.2), with f_t a family of* **complex** *functions (or $f_t := f$ a* **complex** *function, with $\pi_t := \pi$). Fix $\mathcal{F} \in ob(D(X_\bullet))$. Then the isomorphism class of the pair*

$$\Big(\big(\phi_{g_t}(\mathcal{F}|X \cap \{\pi_t = z\})\big)_z\,,\ m_{g_t} \Big) \quad \textit{for} \quad g_t := (f_t - f_t \circ \pi_t)|X \cap \{\pi_t = z\}$$

is **locally independent** *of $t \in T$ (or $z \in S$).*

Proof. By the proof of theorem 5.3.1(3) we know that

- the restriction of

$$\mathcal{G}_t := Rg_{t*}\big(\mathcal{F}|(X \cap \{\pi_t = z,\ |g_t| \leq \epsilon,\ \rho \leq \delta\})\big)$$

  to $\{|w| \leq \epsilon\}$ is **constructible** with respect to the filtration $(\{0\}, \{|w| \leq \epsilon\}$, and its isomorphism class is **locally independent** of $t \in T$ (or $z \in S$).

Here ϵ, δ are chosen as in theorem 5.3.1 or 5.3.2. But by this choice, we can apply the isomorphism (5.84) of lemma 5.4.1 and get

$$\big(\phi_{g_t}(\mathcal{F}|X \cap \{\pi_t = z\})\big)_z \simeq \big(\phi_{id}(\mathcal{G}_t)\big)_0 .$$

Moreover, this isomorphism commutes with the (monodromy) automorphisms m_{g_t} and m_{id} . □

We get especially (as in section 5.3) the counterpart of example 5.3.1 for complex functions (compare also with [GM, p. 140, prop. 12.8]), and in the **complex analytic context** the counterpart of example 5.3.2 for **holomorphic** functions or covectors. This implies in particular the following corollary (compare also with [GM, p. 163, proof of lem. 2.3.2]):

Corollary 5.4.2. *Let $X_\bullet$ be a* **complex analytic** *stratification, which is* **locally C-regular** *in $X \backslash X_{2s}$. Consider an open subset U of X^{2s} such that X^{2j} is Whitney* **b-regular** *over X^{2s} in U for all $j > s$. Let N be a* **complex analytic normal slice** *at $z \in U$, and fix $\mathcal{F} \in ob(D(X_\bullet))$. Then the isomorphism class of the pair*

$$\Big(\big(\phi_{f-f(z)}(\mathcal{F}|X \cap N)\big)_z\,,\ m_{f-f(z)}\Big)$$

depends only on the connected component of the set of **nondegenerate holomorphic** *covectors in $(T^*_{X^{2s}}M)|U$, which contains df_z. Especially, it does* **not** *depend on the choice of the normal slice, or on the holomorphic function f.*

If $X_\bullet$ is Whitney **b-regular**, *then we can take $U = X^{2s}$. In this case the set of* **nondegenerate holomorphic** *covectors in $T^*_S M$ is connected for each stratum S of $X_\bullet$ so that the isomorphism class of this pair only depends on the stratum S.*

Remark 5.4.3. Consider a **complex analytic Whitney b-regular** stratification $X_\bullet$ of a complex space X, which is **not** neccessarily globally embedded into a complex manifold M. Fix $\mathcal{F} \in ob(D(X_\bullet))$. Then one can use a **local holomorphic embedding** $(X, z) \hookrightarrow (M, z)$ into a complex manifold M to define the corresponding "normal Morse data":

$$NMD(\mathcal{F}, f, z) \quad \text{or} \quad \Big(\big(\phi_{f-f(z)}(\mathcal{F}|X \cap N)\big)_z\,,\ m_{f-f(z)}\Big),$$

with f induced from a **smooth or holomorphic function** f on (M, z) such that df_z is **nondegenerate**. Then the isomorphism class of these "normal Morse data" is **independent** of the used local embedding so that it is an invariant of the complex analytic stratum S containing z.

The description of the "normal Morse data" in terms of **vanishing cycles** and **complex links** has an important application in the context of the theory of **characteristic cycles** for a closed complex analytic subset X of the complex analytic manifold M, if X, M are "defined over $\mathbb{R}$" in the following sense:

We assume that M has an **antiholomorphic involution** σ, which is also called a **real structure** such that X is **invariant** under σ (e.g. we consider the algebraic context, with X, M defined over $spec(\mathbb{R})$). If Y is a **σ-invariant** subset of M, then let us denote by $Y(\mathbb{R})$ the **fixed point set** of $\sigma|Y$.

Choose a **complex analytic (algebraic) Whitney b-regular** stratification $X_\bullet$ of X (in the algebraic case as above), which is also **invariant** under σ. Then $X(\mathbb{R})$ is a closed **real analytic (algebraic)** subset of the real analytic (algebraic) manifold $M(\mathbb{R})$. Moreover, the $X_i(\mathbb{R})$ define a **subanalytic (semialgebraic) Whitney b-regular** stratification $X(\mathbb{R})_\bullet$ of $X(\mathbb{R})$.

We now compare the **Euler characteristic of the "normal Morse data"** for

$$\chi(NMD(1_X, \cdot) \quad \text{and} \quad \chi(NMD(1_{X(\mathbb{R})}, \cdot)\,.$$

Let us first compare the corresponding sets of **nondegenerate** covectors (for a fixed s):

$$\Lambda^\circ_{2s} \subset T^*_{X^{2s}}M \quad \text{and} \quad \Delta^\circ_s \subset T^*_{X^{2s}(R)}M(\mathbb{R})\,.$$

Then $\Lambda^\circ_{2s}(\mathbb{R})$ (for the induced real structure of T^*M) is an open **subanalytic (semialgebraic)** subset of Δ°_s, whose complement is of real codimension ≥ 1 in Δ°_s. Especially, $\Lambda^\circ_{2s}(\mathbb{R})$ intersects **all connected components** of Δ°_s.

Consider a covector $\omega \in \Lambda^\circ_{2s}(\mathbb{R})$. Then there exists a holomorphic function germ $f : (M, x) \to (\mathbb{C}, 0)$, with

$$f(M(\mathbb{R})) \subset \mathbb{R} \quad \text{and} \quad d(re(f))_x = \omega\,,$$

e.g. take a **real linearform** in suitable coordinates. Especially

$$dg_x \in \Delta^\circ_s \quad \text{for } g := re(f)|M(\mathbb{R})\,.$$

Take a **complex analytic normal slice** N at x in M, which is also **σ-invariant**. Then $N(\mathbb{R})$ is a **normal slice** at $x \in X^{2s}(\mathbb{R})$ in $M(\mathbb{R})$. Let us calculate the Euler characteristic of the "normal Morse data" with respect to these normal slices:

$$\chi(NMD(1_X, d(re(f))_x) = 1 - \chi(H^*(l_X, \mathbb{Q}))\,,$$
$$\chi(NMD(1_{X(R)}, d(g)_x) = 1 - \chi(H^*(l^-_{X(R)}, \mathbb{Q}))\,.$$

Here l_X is the **complex link** of X in x with respect to f, and $l^-_{X(R)}$ is the **lower halflink** of $X(\mathbb{R})$ in x with respect to g. But by our choices , $l^-_{X(R)}$ is just the **fixed point set** of $\sigma|l_X$. This implies (compare with (3.17) on page 176):

$$\chi(H^*(l^-_{X(R)}, \mathbb{Q})) \equiv \chi(H^*(l_X, \mathbb{Q})) \;\; (\; mod\; 2\;)\,.$$

Therefore we get the following relation between the corresponding **characteristic cycles** of 1_X and $1_{X(R)}$:

$$CC_2(1_{X(R)}) = cl_R(CC(1_X)) \,. \tag{5.86}$$

Here we use the notation

$$CC : CF(X_\bullet, \mathbb{Z}) \to H_{2m}^{BM}(\Lambda, \mathbb{Z})$$

and

$$CC_2 : CF(X(\mathbb{R})_\bullet, \mathbb{Z}_2) \to H_m^{BM}(\Delta, \mathbb{Z}_2)$$

for the corresponding characteristic cycle maps as in (5.22) of the introduction, with Λ and Δ the union of the corresponding conormal bundles for the stratifications $X_\bullet$ and $X(\mathbb{R})_\bullet$. Note that in the last case we do **not** need to assume $M(\mathbb{R})$ oriented, since we work with $\mathbb{Z}_2$-coefficients (compare also with [FuMC]).

Finally $cl_R(\cdot)$ is the **real cycle map** of [BH] induced by

$$[\, cl(T_S^*M)\,] \mapsto [\, cl(T_{S(R)}^*M(\mathbb{R}))\,] \quad \text{for a stratum } S \text{ of } X_\bullet,$$

with $[\cdot]$ the corresponding fundamental class in Borel-Moore homology.

Remark 5.4.4. The relation (5.86) implies especially that $CC_2(1_{X(R)})$ is **invariant** under the action of the **antipodal map** a of $T^*M(\mathbb{R})$. By (5.25) (or [FuMC, thm. 4.3]) this implies the classical result of Sullivan that $X(\mathbb{R})$ is a *mod* 2 **Euler space**:

$$D(1_{X(R)}) = 1_{X(R)} \in CF(X(\mathbb{R})_\bullet, \mathbb{Z}_2) \,,$$

i.e. the **Euler characteristic of a link is even** for all $x \in X(\mathbb{R})$ (as explained on page 176). So one should see this relation (5.86) as a **micro-local refinement** of the result of Sullivan.

Let us come back to the general case of a stratification $X_\bullet$, which is not neccessarily complex analytic. By a small modification of the proof of theorem 5.3.3 we get the

Theorem 5.4.1. *Assume $X_\bullet$ is* **a+d-regular** *in $X \backslash X_s$, and X^j is* **b-regular** *over X^s in the point $z \in X^s$ for all $j > s$. Consider a smooth function germ*

$$f : (M, z) \to (\mathbb{C}, 0), \quad \textit{with } df_z | T_z X^s = 0 \textit{ and } df_z \textbf{ nondegenerate}.$$

Fix $\mathcal{F} \in ob(D(X_\bullet))$, and choose local coordinates

$$(M, X^s, z) \simeq (\mathbb{R}^m, \mathbb{R}^s \times \{0\}, 0) \,, \quad \textit{with} \quad \pi : (\mathbb{R}^m, 0) \to (\mathbb{R}^s \times \{0\}, 0)$$

the linear projection. Suppose that the triple $(f|X^s, \{z\}, \mathbb{Z}_{X^s})$ satisfies the assumptions of definition 5.1.1 about a **cohomological version of a Milnor fibration**, *with $K_n = B_n$ a sequence of* **compact balls** *in $X^s \simeq \mathbb{R}^s \times \{0\}$.*

Let $N := \{\pi = z\}$. Then there exists an isomorphism

$$\left(\phi_f[-1](\mathcal{F})\right)_z \simeq \left(\phi_{f|X^s}[-1](\mathbb{Z}_{X^s})\right)_z \otimes^L_Z \left(\phi_{f|N\cap X}[-1](\mathcal{F}|N \cap X)\right)_z \,, \tag{5.87}$$

which identifies $m_f[-1]$ with $m_{f|X^s}[-1] \otimes^L_Z m_{f|N\cap X}[-1]$.

Proof. We follow the proof of theorem 5.3.3, with the corresponding notation K_n, N_n, B_n and g_n introduced there. Then we only have to change the last argument, where we assumed $\mathbb{K} = \mathbb{R}$. As in that proof, we get by the isomorphism (5.84) of lemma 5.4.1:

$$\left(\phi_f[-1](\mathcal{F})\right)_z \simeq \left(\phi_a[-1](Rg_{n*}(\mathcal{F}|K_n))\right)_0 .$$

Here a is the addition map

$$a : \mathbb{C}^2 \to \mathbb{C}\,;\ (w_1, w_2) \mapsto w_1 + w_2 .$$

Moreover, this isomorphism identifies $m_f[-1]$ and $m_a[-1]$. But we also have

$$Rg_{n*}(\mathcal{F}|K_n) \simeq (Rf_*\mathbb{Z}_{B_n}) \boxtimes_Z^L Rf_*(\mathcal{F}|N_n) .$$

Now we can apply the **Thom-Sebastiani Theorem** 1.2.2 to the map a and get the isomorphism

$$\left(\phi_a[-1](Rg_{n*}(\mathcal{F}|K_n))\right)_0 \simeq \left(\phi_{w_1}[-1](Rf_*\mathbb{Z}_{B_n})\right)_0 \otimes_Z^L \left(\phi_{w_2}[-1](Rf_*(\mathcal{F}|N_n))\right)_0 .$$

Moreover, this isomorphism identifies $m_a[-1]$ and $m_{w_1}[-1] \otimes_Z^L m_{w_2}[-1]$. Finally, by the same argument we have

$$\left(\phi_{w_1}[-1](Rf_*\mathbb{Z}_{B_n})\right)_0 \simeq \left(\phi_{f|X^s}[-1](\mathbb{Z}_{X^s})\right)_z$$

and

$$\left(\phi_{w_2}[-1](Rf_*(\mathcal{F}|N_n))\right)_0 \simeq \left(\phi_{f|N\cap X}[-1](\mathcal{F}|N\cap X)\right)_z ,$$

identifying the corresponding "monodromy isomorphisms" in the right way. □

Again there are two simple but important cases, where the assumptions on $f|X^s$ are true and where we can **calculate** the isomorphism class of the pair

$$\left(\left(\phi_{f|X^s}[-1](\mathbb{Z}_{X^s})\right)_z ,\ m_{f|X^s}[-1]\right) .$$

1. The trivial case that $f|X^s$ is the **zero function**. Then this pair is isomorphic to $(\mathbb{Z}, id)$, with $\mathbb{Z}$ in degree zero.

2. The other case deals with a **complex Morse critical point**: Assume that $(f|X^{2s}, z)$ is in some coordinates of X^{2s} isomorphic to

$$g : (\mathbb{C}^s, 0) \to (\mathbb{C}, 0)\,;\ g(w) := \sum_{i=1}^{s} w_i^2 .$$

For example in the complex analytic context, with f a **holomorphic** function germ such that $f|X^{2s}$ has an **ordinary quadratic point** at z.

- In the case $s = 1$, the **Milnor fiber** consists of two points, which are permuted by the geometric monodromy. This pair is for $s = 1$ therefore isomorphic to $(\mathbb{Z}[-1], -id)$.

- For $s > 1$ we get by iterated use of the Thom-Sebastini Theorem, that this pair is isomorphic to $(\mathbb{Z}[-s], (-1)^s \cdot id)$.

Remark 5.4.5. These two examples together with theorem 5.4.1 show especially, that there is a sign-mistake in [Gi1, cor. 6.19], i.e. the corresponding Euler characteristic of the vanishing cycles in the cases a) and b) of [Gi1, cor. 6.19] have to differ by a factor $(-1)^{dim_{\mathbb{C}}(Z)}$.

Before we can now explain the counterpart of corollary 5.1.2 for the **vanishing cycle functor**, we have to explain another basic distinguished triangle related to the **variation map** $var : \phi_f \to \psi_f$.

We use the notation of the diagram (5.80) so that

$$\psi_f \simeq i^* Rj_* Rp_* p^* j^* ,$$

with $p : \widetilde{X^*} \to \{f \neq 0\} =: U$ the induced **covering map**. Let g be a generator of $\pi_1(\mathbb{C}^*) \simeq \mathbb{Z}$, given by an orientation of $\mathbb{C}^*$, which we interpret now as an **automorphism** of $\widetilde{X^*}$. Let $\mathcal{F}$ be a sheaf on U. We claim that the following sequence of sheaves on U is **exact**:

$$0 \longrightarrow \mathcal{F} \xrightarrow{ad_p} p_* p^* \mathcal{F} \xrightarrow{id-g} p_* p^* \mathcal{F} \longrightarrow 0 .$$

Take a ball B in $\mathbb{C}^*$ so that the restriction of the universal covering map of $\mathbb{C}^*$ to B is isomorphic to the projection $B \times \mathbb{Z} \to B$, with g correponding to the translation $\mathbb{Z} \to \mathbb{Z}, i \mapsto i+1$. If we take an open subset $V \subset f^{-1}(B)$, then $p : p^{-1}(V) \to V$ is isomorphic to the projection $V \times \mathbb{Z} \to V$, with g acting as before on the second factor. Then

$$\Gamma(V, p_* p^* \mathcal{F}) \simeq \prod_{i \in \mathbb{Z}} \Gamma(V, \mathcal{F})$$

such that g acts by the **permutation** $i \mapsto i+1$. Moreover the adjunction map ad_p corresponds to the **diagonal embedding** $\Gamma(V, \mathcal{F}) \to \prod_{i \in \mathbb{Z}} \Gamma(V, \mathcal{F})$. Then it is an exercise that the sequence

$$0 \longrightarrow \Gamma(V, \mathcal{F}) \xrightarrow{diag} \prod_{i \in \mathbb{Z}} \Gamma(V, \mathcal{F}) \xrightarrow{id-g} \prod_{i \in \mathbb{Z}} \Gamma(V, \mathcal{F}) \longrightarrow 0$$

is exact. Since each point of $U = \{f \neq 0\}$ has a **fundamental system** of open neighborhoods V as before, this implies our claim.

Let in addition $\mathcal{F}$ be **flabby**. Since p is a covering and flabbiness is a local condition ([KS, prop. 2.4.10(i)]), $p^*\mathcal{F}$ is flabby and then also $p_* p^* \mathcal{F}$ ([KS, prop. 2.4.6]). By flabby resolution (recall that all our spaces are of bounded cohomological dimension) we get therefore the distinguished triangles (compare also with [Bry, prop. 1.1])

$$id \xrightarrow{ad_p} Rp_* p^* \xrightarrow{id-g} Rp_* p^* \xrightarrow{[1]} ,$$

and

$$i^* Rj_* j^* \xrightarrow{ad_p} \psi_f \xrightarrow{id-m_f} \psi_f \xrightarrow{[1]} . \tag{5.88}$$

Remark 5.4.6. Consider the sitation of lemma 5.4.1. Then m_f corresponds to the **monodromy** action on $Rf_*(\mathcal{F}|K_n)$. If we in addition assume that

$$f : K_n \cap \{|f| = \epsilon_n\} \to \{|z| = \epsilon_n\}$$

is a **stratified fibration**, then the cohomology sequence of the distinguished triangle that we get from application of $R\Gamma(S, \cdot)$ to (5.88), corresponds to the **Wang sequence** of this fibration.

Consider also the distingished triangle

$$i^* \xrightarrow{ad_j} i^*Rj_*j^* \longrightarrow i^![1] \xrightarrow{[1]} . \tag{5.89}$$

Since the diagram

$$\begin{array}{ccc} i^* & \xrightarrow{ad_j} & i^*Rj_*j^* \\ \| & & \downarrow{\scriptstyle ad_p} \\ i^* & \xrightarrow{sp} & \psi_f \end{array}$$

commutes, we get from the distinguished triangles (5.89), (5.88), (5.81) and the **octahedral axiom** ([KS, p. 36,(TR5)]) a distinguished triangle

$$i^![1] \longrightarrow \phi_f \xrightarrow{var} \psi_f \xrightarrow{[1]} , \tag{5.90}$$

with $var \circ can = id - m_f : \psi_f \to \psi_f$.

Remark 5.4.7. Our approach to the important distinguished triangle (5.90) works without any assumptions on the sheaf complexes or the base ring. For another construction see [KS, p. 352, (8.6.7)]. This depends on Poincaré-Verdier duality ([KS, p. 351, (8.6.2)]) so that one has to make suitable restrictions.

Now we can explain the counterpart of corollary 5.1.2 for the **vanishing cycle functor**. For other applications, we work in a more general context and consider first the **nearby cycle functor**:

Lemma 5.4.2. *Let $X_\bullet$ be a stratification of the closed subset X of the manifold M, which is* **a+d-regular**. *Let $f : M \to \mathbb{C}$ be a smooth function such that $X' := X \cap \{f = 0\}$ is a union of strata. Assume that $X_\bullet$ satisfies the following* **$\mathbf{a}_f$-condition** *of Thom:*

1. *f restricted to $X \backslash X'$ is a stratified submersion.*
2. *If x_i is a sequence in $X^j \backslash X'$ converging to $x \in X^s \cap X'$ such that*

$$kern(df(x_i)|T_{x_i}X^j)$$

converges to some limiting plane τ, then $\tau \supset T_xX^s$.

Fix a point $x \in X'$ and assume that the strata of X' are Whitney **b-regular** *over $\{x\}$. Let $\mathcal{F}$ be $X_\bullet$-constructible, and define in local coordinates $(M,x) \simeq (\mathbb{R}^m, 0)$ the distance function r by $r(x) := \sum_{i=1}^m x_i^2$. Let $i_x : \{x\} \to X'$ be the inclusion. Then one has for $0 < |w| << \delta << 1$:*

$$i_x^* \psi_f(\mathcal{F}) \simeq R\Gamma(X \cap \{r \leq \delta, f = w\}, \mathcal{F}) \,. \tag{5.91}$$

$$i_x^! \psi_f(\mathcal{F}) \simeq R\Gamma(X \cap \{r \leq \delta, f = w\}, X \cap \{r = \delta, f = w\}, \mathcal{F}) \,. \tag{5.92}$$

Proof. Since the strata of X' are Whitney **b-regular** over $\{x\}$, there exists a $\delta > 0$ such that r is a **stratified submersion** in

$$X' \cap \{0 < r < \delta\} \,.$$

If we take sequences $\epsilon_n, \delta_n \searrow 0$ with $0 < \epsilon_n << \delta_n << \delta$ and define

$$K_n := X \cap \{|f| \leq \epsilon_n,\, r \leq \delta_n\} \,,$$

then the assumptions of definition 5.1.3 are true for $(f, \{x\}, \mathcal{F})$ and we get a **cohomological version of a local Milnor fibration**. This follows (similarly to example 5.1.4) from lemma 5.1.5 and example 1.1.3 on page 36.

Then (5.91) is a special case of lemma 5.4.1(1) for the case $S = \{x\}$. Now $\psi_f(\mathcal{F})$ is constructible with respect to the induced stratification of X' by proposition 4.2.1 and lemma 4.2.1. Then the proof of (5.92) is the same as in the proof of corollary 5.1.2. This was already noticed in remark 5.1.6(2). □

Remark 5.4.8.

1. The assumptions of lemma 5.4.2 are true in the **complex analytic context**, i.e. X is locally a closed complex subspace of the complex manifold M, f is a holomorphic function and $\mathcal{F}$ is complex analytically constructible. For the existence of a **complex analytic Whitney b-regular a_f-stratification** (refining a given complex analytic stratification) see [Hiro, Cor. 1 to thm. 2, p. 248], [LeTe, cor. 1.3.5.1] and compare also with [KS2, prop. 8.5.9].

2. The reader should compare our proof of (5.92) with [De, (3.1.8.4)], which corresponds to the very special case of the complex analytic context, with (X,x) an isolated singularity, $f|X\backslash\{x\}$ a submersion and $\mathcal{F}$ a constant sheaf.

It is a simple exercise to show that the **a_f-condition** of Thom is true in the context of example 5.1.4, with f a complex function germ. We therefore get the

Corollary 5.4.3. *Consider the situation of example* 5.1.4, *with f a complex function germ. Let $k_x : \{x\} \to X$ be the inclusion. Then there exists for $f' := f - f(x)$ and $0 < |w| << \delta << 1$ two distinguished triangles:*

$$R\Gamma(M_{f'}, \mathcal{F})[-1] \to \left(\phi_{f'}[-1](\mathcal{F})\right)_x \to k_x^* \mathcal{F} \xrightarrow{[1]} \,, \tag{5.93}$$

and

$$k_x^! \mathcal{F} \to \left(\phi_{f'}[-1](\mathcal{F})\right)_x \to R\Gamma(M_{f'}, \partial M_{f'}, \mathcal{F})[-1] \xrightarrow{[1]} . \tag{5.94}$$

Here $(M_{f'}, \partial M_{f'}) := (X \cap \{r \leq \delta, f' = w\}, X \cap \{r = \delta, f' = w\})$ *is a* **local Milnor fiber** *of* f'. *Moreover*

$$\left(R\Gamma_{\{re(f') \geq 0\}}(\mathcal{F})\right)_x \simeq \left(\phi_{f'}[-1](\mathcal{F})\right)_x .$$

Proof. The last isomorphism and the first distinguished triangle are a special case of lemma 5.4.1(1). By remark 4.2.4 on page 248 we have $(\phi_{f'}(\mathcal{F}))_{x'} = 0$ for $x' \in \{f' = 0\} \backslash \{x\}$ (locally near x). This implies

$$i_x^*(\phi_{f'}(\mathcal{F})) \xleftarrow{\sim} i_x^!(\phi_{f'}(\mathcal{F})) ,$$

with $i_x : \{x\} \to \{f' = 0\}$ the inclusion. Then the second distinguished triangle follows from application of $i_x^![-1]$ to the distinguished triangle (5.90), together with (5.92). □

Remark 5.4.9. Let us consider especially the above situation after taking a normal slice N to X^s at x. Then the **local Milnor fiber** of $f'|N \cap X$:

$$(l'_X, \partial l'_X) := (X \cap N \cap \{r \leq \delta, f - f(x) = w\}, X \cap N \cap \{r = \delta, f - f(x) = w\})$$

(with $0 < |w| << \delta << 1$) is the **quarterlink** of [GM, p. 131, def. 12.4.1]. The distinguished triangles of corollary 5.4.3 correspond to [GM, p. 135, prop. 12.5(d)]. Moreover, corollary 5.4.3 and remark 5.4.2(1) imply in the complex analytic context the proposition 5.0.3.

The (relative) cohomology of the quarterlink (modulo its "boundary") with respect to $\mathcal{F} \in ob(D(X_\bullet))$ is also a **special case** of a "normal Morse datum", since we started with an **a+d-regular** stratification:

Let $j : U := X \backslash X^s \to X$ be the inclusion of the locally closed complement. Then $Rj_!j^*\mathcal{F}$ and $Rj_*j^*\mathcal{F}$ are also $X_\bullet$-constructible (corollary 4.2.1), and we want to calculate their stalks in $x \in X^s$. Then one gets from corollary 5.4.3 by the same argument as on page 309:

$$R\Gamma(l'_X, \mathcal{F})[-1] \simeq R\Gamma(l'_X, Rj_!j^*\mathcal{F})[-1] \simeq NMD(Rj_!j^*\mathcal{F}, f, x) , \tag{5.95}$$

and

$$R\Gamma(l'_X, \partial l'_X, \mathcal{F})[-1] \simeq R\Gamma(l'_X, \partial l'_X, Rj_*j^*\mathcal{F})[-1] \simeq NMD(Rj_*j^*\mathcal{F}, f, x) . \tag{5.96}$$

We do not use or explain the general relation between the functor $\psi_f[-1]$ or $\phi_f[-1]$ and **Poincaré-Verdier duality**. Compare for example with [Bry, lem. 1.4] for the **complex analytic context**, with the base ring R a field. Here we remark only the following simpler facts:

Corollary 5.4.4. *Consider the context of lemma 5.4.2, and assume that the base ring is* **commutative**. *Then one has isomorphisms*

1. $D\big(i^!_x\psi_f[-1](\mathcal{F})\big) \simeq i^*_x\psi_f[-1]\big(D(\mathcal{F})\big)$ *for* $\mathcal{F} \in ob(D^-(X_\bullet))$.
2. $D\big(i^*_x\psi_f[-1](\mathcal{F})\big) \simeq i^!_x\psi_f[-1]\big(D(\mathcal{F})\big)$ *for* $\mathcal{F} \in ob(D_{perf}(X_\bullet))$.

Proof. Let $(M_f, \partial M_f) := (X \cap \{r \leq \delta, f = w\}, X \cap \{r = \delta, f = w\})$ be a **local Milnor fiber** of f. Define similarly $M_f^\circ := X \cap \{r < \delta, f = w\}$ for $0 < |w| << \delta << 1$.

1. By lemma 5.4.2 and proposition 4.0.2 we get

$$\begin{aligned} D\big(i^!_x\psi_f[-1](\mathcal{F})\big) &\simeq D\big(R\Gamma(M_f, \partial M_f, \mathcal{F})[-1]\big) \simeq D\big(R\Gamma_c(M_f^\circ, \mathcal{F})[-1]\big) \\ &\simeq R\Gamma\big(M_f^\circ, D(\mathcal{F}|M_f^\circ)\big)[+1] \simeq R\Gamma\big(M_f^\circ, D(\mathcal{F})\big)[-1] \\ &\simeq R\Gamma\big(M_f, D(\mathcal{F})\big)[-1] \simeq i^*_x\psi_f[-1]\big(D(\mathcal{F})\big)\,. \end{aligned}$$

Here we use the isomorphism

$$D(\mathcal{F}|M_f^\circ) \simeq \big(D(\mathcal{F})\big)|M_f^\circ[-2]\,.$$

This isomorphism follows from the proof of lemma 5.1.6 and corollary 5.1.3 on page 314, together with $i^!(\mathbb{Z}_M) \simeq \mathbb{Z}_N[-2]$ for the inclusion $i : N := \{f = w\} \to M$. This isomorphism is in our context **globally** (and not only locally) true, since $T^*_N M$ is **globally trivial** near $N \cap X$ (by the section $x \mapsto (x, df_x)$).

2. Since the induced stratifications of M_f° and M_f are also **d-regular** (compare lemma 4.3.2), we have the **biduality isomorphism** (proposition 4.0.2):

$$\mathcal{F}|M_f^\circ \xrightarrow{\sim} D(D(\mathcal{F}|M_f^\circ))\,.$$

Moreover

$$R\Gamma_c(M_f^\circ, \mathcal{F}) \simeq R\Gamma(M_f, \partial M_f, \mathcal{F})$$

is **perfect** for $\mathcal{F} \in ob(D_{perf}(X_\bullet))$ by proposition 4.0.2 and (4.10) of remark 4.2.2 on page 240. Together this implies

$$D\big(R\Gamma(M_f^\circ, \mathcal{F})\big) \simeq R\Gamma_c\big(M_f^\circ, D(\mathcal{F}|M_f^\circ)\big)\,.$$

Then the proof goes on as in the case before. □

We can use this duality result for example in the complex analytic context or in the context of example 5.1.4 (with f a complex function germ).

Chapter 6

Vanishing theorems for constructible sheaves

Introduction: Results and examples

In this final chapter we explain some important applications of our stratified Morse theory for constructible sheaves in the **complex analytic (algebraic)** context. All topological spaces are locally compact with countable topology and of finite (cohomological) dimension.

6.0.1 (Co)stalk properties

First we recall a basic definition related to an abstract triangulated category:

Definition 6.0.1. *Let D be a triangulated category with translation functor $T = [1]$. A subfamily B of $ob(D)$ satisfies the property "$\geq$" (or "$\leq$"), if we have*

I. $0 \in B$.

II. *Let $X \to Y \to Z \overset{[1]}{\to}$ is a distinguished triangle in D. Then $X, Z \in B \Rightarrow Y \in B$ (i.e. B is stable by extensions [BBD,* 1.2.6*]).*

III. $X \in B \Rightarrow X[-1] \in B$ *(or* $X \in B \Rightarrow X[1] \in B$*).*

In the following, we do not distinguish between a family B of objects of D and the corresponding full subcategory.

Remark 6.0.1. 1. If B is non-empty, then II. and III. imply the first property (just take a shift of the distinguished triangle $X \overset{id}{\to} X \to 0 \overset{[1]}{\to}$, with $X \in B$).

2. I. and II. imply that

(a) B is stable by isomorphisms: If $X \simeq Y$ in D, then $X \in B \Rightarrow Y \in B$ (this follows from the distinguished triangle $X \xrightarrow{\sim} Y \to 0 \xrightarrow{[1]}$).

(b) B is stable by direct sums: $X, Y \in B \Rightarrow X \oplus Y \in B$ (this follows from the distinguished triangle $X \to X \oplus Y \to Y \xrightarrow{[1]}$). Especially, B defines an additive subcategory.

3. These notions are **dual** in the sense that B satisfies the property "$\geq$" if and only if B satisfies the property "$\leq$" in the dual triangulated category D°.

4. B satisfies both properties "$\geq$" **and** "$\leq$" if and only if B is a **strict full triangulated** subcategory (i.e. the full subcategory of D, whose objects belong to B is a triangulated subcategory which is stable by isomorphisms). This corresponds also to the notion of a "null system" used in [KS, def. 1.6.6].

5. All these properties are stable by intersection, pullback by an exact functor of triangulated categories, and by translation with $[n]$ for $n \in \mathbb{Z}$. Moreover, for the last case let us introduce for $n \in \mathbb{Z}$ the notation

$$D^{\geq n} := B[-n] \quad \text{or} \quad D^{\leq n} := B[-n]\,.$$

Especially, for B a "null system" one gets $D^{\geq n} = B = D^{\leq n}$ for all $n \in \mathbb{Z}$.

In the following, we are interested in such a subfamily B of $ob(D(R))$, given by suitable complexes of R-modules. Then we consider constructible complexes of sheaves, whose **(co)stalks** belong to B up to suitable shifts so that also this family of constructible complexes of sheaves has the property "$\geq$" or "$\leq$". We therefore list now some general examples, which we specify in the case $D(R)$:

Example 6.0.1. 1. *A* **thick** *subcategory of D is by definition [Nee, def. 2.1.6] just a "null system" stable by direct summands (by [Ri] this is also equivalent to Verdier's notion [SGA4$\frac{1}{2}$, p. 276, def. 1.1] of an "épaisse" subcategory).*

- *The family $D_{perf}(R)$ of* **perfect** *complexes, i.e. which are isomorphic to a bounded complex of finitely generated projective R-modules, is a thick subcategory of $D(R)$. Assume that R is commutative. Then the duality functor $D = Rhom(\cdot, R)$ induces an automorphism of $D_{perf}(R)$ with $D \circ D \simeq id$. It* **exchanges** *the subfamilies of $D_{perf}(R)$ with the property "$\geq$" or "$\leq$".*

2. *Let $(D^{\leq 0}, D^{\geq 0})$ be a t-***structure** *on D, i.e.:*

(i) *$D^{\leq -1} \subset D^{\leq 0}$ and $D^{\geq 1} \subset D^{\geq 0}$.*

(ii) *$Hom_D(X, Y) = 0$ for $X \in ob(D^{\leq 0})$ and $Y \in ob(D^{\geq 1})$.*

(iii) *For any $X \in ob(D)$ there exists a distinguished triangle $X_0 \to X \to X_1 \xrightarrow{[1]}$ in D, with $X_0 \in ob(D^{\leq 0})$ and $X_1 \in ob(D^{\geq 1})$.*

Then $D^{\geq 0}$ *(or* $D^{\leq 0}$*) satisfies the property "*$\geq$*" (or "*$\leq$*"), since these subcategories are stable by extensions (see [KS2, prop.* 10.1.7*] or [BBD,* 1.3.4*]). Moreover, our notation* $D^{\geq n} := D^{\geq 0}[-n]$ *(or* $D^{\leq n} := D^{\leq 0}[-n]$*) corresponds to the usual notation in this context.*

- *The standard t-structure* $(D^{\leq 0}, D^{\geq 0})$ *on* $D(R)$ *is given by* $D^{\geq 0}$ *(or* $D^{\leq 0}$*) consisting of the complexes with* **vanishing cohomology in negative (or positive) degrees***.*

3. *Let* N *be a null system in* D*,* $F : N \to Ab$ *a cohomology functor with values in the abelian category* Ab*. Consider a subfamily* A *of* $ob(Ab)$ *with the property: if*

$$0 \to M' \to M \to M'' \to 0$$

is a short exact sequence in Ab*, then* $M', M'' \in A \Rightarrow M \in A$*. Assume* A *is also stable by taking a subobject (or quotient object). Let* B *be the family of objects* $X \in ob(N)$*, with*

$$F(X[-n]) = 0 \quad \forall n > 0 \quad (\forall n < 0), \quad \text{and} \quad F(X) \in A\,.$$

Then B *satisfies the property "*$\geq$*" (or "*$\leq$*").*

- *If* $F := h^0(\cdot) : D(R) \to R-mod$ *is the usual cohomology functor to the category of* R*-modules, then we can take for* A *the family of* R*-modules which are* **torsion-free** *(with respect to a given subset* S *of* R*). This is stable by subobjects so that the category* B *has the property "*$\geq$*".*

  Here B *is given by the complexes of* R*-modules with* **vanishing cohomology in negative degrees and** $h^0(\cdot)$ **torsion-free** *(with respect to a given subset* S *of* R*).*

Let us **fix** for the rest of this chapter a subfamily B of $ob(D(R))$ with the property "$\geq$" (or "$\leq$").

During this chapter, we will **illustrate** our general results for the following special "classical case" of a family B (so that one does not need for this case the abstract notions of example 6.0.1):

Example 6.0.2 ("classical case"). 1. $B := D^{\leq 0}$ *is given by the complexes of* R*-modules with* **vanishing cohomology in positive degrees***, or by*

$$B := D^{\leq 0}_{perf} := D^{\leq 0} \cap D_{perf}\,.$$

2. $B := D^{\geq 0}$ *is given by the complexes of* R*-modules with* **vanishing cohomology in negative degrees***, or by*

$$B := D^{\geq 0}_{perf} := D^{\geq 0} \cap D_{perf}\,.$$

3. $B := {}^+D^{\geq 0}$ *is given by the complexes of* R*-modules with* **vanishing cohomology in negative degrees and** $h^0(\cdot)$ **torsion-free**, *or by*

$$B := {}^+D^{\geq 0}_{perf} := {}^+D^{\geq 0} \cap D_{perf} \, .$$

Especially, for R *a (commutative)* **principal ideal domain** *(like* $\mathbb{Z}$ *or a field) one has the following* **duality** *relation (compare [BBD, sec. 3.3]):*

$$D(\, D^{\leq 0}_{perf} \,) = {}^+D^{\geq 0}_{perf} \, . \tag{6.1}$$

Moreover, a complex of R*-modules is* **perfect** *if and only if it is cohomologically bounded with* **finitely generated** *cohomology (this is more generally true for* R *noetherian with finite homological dimension).*

We extend these definitions now to constructible complexes of sheaves of R-modules on a topological stratification $X_\bullet$ of the topological space X (compare [BBD, sec. 2.1] and [KS2, sec. 10.2]). First we recall the following

Definition 6.0.2. *Let* X *be a locally compact topological space.*

1. *A* **filtration** $X_\bullet$ *of* X *is a filtration* $\emptyset = X_{-1} \subset X_0 \subset \cdots \subset X_n = X$ *by closed subsets.*

2. *A complex of sheaves* $\mathcal{F} \in ob(D(X))$ *is called* **constructible** *with respect to* $X_\bullet$ *(or* $X_\bullet$*-constructible), if the restriction of* $\mathcal{F}$ *to all* $X^i := X_i \backslash X_{i-1}$ $(i = 0, \ldots, n)$ *has locally constant cohomology sheaves. Let us denote by* $D(X_\bullet)$ *the corresponding full subcategory of all* $X_\bullet$*-constructible complexes.*

3. *A* **(topological) stratification** *of* X *is a filtration* $X_\bullet$ *of* X *such that* X^i $(i = 0, \ldots, n)$ *is a topological manifold of pure dimension* i *(or empty). We call* $dim(X) := sup\{i | X^i \neq \emptyset\}$ *the* **dimension** *of* X *(with* $dim(\emptyset) := -\infty$*). The connected components of the* X^i *are called the* **strata** *(of this stratification).*

We assume that the stratification $X_\bullet$ satisfies for all i the following condition:

$(*_i)$ Let $j : V' := X \backslash X_i \to X \backslash X_{i-1} =: V$ be the open inclusion. Then $Rj_*\mathcal{G}$ is constructible with respect to the induced stratification of V for each $\mathcal{G} \in ob(D(V'))$, which is constructible with respect to the induced stratification of V'.

By proposition 4.2.1 this is for example the case for a **cohomology cone-like** or **ccs**-stratification in the sense of definition 4.0.6 on page 214. Especially for a **smooth stratification**, which is **b-,w-,d- or locally C-regular** (compare with chapter 4 for more on this). Let A be a **locally closed** subset of X, which is a union of strata. Then the induced stratification of A also has the property $(*_i)$ for all i (see page 235).

These stratifications have the following important stability properties (compare with proposition 4.0.2 and corollary 4.2.1):

Proposition 6.0.1. *Suppose $X_\bullet$ satisfies the condition $(*_i)$ for all i. Let $j : U \to V$ be the inclusion of* **locally closed** *subsets of X, which are unions of strata of $X_\bullet$. Then the corresponding categories of constructible complexes of sheaves (for the induced stratifications)*

$$D(U_\bullet) \quad and \quad D(V_\bullet)$$

are **stable** *under the functors*

$$j^*,\, j^!,\, j_! \quad and\ Rj_*\,.$$

Assume in addition that we consider complexes of sheaves of R-modules over a **commutative** *ring R, and that $X_\bullet$ is a* **ccs**-*stratification. Let $D_{perf}(X_\bullet)$ be the full subcategory of $X_\bullet$-constructible complexes $\mathcal{F}$, whose stalk complexes $\mathcal{F}_x$ are* **perfect** *complexes for all $x \in X$. Then also*

$$D_{perf}(U_\bullet) \quad and \quad D_{perf}(V_\bullet)$$

are stable under these four functors. Moreover, $D_{perf}(X_\bullet)$ is stable under the **duality functor** *D_X on X, with*

$$\mathcal{F} \xrightarrow{\sim} D_X(D_X\mathcal{F}) \quad for \quad \mathcal{F} \in ob(D_{perf}(X_\bullet))\,. \tag{6.2}$$

Consider the **fixed** subfamily B of $ob(D(R))$ with the property "$\geq$" (or "$\leq$"). Also recall, that $D^{\geq n} := B[-n]$ (or $D^{\leq n} := B[-n]$).

Definition 6.0.3. *Suppose that the stratification $X_\bullet$ has for all i the property $(*_i)$. Choose in addition a function*

$$p : \{strata\ of\ X_\bullet\} \to \mathbb{Z}\,.$$

Then ${}^pD^{\geq n}(X_\bullet)$ or ${}^pD^{\leq n}(X_\bullet)$ is the subfamily of $ob(D(X_\bullet))$ given by the complexes $\mathcal{F} \in ob(D(X_\bullet))$ with

$$k_x^* i_S^! \mathcal{F}\,[\,p(S)] \in D^{\geq n} \quad or \quad k_x^* i_S^* \mathcal{F}\,[\,p(S)] \in D^{\leq n} \tag{6.3}$$

for all strata S of $X_\bullet$ and $x \in S$. Here $k_x : \{x\} \to S$ and $i_S \to X$ are the inclusions.

Note that this is a **local** condition, i.e. $\mathcal{F} \in ob(D(X_\bullet))$ belongs to ${}^pD^{\geq n}(X_\bullet)$ or ${}^pD^{\leq n}(X_\bullet)$ if and only if each point $x \in X$ has an open neighborhood U in X such that $\mathcal{F}|U$ belongs to ${}^pD^{\geq n}(U_\bullet)$ or ${}^pD^{\leq n}(U_\bullet)$.

Remark 6.0.2. 1. The cohomology sheaves of $i_S^!\mathcal{F}$ are locally constant for $\mathcal{F} \in ob(D(X_\bullet))$ (by proposition 6.0.1). Therefore

$$k_x^* i_S^! \mathcal{F} \simeq k_x^! i_S^! \mathcal{F}\,[dim(S)]$$

(for example by lemma 5.1.6, with $M = S$), and the above conditions are equivalent to

$$i_x^! \mathcal{F}\,[\,p(S) + dim(S)] \in D^{\geq n} \quad \text{or} \quad i_x^* \mathcal{F}\,[\,p(S)] \in D^{\leq n} \tag{6.4}$$

for all $x \in S$, with $i_x : \{x\} \to X$ the inclusion. Also note, that the isomorphism class of $i_x^* \mathcal{F}$ and $i_x^! \mathcal{F}$ **depends only on the stratum** S, and not on the point $x \in S$.

2. Let $(D^{\leq 0}, D^{\geq 0})$ be the standard t-structure on $D(R)$ as in example 6.0.1(2). Then our definition and notation is the same as in [BBD, sec. 2.1].

3. Suppose that $X_\bullet$ is a **ccs**-stratification, and the base ring R is **commutative**. Assume $D^{\geq 0}$ is a subfamily of $ob(D_{perf}(R))$ with the property "$\geq$". Then $D^{\leq 0} := D(D^{\geq 0})$ (with $D(\cdot) := Rhom(\cdot, R)$ the duality functor) is also a subfamily of $ob(D_{perf}(R))$ with the property "$\leq$". If we define the **dual** function p^* of p by

$$p^* : \{\text{strata of } X_\bullet\} \to \mathbb{Z}, \quad p^*(S) := -p(S) - dim(S)\,,$$

then the duality functor D on $D_{perf}(X_\bullet)$ **exchanges**

$${}^{p}D^{\geq n}(X_\bullet) \quad \text{and} \quad {}^{p^*}D^{\leq -n}(X_\bullet)$$

by corollary 4.2.2 and the reformulation 1. above. If all non-empty strata of $X_\bullet$ are **even-dimensional**, then $p := m$ with $m(S) := -dim(S)/2$ is the only "self-dual function", i.e. with $p = p^*$.

The reformulation 1. above and property III. of definition 6.0.1 imply also the following counterpart of [BBD, prop. 2.1.14]:

Lemma 6.0.1. *Let $X_\bullet, X'_\bullet$ be topological stratification of the topological space X as in definition 6.0.3. Assume $X'_\bullet$ is a* **refinement** *of $X_\bullet$ in the sense that each stratum S of $X_\bullet$ is a union of strata T of $X'_\bullet$. Let*

$$p : \{\textit{strata of } X_\bullet\} \to \mathbb{Z} \quad \textit{and} \quad q : \{\textit{strata of } X'_\bullet\} \to \mathbb{Z}$$

be two functions with

$$q(T) + dim(T) \leq p(S) + dim(S) \quad \textit{or} \quad p(S) \leq q(T) \tag{6.5}$$

for each such pair of strata (S, T). Then

$${}^{p}D^{\geq n}(X_\bullet) \subset {}^{q}D^{\geq n}(X'_\bullet) \quad \textit{or} \quad {}^{p}D^{\leq n}(X_\bullet) \subset {}^{q}D^{\leq n}(X'_\bullet)\,.$$

If moreover

$$q(T) + dim(T) = p(S) + dim(S) \quad \textit{or} \quad p(S) = q(T) \tag{6.6}$$

for each such pair of strata (S, T) with T open in S, then one has for $\mathcal{F} \in ob(D(X_\bullet)) \subset ob(D(X'_\bullet))$:

$$\mathcal{F} \in {}^{p}D^{\geq n}(X_\bullet) \Leftrightarrow \mathcal{F} \in {}^{q}D^{\geq n}(X'_\bullet) \quad \textit{or} \quad \mathcal{F} \in {}^{p}D^{\leq n}(X_\bullet) \Leftrightarrow \mathcal{F} \in {}^{q}D^{\leq n}(X'_\bullet)\,.$$

Remark 6.0.3. In most cases the function p **depends only on the dimension** of the stratum, i.e.

$$p(S) := p(dim(S)), \quad \text{with} \quad p : \mathbb{N}_0 \to \mathbb{Z}\,.$$

In this case one gets by lemma 6.0.1, with $q = p$: The condition

(per) $0 \leq p(n) - p(m) \leq m - n$ for all $n, m \in \mathbb{N}_0$ with $n \leq m$,

implies the **stability** of ${}^pD^{\geq n}(X_\bullet)$ or ${}^pD^{\leq n}(X_\bullet)$ **with respect to refinements** of the stratification. Such a function is called a **perversity**, and the condition (per) can also be reformulated as (compare [KS, p. 419, (10.2.3)]):

$$p(n+1) \geq p(n) \geq p(n+1) - 1 \quad \text{for all } n \in \mathbb{N}_0. \tag{6.7}$$

So the **dual perversity** p^* of p is given by $p(n) := -p(n) - n$ $(n \in \mathbb{N}_0)$. We call p a **normalized** perversity, if in addition $p(0) = 0$. Two simple examples are given by the **zero perversity** $p = o$ and the **minimum perversity** $p = t$, with $o(n) := 0$ and $t(n) := -n$ for all $n \in \mathbb{N}_0$. Note, that $t \leq p \leq o$ for each **normalized perversity**, with $m := (t + o)/2 : 2 \cdot \mathbb{N}_0 \to \mathbb{Z}$ the **middle perversity**.

6.0.2 Intersection (co)homology and perverse sheaves

The notion of a perversity has its origin in the definition of **intersection (co)homology** by Goresky-MacPherson (cf. [GM4, GM5, HS, IC, Klei, MP2]). In this context one assumes $X_{n-2} = X_{n-1}$ for $n = dim(X)$, with $U := X^n$ dense in X. Then a **classical perversity** is a function $\bar{p} : \mathbb{N}_0 \to \mathbb{Z}$, with

$$\bar{p}(0) = \bar{p}(1) = \bar{p}(2) = 0, \quad \text{and} \quad \bar{p}(k+1) \leq \bar{p}(k) \leq \bar{p}(k+1) + 1 \quad \text{for all } k \in \mathbb{N}_0$$

(with k corresponding to the **codimension** of the stratum). So it is enough to consider $\bar{p}$ as a function $\bar{p} : \mathbb{N}_0 \backslash \{0, 1\} \to \mathbb{Z}$. Then there is a one-to-one correspondence between **normalized perversities** and **classical perversities**:

$$\bar{p} \leftrightarrow p : \quad \bar{p}(k) = -p(k-2) \quad \text{for all } k \in \mathbb{N}_0 \backslash \{0, 1\}\,.$$

Especially, $\bar{o}, \bar{t}$ or $\bar{m}$ is the **classical zero, top or middle perversity**, (with $\bar{t}(k) := k - 2$ and $\bar{m}(2k) := k - 1$).

We use in this chapter only **(normalized) perversities** as in remark 6.0.3, and **not** "classical perversities" so that there should be no misunderstanding. The only exception is the following definition of **intersection (co)homology**:

Let $\mathcal{L}$ be a locally constant sheaf on $U = X \backslash X_{n-1} = X \backslash X_{n-2}$. Then Deligne's definition of the **intersection cohomology complex** for the "classical perversity" $\bar{p}$ is

$$\mathcal{IC}^\bullet_{\bar{p}}(X, \mathcal{L}) := \tau_{\leq \bar{p}(n)}\, Rj_{n*} \cdots \tau_{\leq \bar{p}(2)}\, Rj_{2*}\, \mathcal{L}\,, \tag{6.8}$$

with $j_k : X \backslash X_{n-k} \to X \backslash X_{n-k-1}$ the open inclusion for $2 \leq k \leq n$. Here $\tau_{\leq i}$ is the truncation functor that kills the stalk cohomology in degree above i (compare

[KS, p.33, rem. 1.7.6, p.46]). Then the corresponding **intersection (co)homology** (with compact support) is defined by

$$I_{\bar{p}}C_k^{(c)}(X,\mathcal{L}) := H_{(c)}^{n-k}(X,\mathcal{IC}^\bullet_{\bar{p}}(X,\mathcal{L}))\,, \tag{6.9}$$

and

$$I_{\bar{p}}C^k_{(c)}(X,\mathcal{L}) := H^k_{(c)}(X,\mathcal{IC}^\bullet_{\bar{p}}(X,\mathcal{L}))\,. \tag{6.10}$$

Our notation for the "intersection complex" $\mathcal{IC}^\bullet_{\bar{p}}(X,\mathcal{L})$ agrees with that of [HS, IC, Klei], whereas [GM4, GS] us a homological language, i.e. they work in our notation with $\mathcal{IC}^\bullet_{\bar{p}}(X,\mathcal{L})[n]$.

Note that $\mathcal{P} := \mathcal{IC}^\bullet_{\bar{p}}(X,\mathcal{L})$ is $X_\bullet$-constructible by proposition 6.0.1. Moreover, by definition (6.8) it has the following properties:

IC1. $\mathcal{P}|U$ is (quasi-)isomorphic to $\mathcal{L}$.

IC2. All cohomology sheaves $h^i(i_S^*\mathcal{P})$ are zero for $i \geq \bar{p}(k)+1$, if S is a stratum of dimension $n-k$ with $k \geq 2$.

IC3. All cohomology sheaves $h^i(i_S^!\mathcal{P})$ are zero for $i \leq \bar{p}(k)+1$, if S is a stratum of dimension $n-k$ with $k \geq 2$.

In fact, $\mathcal{IC}^\bullet_{\bar{p}}(X,\mathcal{L})$ is the **only** $X_\bullet$-constructible complex with these properties (up to isomorphism). This is a special case of [BBD, prop. 2.1.9]:

$$\mathcal{IC}^\bullet_{\bar{p}}(X,\mathcal{L}) \simeq j_{!*}\mathcal{L}\,, \tag{6.11}$$

with $j_{!*}\mathcal{L}$ the **intermediate extension** of [BBD] for the following function $\tilde{p}$: {strata of $X_\bullet$} $\to \mathbb{Z}$:

$$\tilde{p}(S) := \begin{cases} 0 & \text{for } dim(S) = n. \\ \bar{p}(k)+1 & \text{for } dim(S) = n-k \text{ with } k \geq 2. \end{cases}$$

Assume in addition, that $X_\bullet$ is a **ccs**-stratification. Let the base ring R be a field, and consider a locally constant sheaf $\mathcal{L}$ on U, whose stalks are finite dimensional. Then

$$\mathcal{IC}^\bullet_{\bar{p}}(X,\mathcal{L}) \in ob(D_{perf}(X_\bullet))$$

by proposition 6.0.1. Then one gets by remark 6.0.2(3) and this **characterization** the following **duality result**:

$$D_X(\,\mathcal{IC}^\bullet_{\bar{p}}(X,\mathcal{L})\,) \simeq \mathcal{IC}^\bullet_{\bar{q}}(X,\mathcal{L}^*)\,. \tag{6.12}$$

Here

$$\mathcal{L}^* := hom_{R_U}(\mathcal{L}, or_U) \simeq Rhom_{R_U}(\mathcal{L}, or_U[n])\,[-n] \simeq D_U(\mathcal{L})[-n]$$

is the "dual local system", with or_U the "orientation sheaf" of U, and $\bar{q}$ the **dual classical perversity** of $\bar{p}$:

$$\bar{q}(k) := -\bar{p}(k) + k - 2 \quad \text{for } k \geq 2.$$

The last definition comes from

$$\bar{q}(k) + 1 = \tilde{q}(S) = n + \tilde{p}^*(S) = n - (\bar{p}(k) + 1) - (n - k) = -\bar{p}(k) + k - 1$$

for S a stratum of dimension $n - k$ for $k \geq 2$.

This duality property (6.12) was of course in some sense the starting point for the development of intersection (co)homology (compare [GM4, HS, IC, MP2], and especially with [Klei]). Also note, that it is important to use field coefficients in this result.

Consider more generally the case of a **principal ideal domain** R, with $\mathcal{L}$ a locally constant sheaf on U, whose stalks are **torsion-free** and finitely generated. Then one also gets $\mathcal{L}^* := hom_{R_U}(\mathcal{L}, or_U) \simeq D_U(\mathcal{L})[-n]$. But (6.12) need **not** be true. By (6.1) one has the following necessary condition for $\mathcal{P} := \mathcal{IC}^\bullet_{\bar{p}}(X, \mathcal{L})$:

IC3$_{tor}$. The cohomology sheaves $h^i(i^!_S\mathcal{P})$ are **torsion-free** for $i = \bar{p}(k)+2$, if S is a stratum of dimension $n-k$ with $k \geq 2$. This is also equivalent to the costalk condition $h^i(i^!_x\mathcal{P})$ is **torsion-free** for $i = n + \bar{p}(k) - k + 2 = n - \bar{q}(k)$ and all x in a stratum of dimension $n - k$ with $k \geq 2$.

In fact, one can show that this condition is also sufficient for (6.12) (compare with [GS] and [BBD, sec. 3.3]).

Another important fact is, that in the case of a **locally cone-like** stratification the intersection cohomology complex

$$\mathcal{IC}^\bullet_{\bar{p}}(X, \mathcal{L})$$

does **not** depend on the choice of this stratification. For the details we refer again to [GM4, HS, IC, MP2]. Let us only remark, that this property is closely related to lemma 6.0.1.

Similarly for the following example, where we use again a **fixed** subfamily B of $ob(D(R))$ with the property "$\geq$" (or "$\leq$").

Example 6.0.3. *Let p be a* **perversity***, and consider one of the following cases:*

1. *X is a closed subset of $\mathbb{R}^n$ or a real analytic (Nash-)manifold M, and X is definable with respect to an o-minimal structure $\mathcal{S}$ on $(\mathbb{R}, +, \cdot)$ or an analytic (Nash) geometric category (compare with chapter 2).*

2. *X is a complex analytic (algebraic) space.*

Then we can define ${}^pD^{\geq n}(X)$ *or* ${}^pD^{\leq n}(X)$ *as the* **limit** *of the families*

$$ {}^pD^{\geq n}(X_\bullet) \quad or \quad {}^pD^{\leq n}(X_\bullet)\,, $$

with $X_\bullet$ *a* **filtration by closed definable or complex analytic (algebraic)** *subsets (in addition to the assumptions as before). Note that these stratifications build a directed system (with respect to refinements) by the well-known existence results about* **Whitney b-regular stratifications** *of this type (compare [vDrMi, Te, Ver]).*

Especially, ${}^pD^{\geq n}(X)$ *or* ${}^pD^{\leq n}(X)$ *has the property "$\geq$" or "$\leq$". Moreover, one has for such a* **constructible** *complex of sheaves* $\mathcal{F}$ *the following reformulation in terms of a* **(co)support condition***:*

$$ \mathcal{F} \in {}^pD^{\geq n}(X) \Leftrightarrow dim_R\big(\{x \in X \mid i_x^!\mathcal{F}\,[\,p(s)+s] \notin D^{\geq n}\,\}\big) \;<\; s \;\;\forall\, s \in \mathbb{N}_0\,, $$

and

$$ \mathcal{F} \in {}^pD^{\leq n}(X) \Leftrightarrow dim_R\big(\{x \in X \mid i_x^*\mathcal{F}\,[\,p(s)] \notin D^{\leq n}\,\}\big) \;<\; s \;\;\forall\, s \in \mathbb{N}_0\;. $$

Note that for $\mathcal{F} \in ob(D(X_\bullet))$ the above sets are unions of strata. Especially, they are definable (or constructible) and have a well-defined dimension. Before we switch to the complex analytic context, let us notice the simple

Lemma 6.0.2. *Suppose that the stratification* $X_\bullet$ *of* X *has for all* k *the property* $(*_k)$*. Let* $j : U \to X$ *be the inclusion of an open subset, which is a union of strata, and denote by* $i : Y := X\backslash U \to X$ *the inclusion of the closed complement (which is also a union of strata). Then one has for* $\mathcal{F} \in ob(D(X_\bullet))$ *and the induced stratifications* $U_\bullet$ *and* $Y_\bullet$ *the following properties:*

1. $\mathcal{F} \in {}^pD^{\geq n}(X_\bullet)$ *if and only if* $j^*\mathcal{F} \in {}^pD^{\geq n}(U_\bullet)$ *and* $i^!\mathcal{F} \in {}^pD^{\geq n}(Y_\bullet)$*.*
2. $\mathcal{F} \in {}^pD^{\leq n}(X_\bullet)$ *if and only if* $j^*\mathcal{F} \in {}^pD^{\leq n}(U_\bullet)$ *and* $i^*\mathcal{F} \in {}^pD^{\leq n}(Y_\bullet)$*.*
3. $Ri_* = i_!$ *maps* ${}^pD^{\geq n}(Y_\bullet)$ *into* ${}^pD^{\geq n}(X_\bullet)$*, and* ${}^pD^{\leq n}(Y_\bullet)$ *into* ${}^pD^{\leq n}(X_\bullet)$*.*
4. Rj_* *maps* ${}^pD^{\geq n}(U_\bullet)$ *into* ${}^pD^{\geq n}(X_\bullet)$*.*
5. $Rj_!$ *maps* ${}^pD^{\leq n}(U_\bullet)$ *into* ${}^pD^{\leq n}(X_\bullet)$*.*

Assume now that all strata are **even dimensional** (e.g. a complex analytic or algebraic stratification). Especially, $dim(X) =: 2n$ is even.

Let us for example consider the standard t-structure $(D^{\leq 0}, D^{\geq 0})$ on $D(R)$ as in example 6.0.1(2). Then our definitions and notations for $B := D^{\geq 0}$ (or $B := D^{\leq 0}$) are the **same** as the one used in the context of the **perverse t-structure with respect to the middle perversity** m (compare [BBD, sec. 2.1] and [KS, prop. 10.2.4]). Especially, the objects of

$$ {}^mD^{\leq 0}(X_\bullet) \cap {}^mD^{\geq 0}(X_\bullet) $$

are the famous **perverse sheaves** on X, which are $X_\bullet$-constructible. Of course, in the complex analytic (algebraic) context we can also consider

$$^m D^{\leq 0}(X) \cap {}^m D^{\geq 0}(X)\,,$$

whose objects are the **complex analytically (algebraically) constructible perverse sheaves**. The **support condition** can in both cases easily be formulated in terms of the complex dimension $dim(\cdot) := dim_R(\cdot)/2$ of the supports of the cohomology sheaves:

$$\mathcal{F} \in {}^m D^{\leq 0}(X_\bullet) \text{ or } {}^m D^{\leq 0}(X) \Leftrightarrow dim(\,supp(h^k(\mathcal{F}))\,) \leq -k \text{ for all } k.$$

The most important example is given by the **shifted intersection cohomology complex**

$$\mathcal{IC}^\bullet_{\bar{m}}(X, \mathcal{L})\,[n] \simeq j_{!*}(\mathcal{L}[n])\,,$$

with $j_{!*}(\mathcal{L}[n])$ the **intermediate extension** of [BBD] for the middle perversity m. Note that $\mathcal{L}[n]$ is a **perverse sheaf** on $U = X^{2n}$, and $-n + \tilde{m} : \{\text{strata of } X_\bullet\} \to \mathbb{Z}$ is the **middle perversity** m:

$$-n + \tilde{m}(S) := \begin{cases} -n + 0 = -n & \text{for } dim(S) = 2n. \\ -n + \bar{m}(2k) + 1 = -(n-k) & \text{for } dim(S) = 2(n-k) \text{ with } k \geq 1. \end{cases}$$

More generally,

$$\mathcal{IC}^\bullet_{\bar{p}}(X, \mathcal{L})\,[n] \in {}^m D^{\leq 0}(X_\bullet) \quad \text{or} \quad \mathcal{IC}^\bullet_{\bar{p}}(X, \mathcal{L})\,[n] \in {}^m D^{\geq 0}(X_\bullet)\,,$$

if $\bar{p}(2k) \leq \bar{m}(2k) + 1 = k$ or $\bar{p}(2k) \geq \bar{m}(2k) - 1 = k - 2$ for all $k \geq 1$. In particular

$$\mathcal{IC}^\bullet_{\bar{p}}(X, \mathcal{L})\,[n] \in {}^m D^{\leq 0}(X_\bullet) \cap {}^m D^{\geq 0}(X_\bullet)$$

is a perverse sheaf, if $k - 2 \leq \bar{p}(2k) \leq k$ for all $k \geq 1$. Finally

$$\mathcal{IC}'_X := \mathcal{IC}^\bullet_{\bar{m}}(X, R_X)\,[n]$$

is **selfdual** for R a field, i.e.

$$D_X(\,\mathcal{IC}'_X\,) \simeq \mathcal{IC}'_X\,,$$

if $X_\bullet$ is in addition a **ccs**-stratification with $U = X^{2n}$ an oriented manifold (e.g. for a complex analytic or algebraic Whitney b-regular stratification).

6.0.3 Vanishing results in the complex context

From now on we consider only the **complex analytic (algebraic)** context, with $p := m$ the **middle perversity**, and B a **fixed** subfamily of $ob(D(R))$ with the property "$\geq$" (or "$\leq$"). Moreover, dimension always refers to the **complex dimension**.

If $(D^{\leq 0}, D^{\geq 0})$ is the standard t-structure on $D(R)$, then some of the following results for $B := D^{\geq 0}$ (or $B := D^{\leq 0}$) are well known (compare [BBD, 4.1], [GM, II. chapter 6] and [KS, chapter X]). But in our approach we do **not** need the general theory of t-structures. The only important point is:

- The corresponding "(co)stalk property" $B \subset ob(D(R))$ satisfies the condition "$\leq$" (or "$\geq$") of definition 6.0.1, and for a constructible complex of sheaves we measure these "(co)stalk properties" with respect to the **middle perversity** m.

Lemma 6.0.3. *Let $f : X \to Y$ be a holomorphic (algebraic) map of complex analytic (algebraic) spaces, whose fiber dimension is bounded by d.*

1. *$f^!$ maps ${}^mD^{\geq n}(Y)$ into ${}^mD^{\geq n-d}(X)$.*
2. *f^* maps ${}^mD^{\leq n}(Y)$ into ${}^mD^{\leq n+d}(X)$.*

Suppose in addition that $f : X \to Y$ is a **smooth map of constant relative dimension** *d. Then $f^! \simeq f^*[2d]$ by* **Poincaré-Verdier duality** *(compare with remark 4.2.3) so that*

3. *f^* maps ${}^mD^{\geq n}(Y)$ into ${}^mD^{\geq n+d}(X)$.*
4. *$f^!$ maps ${}^mD^{\leq n}(Y)$ into ${}^mD^{\leq n-d}(X)$.*

Note that the complex analytically (algebraically) constructible complexes are stable under f^* and $f^!$ by theorem 4.0.2. One also gets similar results for a **transversal intersection** (compare with [KS, lem. 10.3.9]):

Lemma 6.0.4. *Let $Y_\bullet$ be a complex analytic Whitney* **b-regular** *stratification of the closed complex analytic subset Y of the complex manifold M. Suppose the complex submanifold $N \subset M$ of* **constant codimension** *$dim(M) - dim(N) =: d$ is* **transversal** *to all strata of $Y_\bullet$. Let $i' : X := Y \cap N \to Y$ be the inclusion, and denote by $X_\bullet$ the induced b-regular stratification of X. Then*

1. *$i'^!$ and $i'^*[-2d]$ map ${}^mD^{\geq n}(Y_\bullet)$ into ${}^mD^{\geq n+d}(X_\bullet)$.*
2. *i'^* and $i'^![2d]$ map ${}^mD^{\leq n}(Y_\bullet)$ into ${}^mD^{\leq n-d}(X_\bullet)$.*

For $c : X \to \{pt\}$ a constant map we get by lemma 6.0.3 especially the

Corollary 6.0.1. *Suppose the base ring R, viewed as a complex in degree zero, belongs to our fixed family $B = D^{\geq 0}$ or $B = D^{\leq 0}$. Then*

$$D(R_X) = c^! R \in {}^mD^{\geq -d}(X) \quad or \quad R_X = c^* R \in {}^mD^{\leq d}(X)\,,$$

with $D(R_X) = c^! R$ the **dualizing complex** *on X and $d := dim(X)$. Assume in addition that X is* **smooth of constant dimension** *d. Then*

$$R_X = c^* R \in {}^mD^{\geq d}(X) \quad or \quad D(R_X) = c^! R \in {}^mD^{\leq -d}(X)\,.$$

This corollary applies in particular to B as in the "classical case" of example 6.0.2, with

$$R \in {}^+D^{\geq 0}_{perf} \cap D^{\leq 0}_{perf}\,.$$

Moreover, it is closely related to the following

Definition 6.0.4. *Let X be a complex space. Then the* **rectified homological depth** *$rHd(X,R)$ of X with respect to the commutative base ring R is $\geq n$ (for $n \in \mathbb{Z}$) if*

$$D(R_X) = c^! R \in {}^m D^{\leq -n}(X) \tag{6.13}$$

for $B := D^{\leq 0}$ as in example 6.0.2(1). If R is a principal ideal domain (e.g. $\mathbb{Z}$ or a field), then this is also equivalent to

$$R_X = c^* R \in {}^m D^{\geq n}(X)$$

for $B := {}^+D^{\geq 0}$ as in example 6.0.2(3). The **rectified homological depth** *$rHd(X,R)$ of $X \neq \emptyset$ with respect to R is the maximum of the integers n such that (6.13) holds.*

For example $rHd(X,R) \leq dim(X)$ for any X, with $rHd(X,R) = dim(X)$, if $X \neq \emptyset$ is smooth of pure dimension. For $X \neq \emptyset$ one has more generally:

$$\begin{gathered} rHd(X,R) = dim(X) \Leftrightarrow \\ R_X[dim(X)] \text{ is perverse and } X \text{ is pure dimensional.} \end{gathered} \tag{6.14}$$

Choose a **complex analytic Whitney b-regular** stratification $X_\bullet$ of X so that $D(R_X) = c^! R \in ob(D(X_\bullet))$. Fix a point x in a stratum S of dimension s, with $k_x : \{x\} \to S$ and $i_S \to X$ the inclusions. Then one gets:

$$\begin{aligned} &k_x^* i_S^* D(R_X)[-2s] \simeq k_x^! i_S^* D(R_X) \simeq R\Gamma_c(B, i_S^* D(R_X)), \\ &\text{and} \quad H_c^{-k}(B, i_S^* D(R_X)) \simeq H_k^c(B, B \backslash S). \end{aligned} \tag{6.15}$$

Here B is the intersection of X with a small open ball around x (in a local embedding of (X,x)). Therefore we get the following equivalence for the vanishing of the relative homology (with compact support):

$$\begin{aligned} H_k^c(B, B\backslash S) = 0 \quad \text{for } k < n - s \quad &\Leftrightarrow \quad H^{k'}(\, k_x^* i_S^* D(R_X)\,) = 0 \quad \text{for } k' > -n - s \\ &\Leftrightarrow \quad k_x^* i_S^* D(R_X)[-s] \in D^{\leq -n} := B[n], \end{aligned}$$

with $k' \leftrightarrow -k - 2s$. So our definition 6.0.4 agrees with that of [HL, def. 1.1.3, p. 313] (one also has $rHd(X,R) = dim(X) - Lh(X,R)$, with $Lh(X,R)$ as in [Kaup, sec. 5]).

In fact, by [HL, thm. 1.9] this notion of **rectified homological depth** is also equivalent to the one of Grothendieck in [SGA2], who introduced this notion, together with its "homotopy version" of **rectified homotopical depth**, long before the creation of "intersection (co)homology and perverse sheaves".

He conjectured that it gives the level of comparison for the homology (or homotopy) in the **weak Lefschetz theorem** for singular varieties. As we will see soon (for the case of homology), this turns out to be true! For the corresponding "homotopy results", we refer to [HL].

The first **main result** in the complex analytic context is the

Theorem 6.0.1. *Let $X_\bullet$ be a* **complex analytic Whitney b-regular stratification** *of the complex analytic space X, and consider a C^∞-function $f : X \to \mathbb{R}$. Assume $a < b$ are* **regular values** *of f (with respect to $X_\bullet$) such that $\{a \le f \le b\}$ is* **compact**.

1. *Suppose f is* **q-convex** *in $\{a \le f \le b\}$, and let $q' := min\{q, dim(X)\}$.*

$$\mathcal{F} \in {}^m D^{\le n}(X_\bullet) \Rightarrow R\Gamma(\{a \le f \le b\}, \{f = a\}, \mathcal{F}) \in D^{\le n+q'} .$$

2. *Suppose $-f$ is* **q-convex** *in $\{a \le f \le b\}$, and let $q' := min\{q, dim(X)\}$.*

$$\mathcal{F} \in {}^m D^{\ge n}(X_\bullet) \Rightarrow R\Gamma(\{a \le f \le b\}, \{f = a\}, \mathcal{F}) \in D^{\ge n-q'} .$$

Here we use the following

Definition 6.0.5. *A function $f : X \to \mathbb{R}$ on the complex space X is a* **C^k-function** *(with $k \ge 2$), if for any $x \in X$ there exists a local model $(X, x) \hookrightarrow (\mathbb{C}^n, x)$ such that f is induced by restriction of a C^k-function germ $\hat{f} : (\mathbb{C}^n, x) \to (\mathbb{R}, f(x))$.*

f is called **q-convex in** *$x \in X$, if one can choose this $\hat{f}$ so that its* **Levi form** *$L_{\hat{f}}(x)$ at the point x has at most q non-positive eigenvalues (cf. [SV], where such a function is called "strongly q-plurisubharmonic"). Here the* **Levi form** *of $\hat{f}$ at the point x is the Hermitian form given by the matrix of partial derivatives:*

$$L_{\hat{f}}(x) := \left(\frac{\partial^2 \hat{f}}{\partial z_j \, \partial \bar{z}_k}(x) \right)_{1 \le j,k \le n} .$$

f is called **q-convex** *(in a subset $S \subset X$), if it is q-convex in all points (of S).*

For example a **strongly plurisubharmonic** C^k-function on an open subset U of $\mathbb{C}^n$ (with $k \ge 2$) is 0-convex in all of U. Especially the distance function

$$r : \mathbb{C}^n \to \mathbb{R}^{\ge 0} ;\ r(z) := \sum_{i=1}^n z_i \bar{z}_i \quad \text{is 0-}\textbf{convex}\text{ in } \mathbb{C}^n .$$

The proof of theorem 6.0.1 given in the next section depends essentially on our **Morse theory for constructible sheaves** as developed in chapter 5 ! After approximation we can assume that f has only **finitely** many **stratified Morse critical points** in $\{a \le f \le b\}$. Then the assumption $\pm f$ is **q-convex** implies the following important **estimate for the Morse index** $\tau(f|S)$ of $f|S$ for a stratum S in terms of $q' := min\{q, dim(X)\}$ (compare [GM, p. 191, lem. 4.A.2]):

$$\begin{cases} \tau(f|S) \le dim_C(S) + q', & \text{if } f \text{ is q-convex.} \\ \tau(f|S) = dim_R(S) - \tau(-f|S) \ge dim_C(S) - q', & \text{if } -f \text{ is q-convex.} \end{cases} \tag{6.16}$$

Besides this estimate of the "Morse index", we also need some information about the **normal Morse datum**

$$NMD(\mathcal{F}, f, x) =: NMD(\mathcal{F}, x) \in ob(D(R)) \quad \text{of } \mathcal{F} \text{ in } x \in S.$$

Its **isomorphism class depends only on the stratum** S containing x, but not on the function f (by theorem 5.0.1 and remark 5.4.3). Here one gets by induction over $dim(X)$ the following "micro-local" characterization of our "generalized vanishing condition" for the (co)stalks:

Corollary 6.0.2. *Let* $X_\bullet$ *be a* **complex analytic Whitney b-regular stratification**, *and consider* $\mathcal{F} \in ob(D(X_\bullet))$. *Then we have:*

1. $\mathcal{F} \in {}^mD^{\leq 0}(X_\bullet) \Leftrightarrow NMD(\mathcal{F}, x)[-dim(S)] \in D^{\leq 0}$ *for all strata* S *and* $x \in S$.
2. $\mathcal{F} \in {}^mD^{\geq 0}(X_\bullet) \Leftrightarrow NMD(\mathcal{F}, x)[-dim(S)] \in D^{\geq 0}$ *for all strata* S *and* $x \in S$.

Remark 6.0.4. This corresponds in the "classical case" of example 6.0.2(1),(2) to [KS, thm. 10.3.12]. Especially, $\mathcal{F} \in ob(D(X_\bullet))$ is a **perverse sheaf** if and only if the cohomology of the "normal Morse data"

$$NMD(\mathcal{F}, x)[-dim(S)] \quad \textbf{vanishes in all degrees except zero}.$$

This corresponds to the "dimension axiom" in the Morse theoretic approach to perverse sheaves as outlined in [MP2, GrM]. It implies also the **purity result** stated in [GM, p. 223, 6.A.3].

As an example, let us explain corollary 6.0.2 in the "classical case" of example 6.0.2(1),(2) for $\mathcal{F} := D_X := D(R_X)$ the **dualizing complex** of X. Fix a point x in the stratum S of $X_\bullet$, and a local embedding $(X, x) \hookrightarrow (\mathbb{C}^n, 0)$. Then one gets by proposition 5.0.3 a distinguished triangle:

$$R\Gamma(l_X, \partial l_X, \mathcal{F})[-2] \longrightarrow k_x^!(\mathcal{F}|N \cap X) \longrightarrow NMD(\mathcal{F}, x) \xrightarrow{[1]} .$$

Here $k_x : \{x\} \to N \cap X$ is the inclusion into the intersection of X with a complex submanifold N of $\mathbb{C}^n$ such that N is transversal to S, with $X \cap N = \{x\}$. Finally the **complex link** of X in $x \in S$ is given by

$$(l_X, \partial l_X) := X \cap N \cap \{g = w\} \cap (B, \partial B) \quad \text{for } 0 < |w| << 1,$$

with $g : (N, x) \to (\mathbb{C}, 0)$ a **generic** holomorphic or $\mathbb{C}$-linear function, and B a small closed ball around x. By corrollary 5.1.3 we get

$$D_X|(X \cap N) \simeq D_{X\cap N}\,[2 \cdot dim(S)] \quad \text{and} \quad D_X|l^\circ \simeq D_{l^\circ}\,[2 \cdot dim(S) + 2]\,,$$

with $l^\circ := l_X \backslash \partial l_X$. Therefore

$$H^k(l_X, \partial l_X, D_X)[-2] \simeq H^c_{-2\cdot dim(S)-k}(l_X, R)$$

and

$$H^k(k_x^!(D_X|X \cap N)) \simeq H^c_{-2\cdot dim(S)-k}(X \cap N \cap B, R)$$

so that the cohomology of $NMD(D_X, x)\,[-dim(S)]$ is given by

$$H^k(\,NMD(D_X, x)\,[-dim(S)]\,) \simeq H^c_{-dim(S)-k}(X \cap N \cap B, l_X, R)\,.$$

By definition 6.0.4 and corollary 6.0.2 we get

$$\begin{aligned} &rHd(X,R) \leq n \Leftrightarrow D(R_X) \in D^{\leq -n} \Leftrightarrow D(R_X)[-n] \in D^{\leq 0} \Leftrightarrow \\ &H^c_k(X \cap N \cap B, l_X, R) = 0 \quad \text{for } k < n - dim(S) \text{ and all strata } S. \end{aligned} \tag{6.17}$$

Compare also with [HL, thm. 4.1.2] for the corresponding homotopy result.

All the remaining applications are based on theorem 6.0.1, and **no further use of Morse theory** for constructible sheaves is needed!

Corollary 6.0.3. *Let $f : X \to \mathbb{R}^{\geq 0}$ a* **proper C^∞-function** *on the complex space X. Assume f and X have a* **(sub)analytic compactification**, *i.e. there exists a compact (sub)analytic space $\overline{X}$, with $j : X \to \overline{X}$ the inclusion of a locally closed (sub)analytic subset of $\overline{X}$ such that f is the restriction of a continuous subanalytic function $\bar{f} : \overline{X} \to [0, \infty]$. Fix a complex analytically constructible complex $\mathcal{F}$ on X such that $j_!\mathcal{F}$ (or equivalently $Rj_*\mathcal{F}$) is (sub)analytically constructible. Then one gets for $d := dim(X)$:*

1. $\mathcal{F} \in {}^mD^{\leq n}(X) \;\Rightarrow\; R\Gamma_c(X, \mathcal{F}) \in D^{\leq n+d}\,.$
2. $\mathcal{F} \in {}^mD^{\geq n}(X) \;\Rightarrow\; R\Gamma(X, \mathcal{F}) \in D^{\geq n-d}\,.$

Assume moreover that f is also **q-convex**. *Then one gets for $q' := min\{q, dim(X)\}$:*

3. $\mathcal{F} \in {}^mD^{\leq n}(X) \;\Rightarrow\; R\Gamma(X, \mathcal{F}) \in D^{\leq n+q'}\,.$
4. $\mathcal{F} \in {}^mD^{\geq n}(X) \;\Rightarrow\; R\Gamma_c(X, \mathcal{F}) \in D^{\geq n-q'}\,.$

The assumptions on the compactification are used to get for all $r >> 0$:

$$R\Gamma(X, \mathcal{F}) \simeq R\Gamma(\{f \leq r, \mathcal{F}) \simeq R\Gamma(\{-1 \leq f \leq r\}, \{f = -1\}, \mathcal{F})$$

and

$$R\Gamma_c(X, \mathcal{F}) \simeq R\Gamma_c(\{f < r, \mathcal{F}) \simeq R\Gamma(\{-r \leq -f \leq 1\}, \{-f = -r\}, \mathcal{F})$$

so that we can apply theorem 6.0.1 to $\pm f$. Note that a C^2-function f on a complex space X is always q-convex on compact subset $K \subset X$ for some integer $q \in \mathbb{N}_0$.

Remark 6.0.5. The assumption on $\mathcal{F}$ in corollary 6.0.3 are satisfied, if $\mathcal{F}$ is constructible with respect to a complex analytic Whitney b-regular stratification $X_\bullet$ of X such that all X_i are locally closed subanalytic subsets of the compactification $\overline{X}$.

A simple example comes from a complex analytic Whitney b-regular stratification $\overline{X}_\bullet$ of the compact complex space $\overline{X}$, which one restricts to the complement $X := \overline{X} \backslash Y$ of a closed (sub)analytic subset $Y \subset \overline{X}$.

The most important special case is given by a **complex algebraic Whitney b-regular stratification** of the **affine** complex algebraic variety X. If we consider X as a closed algebraic subspace of some $\mathbb{C}^N$, then we can take for f the **strongly plurisubharmonic** distance function

$$f := r : \mathbb{C}^n \to \mathbb{R}^{\geq 0}\,;\ r(z) := \sum\nolimits_{i=1}^{n} z_i \bar{z}_i\ .$$

This extends of course to a continuous semialgebraic function on the projective space $\overline{X} := \mathbb{P}^n(\mathbb{C})$. In fact $1/(1+r)$ is the restriction to $\mathbb{C}^n = \{z_0 \neq 0\}$ of

$$[z_0, \ldots, z_n] \mapsto\ z_0\bar{z}_0 / \sum\nolimits_{i=0}^{n} z_i \bar{z}_i\ .$$

So we can take $q = 0$ for this f and get the **Artin-Grothendieck type result**:

Corollary 6.0.4. *Let X be an* **affine** *complex algebraic variety, with $\mathcal{F} \in ob(D(X))$* **complex algebraically** *constructible. Then*

$$\mathcal{F} \in {}^{m}D^{\leq n}(X) \Rightarrow R\Gamma(X, \mathcal{F}) \in D^{\leq n}\ \ \textit{and}\ \mathcal{F} \in {}^{m}D^{\geq n}(X) \Rightarrow R\Gamma_c(X, \mathcal{F}) \in D^{\geq n}\ .$$

The classical theorem of Artin-Grothendieck [SGA4, p. 159, thm. 3.1] is the relative counterpart for an affine morphism in the context of the **perverse t-structure** (for the middle perversity) in l-adic cohomology (compare [BBD, sec. 4], especially [BBD, p. 102, thm. 4.1.1]). The absolute case of an affine algebraic variety corresponds to [SGA4, p. 160, cor. 3.2].

The relative version of corollary 6.0.4 for an **affine complex algebraic morphism** is also true, as we will see in the end of this introduction. At the moment, let us only illustrate this corollary 6.0.4 by the following (well-known)

Example 6.0.4 (weak Lefschetz theorem for singular spaces). *Let X be a closed algebraic subvariety of the complex projective space, with H a hyperplane. Consider the open inclusion $j : U := X\backslash H \to X$, with $\mathcal{F} \in ob(D(X))$* **complex algebraically** *constructible. Then*

$$j^*\mathcal{F} \in {}^{m}D^{\leq n}(U) \Rightarrow R\Gamma(X, Rj_*j^*\mathcal{F}) \simeq R\Gamma(U, j^*\mathcal{F}) \in D^{\leq n}$$

and

$$j^*\mathcal{F} \in {}^{m}D^{\geq n}(U) \Rightarrow R\Gamma(X, X \cap H, \mathcal{F}) \simeq R\Gamma_c(U, j^*\mathcal{F}) \in D^{\geq n}\ .$$

In particular, one gets for B as in the "classical case" of example 6.0.2:

1. *$H_k^c(X, X \cap H, R) = H^{-k}(U, j^*D_X) = 0$ for $k < rHd(U, R)$.*
2. *If R is a principal ideal domain, then*

$$H^k(X, X \cap H, R) = \begin{cases} 0, & \textit{for } k > rHd(U,R). \\ \textit{(torsion)free} & \textit{for } k = rHd(U,R). \end{cases}$$

3. *Suppose X is purely n-dimensional, and H is a* **generic** *hyperplane, i.e. H is transversal to a Whitney b-regular stratification of X. By this transversality one gets*
$$i^! \mathcal{IC}^\bullet_{\bar{p}}(X, R_X) \simeq \mathcal{IC}^\bullet_{\bar{p}}(X \cap H, R_{X\cap H})\,[2]$$
for $i : X \cap H \to X$ the closed inclusion. Therefore
$$I_{\bar{p}}C^c_k(X, X \cap H, R) \simeq H^{2n-k}(U, j^*\mathcal{IC}^\bullet_{\bar{p}}(X, R_X)) = 0 \quad \textit{for } k < n,$$
and $\bar{p}$ any **classical perversity** *with $\bar{p}(2k) \le k$ for all $k \ge 1$.*

Here (1.) and (2.) show the role of the **rectified homological depth** for the "weak Lefschetz theorem for singular spaces" as conjectured by Grothendieck in [SGA2]. Compare also with [HL, conj. D, thm. 3.4.1] for the corresponding homotopy result.

The "weak Lefschetz theorem for intersection homology" (3.) is due to [GM4, thm. 7.1], and compare also with [GM, p. 217, thm. 6.10]. Here we only cite the following remark of Goresky-MacPherson in the beginning of [GM, p. 217, sec. 6.10]: "...The following Lefschetz hyperplane theorem was our original motivation for developing Morse theory on singular spaces ...". The proof given in [GM4, sec. 7] is different, and corresponds to the proof of the absolute version of the Artin-Grothendieck theorem given in [SGA4, p. 160, cor. 3.2].

The "weak Lefschetz theorem" of example 6.0.4 can be generalized in many different directions. Especially, it is enough to assume that X is a **quasiprojective** algebraic subvariety $X = X'\backslash A$, with $A \subset X' \subset \mathbb{P}^N(\mathbb{C})$ closed subvarieties, if H is a **generic** hyperplane. Just take a Whitney b-regular stratification of X' such that A is a union of strata, and H is **transversal** to all strata. Then one can apply the following

Lemma 6.0.5 (base change isomorphism). *Let $A \subset X'$ be closed analytic subvarieties of the complex manifold M. Fix a Whitney b-regular stratification $X'_\bullet$ of X', with A a union of strata. Suppose H is a closed complex submanifold of M, which is transversal to all strata, and consider the following cartesian diagram of inclusions:*
$$\begin{array}{ccc} X := X'\backslash A & \xleftarrow{\;j\;} & U := X\backslash H \\ \big\downarrow{\scriptstyle k'} & & \big\downarrow{\scriptstyle k} \\ X' & \xleftarrow[j']{} & U' := X'\backslash H\,. \end{array}$$
Then one has for $\mathcal{F} \in ob(D(X_\bullet))$ constructible with respect to the induced stratification of X natural isomorphisms:
$$j'_! Rk_* j^*\mathcal{F} \simeq Rk'_* j_! j^*\mathcal{F} \quad \textit{and} \quad Rj'_* k_! j^*\mathcal{F} \simeq k'_! Rj_* j^*\mathcal{F}\,. \tag{6.18}$$
If X' is also **compact***, then one gets for $\mathcal{F} \in ob(D(X_\bullet))$:*
$$R\Gamma_c(X, Rj_*j^*\mathcal{F}) \simeq R\Gamma(X', k'_! Rj_*j^*\mathcal{F}) \simeq R\Gamma(U', k_! j^*\mathcal{F})\,,$$

and

$$R\Gamma(X, X \cap H, \mathcal{F}) \simeq R\Gamma(X', Rk'_* j_! j^* \mathcal{F}) \simeq R\Gamma_c(U', Rk_* j^* \mathcal{F}) \,.$$

Note that the sheaf complexes of (6.18) are **complex analytically constructible** on X' for $\mathcal{F} \in ob(D(X_\bullet))$. Moreover

$$j^*\mathcal{F} \in {}^m D^{\leq n}(U) \Rightarrow k_! j^* \mathcal{F} \in {}^m D^{\leq n}(U')$$

and

$$j^*\mathcal{F} \in {}^m D^{\geq n}(U) \Rightarrow Rk_* j^* \mathcal{F} \in {}^m D^{\geq n}(U')$$

by lemma 6.0.2. So we can apply corollary 6.0.3, if there exists a **proper q-convex** $\mathbf{C}^\infty$**-function** $f : U' \to \mathbb{R}^{\geq 0}$, which is the restriction of a continuous subanalytic function $\bar{f} : X' \to [0, \infty]$. This applies in particular to the **quasiprojective context**, with H a generic hyperplane so that $\mathbb{P}^N(\mathbb{C}) \backslash H \simeq \mathbb{C}^N$.

So from this point of view, "weak Lefschetz theorems" are just special cases of **Artin-Grothendieck type theorems** as in corollary 6.0.3 for the complement $U' = X' \backslash H$. In particular, for B as in the "classical case" of example 6.0.2 (without the property *perf*), we do **not** need the additional assumption on a **(sub)analytic compactification** of the q-convex exhaustion function as in corollary 6.0.3.

Then we are allowed to consider a stratified Morse function with possible **infinitely** many critical points, since the corresponding "vanishing conditions" are stable under "limits" (compare [Sch3]). So one gets corresponding vanishing results on **q-complete** complex spaces, which are defined as follows:

Definition 6.0.6. *A complex space X is called* **q-complete** *($q \in \mathbb{N}_0$), if there exists a* **proper q-convex** $\mathbf{C}^\infty$**-function** $f : X \to \mathbb{R}^{\geq 0}$. *In particular, the* **0-complete** *complex spaces are just the complex analytic* **Stein spaces** *(cf. [Na]), and a closed complex subspace of a q-complete space is again q-complete.*

But if we work with the more general (co)stalk properties as in definition 6.0.1, then these are in general not "stable under limits", e.g. direct sums or products. Consider for example X a **countable discrete set**, which is **0-complete**. Then

$$R\Gamma(X, R) \simeq \prod\nolimits_{x \in X} R \quad \text{or} \quad R\Gamma_c(X, R) \simeq \bigoplus\nolimits_{x \in X} R$$

is **perfect** if and only if X is a **finite** set. So in general it is necessary to make a suitable "compactness assumption" as in corollary 6.0.3. To cover all the different cases, it is useful to introduce the

Definition 6.0.7. *Let X be a complex space, and fix a family $C(X) \subset ob(D(X))$ of complex analytically constructible complexes. Consider a* **fixed** *subfamily B of $ob(D(R))$ with the property "$\leq$" (or "$\geq$"). Then X is* **topologically q-complete** *with respect to the family $C(X)$ and the (co)stalk property B ($q \in \mathbb{Z}$), if*

$$\mathcal{F} \in {}^m D^{\leq n}(X) \Rightarrow R\Gamma(X, \mathcal{F}) \in D^{\leq n+q} \quad \textit{or}$$

$$\mathcal{F} \in {}^m D^{\geq n}(X) \Rightarrow R\Gamma_c(X, \mathcal{F}) \in D^{\geq n-q} \quad \textit{for } \textbf{all } \mathcal{F} \in C(X).$$

If X has this property for an integer $q \in \mathbb{Z}$, then the same is true for any $q' \geq q$ (by III. of definition 6.0.1*).*

Let us explain some important special cases:

Example 6.0.5.

1. *Assume X is a complex algebraic variety, and take for $C(X)$ the family of all complex algebraically constructible complexes of sheaves. If X is topologically q-complete with respect to $C(X)$ and a (co)stalk property B as above, then we call X* **algebraically topological q-complete** *with respect to the (co)stalk property B.*

   In particular, any closed algebraic subspace of X is then also **algebraically topological q-complete** *with respect to the (co)stalk property B. If X is* **affine**, *then X is* **algebraically topological 0-complete** *with respect to* **any** *(co)stalk property B as above (corollary* 6.0.4*).*

2. *Assume the complex algebraic variety X is the union $X = U_1 \cup U_2$ of two open algebraic subsets, with U_1, U_2 and $U_1 \cap U_2$* **algebraically topological q-complete** *with respect to a (co)stalk property B. Then X is* **algebraically topological (q+1)-complete** *with respect to B. This follows from a corresponding* **Mayer-Vietoris triangle** *([KS,* (2.6.30), (2.6.28), *p.* 114*]).*

   If for example X is the union of $k+1$ **affine** *open algebraic subsets, then X is* **algebraically topological k-complete** *with respect to* **any** *(co)stalk property B as above.*

3. *Let X be a complex space, and consider for $C(X)$ the family of all complex analytically constructible complexes of sheaves. If X is topologically q-complete with respect to $C(X)$ and a (co)stalk property B as above, then we call X* **topologically q-complete** *with respect to the (co)stalk property B.*

   Again, this remains true for a closed complex subspace of X. This notion corresponds for the (co)stalk property $B = D^{\leq 0}, D^{\geq 0}$ or ${}^{+}D^{\geq 0}$ as in the "classical case" of example 6.0.2 *to the "(dual) topological q-completeness" studied in [HL2, HL3, HL4]. As remarked before, any* **q-complete** *complex space is* **topologically q-complete** *in this special case (cf. [Sch3]).*

4. *Let $X_\bullet$ be a* **complex analytic stratification** *of X, satisfying the property $(*_i)$ for all i (e.g. a Whitney b-regular stratification). Take for $C(X)$ the family of all $X_\bullet$-constructible complexes of sheaves. If X is topologically q-complete with respect to $C(X)$ and a (co)stalk property B as above, then we call the* **stratification** *$X_\bullet$* **topologically q-complete** *with respect to the (co)stalk property B.*

   Again, this remains true for the induced stratification of a closed complex subspace of X, which is a union of strata. For the (co)stalk property $B = D^{\leq 0}, D^{\geq 0}$ or ${}^{+}D^{\geq 0}$ as in the "classical case" of example 6.0.2*, this definition is closely related to a similar notion studied in [FK1, FK2, FK3, Kaup].*

For example a compactifiable stratification as in corollary 6.0.3 *and remark* 6.0.5 *is* **topologically q-complete** *with respect to* **any** *(co)stalk property* B *as above.*

5. *Assume the (co)stalk property* B *is*

$$\text{stable under countable} \quad \begin{cases} \text{direct sums in the case ``}\geq\text{''}. \\ \text{products in the case ``}\leq\text{''}. \end{cases}$$

Then any **q-complete** *complex space is* **topologically (q+1)-complete** *with respect to the (co)stalk property* B.

As seen above, the "q-completeness" of a complex space X implies the same property for any closed complex subspace Y. But often a given Y is "q'-complete" for a $q' < q$. Therefore it is natural to **refine** given "vanishing theorems" by working with a (finite) filtration

$$\emptyset =: X_{M-1} \subset X_M \cdots \subset X_i \subset \cdots \subset X_N = X$$

of X by closed complex subsets X_i ($M \leq i \leq N$ for some $M, N \in \mathbb{Z}$). Note that $X^i := X_i \backslash X_{i-1}$ is a locally closed constructible subset of X. Especially, its closure $cl(X^i)$ is a closed complex analytic subset of X. Let

$$j_i : X^i \to Z_i := cl(X^i) \quad \text{and} \quad k_i : X^i \to X \quad \text{be the inclusions for } M \leq i \leq N.$$

Then one has the following simple

Lemma 6.0.6. *Fix a complex analytically constructible complex* $\mathcal{F}$ *on* X, *and a "(co)stalk family"* $B \subset ob(D(R))$ *with the property "*$\leq$*" (or "*$\geq$*"). Assume* Z_i *is for* $M \leq i \leq N$ **topologically q(i)-complete** *with respect to* B *and*

$$C(Z_i) := \{ Rj_{i!}k_i^*\mathcal{F} \} \quad \text{or} \quad C(Z_i) := \{ Rj_{i*}k_i^!\mathcal{F} \} .$$

Then

$$k_i^*\mathcal{F} \in {}^m D^{\leq n(i)}(X^i) \quad \text{for all } M \leq i \leq N \;\Rightarrow\; R\Gamma(X, \mathcal{F}) \in D^{\leq n}$$

for $n := max\{n(i) + q(i) \mid M \leq i \leq N\}$, *and*

$$k_i^!\mathcal{F} \in {}^m D^{\geq n(i)}(X^i) \quad \text{for all } M \leq i \leq N \;\Rightarrow\; R\Gamma_c(X, \mathcal{F}) \in D^{\geq n}$$

for $n := min\{n(i) - q(i) \mid M \leq i \leq N\}$.

A very natural example is given by $\mathcal{F} := \mathcal{IC}^\bullet_{\bar{p}}(X, \mathcal{L})$ an **intersection cohomology complex** on the purely n-dimensional complex space X, with the "(co)stalk property" B a vanishing condition as in example 6.0.2.

Let $X_\bullet$ be a **complex analytic stratification** of X, satisfying the property $(*_i)$ for all i. Choose a locally constant sheaf $\mathcal{L}$ on $U := X^{2n}$. Then we consider for lemma 6.0.6 the two-step filtration $X_{n-2} \subset X$, because

$$\mathcal{F}|U \simeq \mathcal{L} \in {}^m D^{\leq n}(U) \cap {}^m D^{\geq n}(U)$$

for **any** "classical perversity" $\bar{p}$, with $B = D^{\leq 0}$ or $D^{\geq 0}$ as in example 6.0.2(1),(2). If the stalks of $\mathcal{L}$ are **torsionfree**, then this is also true for $B = {}^+D^{\geq 0}$ as in example 6.0.2(3).

Example 6.0.6. *Assume X is* **topologically q-complete**, *and X_{n-2} is* **topologically s-complete** *for the "(co)stalk properties" $B = D^{\leq 0}$ and $B = D^{\geq 0}, {}^+D^{\geq 0}$ as before. Especially, for X* **q'-complete**, *we can take $q := min\{q', dim(X)\}$ and $s := min\{q', dim(X_{n-2})\}$. Then*

$$H^k(X, \mathcal{IC}^\bullet_{\bar{p}}(X, \mathcal{L})) = 0 \quad for \quad k > max\{n+q, n-1+\beta+s\}\,,$$

if $\bar{p}(2k) \leq \bar{m}(2k) + \beta = k - 1 + \beta$ for all $k \geq 1$.

$$H^k_c(X, \mathcal{IC}^\bullet_{\bar{p}}(X, \mathcal{L})) = 0 \quad for \quad k < min\{n-q, n+1-\beta-s\}\,,$$

if $\bar{p}(2k) \geq \bar{m}(2k) - \beta = k - 1 - \beta$ for all $k \geq 1$.

In terms of **intersection homology** *this can be rewritten as*

$$I_{\bar{p}}C^c_k(X, \mathcal{L}) := H^{2n-k}_c(X, \mathcal{IC}^\bullet_{\bar{p}}(X, \mathcal{L})) = 0$$

for $k > max\{n+q, n-1+\beta+s\}$. If the stalks of $\mathcal{L}$ are also **torsionfree**, *then*

$$H^k_c(X, \mathcal{IC}^\bullet_{\bar{p}}(X, \mathcal{L})) \quad or \quad I_{\bar{p}}C^c_k(X, \mathcal{L}) \quad is\ \textbf{torsionfree}$$

for $k = min\{n-q, n-\beta-s\}$ or $k = max\{n+q, n+\beta+s\}$. If one assumes in addition the condition **(IC3$_{tor}$)**, *this is already true for $k = min\{n-q, n+1-\beta-s\}$ or $k = max\{n+q, n-1+\beta+s\}$.*

Compare also with [FK1, FK2, FK3, Kaup] for similar results.

For general "(co)stalk properties" B and $\mathcal{F}$ a complex analytically constructible complex on the complex space X, one can consider the **open** subsets

$$U_k := \{x \in X \,|\, \exists \text{open neighb. } U \text{ of } x, \text{ with } \mathcal{F}|U \in \begin{cases} {}^m D^{\geq 1+k}(U) \\ {}^m D^{\leq -1-k}(U) \end{cases} . \tag{6.19}$$

Note that these are **decreasing** sequences of open subsets (by property III. of definition 6.0.1). Moreover, if $\mathcal{F}$ is constructible with respect to a complex analytic Whitney b-regular stratification, then these open subsets are unions of strata (by the frontier condition). Especially, their complements

$$X_k := X \backslash U_k$$

are closed analytic subsets, which are also unions of strata. If this increasing sequence of closed analytic subsets is **finite**, then we can apply lemma 6.0.6 with $n(k) = k$ or $n(k) = -k$, since

$$X^k = U_{k-1} \backslash U_k \hookrightarrow U_{k-1} .$$

Example 6.0.7. *Consider the* **dualizing complex** $\mathcal{F} := D_X := c^! R$ *on* X *(with* $c : X \to \{pt\}$ *a constant map), and take* $B = D^{\leq 0}$ *the vanishing condition of example* 6.0.2(1). *Then one has for* $x \in X$ *by definition:*

$$rHd_x(X, R) \geq k \;:\Leftrightarrow\; \exists \text{ open neighb. } U \text{ of } x, \text{ with } D_X|U \simeq D_U \in {}^m D^{\leq -k}(U)$$

so that $X^k = \{x \in X \,|\, rHd_x(X, R) = k\}$. *Moreover, the sequence* X_k *is finite, since we assume that* X *is finite dimensional. Suppose* $Z_k := cl(X^k)$ *is* **topologically q(k)-complete** *with respect to* $B = D^{\leq 0}$ $(k \in \mathbb{Z})$. *Especially, if* X *is* **q-complete**, *we can take* $q(k) := min\{q, dim(Z_k)\}$. *Then*

$$H_i^{BM}(X, R) = H^{-i}(X, D_X) = 0 \quad \text{for } i < -max\{-k + q(k)\} = min\{k - q(k)\} .$$

Compare also with [Kaup, thm. 5.1, cor. 5.6], where the invariant $Lh_x(X, R) = dim(X) - rHd_x(X, R)$ is used.

Another useful application of corollary 6.0.3(1),(2) is the

Corollary 6.0.5. *Let* $\bar{f} : \overline{X} \to Y$ *be a* **proper** *holomorphic map of complex spaces, with* $Z \subset \overline{X}$ *a closed complex subspace. Consider the induced holomorphic map*

$$f := \bar{f}|X : \; X := \overline{X} \backslash Z \to Y .$$

Suppose the **fiber dimension of** f **is bounded by** d, *and consider a complex analytically constructible complex* $\mathcal{F}$ *on* $\overline{X}$.

1. $\mathcal{F}|X \in {}^m D^{\geq n}(X) \;\Rightarrow\; Rf_*(\mathcal{F}|X) \in {}^m D^{\geq n-d}(Y)$.
2. $\mathcal{F}|X \in {}^m D^{\leq n}(X) \;\Rightarrow\; Rf_!(\mathcal{F}|X) \in {}^m D^{\leq n+d}(Y)$.

Here we allow any "(co)stalk family" $B \subset ob(D(R))$ with the property "$\leq$" or "$\geq$". Also note, that $Rf_*(\mathcal{F}|X)$ and $Rf_!(\mathcal{F}|X)$ are complex analytically constructible by proposition 6.0.1 and theorem 4.0.2. Moreover, in the **algebraic** context one can always find such a compactification of the algebraic map f. Finally the "absolute case" $Y = pt$ a point is just a special case of corollary 6.0.3.

Consider in addition a **locally closed constructible** subset Y^i of Y. Then one has for the induced cartesian diagram

$$\begin{array}{ccc} X^i := f^{-1}(Y^i) & \xrightarrow{k'_i} & X \\ {\scriptstyle f_i}\downarrow & & \downarrow{\scriptstyle f} \\ Y^i & \xrightarrow{k_i} & Y \end{array} \tag{6.20}$$

the **base change isomorphisms** (cf. [KS, prop. 2.6.7, prop. 3.1.9]):

$$k_i^* Rf_! \simeq Rf_{i!}k_i'^* \quad \text{and} \quad k_i^! Rf_* \simeq Rf_{i*}k_i'^! \,.$$

So one can easily combine corollary 6.0.5 and lemma 6.0.6.

Example 6.0.8. *Let $f: X \to Y$ be a* **proper** *holomorphic map of complex spaces, and consider the set of points*

$$Y_i := \{y \in Y \,|\, dim(f^{-1}(y)) \geq i\} \quad for\ i \in \mathbb{N}_0 \,.$$

This is a closed analytic subset of Y, since f is proper. Let $d(i) := dim(Y^i)$, with

$$Y^i := Y_i \backslash Y_{i-1} = \{y \in Y \,|\, dim(f^{-1}(y)) = i\} \,.$$

Then $dim(X^i) \leq d(i) + i$ so that

$$R_{X^i} \in {}^m D^{\leq d(i)+i}(X^i) \;\Rightarrow\; Rf_{i!}(R_{X^i}) \in {}^m D^{\leq d(i)+2i}(Y^i) \,,$$

and

$$D_{X^i} \in {}^m D^{\geq -(d(i)+i)}(X^i) \;\Rightarrow\; Rf_{i!}(D_{X^i}) \in {}^m D^{\geq -(d(i)+2i)}(Y^i) \,.$$

Here we take $B = D^{\leq 0}$ or $B = {}^+D^{\geq 0}$ as in example 6.0.2(1),(3)*. Suppose $Z_i := cl(Y^i)$ is* **topologically q(i)-complete** *with respect to this B. Especially, if Y is* **q-complete**, *we can take $q(i) := min\{q, d(i)\}$. Then*

$$H^k(X,R) \simeq H^k(Y, Rf_!(R_X)) = 0 = H_c^{-k}(Y, Rf_!(D_X)) \simeq H_k^c(X,R)$$

for $k > max\{2i + d(i) + q(i)\}$. Moreover,

$$H_k^c(X,R) \quad is\ torsionfree\ for\ k = max\{2i + d(i) + q(i)\}.$$

Suppose X is purely n-dimensional, and also an **oriented R-homology manifold**, *i.e. $D_X \simeq R_X[2n]$. Then*

$$H_k^{BM}(X,R) \simeq H^{-k}(X, D_X) \simeq H^{2n-k}(X,R) = 0$$

for $k < 2n - max\{2i + d(i) + q(i)\}$.

Compare with [GM, p. 150, thm. 1.1, p. 152, thm. 1.1*] for the corresponding homotopy results.

In the last example, we have taken into account the **global** dimension of the fibers of the holomorphic map $f: X \to Y$. But corollary 6.0.5 can also be used in another way. Consider directly a locally closed (connected) complex manifold S of X, with $f|S$ of constant rank

$$rk_x(f|S) = dim_x(S) - dim_x(\{x' \in S \,|\, f(x') = f(x)\}) \quad \text{for } x \in S.$$

So the fiber dimension of $f|S$ is of constant dimension $dim(S) - rk(f|S)$, and in view of corollary 6.0.5 it is natural to measure a "(co)stalk property" with respect to the "shifted middle perversity" $m(S) \mp (dim(S) - rk(f|S)) =$

$$\begin{cases} -dim(S) - (dim(S) - rk(f|S)) & = -2 \cdot dim(S) + rk(f|S)\,, \\ -dim(S) + (dim(S) - rk(f|S)) & = -rk(f|S)\,. \end{cases}$$

Note that for a given holomorphic map $f : X \to Y$, there always exists a **complex analytic Whitney b-regular stratification** $X_\bullet$, refining a given complex analytic stratification or partition such that the restriction $f|S$ of f to any stratum S is of **constant rank** $rk(f|S)$ (compare [HL5, lem. 1.2]). For such a stratification we introduce the **(dual) relative middle perversity** $m(f)$ (or $m(f)^*$):

$$m(f) : \{\text{strata of } X_\bullet\} \to \mathbb{Z}\,;\ S \mapsto -2 \cdot dim(S) + rk(f|S)\,, \tag{6.21}$$

$$m(f)^* : \{\text{strata of } X_\bullet\} \to \mathbb{Z}\,;\ S \mapsto -rk(f|S)\,. \tag{6.22}$$

Note that $m(id) = m(id)^*$ is just the middle perversity m. More generally, $m(f) = m(f)^* = m$ for f a **discrete** holomorphic map, i.e. all fibers of f are discrete or zero-dimensional.

Fix a "(co)stalk family" $B \subset ob(D(R))$ with the property "$\leq$" or "$\geq$". Then we consider

$${}^{m(f)}D^{\leq n}(X_\bullet) \quad \text{or} \quad {}^{m(f)^*}D^{\geq n}(X_\bullet)$$

for such a complex Whitney stratification of X. Suppose the complex analytic Whitney b-regular stratification $X'_\bullet$ is a refinement of $X_\bullet$ such that also $f|T$ is of **constant rank** $rk(f|T)$ for each stratum T of $X'_\bullet$. If T is contained in a stratum S of $X_\bullet$, then one gets:

$$dim(T) \leq dim(S) \quad \text{and} \quad dim(T) - rk(f|T) \leq dim(S) - rk(f|S)\,,$$

with equality for T open in S. So the condition

$$\mathcal{F} \in {}^{m(f)}D^{\leq n}(X_\bullet) \quad \text{or} \quad \mathcal{F} \in {}^{m(f)^*}D^{\geq n}(X_\bullet)$$

for a complex analytically constructible complex $\mathcal{F}$ on X depends by lemma 6.0.1 **not** on the choice of such a Whitney stratification $X_\bullet$, with $\mathcal{F}$ $X_\bullet$-constructible and $rk(f|S)$ constant on the strata. We denote the corresponding families of complex analytically constructible complexes on X by

$${}^{m(f)}D^{\leq n}(X) \quad \text{or} \quad {}^{m(f)^*}D^{\geq n}(X)\,.$$

Then one gets the following counterpart of corollary 6.0.5:

Corollary 6.0.6. *Let* $\bar{f} : \overline{X} \to Y$ *be a* **proper** *holomorphic map of complex spaces, with* $Z \subset \overline{X}$ *a closed complex subspace. Consider the induced holomorphic map*

$$f := \bar{f}|X : X := \overline{X} \backslash Z \to Y\,,$$

and a complex analytically constructible complex $\mathcal{F}$ *on* $\overline{X}$.

1. $\mathcal{F}|X \in {}^{m(f)^*}D^{\geq n}(X) \Rightarrow Rf_*(\mathcal{F}|X) \in {}^{m}D^{\geq n}(Y)$.

2. $\mathcal{F}|X \in {}^{m(f)}D^{\leq n}(X) \Rightarrow Rf_!(\mathcal{F}|X) \in {}^{m}D^{\leq n}(Y)$.

So for f a **discrete** holomorphic map, this gives nothing new compared to corollary 6.0.5. But otherwise, it contains more information. Let $f : X \to Y$ be a **proper** holomorphic map, and choose complex analytic Whitney b-regular stratifications $X_\bullet$ and $Y_\bullet$ of X and Y such that f is a **stratified submersion**, i.e. f maps each stratum S of $X_\bullet$ into a stratum T of $Y_\bullet$, with $f : S \to T$ a submersion. In particular, $rk_x(f|S)$ is constant on S with $rk(f|S) = dim(T)$.

Example 6.0.9 (semismall map). *Let $f : X \to Y$ be a* **proper** *holomorphic map, and stratify f as above. Assume*

$$m(f)(S) = -2 \cdot dim(S) + rk(f|S) \geq -dim(X)$$

for each stratum S of $X_\bullet$. This is also equivalent to

$$2 \cdot dim(f^{-1}(x)) \leq dim(X) - dim(T)$$

for all x in a stratum T of $Y_\bullet$. If $R \in B = D^{\leq 0}$ for the "stalk property" B, then $R_X[dim(X)] \in {}^{m(f)}D^{\leq 0}(X)$ so that

$$Rf_*(R_X[dim(X)]) \simeq Rf_!(R_X[dim(X)]) \in {}^{m}D^{\leq 0}(Y)\,.$$

The assumption on R is in particular true for $B = D^{\leq 0}_{perf}$ as in example 6.0.2(1).

Consider this special case for R a field, with X pure dimensional and an **R-homology manifold**, *i.e. $D_X \simeq R_X[2 \cdot dim(X)]$. Then $R_X[dim(X)]$ is* **self-dual**, *i.e.*

$$D(R_X[dim(X)]) \simeq R_X[dim(X)] \quad \text{so that}$$

$$Rf_*(R_X[dim(X)]) \simeq Rf_!(R_X[dim(X)]) \in {}^{m}D^{\leq 0}(Y) \cap {}^{m}D^{\geq 0}(Y)$$

is a **perverse sheaf** *on Y by duality.*

This result is due to [BoM, p. 38] for the case Y irreducible with $dim(X) = dim(Y)$. f is called **semismall** *in this special case, and the above conditions are then equivalent to*

$$dim(\{y \in Y \mid dim(f^{-1}(y)) \geq i\} \leq dim(Y) - 2 \cdot i \quad \text{for all } i. \tag{6.23}$$

Especially, they do **not** *depend on the stratification of the map f. Moreover, there is a constructible open subset $U \subset Y$ such that $f_U := f : f^{-1}(U) \to U$ is a finite covering, with $\mathcal{L} := (f_* R_X)|U$ locally constant. The map f is called* **small**, *if we have in* (6.23) *the strict inequalities for all $i > 0$. Then one can refine outside $f^{-1}(U)$ our starting estimate by one, and get in the same way (compare [GM4, sec. 6.2]):*

$$Rf_*(R_X) \simeq \mathcal{IC}^\bullet_{\bar{m}}(X, \mathcal{L})\,.$$

Note that $\mathcal{L} = R_U$, if f is also birational, i.e. f_U is an isomorphism.

One can also combine corollary 6.0.6 and lemma 6.0.6. For general "(co)stalk properties" B and $\mathcal{F}$ a complex analytically constructible complex on the complex space X, one can consider the **open** subsets

$$U_k := \{x \in X \mid \exists \,\text{open neighb. } U \text{ of } x, \text{ with } \mathcal{F}|U \in \begin{cases} {}^{m(f)^*}D^{\geq 1+k}(U) \\ {}^{m(f)}D^{\leq -1-k}(U) \end{cases} .$$

Again these are **decreasing** sequences of open subsets, which are unions of strata for any complex Whitney stratification $X_\bullet$ of X as before such that $\mathcal{F}$ is $X_\bullet$-constructible. Suppose f is **proper**. Then $Y_i := f(X \backslash U_i)$ is an increasing sequence of closed analytic subsets of Y such that

$$X^i := f^{-1}(Y^i) \hookrightarrow X \backslash f^{-1}(Y_{i-1}) \subset U_{i-1} \quad \text{for } Y^i := Y_i \backslash Y_{i-1}.$$

So for the induced proper map $f = f_i : X^i \to Y^i$ as in (6.20) we can apply corollary 6.0.6 to

$$k_i'^! \mathcal{F} \in {}^{m(f)^*}D^{\geq i}(X^i) \quad \text{or} \quad k_i'^* \mathcal{F} \in {}^{m(f)}D^{\leq -i}(X^i) .$$

If these filtrations are **finite**, then we can apply the corresponding base change isomorphisms as in (6.20) together with lemma 6.0.6. Let us illustrate these things for $\mathcal{F} := D_X$ the **dualizing complex** on X, with $B = D^{\leq 0}$ the "stalk property" as in example 6.0.2(1).

Definition 6.0.8. *Let $f : X \to Y$ be a holomorphic map of complex spaces. Then the* **relative rectified homological depth** $rHd(f, R)$ *of f with respect to the commutative base ring R is $\geq n$ (for $n \in \mathbb{Z}$) if*

$$D_X = D(R_X) \in {}^{m(f)}D^{\leq -n}(X) . \tag{6.24}$$

The **relative rectified homological depth** $rHd(f, R)$ *of f for $X \neq \emptyset$ is the maximum of the integers n such that* (6.24) *holds.*

Choose a **complex analytic Whitney b-regular** stratification $X_\bullet$ of X such that $rk(f|S)$ is constant on each stratum S. Let B be the intersection of X with a small open ball around $x \in S$. Then one gets by (6.15) the following equivalence for the vanishing of the relative homology (with compact support):

$$\begin{aligned} & H_k^c(B, B \backslash S) = 0 \quad \text{for } k < n - rk(f|S) \\ \Leftrightarrow \; & H^{k'}(\, D(R_X)_x \,) = 0 \quad \text{for } k' > -n - 2s + rk(f|S) \\ \Leftrightarrow \; & D(R_X)_x \, [-2s + rk(f|S)] \in D^{\leq -n} := B[n] , \end{aligned}$$

with $k' \leftrightarrow -k - 2s$ and $s := dim(S)$. So our definition 6.0.8 is the "homology version" of the corresponding definition for the **relative rectified homotopical depth** given in [HL5, def. 1.1.2].

By definition one has for $x \in X$: $rHd_x(f, R) \geq k$ if and only if there exists an open neighborhood U of x with

$$D_X|U \simeq D_U \in {}^{m(f)}D^{\leq -k}(U)\,.$$

Assume now that $f : X \to Y$ is **proper**. Then the corresponding filtrations U_k and Y_k are finite (since all spaces are finite dimensional), with

$$Y^k = \{y \in Y \,|\, min\{rHd_x(f, R) \,|\, f(x) = y\} = k\}\,.$$

Therefore we get the following "relative version" of example 6.0.7:

Example 6.0.10. *Let $f : X \to Y$ be a* **proper** *holomorphic map, with*

$$Y^k := \{y \in Y \,|\, min\{rHd_x(f, R) \,|\, f(x) = y\} = k\}\,.$$

Suppose that $Z_k := cl(Y^k)$ is **topologically q(k)-complete** *($k \in \mathbb{Z}$) with respect to $B = D^{\leq 0}$ as in example 6.0.2(1). Especially, if Y is* **q-complete**, *we can take $q(k) := min\{q, dim(Z_k)\}$. Then*

$$H_i^{BM}(X, R) = H^{-i}(Y, Rf_*(D_X)) = 0$$

for $i < -max\{-k + q(k)\} = min\{k - q(k)\}$.

Compare also with [HL5, thm. 2.1.3] for a corresponding homotopy result.

6.0.4 Nearby and vanishing cycles

Now we come to our second **main result** in the complex analytic context, which deals with the famous **nearby and vanishing cycle functors** of Deligne.

Let $f : X \to \mathbb{C}$ be a holomorphic (algebraic) function on the complex analytic (algebraic) space X, with $i : X_0 := \{f = 0\} \to X$ and $j : U := \{f \neq 0\} \to X$ the inclusions. Denote by

$$\psi_f, \phi_f : D(X) \to D(X_0)$$

the **nearby and vanishing cycle functors** of f (compare chapter 1 for the definitions). These respect the corresponding categories of complex analytically (algebraically) constructible complexes (compare theorem 4.0.2). Similarly for the functor Rj_*j^*.

These functors are related by three basic distinguished triangles (compare with section 5.4):

$$i^* \xrightarrow{sp} \psi_f \xrightarrow{can} \phi_f \xrightarrow{[1]}\,, \tag{6.25}$$

$$i^![1] \longrightarrow \phi_f \xrightarrow{var} \psi_f \xrightarrow{[1]}\,, \tag{6.26}$$

$$i^*Rj_*j^* \longrightarrow \psi_f \xrightarrow{id-m_f} \psi_f \xrightarrow{[1]}\,. \tag{6.27}$$

Theorem 6.0.2. *Assume $\mathcal{F} \in ob(D(X))$ is complex analytically (algebraically) constructible. Then we have:*

1. $j^*\mathcal{F} \in {}^mD^{\leq n}(U) \Rightarrow \psi_f(\mathcal{F})[-1] \in {}^mD^{\leq n}(X_0)$.
2. $j^*\mathcal{F} \in {}^mD^{\geq n}(U) \Rightarrow \psi_f(\mathcal{F})[-1] \in {}^mD^{\geq n}(X_0)$.
3. $\mathcal{F} \in {}^mD^{\leq n}(X) \Rightarrow \phi_f(\mathcal{F})[-1] \in {}^mD^{\leq n}(X_0)$.
4. $\mathcal{F} \in {}^mD^{\geq n}(X) \Rightarrow \phi_f(\mathcal{F})[-1] \in {}^mD^{\geq n}(X_0)$.

Here we allow any "(co)stalk family" $B \subset ob(D(R))$ with the property "$\leq$" or "$\geq$"! Note that the result (3.) or (4.) for the **vanishing cycle functor** follows directly from the corresponding result (1.) or (2.) for the nearby cycle functor, if one uses the distinguished triangle (6.25) or (6.26).

Remark 6.0.6. In the classical case of example 6.0.1(1),(2), we get especially that the (shifted) **nearby and vanishing cycle functors** $\psi_f[-1]$ and $\phi_f[-1]$ are **exact** functors with respect to the **perverse t-structure** (for the middle perversity).

This is a result of Gabber in the algebraic context for l-adic cohomology (unpublished, but compare with [BBD, prop. 4.4.2], [Bry, thm. 1.2] and [Ill, cor. 4.5, 4.6]). In the complex analytic context it is due to Goresky-MacPherson [GM3, thm.6.5], and [GM, p. 222, II cor. 6.13.6, p. 224 6.A.5]. For another proof of this classical case see [KS, cor. 10.3.11, 10.3.13].

Even if we apply our generalization above only to the simple case of example 6.0.1(3), then we get additional torsion information, which was not covered before.

As explained above, the deep part of theorem 6.0.2 lies in the proof of the results for the nearby cycle functor. But most of the applications of theorem 6.0.2 use the results for the **vanishing cycle functor**. This comes from the fact, that the support of the vanishing cycle functor is in general much smaller (compare with remark 4.2.4 on page 248):

- If $\mathcal{F}$ is constructible with respect to a complex analytic stratification with a suitable regularity condition (like Whitney **b-regularity**), then the support of $\phi_f(\mathcal{F})$ is contained in the set of stratified critical points of f in $\{f = 0\}$.

A nice application is the following characterization:

Corollary 6.0.7. *Let X be a complex analytic space. Consider a complex analytically constructible complex $\mathcal{F} \in ob(D(X))$. Then we have for ? = "$\leq$" or ? = "$\geq$":*

$$\mathcal{F} \in {}^mD^{?n}(X) \quad \Leftrightarrow \quad \left(\phi_f(\mathcal{F})[-1]\right)_x \in D^{?n}$$

for **all** *holomorphic function germs $f : (X, x) \to (\mathbb{C}, 0)$ such that x is* **isolated in the support of** $\phi_f(\mathcal{F})$.

Note that the implication $\Rightarrow$ is just a special case of theorem 6.0.2(3),(4). Here we use of course

$$i_x^* \phi_f(\mathcal{F})[-1] \simeq i_x^! \phi_f(\mathcal{F})[-1] \tag{6.28}$$

for the inclusion $i_x : \{x\} \to X_0 = \{f = 0\}$, if x is **isolated** in the support of $\phi_f(\mathcal{F})$.

Before we explain some examples, let us notice the following application of theorem 6.0.2:

Proposition 6.0.2. *Let $i : Y \to X$ be the inclusion of a closed complex analytic subset, with $j : U := X \backslash Y \to X$ the inclusion of the open complement. Suppose that Y can locally be described (at each point $x \in Y$) as the common zero-set of at most k holomorphic functions on X ($k \geq 1$). Then one has:*

1. $\mathcal{F} \in {}^m D^{\leq n}(X) \Rightarrow i^! \mathcal{F}[k] \in {}^m D^{\leq n}(Y)$.
2. $\mathcal{F} \in {}^m D^{\geq n}(X) \Rightarrow i^* \mathcal{F}[-k] \in {}^m D^{\geq n}(Y)$.
3. $j^*\mathcal{F} \in {}^m D^{\leq n}(U) \Rightarrow i^* Rj_* j^* \mathcal{F}[k-1] \simeq i^! Rj_! j^* \mathcal{F}[k] \in {}^m D^{\leq n}(Y)$ *and* $Rj_* j^* \mathcal{F}[k-1] \in {}^m D^{\leq n}(X)$.
4. $j^*\mathcal{F} \in {}^m D^{\geq n}(U) \Rightarrow i^! Rj_! j^* \mathcal{F}[-(k-1)] \simeq i^* Rj_* j^* \mathcal{F}[-k] \in {}^m D^{\geq n}(Y)$ *and* $Rj_! j^* \mathcal{F}[-(k-1)] \in {}^m D^{\geq n}(X)$.

Note that these are local results, since all complexes are complex analytically constructible. So we can assume that Y is the common zero-set of k' holomorphic functions (with $k' \leq k$). Then the claim for (1.) or (2.) follows by induction from the case $k' = 1$, which is a direct application of theorem 6.0.2 and the distinguished triangle (6.26) or (6.25). Finally (3.) or (4.) is by $i^* Rj_* j^* \simeq i^! Rj_! j^*[1]$ a special case of (1.) or (2.), since

$$j^*\mathcal{F} \in {}^m D^{\leq n}(U) \Rightarrow Rj_! j^* \mathcal{F} \in {}^m D^{\leq n}(X),$$

and

$$j^*\mathcal{F} \in {}^m D^{\geq n}(U) \Rightarrow Rj_* j^* \mathcal{F} \in {}^m D^{\geq n}(X).$$

Example 6.0.11 (purity). *Let $i : Y \to X$ be the inclusion of a closed complex analytic subset, which can locally be described as the common zero-set of at most k holomorphic functions on X. Then $i^! D_X \simeq D_Y$ implies by proposition 6.0.2(1):*

$$rHd(X, R) \geq n \Rightarrow rHd(Y, R) \geq n - k. \tag{6.29}$$

Especially for X smooth of pure dimension n, we get

$$rHd(Y, R) \geq n - k = dim(Y) \geq rHd(Y, R),$$

if Y is locally a **set-theoretical complete intersection** *of codimension k in X. In particular*

$$rHd(Y, R) = dim(Y) \quad \text{and} \quad R_Y[dim(Y)] \quad \text{is a perverse sheaf}$$

for Y a **pure-dimensional local complete intersection.**

Compare also with [HL, thm. 3.2.1, cor. 3.2.2] for the corresponding homotopy results. Another important special case of proposition 6.0.2(2) is the case $Y = \{x\}$ a point, if we factorize the inclusion $i : \{x\} \to X$ by

$$\{x\} \xrightarrow{\ i_x\ } S := supp(\mathcal{F}) \xrightarrow{\ \kappa\ } X\,,$$

with $supp(\mathcal{F})$ the **support** of the constructible complex $\mathcal{F}$. Then $\{x\}$ is locally on S the zero-set of $k := dim_x(S)$ holomorphic functions so that

$$\begin{aligned} \mathcal{F} \in {}^mD^{\geq n}(X) &\Rightarrow \kappa^*\mathcal{F} \simeq \kappa^!\mathcal{F} \in {}^mD^{\geq n}(S) \\ &\Rightarrow \mathcal{F}_x \simeq i_x^*\kappa^*\mathcal{F} \in D^{\geq n - dim_x(S)}\,. \end{aligned} \tag{6.30}$$

Example 6.0.12 (connectivity of a Milnor fiber). *Consider the "(co)stalk property"*

$$B = D^{\leq 0}_{perf} \quad and \quad B = D^{\geq 0}_{perf} \quad as\ in\ example\ 6.0.2(1),(3),$$

with the base ring R a principal ideal domain. Let $f : X \to \mathbb{C}$ be a holomorphic function on the complex space X. Then one gets for $x \in \{f = 0\}$ and $d := dim_x(X)$:

$$\phi_f(R_X)_x[-1] \in D^{\leq d} \cap D^{\geq n}\,, \quad for \quad n := rHd_x(X, R) - dim_x(\,supp(\phi_f(R_X))\,)\,.$$

As explained in chapter 1, $\phi_f(R_X)_x$ calculates the **reduced cohomology of a Milnor fiber** *of f in x. This is given by*

$$M_{f,x} := \{z \in X|\ f = w, |z| \leq \delta\} \quad (0 < |w| << \delta << 1)$$

for a local model $(X, x) \hookrightarrow (\mathbb{C}^N, 0)$. So we get $\tilde{H}^*(M_{f,x}, R)$ is **finitely generated**, *with*

$$\tilde{H}^*(M_{f,x}, R) = \begin{cases} 0 & for\ k < n-1\ and\ k > dim_x(X) - 1, \\ (torsion)free & for\ k = n-1. \end{cases} \tag{6.31}$$

Assume $rHd_x(X, R) = dim_x(X)$, e.g. X is smooth or a local complete intersection of pure dimension (near x), with x an **isolated critical point of** *f in the sense that x is an isolated point in $supp(\phi_f(R_X))$, e.g. an isolated stratified critical point of f with respect to a Whitney b-regular stratification of X. Then*

$$\tilde{H}^*(M_{f,x}, R) = \begin{cases} 0 & for\ k \neq dim_x(X) - 1, \\ finitely\ generated\ (torsion)free & for\ k = dim_x(X) - 1, \end{cases}$$

i.e. the Milnor fiber of f in x has the (co)homology of a "finite bouquet of spheres" of real dimension $dim_x(X) - 1$.

So corollary 6.0.7 corresponds in the very special case of example 6.0.12 to [Le, thm. 4.12, cor. 5.5], where also a corresponding homotopy result is discussed!

Let us also recall, that the case X smooth of example 6.0.12 is the cohomological version of a classical result of Kato-Matsumoto [KM]:

- The Milnor fiber of $f : (\mathbb{C}^{n+1}, 0) \to (\mathbb{C}, 0)$ is $(n-s-1)$-connected, with s the dimension in x of the singularity set of f. The special case $s = 0$ of an isolated singularity is due to Milnor [Mi].

This estimate is optimal, as the simple example

$$f(z_0, \dots, z_n) := z_0 \cdot \dots \cdot z_n$$

shows. Then the singularity set is of dimension $n-1$, and the Milnor fiber in 0 is homeomorphic to $(\mathbb{C}^*)^n$. Using this example together with our **Thom-Sebastiani results** of chapter 1, we see that also the estimate of example 6.0.12 is optimal in general. Moreover, the "shift by the dimension of the singularity set"

$$s := dim_x(\, supp(\phi_f(\mathcal{F}))\,)$$

in such "(co)homologically connectivity results" is a general effect due to (6.30).

Example 6.0.13 (connectivity of a global fiber). *Consider the "(co)stalk properties" as in example* 6.0.12. *Let $f : X \to \mathbb{C}$ be a* **proper** *holomorphic function on the complex space X, with $Y \subset X$ a closed complex analytic subset. Denote by $j : U := X\backslash Y \to X$ the open inclusion, and by $g := f : U \to \mathbb{C}$ the induced function. Then*

$$R\Gamma(\{f=0\}, \phi_f[-1](Rj_* R_U)) \simeq R\Gamma(\{|g| \le \epsilon\}, \{g = w\}, R) \quad \textit{for } 0 < |w| < \epsilon << 1$$

calculates the **relative cohomology of a tube modulo a generic fiber** *for the induced function g (compare section 1.1).*

Assume Y can locally be described on X as the common zero-set of at most k holomorphic functions. Then we get for $d := dim(U)$ and $n := rHd(U, R)$:

$$R_U \in {}^mD^{\le d}(U) \cap {}^mD^{\ge n}(U) \Rightarrow Rj_* R_U \in {}^mD^{\le d+k-1}(X) \cap {}^mD^{\ge n}(X)$$

$$\Rightarrow \phi_f[-1](Rj_* R_U) \in {}^mD^{\le d+k-1}(\{f=0\}) \cap {}^mD^{\ge n}(\{f=0\})$$

$$\Rightarrow R\Gamma(\{f=0\}, \phi_f[-1](Rj_* R_U)) \in D^{\le d+k+s-1} \cap D^{\ge n-s}\,,$$

with $s := dim(\, supp(\phi_f(Rj_* R_U))\,)$. So we get

$$H^i(\{|g| \le \epsilon\}, \{g=w\}, R) = \begin{cases} 0 & \textit{for } i > d+k+s-1, \\ 0 & \textit{for } i < n-s, \\ \textit{(torsion)free} & \textit{for } i = n-s. \end{cases} \tag{6.32}$$

Also note, that

$$H^i(\{|g| \le \epsilon\}, \{g=w\}, R) \simeq \tilde{H}^{i-1}(\{g=w\}, R)\,,$$

if the "tube" $\{|g| \le \epsilon\}$ has the cohomology of a point, e.g. it is contractible.

Assume $rHd(U, R) = dim(U)$, e.g. U is smooth or a local complete intersection of pure dimension, with Y locally given by **one** *equation. Then*

$$H^i(\{|g| \leq \epsilon\}, \{g = w\}, R) = \begin{cases} 0 & \textit{for } i > dim(U) + s, \\ 0 & \textit{for } i < dim(U) - s, \\ \textit{(torsion)free} & \textit{for } i = dim(U) - s. \end{cases}$$

So this relative cohomology is concentrated in one degree, if moreover f has globally **isolated** *singularities with respect to Rj_*R_U, i.e. the dimension s of the support of the corresponding vanishing cycles is zero.*

As a final example, let us mention the

Example 6.0.14 (local Lefschetz and vanishing theorems). *Consider the "(co)stalk properties" as in example 6.0.12. Let $f : (X, x) \to (\mathbb{C}, 0)$ be a holomorphic function germ, with $Y \subset X$ a closed complex analytic subset. Denote by $j : U := X \backslash Y \to X$ the open inclusion. Then*

$$(\phi_f[-1](Rj_*R_U))_x \simeq R\Gamma(B_\epsilon \cap X \backslash Y, B_\epsilon \cap X \cap \{g = w\} \backslash Y, R)$$

for $0 < |w| << \epsilon$, with B_ϵ a closed ball of radius $0 < \epsilon << 1$ around x (in local coordinates). So we get for $n := rHd(U, R)$ and $s := dim_x(\, supp(\phi_f(Rj_*R_U))\,)$:

$$H^i(B_\epsilon \cap X \backslash Y, B_\epsilon \cap X \cap \{g = w\} \backslash Y, R) = \begin{cases} 0 & \textit{for } i < n - s, \\ \textit{(torsion)free} & \textit{for } i = n - s. \end{cases} \tag{6.33}$$

Assume $x \in Y$, and Y is locally near x the zero-set of k holomorphic functions on X. Let $\kappa : X \backslash \{x\} \to X$ be the inclusion, with $\mathcal{F}$ a complex analytically constructible complex on X. Then one gets for $0 < \epsilon << 1$ by **transversality** *(compare proposition 4.0.3):*

$$(Rj_*j^*\mathcal{F})_x \simeq (R\kappa_*\kappa^*Rj_*j^*\mathcal{F})_x \simeq R\Gamma(\partial B_\epsilon \cap X \backslash Y, \mathcal{F})\,.$$

So we get for $\mathcal{F} := R_X$ and $d := dim_x(X) \geq dim(U)$:

$$H^i(\partial B_\epsilon \cap X \backslash Y, R) = 0 \quad \textit{for } i > d + k - 1. \tag{6.34}$$

Assume X is purely d-dimensional, and take $\mathcal{F} := \mathcal{IC}^\bullet_{\bar{p}}(X, R_X)$ the **intersection cohomology complex** *for $\bar{p}$ a classical perversity. Then*

$$\mathcal{IC}^\bullet_{\bar{p}}(X, R_X)|L \simeq \mathcal{IC}^\bullet_{\bar{p}}(L, R_L)$$

for $L := \partial B_\epsilon \cap X \backslash Y$ by transversality. Suppose $\bar{p}(2k') \leq \bar{m}(2k') + \beta = k' - 1 + \beta$ for all $k' \geq 1$. Then one gets

$$IH^i_{\bar{p}}(\partial B_\epsilon \cap X \backslash Y, R) \simeq H^i(L, \mathcal{IC}^\bullet_{\bar{p}}(X, R_X)) = 0 \tag{6.35}$$

for $i > d + k - 1 + max\,\{0, \beta - 1\}$.

Compare with [GM, p. 157, thm. 2, p. 188, prop. 4.6*] for the homotopy result corresponding to (6.34). Notice especially the similarities between the inductive proof of [GM, p. 188, prop. 4.6*] and our inductive proof of proposition 6.0.2(1),(2). The result (6.35) for the intersection cohomology is missing (!) in [GM] (cf. the remarks on [GM, p. 158]).

Compare also with [HL, thm. 2.12] for the homotopy result corresponding to (6.33) in the case of a **generic** holomorphic function germ, for which one has $s \leq 0$. More generally, suppose that $S \subset \{f = 0\}$ is an analytic subset such that f is **outside** S **transversal** to a Whitney b-regular stratification of X, with Y a union of strata. Then $supp(\phi_f(Rj_* R_U)) \subset S$ so that $s \leq dim(S)$.

6.0.5 Artin-Grothendieck type theorems

As we have seen, it is important to get information as in proposition 6.0.2(1),(2) about the stability of our "(co)stalk properties" with respect to the functor Rj_* (or $Rj_!$).

Definition 6.0.9. *Let $f : X \to Y$ be a holomorphic map of complex spaces, and fix a family $C(X) \subset ob(D(X))$ of complex analytically constructible complexes on X. Consider a* **fixed** *subfamily B of $ob(D(R))$ with the property "$\leq$" (or "$\geq$"). Then f is* **topologically q-complete** *with respect to the family $C(X)$ and the (co)stalk property B ($q \in \mathbb{Z}$), if one has for* **all** *$\mathcal{F} \in C(X)$:*

$$\mathcal{F} \in {}^m D^{\leq n}(X) \quad \textbf{and} \quad Rf_*\mathcal{F} \quad \textit{complex analytically constructible on } Y$$
$$\Rightarrow \quad Rf_*\mathcal{F} \in {}^m D^{\leq n+q}(Y)\,,$$

or

$$\mathcal{F} \in {}^m D^{\geq n}(X) \quad \textbf{and} \quad Rf_!\mathcal{F} \quad \textit{complex analytically constructible on } Y$$
$$\Rightarrow \quad Rf_!\mathcal{F} \in {}^m D^{\geq n-q}(Y)\,.$$

If f has this property for an integer $q \in \mathbb{Z}$, then the same is true for any $q' \geq q$ (by III. of definition 6.0.1*).*

If we can take for $C(X)$ the family of **all** *complex analytically constructible complexes, then we call f* **topologically q-complete** *with respect to the (co)stalk property B.*

For a constant map $f : X \to \{pt\}$ we get back our definition 6.0.7.

Example 6.0.15. 1. *Consider the open inclusion $j : U := X \backslash Y \to X$ of proposition* 6.0.2*, with $C(X)$ the family of all complex analytically constructible complexes $\mathcal{F}$ such that $Rj_*\mathcal{F}$ (or equivalently $Rj_!\mathcal{F}$) is also complex analytically constructible. Then j is* **topologically (k-1)-complete** *with respect to $C(X)$ and* **any** *(co)stalk property B.*

2. *Consider a* **proper** *holomorphic map $f : X \to Y$ of complex spaces, whose fiber dimension is bounded by d. Then f is* **topologically d-complete** *with respect to* **any** *(co)stalk property B (corollary 6.0.5).*

The most important example is contained in the following "relative version" of corollary 6.0.3. Here we allow **any fixed** subfamily B of $ob(D(R))$ with the property "$\leq$" (or "$\geq$") !

Theorem 6.0.3. *Let $f : X \to Y$ be a holomorpic map of complex spaces, and fix a complex analytically constructible complex $\mathcal{F}$ on X. Suppose each point $y \in Y$ has an open neighborhood U in Y with the following properties:*

a. *There exists a* **q-convex $\mathbf{C}^\infty$-function**

$$g : V := f^{-1}(U) \to \mathbb{R}^{\geq 0} \quad \textit{with} \quad (f,g) : V \to U \times \mathbb{R}^{\geq 0} \quad \textbf{proper}.$$

b. *$(f,g), V$ and $\mathcal{F}|V$ have a* **(sub)analytic compactification**, *i.e. there exists a (sub)analytic space $\overline{V}$, with $j : V \to \overline{V}$ the inclusion of a locally closed (sub)analytic subset of $\overline{V}$ such that (f,g) is the restriction of a proper subanalytic map $(\bar{f},\bar{g}) : \overline{V} \to U \times [0,\infty]$, with $j_!(\mathcal{F}|V)$ (or equivalently $Rj_*(\mathcal{F}|V)$) (sub)analytically constructible.*

Then one gets for $q' := min\{q, dim(X)\}$:

1. *$\mathcal{F} \in {}^m D^{\leq n}(X)$ and $Rf_*\mathcal{F}$ complex analytically constructible on Y*

$$\Rightarrow \quad Rf_*\mathcal{F} \in {}^m D^{\leq n+q'}(Y)\,.$$

2. *$\mathcal{F} \in {}^m D^{\geq n}(X)$ and $Rf_!\mathcal{F}$ complex analytically constructible on Y*

$$\Rightarrow \quad Rf_!\mathcal{F} \in {}^m D^{\geq n-q'}(Y)\,.$$

In the complex algebraic context we get especially the following **Artin-Grothendieck type theorem**:

Theorem 6.0.4. *Let $f : X \to Y$ be a complex algebraic* **affine** *map of complex algebraic spaces, and choose for $C(X)$ the family of all algebraically constructible complexes on X. Then f is* **topologically 0-complete** *with respect to $C(X)$ and* **any** *(co)stalk property B.*

Let us come back to the complex analytic context. Then the proof of theorem 6.0.3 sometimes applies to a more general context.

Definition 6.0.10. *Let $f : X \to Y$ be a holomorpic map of complex spaces. Then f is called* **q-complete**, *if it satisfies the assumption* (a.) *of theorem* 6.0.3. *This is also equivalent to the following property:*

a$'$. *Each point* $y \in Y$ *has an open neighborhood* U *in* Y *such that* $f^{-1}(U)$ *is* q*-complete.*

Then any $y \in Y$ *also has a fundamental system of open neighborhoods* U *in* Y *such that* $f^{-1}(U)$ *is* q*-complete. A 0-complete holomorphic map is also called a* **Stein map**.

Corollary 6.0.8 (corollary of the proof of theorem 6.0.3). *Suppose the holomorphic map* $f : X \to Y$ *is* q*-complete.*

1. f *is* **topologically q-complete** *with respect to the (co)stalk property* $B = D^{\leq 0}, D^{\geq 0}$ *or* $^{+}D^{\geq 0}$ *as in the "classical case" of example* 6.0.2.

2. *Assume the (co)stalk property* B *is stable under countable direct sums in the case "*$\geq$*", or under countable products in the case "*$\leq$*". Then* f *is* **topologically (q+1)-complete** *with respect to the (co)stalk property* B.

Remark 6.0.7. Corollary 6.0.8(1) for the "classical case" $B = D^{\leq 0}$ or $B = D^{\geq 0}$ is just a special case of [HL2, thm. 3.3.1] and [HL3, thm. 3.1.4] (compare also with [HL4, thm. 1.2.3]). Moreover, the case $q = 0$ of a Stein map is due to [KS, prop. 10.3.17]. Our proof of theorem 6.0.3 is a slight modification of the proof of Kashiwara-Schapira, using our corollary 6.0.3 and proposition 6.0.2(1),(2).

Theorem 6.0.3 applies also in the complex analytic context to a **suitable compactifiable holomorphic Stein map**. Consider a **proper** holomorphic map $\bar{f} : \overline{X} \to Y$ of complex spaces, with $Z \subset \overline{X}$ a closed complex analytic subset. Let $j : X := \overline{X} \backslash Y$ be the open inclusion of the complement, with $f := \bar{f} \circ j : U \to Y$ the induced holomorphic map. Suppose f is an **affine** map in the following sense: for each point $y \in Y$ there exists a commutative diagram

$$\begin{array}{ccc} f^{-1}(U) & \xrightarrow{i'} & U \times \mathbb{C}^N \\ {\scriptstyle j}\downarrow & & \downarrow{\scriptstyle j'} \\ \bar{f}^{-1}(U) & \xrightarrow[i]{} & U \times \mathbb{P}^N(\mathbb{C}) \\ {\scriptstyle \bar{f}}\downarrow & & \downarrow{\scriptstyle pr} \\ U & = & U\,, \end{array} \tag{6.36}$$

with U an open neighborhood of y in Y and pr the projection such that i, i' are closed and j, j' are the obvious open embeddings.

Corollary 6.0.9. *Consider an* **affine** *holomorphic map as before, with* $C(X)$ *the family of all complex analytically constructible complexes* $\mathcal{F}$ *on* X *such that* $Rj_*\mathcal{F}$ *(or equivalently* $Rj_!\mathcal{F}$*) is also complex analytically constructible. Then* f *is* **topologically 0-complete** *with respect to* $C(X)$ *and* **any** *(co)stalk property* B.

Also remark, that in the context of this corollary

$$Rf_!\mathcal{F} \simeq R\bar{f}_! Rj_! \mathcal{F} \quad \text{and} \quad Rf_*\mathcal{F} \simeq R\bar{f}_* Rj_* \mathcal{F}$$

are complex analytically constructible on Y for $\mathcal{F} \in C(X)$.

Example 6.0.16 (Fourier-Sato transformation). *Let E be a holomorphic vector bundle of rank r over the complex space X, with E^* the dual holomorphic vector bundle. Consider the following holomorphic maps:*

$$E \xleftarrow{\ p_1\ } E \times_X E^* \xrightarrow{\ \gamma\ } E^* \times \mathbb{C} \xrightarrow{\ t\ } \mathbb{C}\,,$$

with p_1, t the corresponding projections, and γ the map $(x, y) \mapsto (y, y(x))$. Then γ is an **affine** *map in the sense above, with $p_1^*\mathcal{F} \in C(E \times_X E^*)$ in the corresponding family for any $\mathcal{F} \in D_{mon}(E)$.*

Especially, the **Fourier-Sato transformation**

$$(\,\cdot\,)^\wedge \simeq \phi_t[-1]R\gamma_*\, p_1^* \simeq \phi_t[-1]R\gamma_!\, p_1^* : \quad D_{mon}(E) \to D(E^*)$$

maps

$${}^m D^{\leq n}_{mon}(E) \quad \textit{into} \quad {}^m D^{\leq n+r}(E^*) \quad \textit{and} \quad {}^m D^{\geq n}_{mon}(E) \quad \textit{into} \quad {}^m D^{\geq n+r}(E^*)\,,$$

since p_1 is a smooth map of constant relative dimension r.

Here we denote by $D_{mon}(\cdot)$ the complex analytically constructible complexes on a holomorphic vector bundle, which are **monodromic** *with respect to the corresponding $\mathbb{C}^*$ action on this bundle, i.e. whose cohomology sheaves are locally constant on the $\mathbb{C}^*$-orbits.*

Remark 6.0.8. This example generalizes [KS, prop. 10.3.18] to the case of a **general (co)stalk property** B. Compare also with [KS, (10.3.31), (10.3.32)] for our description of the Fourier-Sato transformation for **monodromic** complexes in terms of the vanishing cycle functor. But be aware that our $\phi_t[-1]$ corresponds to the vanishing cycle functor used in [KS].

Moreover, theorem 6.0.2(1),(2) and example 6.0.16 imply for a **general (co)-stalk property** B the "stability" of

$${}^m D^{\leq n}(\cdot) \quad \text{and} \quad {}^m D^{\geq n}(\cdot)$$

under the **specialization and microlocalization functors** ν_Y and $\mu_Y[codim(Y)]$ as in [KS, prop. 10.3.19].

See also [BMV1, BMV2] for another presentation of the Fourier transformation, and compare especially with [BMV1, prop. 5.1] and [BMV2, thm. 4.1].

6.0.6 Applications to constructible functions

We finish this introduction with an application of theorem 6.0.2 to the theory of complex analytically (algebraically) constructible functions.

Suppose that our fixed family $B := D^{\leq 0} = D^{\geq 0}$ is a **"null system"** in $D(R)$. Then

$$D_B(X) := {}^m D^{\leq 0}(X) = {}^m D^{\geq 0}(X)$$

by corollary 6.0.2. This is a **triangulated** category so that we can consider its **Grothendieck group** $K_0(D_B(X))$.

Let $f : X \to \mathbb{C}$ be a holomorphic (algebraic) function on the complex analytic (algebraic) space X, with $i : X_0 := \{f = 0\} \to X$ and $j : U := \{f \neq 0\} \to X$ the inclusions. Then the distinguished triangle (6.27) together with theorem 6.0.2(1),(2) implies

$$\{i^* Rj_* j^* \mathcal{F}\} = 0 \in K_0(D_B(X_0)), \tag{6.37}$$

with $\{\cdot\}$ the corresponding class in the Grothendieck group of all complex analytically (algebraically) constructible complexes with the (co)stalk property B.

Let more generally $X_0 := \{f_i = 0,\ i = 1, \ldots, k+1\}$ be the common zero-set of the holomorphic (algebraic) functions $f_i : X \to \mathbb{C}$ $(i = 1, \ldots, k+1)$. Let $j : U := X \backslash X_0 \to X$ be again the inclusion. Then the equality (6.37) is also true. This follows by induction from a **Mayer-Vietoris triangle** (compare [KS, p.114,(2.6.28)], with $R\Gamma_U(\cdot) \simeq Rj_* j^*$) associated to the open covering $\{U_1, U_2\}$ of U, with

$$U_1 := \{f_{k+1} \neq 0\} \quad \text{and} \quad U_2 := X \backslash \{f_i = 0,\ i = 1, \ldots, k\}.$$

Note that

$$U_1 \cap U_2 = X \backslash \{f_i \cdot f_{k+1} = 0,\ i = 1, \ldots, k\}$$

is the complement of the common zero-set of k holomorphic (algebraic) functions.

Let $i : X_0 \to X$ be the inclusion. By the distinguished triangle

$$Rj_! j^* \longrightarrow Rj_* j^* \longrightarrow i_* i^* Rj_* j^* \xrightarrow{[1]}$$

we get

$$Rj_! j^* = Rj_* j^* : K_0(D_B(X)) \to K_0(D_B(X)). \tag{6.38}$$

Similarly, the distinguished triangle

$$i^! \longrightarrow i^* \longrightarrow i^* Rj_* j^* \xrightarrow{[1]}$$

implies

$$i^! = i^* : K_0(D_B(X)) \to K_0(D_B(X_0)). \tag{6.39}$$

Consider in addition a proper holomorphic map $h : X \to Y$ of complex analytic spaces. Then $Rh_! = Rh_*$ maps $D_B(X)$ into $D_B(Y)$ (corollary 6.0.5), and we get the following equality for the induced maps on the level of Grothendieck groups:

$$R(h \circ j)_! j^* = R(h \circ j)_* j^* : K_0(D_B(X)) \to K_0(D_B(Y)) . \tag{6.40}$$

Let $f : X \to Y$ be a morphism of complex algebraic varieties (corresponding to seperated schemes of finite type over $spec(\mathbb{C})$). This can be compactified to $f = \bar{f} \circ j$, with $j : X \to \bar{X}$ an open inclusion and $\bar{f} : \bar{X} \to Y$ a proper morphism of complex algebraic varieties. Then the functors $Rf_!, Rf_*$ map $D_B(X)$ into $D_B(Y)$. Here we consider of course complex **algebraically constructible complexes** of sheaves. Using a **Mayer-Vietoris argument** as before (i.e. a suitable finite open covering), one gets as in [La, p. 210] the equality

$$Rf_! = Rf_* : K_0(D_B(X)) \to K_0(D_B(Y)) . \tag{6.41}$$

Let $CF(X, \mathbb{Z})$ be the abelian group of complex **algebraically constructible** $\mathbb{Z}$-valued functions. Then (6.41) implies the equality (compare section 2.3):

$$f_! = f_* : CF(X, \mathbb{Z}) \to CF(X, \mathbb{Z}) \tag{6.42}$$

for the corresponding group homomorphisms on the level of constructible functions.

Let us go back to the complex analytic context with $CF(X, \mathbb{Z})$ the abelian group of complex **analytically constructible** $\mathbb{Z}$-valued functions.

Example 6.0.17. *Let $i : X_0 \to X$ be the inclusion of a closed complex analytic subset, with $j : U := X \backslash X_0 \to X$ the inclusion of the open complement. Fix also a proper holomorphic map $f : X \to Y$ of complex analytic spaces. Then we get the following equalities for the corresponding group homomorphisms on the level of constructible functions:*

1. $j_! j^* = j_* j^* : CF(X, \mathbb{Z}) \to CF(X, \mathbb{Z})$, *especially* $i^* j_* j^* = 0$.
2. $i^! = i^* : CF(X, \mathbb{Z}) \to CF(X_0, \mathbb{Z})$.
3. $(f \circ j)_! j^* = (f \circ j)_* j^* : CF(X, \mathbb{Z}) \to CF(Y, \mathbb{Z})$.

Note that (1.) implies (3.). Moreover, (1.) and (2.) are local results so that we can assume that X_0 is defined by finitely many holomorphic functions. Then they follow from the corresponding result on the level of Grothendieck groups (compare section 2.3).

A typical application is the following generalization of a **result of Sullivan**: assume in addition that X_0 is **compact**. Consider as a base-ring a field k, and let the "null system" $B := D^{\leq 0} = D^{\geq 0}$ in $D(k)$ be given by the **bounded complexes with finite-dimensional cohomology**. Then one gets for $\mathcal{F} \in D_B(X)$:

$$\chi(R\Gamma(X_0, i^* Rj_* j^* \mathcal{F})) = f_* i^* j_* j^* (\chi_{\mathcal{F}}) = 0 . \tag{6.43}$$

Here χ is the usual **Euler characteristic**, $\chi_{\mathcal{F}}$ is the **constructible function** $x \mapsto \chi(\mathcal{F}_x)$ and f is a constant map on X_0. Then f is proper, since X_0 is compact. The first equality above follows again from section 2.3.

Remark 6.0.9. 1. The equality $\chi(R\Gamma(X_0, i^*Rj_*j^*\mathcal{F})) = 0$ is more generally true (by the same argument) for $\mathcal{F} \in D_B(X)$, with $B := D^{\leq 0} = D^{\geq 0}$ **any** null system in $D(R)$ and $\chi : B \to G$ an **additive** function with values in the abelian group G.

2. $R\Gamma(X_0, i^*Rj_*j^*\mathcal{F})$ calculates the **global link cohomology** of $\mathcal{F}$ as defined in [DS] (compare also with section 1.1). For example for $X_0 = \{x\}$ a point one has (for $0 < \epsilon << 1$):

$$R\Gamma(X_0, i^*Rj_*j^*\mathcal{F}) \simeq R\Gamma(X \cap \{|z| = \epsilon\}, \mathcal{F}) ,$$

with $|\cdot|$ defined by some local embedding $(X, x) \hookrightarrow (\mathbb{C}^n, 0)$. So for $\mathcal{F} := k_X$ the constant sheaf we get that the **link**

$$X \cap \{|z| = \epsilon\} \quad \text{of } X \text{ in } x$$

has a **vanishing Euler characteristic**. For a proof of this classical result of Sullivan [Su], based on the **fibration theorem of Milnor**, compare with [BV, prop.4.1]. Note that we deduced our general result from the distinguished triangle (6.27), which is an abstract counterpart of a "cohomological version of a Milnor fibration" (compare with page 369).

6.1 Proof of the results

In this section we give the proof of the results stated in the introduction. Let us start with the simple

Proof of the lemma 6.0.2. We use the notation of this lemma, together with i_S for the inclusion of a stratum S of $X_\bullet$ into the open (or closed) subset U (or Y). Let $p : \{\text{strata of } X_\bullet\} \to \mathbb{Z}$ be given. The statements (1.),(2.) follow by definition from the functoriality of $(\cdot)^*$ and $(\cdot)^!$. Then one gets the other claims by:

(3.) $(i \circ i_S)^! i_! = i_S^! i^! i_! = i_S^!$ and $(j \circ i_S)^! i_! = i_S^! j^* i_! \simeq 0$.

$(i \circ i_S)^* i_! = i_S^* i^* i_! = i_S^*$ and $(j \circ i_S)^* i_! = i_S^* j^* i_! \simeq 0$.

(4.) $(j \circ i_S)^! Rj_* = i_S^! j^! Rj_* = i_S^!$ and $(i \circ i_S)^! Rj_* = i_S^! i^! Rj_* \simeq 0$.

(5.) $(j \circ i_S)^* Rj_! = i_S^* j^* Rj_! = i_S^*$ and $(i \circ j_S)^* Rj_! = i_S^* i^* Rj_! \simeq 0$. □

Similarly one gets the proof of lemma 6.0.3(1),(2). Let $f : X \to Y$ be a holomorphic map, whose fiber dimension is bounded by d. Start with a complex analytic stratification $Y_\bullet$ of Y, which satisfies the condition $(*_k)$ for all k (e.g. a Whitney b-regular stratification). Choose a similar stratification $X_\bullet$ of X, which refines the induced analytic filtration $f^{-1}(Y_i)$ of X. Then for each stratum S of $X_\bullet$ there exists a stratum T of $Y_\bullet$, with

$$f(S) \subset T \quad \text{and} \quad dim(S) \leq dim(f^{-1}(T)) \leq dim(T) + d .$$

Let $i_x : \{x\} \to X$ and $i_y : \{y\} \to Y$ be the inclusions of $x \in S$ and $y := f(x) \in T$. Then

$$i_x^! f^! = (f \circ i_x)^! \simeq i_y^! \quad \text{and} \quad i_x^* f^* = (f \circ i_x)^* \simeq i_y^*$$

imply the claim of lemma 6.0.3(1),(2) for a $Y_\bullet$-constructible complex by the reformulation (6.4) and property III. of definition 6.0.1.

The same argument applies to lemma 6.0.4, with

$$X^{2(j-d)} = Y^{2j} \cap N \quad \text{and} \quad dim(X^{2(j-d)}) = j - d = dim(Y^{2j}) - d\,.$$

Here $f := i' : X := Y \cap N \to Y$ is the inclusion, with $X_{j-2d} := f^{-1}(Y_j) = Y_j \cap N$ the induced (Whitney b-regular) stratification of X, which one gets by the transversal intersection with the submanifold $N \subset M$ of constant complex codimension d. Then one gets as before the claim (1) for $i'^!$ and (2) for i'^*.

The remaining two cases follow from the isomorphism

$$i'^*\mathcal{F}[-2d] \simeq i'^!\mathcal{F} \quad \text{locally on } X \text{ for } \mathcal{F} \in ob(D(Y_\bullet)). \tag{6.44}$$

This isomorphism is an easy application of (the proof of) lemma 5.1.6 on page 314, if we use the following cartesian diagram of inclusions:

$$\begin{array}{ccc} X := Y \cap N & \xrightarrow{k'} & N \\ \downarrow{\scriptstyle i'} & & \downarrow{\scriptstyle i} \\ Y & \xrightarrow{k} & M\,. \end{array}$$

Then we get by (the proof of) lemma 5.1.6:

$$i'^*\mathcal{F}[-2d] \simeq i'^*k^*Rk_*\mathcal{F}[-2d] \simeq k'^*i^*Rk_*\mathcal{F}[-2d]$$
$$\simeq k'^*i^!Rk_*\mathcal{F} \simeq k'^*Rk'_*i'^!\mathcal{F} \simeq i'^!\mathcal{F}$$

locally on X for $\mathcal{F} \in ob(D(Y_\bullet))$. In fact in our complex context it holds globally on X, since the (complex) normal bundle of N in M is canonical oriented. Moreover, this argument also applies to an **a-regular** stratification with the property **LSA**. (Compare with chapter 4 for all the different regularity conditions of a stratification that we use.)

But as already explained in remark 5.1.8, the proof of this result depends on [KS, cor. 5.4.11(ii)], which is a **non-trivial** application of the "micro-local sheaf theory" of Kashiwara-Schapira (or of Thom's first isotopy lemma). At least for the result about i'^* in lemma 5.1.6(1) we can also give a direct proof (the other result about $i'^!$ will not be used in this book). Then the claim follows from definition 6.0.3 and the second **base change isomorphism** of proposition 4.3.1 (together with remark 4.3.6), if we apply it for a fixed j to the **closed** inclusion $Y^{2j} \hookrightarrow Y \backslash Y_{2j-1}$.

Here we do **not** need b-regularity of the stratification $Y_\bullet$, but only the weaker **d^0-regularity** (or the property **LST**), which we introduced in chapter 4. At the same time, we also get the following more general version of lemma 6.0.5:

Proposition 6.1.1 (base change isomorphism). *Let $X'_\bullet$ be a smooth stratification of the closed subset X' of the manifold M, with $A \subset X'$ a closed union of strata. Suppose H is a closed submanifold of M, which is transversal to all strata, and consider the following cartesian diagram of inclusions:*

$$\begin{array}{ccccc} X \cap H & \xrightarrow{i} & X := X'\backslash A & \xleftarrow{j} & U := X\backslash H \\ \downarrow k'' & & \downarrow k' & & \downarrow k \\ X' \cap H & \xrightarrow[i']{} & X' & \xleftarrow[j']{} & U' := X'\backslash H\,. \end{array}$$

Assume $X'_\bullet$ satisfies the property **LST**, *or is* **d^0-regular**. *Then one has for $\mathcal{F} \in D(X_\bullet)$ the following isomorphisms:*

$$i'^* Rk'_* \mathcal{F} \xrightarrow{\sim} Rk''_* i^* \mathcal{F}\,, \tag{6.45}$$

$$j'_! Rk_* j^* \mathcal{F} \xrightarrow{\sim} Rk'_* j_! j^* \mathcal{F}\,. \tag{6.46}$$

If $X'_\bullet$ is **in addition also Whitney a-regular**, *then one has for $\mathcal{F} \in D(X_\bullet)$ also the isomorphisms:*

$$k''_! i^! \mathcal{F} \xrightarrow{\sim} i'^! k'_! \mathcal{F}\,, \tag{6.47}$$

$$k'_! Rj_* j^* \mathcal{F} \xrightarrow{\sim} Rj'_* k_! j^* \mathcal{F}\,. \tag{6.48}$$

Proof. The isomorphism (6.45) follows from proposition 4.3.1 and remark 4.3.6.

(6.46): The morphism

$$Rk_* \xrightarrow{ad_j} Rk_* j^! j_! \simeq j'^! Rk'_* j_!$$

induces by adjunction a functorial morphism $\alpha : j'_! Rk_* \to Rk'_* j_!$ such that $j'^*(\alpha)$ is isomorphic to the identity morphism

$$Rk_* \simeq j'^* j'_! Rk_* \to j'^* Rk'_* j_! \simeq Rk_* j^* j_! \simeq Rk_*\,.$$

So it is enough to show that $i'^*(\alpha)$ is an isomorphism for $\mathcal{F} \in D(X_\bullet)$. But this is true by $i'^* j'_! \simeq 0$ and

$$i'^* Rk'_* j_! j^* \mathcal{F} \simeq Rk''_* i^* j_! j^* \mathcal{F} \simeq 0\,,$$

since the functorial morphism of (6.45) is an isomorphism for $\mathcal{F}$, $Ri_* i^* \mathcal{F}$ and therefore also for $j_! j^* \mathcal{F}$. More precisely, this morphism applied to $Ri_* i^*$ is isomorphic to the identity morphism

$$Rk''_* i^* \simeq i'^* Ri'_* Rk''_* i^* \simeq i'^* Rk'_* Ri_* i^* \to Rk''_* i^* Ri_* i^* \simeq Rk''_* i^*\,.$$

(6.47): The morphism

$$i^! \xrightarrow{ad_{k'}} i^! k'^! k'_! \simeq k''^! i'^! k'_!$$

induces by adjunction a functorial morphism $\beta : k''_! i^! \to i'^! k'_!$ such that $k''^*(\beta)$ is isomorphic to the identity morphism

$$i^! \simeq k''^* k''_! i^! \to k''^* i'^! k'_! \simeq i^! k'^* k'_! \simeq i^! .$$

So it is enough to show that $i''^*(\beta)$ is an isomorphism for $\mathcal{F} \in D(X_\bullet)$, with $i'' : A \cap H \to X' \cap H$ the closed inclusion. But this is true by $i''^* k''_! \simeq 0$ and

$$i''^* i'^! k'_! \mathcal{F} \simeq i''^* i'^* k'_! \mathcal{F}[-2d] \simeq (i' \circ i'')^* k'_! \mathcal{F}[-2d] \simeq 0 .$$

Here we used (locally) the isomorphism (6.44).

(6.48): The morphism

$$Rj_* \xrightarrow{ad_k} Rj_* k^! k_! \simeq k'^! Rj'_* k_!$$

induces by adjunction a functorial morphism $\gamma : k'_! Rj_* \to Rj'_* k_!$ such that $j'^*(\gamma)$ is isomorphic to the identity morphism

$$k_! \simeq k_! j^* Rj_* \simeq j'^* k'_! Rj_* \to j'^* Rj'_* k_! \simeq k_! .$$

So it is enough to show that $i'^!(\gamma)$ is an isomorphism for $\mathcal{F} \in D(X_\bullet)$. But this is true by $i'^! Rj'_* \simeq 0$ and

$$i'^! k'_! Rj_* j^* \mathcal{F} \simeq k''_! i^! Rj_* j^* \mathcal{F} \simeq 0 ,$$

since the functorial morphism of (6.46) is an isomorphism for $\mathcal{F}$, $Ri_* i^! \mathcal{F}$ and therefore also for $Rj_* j^* \mathcal{F}$. More precisely, this morphism applied to $Ri_* i^!$ is isomorphic to the identity morphism

$$k''_! i^! \simeq k''_* i^! Ri_* i^! \to i'^! k'_! Ri_* i^! \simeq i'^! Ri'_* k''_! i^! \simeq k''_! i^! . \qquad \square$$

After this discussion about suitable base change isomorphisms, let us come back to the proof of the first main result of the introduction. For the following argument, the reader should also look at our proof of theorem 5.0.4 on page 320.

Proof of theorem 6.0.1. We prove the claim by induction over $dim(X)$, where the case $dim(X) = 0$ is trivial (since our families are stable by finite direct sums).

Since $\{a \leq f \leq b\}$ is compact, there exists an open neighborhood U of $\{a \leq f \leq b\}$ in X, and $a' < b'$ with $[a, b] \subset]a', b'[$ such that $f : U \to]a', b'[$ is proper. By approximation we can assume that f has in $\{a < f < b\}$ only **stratified Morse critical points** x with respect to $X_\bullet$, i.e. $x \in X^{2s}$ is a **classical Morse critical** point of $f|X^{2s}$, with df_x **nondegenerate** (i.e. f is induced in some local model $(X, x) \hookrightarrow (\mathbb{C}^n, x)$ by a smooth function germ $f : (\mathbb{C}^n, x) \to (\mathbb{R}, f(x))$ such that df_x is a nondegenerate covector). See for example [Ben] for such an approximation result. If this approximation is close to our given f, then this approximation (or its negative in the second case of the theorem) is also q-convex (since the assumption on the Levi form is an open condition). These critical points are **isolated** so that there are only **finitely** many critical points in $\{a < f < b\}$. To simplify the notation, we can also assume that f takes different values at different critical points.

Then we apply our results about **(stratified) Morse theory for constructible sheaves**. In our context, it is enough to use the results stated in the introduction to chapter 5. Since our families are stable by extensions we only have to show that the **"local Morse datum"**

$$LMD(\mathcal{F}, f, x) := (R\Gamma_{\{f \geq f(x)\}}(\mathcal{F}))_x$$

of f in any such critical point $x \in X^{2s}$ belongs to $D^{\leq n+q'}$ (or $D^{\geq n-q'}$).

The calculation of this "local Morse datum" is a local problem. Therefore we can assume

$$(X, x) \hookrightarrow (\mathbb{C}^n, x)$$

such that f is induced from a smooth function germ on $(\mathbb{C}^n, x)$ with df_x nondegenerate. To simplify the notation we can assume $f(x) = 0$. Since $f|X^{2s}$ has a classical Morse critical point, we can apply theorem 5.0.1(1) and get

$$(R\Gamma_{\{f \geq 0\}}(\mathcal{F}))_x \simeq NMD(\mathcal{F}, f, x)[-\tau],$$

with $NMD(\mathcal{F}, f, x)$ the **"normal Morse datum"** of $\mathcal{F}$ in x and τ the **Morse index** of $f|X^{2s}$.

Since f (or $-f$) is **q-convex**, we can use the estimate (6.16) for this Morse index. By property III. of definition 6.0.1, we only have to show

$$NMD(\mathcal{F}, f, x) \in \begin{cases} D^{\leq n-s} & \text{in the first case of the theorem.} \\ D^{\geq n-s} & \text{in the second case of the theorem.} \end{cases} \tag{6.49}$$

Since we work with a Whitney **b-regular** stratification, we know that the "normal Morse datum"

$$NMD(\mathcal{F}, f, x) = NMD(\mathcal{F}, df_x)$$

depends only on the covector df_x (theorem 5.0.1(1)). In our **complex analytic context** we can therefore assume that f is the **real part of a holomorphic function germ**

$$g : (\mathbb{C}^n, x) \to (\mathbb{C}, 0)\,, \quad \text{with } d(re(g))_x \text{ nondegenerate.}$$

By proposition 5.0.3 we have two distinguished triangles:

$$R\Gamma(l_X, \mathcal{F})[-1] \to NMD(\mathcal{F}, re(g), x) \to k_x^*(\mathcal{F}|N \cap X) \overset{[1]}{\to}, \tag{6.50}$$

and

$$k_x^!(\mathcal{F}|N \cap X) \to NMD(\mathcal{F}, re(g), x) \to R\Gamma(l_X, \partial l_X, \mathcal{F})[-1] \overset{[1]}{\to}. \tag{6.51}$$

Here we use the notation (with $0 < |w| << \delta << 1$)

$$(l_X, \partial l_X) := (X \cap N \cap \{r \leq \delta, g = w\}, X \cap N \cap \{r = \delta, g = w\})$$

for the **complex link** of X with respect to g. N is a complex **normal slice** at x, i.e. a locally closed complex submanifold of $\mathbb{C}^n$ which is **transversal** to $X_\bullet$, with $N \cap X^{2s} = \{x\}$. Finally $k_x : \{x\} \to N \cap X$ is the inclusion and r is the **0-convex distance function** $r(z) := \sum_{i=1}^n z_i \bar{z}_i$ in our fixed local coordinates.

But $N \cap \{g = w\}$ is **transversal** to $X_\bullet$ near x (for $0 < |w| << 1$). So

$$L := X \cap N \cap \{g = w\}$$

gets an induced Whitney b-regular complex analytic stratification $L_\bullet$. Moreover, δ is for $0 < |w| << \delta << 1$ a **regular value** of r with respect to $L_\bullet$, since $d(re(g))_x$ is nondegenerate, and $X_\bullet$ is Whitney b-regular (compare with chapter 5 for the details). Then we can use the induction hypothesis with $q = 0$ for

$$\begin{cases} \mathcal{F}|L,\ r & \text{and } [a,b] := [-1, \delta] \text{ in the first case of the theorem.} \\ \mathcal{F}|L,\ -r & \text{and } [a,b] := [-\delta, 1] \text{ in the second case of the theorem.} \end{cases}$$

Note that

$$\mathcal{F}|L[-1] \in {}^m D^{\leq n-s}(L_\bullet) \quad \text{or} \quad {}^m D^{\geq n-s}(L_\bullet)$$

by transversality and lemma 6.0.4. By induction hypothesis we get

$$R\Gamma(l_X, \mathcal{F})[-1] \in D^{\leq n-s} \quad \text{or} \quad R\Gamma(l_X, \partial l_X, \mathcal{F})[-1] \in D^{\geq n-s} .$$

But by definition $k_x^*(\mathcal{F}|N \cap X) \in D^{\leq n-s}$, and together with (6.50) this implies (6.49). Similarly

$$\mathcal{F}|N \cap X \in {}^m D^{\geq n-s}(L_\bullet)$$

again by lemma 6.0.4 so that $k_x^!(\mathcal{F}|N \cap X) \in D^{\geq n-s}$. Together with (6.51) this implies (6.49) also in the second case. □

Remark 6.1.1.

1. The statement of theorem 6.0.1 is also true for a C^k-function f (with f or $-f$ q-convex) and $k \geq 2$. The proof is the same, if one extends the approximation result of [Ben] about (stratified) Morse functions into this context. Since we need such a result only on a relative compact subset, we can also use a proper real analytic embedding into some $\mathbb{R}^N$. In the embedded context this follows for example from [O2, thm. 1, thm. 3].

2. The statement of theorem 6.0.1 is also true (with the same proof), if we only assume that f can be approximated by a (stratified) Morse function with **"defect"**

$$\lambda - s \leq q \quad \text{or} \quad s - \lambda \leq q$$

for all Morse critical points $x \in X^{2s}$ of f ($s \in \mathbb{N}_0$).

3. Theorem 6.0.1 implies especially the results used in our proof of [Sch3, Satz 1.2] (in that paper we referred to an unpublished preprint, which was a preliminary version of some aspects of this book).

In the classical case of the perverse t-structure (for the middle perversity), theorem 6.0.1 goes back (in special cases) to [GM3], [GM, II, thm. 6.4, prop. 6.7] and [Ha, thm. 5.2].

Our approach to (the proof of) theorem 6.0.1 is an abstraction of the arguments used in these papers. But the distinguished triangles (6.50) and (6.51) (or the base change properties) are **not** used in these papers. Our proof explains, why one has to use **(co)stalk properties** of $\mathcal{F}$, and that one has to **measure them with respect to the middle perversity**:

- In the induction step we use stratified Morse theory on the complex link for the distance function r or $-r$, with r strongly plurisubharmonic (i.e. 0-convex). So the estimate (6.16) for the Morse index (with $q' = 0$) has to be **corrected** by the shift "$[-dim(S)]$" on the stratum S, which is precisely given by the middle perversity!

Of course, this strategy goes back to the work of Goresky-MacPherson (and underlies all applications in [GM, part II]). We realized, that the result of theorem 6.0.1 does **not** depend on the perverse t-structure, but only on **"half of its abstract properties"** as given in our definitions 6.0.1 and 6.0.3. Moreover, the first and the second part of theorem 6.0.1 are completely independent of each other.

Another approach to theorem 6.0.1 in the case $q = 0$ and the classical case of the perverse t-structure (for the middle perversity) is taken up in [KS, thm. 10.3.8, 10.3.32]. They prove these results in an order different from the one that we presented in the introduction. The induction proof given in the case $(b) \Rightarrow (a)$ in [KS, p. 434, thm. 10.3.12] is more or less the same as our proof of theorem 6.0.1, if we use the translation of the **"normal Morse datum"** into the **"local type"** used by Kashiwara-Schapira (compare with the isomorphism (5.52) on page 311).

The same argument as in our proof applies also to the **"real context"**. But here one has to use the distinguished triangles of proposition 5.0.2, with the **"upper/lower halflink"** instead of the "complex link". Since one has no "convexity estimate", one can only use the **trivial estimate** for the Morse index:

$$0 \leq \tau(f|S) \leq dim_R(S).$$

Therefore one has to use the shift "$[-dim_R(S)]$" (or "0") on the stratum S, i.e.:

- The **stalk properties** are measured with respect to the **minimum perversity t** ($t(S) := -dim_R(S)$), and the **costalk properties** are measured with respect to the **zero perversity o** ($o(S) := 0$).

We leave the proof of the following result (which we don't use) as an exercise to the reader. Again we allow **any** "(co)stalk property" $B \subset ob(D(R))$ satisfying the condition "$\leq$" (or "$\geq$") of definition 6.0.1.

Proposition 6.1.2. *Let X be a closed subset of $\mathbb{R}^n$ or a real analytic (Nash-)manifold M, which is definable with respect to an o-minimal structure $\mathcal{S}$ on $(\mathbb{R}, +, \cdot)$ or an analytic (Nash) geometric category (compare with chapter 2).*

Let $X_\bullet$ be a **definable Whitney b-regular stratification** *of X, and consider a C^∞-function $f : M \to \mathbb{R}$. Assume $a < b$ are* **regular values** *of f (with respect to $X_\bullet$) such that $\{a \le f \le b\} \cap X$ is* **compact**. *Then*

$$\mathcal{F} \in {}^tD^{\le n}(X_\bullet) \Rightarrow R\Gamma(\{a \le f \le b\} \cap X, \{f = a\} \cap X, \mathcal{F}) \in D^{\le n} . \tag{6.52}$$

$$\mathcal{F} \in {}^oD^{\ge n}(X_\bullet) \Rightarrow R\Gamma(\{a \le f \le b\} \cap X, \{f = a\} \cap X, \mathcal{F}) \in D^{\ge n} . \tag{6.53}$$

Let us come back to the complex context and the

Proof of corollary 6.0.2. The implication $\Rightarrow$ follows already from the proof of theorem 6.0.1. For the other implication fix a point $x \in X^{2s}$ and take a **complex analytic normal slice** N at x (in some local embedding). Then we have

$$i_x^*(\mathcal{F}) \simeq R\Gamma(N \cap X \cap \{r \le \delta\}, \mathcal{F}) \quad \text{and}$$

$$i_x^!(\mathcal{F}) \simeq R\Gamma(N \cap X \cap \{r \le \delta\}, N \cap X \cap \{r = \delta\}, \mathcal{F})$$

for $0 < \delta << 1$. Here $i_x : \{x\} \to X$ is the inclusion and r the 0-convex distance function

$$r(z) := \sum_{i=1}^{n-s} z_i \bar{z}_i \quad \text{in local coordinates} \quad (N, x) \simeq (\mathbb{C}^{n-s}, x) .$$

For the last isomorphism we use the second base change isomorphism of proposition 4.3.1 (together with remark 4.3.6). Moreover, the "normal Morse datum" doesn't change if we take a transversal intersection:

$$NMD(\mathcal{F}|N \cap X, x')[-(j-s)] \simeq NMD(\mathcal{F}, x')[-j+s] \in D^{\le -s} \quad \text{or} \quad D^{\ge -s}$$

for $x' \in N \cap X^{2j}$ $(j \ge s)$. Then the claim follows from the proof of theorem 6.0.1 for the function r or $-r$ (with $q = 0$). More precisely, by this information about the "normal Morse datum" we get as in that proof:

$$R\Gamma(N \cap X \cap \{r \le \delta\}, \mathcal{F}) \in D^{\le -s} \quad \text{and}$$

$$R\Gamma(N \cap X \cap \{r \le \delta\}, N \cap X \cap \{r = \delta\}, \mathcal{F}) \in D^{\ge -s} . \qquad \square$$

Remark 6.1.2. In the above proof we used the fact, that the "normal Morse datum" doesn't change by a transversal intersection with a submanifold N. This gives us control about the "normal Morse datum" corresponding to a nondegenerate covector $\omega = d(f|N)_{x'}$, which is **induced** from a function f such that df_x is nondegenerate with respect to the stratification $X_\bullet$. This is enough, since we know that the "normal Morse datum" depends only on the corresponding **connected component of the set of nondegenerate covectors** in

$$T^*_{X^j}M \quad \text{or} \quad T^*_{N \cap X^j}N .$$

Especially in our complex analytic context, it depends only on the stratum.

Since the statements are independent of the chosen stratification (compare with the argument after remark 5.2.3 on page 326), we can even refine our stratification such that the **nondegenerate covectors are dense in all fibers**

$$(T^*_{X^s}M)_x \quad \text{for } x \in X^s \quad (s \in \mathbb{N}_0)\,.$$

Compare with remark 5.1.9 on page 320. Moreover, in this case our argument above is similar to (the proof of) [KS, lem. 10.3.9].

Let us point out, that the same reasoning applies by the last remark also to the **real context**, if we use proposition 6.1.2 (instead of theorem 6.0.1):

Corollary 6.1.1. *Let $X_\bullet$ be a* **definable Whitney b-regular stratification** *as in proposition* 6.1.2. *Then we have for $\mathcal{F} \in ob(D(X_\bullet))$:*

1. $\mathcal{F} \in {}^tD^{\leq 0}(X_\bullet) \Leftrightarrow NMD(\mathcal{F}, df_x)[-dim_{\mathbb{R}}(S)] \in D^{\leq 0}$ *for all nondegenerate covectors df_x, with x a point in the stratum S.*
2. $\mathcal{F} \in {}^0D^{\geq 0}(X_\bullet) \Leftrightarrow NMD(\mathcal{F}, df_x) \in D^{\geq 0}$ *for all nondegenerate covectors df_x, with x a point in the stratum S.*

Some other results from **real geometry** are also used in the proof of corollary 6.0.3, where we assumed a **subanalytic compactification**

$$\bar{f} : \overline{X} \to [0, \infty]$$

of the proper (q-convex) function $f : X \to \mathbb{R}^{\geq 0}$, with $j : X \to \overline{X}$ the inclusion of a locally closed (sub)analytic subset.

Assume $j_!\mathcal{F}$ (or equivalently $Rj_*\mathcal{F}$) is subanalytically constructible. Then

$$Rf_!\mathcal{F} \simeq R\bar{f}_!j_!\mathcal{F} \quad \text{or} \quad Rf_*\mathcal{F} \simeq R\bar{f}_*Rj_*\mathcal{F}$$

is subanalytically constructible on $[0, \infty]$, since $\bar{f}$ is a proper subanalytic map (for example by theorem 4.2.1, but compare also with chapter 2). Especially, there is an $r' >> 0$ such that all cohomology sheaves of these direct image complexes are locally constant on $[r', \infty[$. But then we get

$$R\Gamma([0, \infty[\,, Rf_*\mathcal{F}) \simeq R\Gamma([0, r], Rf_*\mathcal{F})$$

and

$$R\Gamma_c([0, \infty[\,, Rf_!\mathcal{F}) \simeq R\Gamma_c([0, r[\,, Rf_!\mathcal{F})$$

for all $r > r'$. This implies the claim used after corollary 6.0.3, since f is proper.

Let us now explain the cases mentioned in example 6.0.5(3-5), where we can **avoid** the use of such a compactification of the **proper q-convex** function

$$f : X \to \mathbb{R}^{\geq 0}\,.$$

Fix a complex analytic Whitney b-regular stratification $X_\bullet$ of X, and a sequence $r_n \nearrow \infty$ of regular values of f (with respect to $X_\bullet$). Then we consider the following exhaustions of X by (relatively) compact subsets:

$$K_n := \{f \leq r_n\} \quad \text{and} \quad U_n := \{f < r_n\} \quad (n \in \mathbb{N}) \ .$$

Let us first look at the (co)stalk property $B = D^{\leq 0}, D^{\geq 0}$ or ${}^{+}D^{\geq 0}$ as in the "classical case" of example 6.0.2. Then we get by theorem 6.0.1 (compare also with [Sch3]):

Corollary 6.1.2. *Let $f : X \to \mathbb{R}^{\geq 0}$ be a* **proper q-convex** *function, with $q' := max\,\{q, dim(X)\}$ and K_n, U_n $(n \in \mathbb{N})$ defined as before.*

1. *Consider the stalk property $B = D^{\leq 0}$, with $\mathcal{F} \in {}^{m}D^{\leq 0}(X_\bullet)$. Then*

$$H^k(K_{n+1}, K_n, \mathcal{F}) \simeq H^k(\{r_n \leq f \leq r_{n+1}\}, \{f = r_n\}, \mathcal{F}) = 0 \quad \textit{for } k > q'$$

so that the restriction map $H^k(K_{n+1}, \mathcal{F}) \to H^k(K_n, \mathcal{F})$ is **surjective** *for $k \geq q'$. Especially, the projective system*

$$H^k(K_n, \mathcal{F}) \quad (n \in \mathbb{N})$$

satisfies the so called **Mittag-Leffler condition** *for $k \geq q'$ so that*

$$H^k(X, \mathcal{F}) \simeq \lim_{\leftarrow} H^k(K_n, \mathcal{F}) = 0 \quad \textit{for } k > q'. \tag{6.54}$$

Compare with [KS, prop. 2.7.1] for the first isomorphism.

2. *Consider the stalk property $B = D^{\geq 0}$, ${}^{+}D^{\geq 0}$, with $\mathcal{F} \in {}^{m}D^{\geq 0}(X_\bullet)$. Then*

$$H^k_c(X, \mathcal{F}) \simeq \lim_{\rightarrow} H^k_c(U_n, \mathcal{F}) = \begin{cases} 0 & \textit{for } k > q', \\ \textit{torsionfree} & \textit{for } k = q'. \end{cases} \tag{6.55}$$

In other words, a **q-complete complex space is topologically q-complete** with respect to the (co)stalk properties $B = D^{\leq 0}, D^{\geq 0}$ and ${}^{+}D^{\geq 0}$ as in the "classical case" of example 6.0.2.

Corollary 6.1.3. *Assume more generally that the (co)stalk property B is*

$$\textit{stable under countable} \quad \begin{cases} \textit{products in the case "}\leq\textit{"}, \\ \textit{direct sums in the case "}\geq\textit{".} \end{cases}$$

Then a **q-complete complex space is topologically (q+1)-complete** *with respect to the (co)stalk property B.*

Using the notation as before, this is a direct application of theorem 6.0.1 by the following two distingished triangles:

$$R\Gamma(X,\mathcal{F}) \longrightarrow \prod_{n=1}^{\infty} R\Gamma(K_n,\mathcal{F}) \longrightarrow \prod_{n=1}^{\infty} R\Gamma(K_n,\mathcal{F}) \xrightarrow{[1]}, \tag{6.56}$$

$$\bigoplus_{n=1}^{\infty} R\Gamma_c(U_n,\mathcal{F}) \longrightarrow \bigoplus_{n=1}^{\infty} R\Gamma_c(U_n,\mathcal{F}) \longrightarrow R\Gamma_c(X,\mathcal{F}) \xrightarrow{[1]}. \tag{6.57}$$

These distingished triangles follow by the use of a flabby resolution from the following short exact sequences for a **flabby** sheaf $\mathcal{F}$ on X:

$$0 \to \Gamma(X,\mathcal{F}) \xrightarrow{can} \prod_{n=1}^{\infty} \Gamma(K_n,\mathcal{F}) \xrightarrow{id-shift} \prod_{n=1}^{\infty} \Gamma(K_n,\mathcal{F}) \to 0\,, \tag{6.58}$$

$$0 \to \bigoplus_{n=1}^{\infty} \Gamma_c(U_n,\mathcal{F}) \xrightarrow{id-shift} \bigoplus_{n=1}^{\infty} \Gamma_c(U_n,\mathcal{F}) \xrightarrow{can} \Gamma_c(X,\mathcal{F}) \to 0\,. \tag{6.59}$$

Here "*can*" is induced by the restriction (or extension) of sections (with compact support). Similarly for the maps

$$\phi_n : \Gamma(K_{n+1},\mathcal{F}) \to \Gamma(K_n,\mathcal{F}) \quad \text{and} \quad \psi_n : \Gamma_c(K_n,\mathcal{F}) \to \Gamma_c(K_{n+1},\mathcal{F})$$

of the corresponding projective or inductive system. Then the map "*shift*" is defined by

$$(s_n) \mapsto (\phi_n(s_{n+1})) \quad \text{or} \quad (s_n) \mapsto (\psi_{n-1}(s_{n-1}))\,,$$

with $\psi_0(s_0) := 0$. We leave the proof of the exactness of the sequences as an exercise. Compare also with [Nee, sec. 1.6] for other applications of such "homotopy (co)limits".

Remark 6.1.3. The proof of corollary 6.1.3 applies more generally also to a complex space X, which has an **exhaustion** X_n **by q-complete open subspaces** (such a space need not to be q-complete in general).

Next we explain the lemma 6.0.6. Consider a (finite) filtration

$$\emptyset =: X_{M-1} \subset X_M \cdots \subset X_i \subset \cdots \subset X_N = X$$

of X by closed complex subsets X_i ($M \leq i \leq N$ for some $M, N \in \mathbb{Z}$). Then $X^i := X_i \backslash X_{i-1}$ is a locally closed constructible subset of X so that its closure $cl(X^i) \subset X_i$ is a closed complex analytic subset of X. Let

$$j_i : X^i \to Z_i := cl(X^i)\,, \quad k_i : X^i \to X \quad \text{and} \quad l_i : X_i \to X$$

be the inclusions for $M \leq i \leq N$. Under the assumptions of lemma 6.0.6, one shows by an induction argument

$$R\Gamma(X_i, l_i^*\mathcal{F}) \in D^{\leq n} \quad \text{and} \quad R\Gamma_c(X_i, l_i^!\mathcal{F}) \in D^{\geq n}\,.$$

For a fixed i, let $k : X_{i-1} \to X_i$ be the closed and $j : X^i \to X_i$ be the open inclusion. Then one has for $\mathcal{F} \in ob(D(X))$ two distinguished triangles

$$Rj_!j^*l_i^*\mathcal{F} \longrightarrow l_i^*\mathcal{F} \longrightarrow Rk_*k^*l_i^*\mathcal{F} \xrightarrow{[1]} ,$$
$$Rk_*k^!l_i^!\mathcal{F} \longrightarrow l_i^!\mathcal{F} \longrightarrow Rj_*j^*l_i^!\mathcal{F} \xrightarrow{[1]} .$$

Using the first distinguished triangle one gets the claim about $R\Gamma(X_i, l_i^*\mathcal{F})$, since

$$R\Gamma(X_{i-1}, l_{i-1}^*\mathcal{F}) \simeq R\Gamma(X_i, Rk_*k^*l_i^*\mathcal{F}) \quad \text{and}$$
$$R\Gamma(Z_i, Rj_{i!}k_i^*\mathcal{F}) \simeq R\Gamma(X_i, Rj_!j^*l_i^*\mathcal{F}) .$$

Note that

$$k_i^*\mathcal{F} \in {}^mD^{\leq n(i)}(X^i) \Rightarrow Rj_{i!}k_i^*\mathcal{F} \in {}^mD^{\leq n(i)}(Z_i) .$$

Using the second distinguished triangle one gets the claim about $R\Gamma_c(X_i, l_i^!\mathcal{F})$, since

$$R\Gamma_c(X_{i-1}, l_{i-1}^!\mathcal{F}) \simeq R\Gamma_c(X_i, Rk_*k^!l_i^!\mathcal{F}) \quad \text{and}$$
$$R\Gamma_c(Z_i, Rj_{i*}k_i^!\mathcal{F}) \simeq R\Gamma_c(X_i, Rj_*j^*k'^!\mathcal{F}) .$$

Again, we have

$$k_i^!\mathcal{F} \in {}^mD^{\geq n(i)}(X^i) \Rightarrow Rj_{i*}k_i^!\mathcal{F} \in {}^mD^{\geq n(i)}(Z_i) .$$

Of course we make use of the fact, that the "(co)stalk family" $B \subset ob(D(R))$ satisfies the property "$\leq$" (or "$\geq$").

A similar induction argument can be used, to reduce corollary 6.0.6 to corollary 6.0.5. Consider the context of corollary 6.0.6. Choose a complex analytic Whitney b-regular stratification $\bar{X}_\bullet$ of $\bar{X}$ such that Z is a (closed) union of strata, with $\mathcal{F}$ $\bar{X}_\bullet$-constructible and $rk(\bar{f}|S)$ constant on any stratum S. We can even assume, that $\bar{X}_\bullet$ has only **finitely** many strata (since the strata are locally finite and the result is local on Y, with $\bar{f}$ proper). By induction over the number of strata of $X_\bullet$, one only has to show (similarly as in the proof of lemma 6.0.6):

$$i_S^!\mathcal{F} \in {}^{m(f)}D^{\geq n}(S) \Rightarrow Rf_*i_S^!\mathcal{F} \in {}^mD^{\geq n}(Y) ,$$

and

$$i_S^*\mathcal{F} \in {}^{m(f)}D^{\leq n}(S) \Rightarrow Rf_!i_S^*\mathcal{F} \in {}^mD^{\leq n}(Y) .$$

Here $i_S : S \to \bar{X}$ is the inclusion of the stratum, with $f := \bar{f}|S : S \to Y$. But $cl(S)$ and $\partial S := cl(S)\backslash S$ are closed complex analytic subsets of $\bar{X}$ (by the frontier condition) so that we can apply corollary 6.0.5 to this map $f : S \to Y$, whose fiber dimension d is given by $d := dim(S) - rk(f|S)$. Then the claim follows from

$$i_S^!\mathcal{F} \in {}^{m(f)}D^{\geq n}(S) \Leftrightarrow i_S^!\mathcal{F} \in {}^mD^{\geq n+d}(S) ,$$

and

$$i_S^*\mathcal{F} \in {}^{m(f)}D^{\leq n}(S) \Leftrightarrow i_S^*\mathcal{F} \in {}^mD^{\leq n-d}(S) .$$

So we have reduced corollary 6.0.6 to the

Proof of corollary 6.0.5. Choose a complex analytic Whitney b-regular stratification $Y_\bullet$ of Y such that $Rf_!(\mathcal{F}|X)$ and $Rf_*(\mathcal{F}|X)$ are constructible with respect to $Y_\bullet$. Choose a point x in the stratum S of dimension s, and consider the cartesian diagram

$$\begin{array}{ccccc} f^{-1}(U) & \xrightarrow{k} & f^{-1}(S) & \xrightarrow{i} & X \\ \downarrow & & \downarrow{\scriptstyle f'} & & \downarrow{\scriptstyle f} \\ U & \longrightarrow & S & \xrightarrow[i_S]{} & Y\,, \end{array}$$

with $U = \{r < r_0\}$ a small open ball around x in S in some local coordinates $(S, x) \simeq (\mathbb{C}^s, 0)$. Let also

$$j : f^{-1}(U) \to \overline{V} := cl(f^{-1}(U)) \subset \bar{f}^{-1}(U) \subset \overline{X}$$

be the **open** inclusion into the closure of $f^{-1}(U)$ in $\bar{f}^{-1}(U)$. Then $\bar{f} : \overline{V} \to U$ is **proper**, with $dim(\overline{V}) = dim(f^{-1}(U)) \leq dim(S) + d$. Finally, the proper C^∞-function

$$\frac{1}{r_0 - r} \circ \bar{f} : \overline{V} \to \mathbb{R}^{\geq 0}$$

has a suitable subanalytic extension so that we can apply corollary 6.0.3(1),(2).

Assume $\mathcal{F}|X \in {}^mD^{\geq n}(X)$. Then we also have

$$Rj_*k^*i^!(\mathcal{F}|X) \in {}^mD^{\geq n}(\overline{V})\,.$$

Moreover, we can apply corollary 6.0.3(2) to this complex, with

$$\begin{aligned} \left(i_S^! Rf_*(\mathcal{F}|X)\right)_x &\simeq R\Gamma(U, i_S^! Rf_*(\mathcal{F}|X)) \simeq R\Gamma(U, Rf'_*i^!(\mathcal{F}|X)) \\ &\simeq R\Gamma(\overline{V}, Rj_*k^*i^!(\mathcal{F}|X))\,. \end{aligned}$$

So we get

$$\left(i_S^! Rf_*(\mathcal{F}|X)\right)_x \in D^{\geq n - dim(S) - d}\,.$$

Assume $\mathcal{F}|X \in {}^mD^{\leq n}(X)$. Then we also have

$$Rj_!k^*i^*(\mathcal{F}|X) \in {}^mD^{\leq n}(\overline{V})\,.$$

Moreover, we can apply corollary 6.0.3(1) to this complex, with

$$\begin{aligned} \left(i_S^* Rf_!(\mathcal{F}|X)\right)_x[-2s] &\simeq R\Gamma_c(U, i_S^* Rf_!(\mathcal{F}|X)) \simeq R\Gamma_c(U, Rf'_!i^*(\mathcal{F}|X)) \\ &\simeq R\Gamma_c(\overline{V}, Rj_!k^*i^*(\mathcal{F}|X))\,. \end{aligned}$$

So we get

$$\left(i_S^* Rf_!(\mathcal{F}|X)\right)_x \in D^{\leq n - dim(S) + d}\,.$$ □

Following the presentation of the introduction, we come to the proof of our second main theorem for the nearby cycle functor.

Proof of theorem 6.0.2(1),(2). The argument is similar to the proof of theorem 6.0.1. But this time we use the corresponding description of lemma 5.4.2 for the **(co)stalk of the nearby cycle functor**. Choose a complex analytic Whitney b-regular stratification $X_\bullet$ of X with the following properties (compare with the references given in remark 5.4.8 on page 371):

1. $\{f = 0\}$ is a union of strata.
2. $j^*\mathcal{F} \in {}^mD^{\leq 0}(U_\bullet)$ or $j^*\mathcal{F} \in {}^mD^{\geq 0}(U_\bullet)$, with $U_\bullet$ the induced stratification of $U = \{f \neq 0\}$.
3. The stratification satisfies the a_f**-condition of Thom** as in lemma 5.4.2.

Consider a point $x \in S$ for a stratum $S \subset \{f = 0\}$ of dimension s. First we assume that $S = \{x\}$ is a point stratum. By lemma 5.4.2 we get

$$i_x^*\psi_f(\mathcal{F}) \simeq R\Gamma(X \cap \{r \leq \delta, f = w\}, \mathcal{F})\,, \tag{6.60}$$

and

$$i_x^!\psi_f(\mathcal{F}) \simeq R\Gamma(X \cap \{r \leq \delta, f = w\}, X \cap \{r = \delta, f = w\}, \mathcal{F}) \tag{6.61}$$

for $0 < |w| << \delta << 1$, with $i_x : \{x\} \to \{f = 0\}$ the inclusion and

$$r(z) := \sum\nolimits_{i=1}^m z_i\bar{z}_i \quad \text{in a local embedding} \quad (X, x) \hookrightarrow (\mathbb{C}^m, 0)\,.$$

Then $L := \{f = w\}$ is **transversal** to $X_\bullet$ near x for $0 < |w| << 1$ (by the curve selection lemma). So we get by lemma 6.0.4:

$$\mathcal{F}|L[-1] \in {}^mD^{\leq n}(L_\bullet) \quad \text{or} \quad \mathcal{F}|L[-1] \in {}^mD^{\geq n}(L_\bullet)\,.$$

Moreover, δ is for $0 < |w| << \delta << 1$ a **regular value** of r with respect to $L_\bullet$. This follows from the a_f**-condition** (compare with example 1.1.3 on page 36). If we apply theorem 6.0.1 with $q = 0$ to r (or $-r$), then we get by (6.60) (or (6.61)):

$$i_x^*\psi_f(\mathcal{F})[-1] \in D^{\leq n} \quad \text{or} \quad i_x^!\psi_f(\mathcal{F})[-1] \in D^{\geq n}\,.$$

This proves our claim for a point stratum.

We reduce the general case to the first case by taking a **complex analytic normal slice** N at x (in some local embedding). Consider the cartesian diagram

$$\begin{array}{ccccc}
\{x\} = N \cap S & \xrightarrow{\kappa_x} & N \cap \{f = 0\} & \xrightarrow{i'} & N \cap X \\
\downarrow{\scriptstyle k_x} & & \downarrow{\scriptstyle k'} & & \downarrow{\scriptstyle k} \\
S & \xrightarrow{i_S} & \{f = 0\} & \xrightarrow{i} & X\,.
\end{array}$$

Then we have

$$k_x^* i_S^* \psi_f(\mathcal{F}) \simeq \kappa_x^* k'^* \psi_f(\mathcal{F})\,.$$

Since $\psi_f(\mathcal{F})$ is constructible with respect to the induced stratification of $\{f = 0\}$ (lemma 4.2.1), we get by proposition 4.3.1 and remark 4.3.6 :

$$k_x^* i_S^! \psi_f(\mathcal{F}) \simeq \kappa_x^! k'^* \psi_f(\mathcal{F}) \,.$$

But we also have the **base change isomorphism** (compare with lemma 4.3.4 on page 265):

$$k'^* \psi_f(\mathcal{F}) \simeq \psi_{f'}(k^* \mathcal{F}) \,, \quad \text{with } f' := f \circ k.$$

Then the claim follows from the first case for f', $X' := N \cap X$ and $k^*\mathcal{F}$, since N is transversal to $X_\bullet$ near x so that

$$k^*\mathcal{F} \in {}^m D^{\leq n-s}(X'_\bullet) \quad \text{or} \quad k^*\mathcal{F} \in {}^m D^{\geq n-s}(X'_\bullet) \,. \qquad \square$$

Using the description (5.13) of the **"local Morse data"** in terms of the **vanishing cycle functor**, we can also give the missing proof of the implication $\Leftarrow$ of corollary 6.0.7:

Choose a complex analytic Whitney b-regular stratification $X_\bullet$ of X such that $\mathcal{F}$ is $X_\bullet$-constructible. Fix a point $x \in X^{2s}$, and choose (in some local coordinates) a holomorphic function germ $f : (X, x) \to (\mathbb{C}, 0)$ such that $f|X^{2s}$ has in x a **complex Morse critical point**, with $d(re(f))_x$ a **nondegenerate covector**. Especially, x is an isolated stratified critical point of f so that x is **isolated** in the support of $\phi_f(\mathcal{F})$.

Then $re(f)|X^{2s}$ has in x a classical Morse critical point of Morse index s, and the "normal Morse datum" corresponding to $\omega := d(re(f))_x$ is isomorphic to

$$NMD(\mathcal{F}, \omega)[-s] \simeq LMD(\mathcal{F}, \omega) \simeq \left(\phi_f[-1](\mathcal{F})[-1]\right)_x \,.$$

Compare with theorem 5.0.1(1) and (5.13) on page 283. But any nondegenerate covector ω is of the form $\omega = d(re(f))_x$, with f a holomorphic function germ as before. Therefore our claim follows from corollary 6.0.2.

Let us finally come to the proof of the **Artin-Grothendieck type theorems**.

Proof of theorem 6.0.3. Fix a complex analytically constructible complex $\mathcal{F}$ on X such that $\mathcal{G} := Rf_*\mathcal{F}$ or $\mathcal{G} := Rf_!\mathcal{F}$ is also complex analytically constructible on Y. Choose a complex analytic Whitney b-regular stratification $Y_\bullet$ of Y such that $\mathcal{G}$ is $Y_\bullet$-constructible.

For the following argument, compare also with the proof of [KS, prop.10.3.17]. Fix a point $y \in S := Y^{2s}$, and consider the intersection with a **complex normal slice** N at y (in some local coordinates) so that one gets the following cartesian diagram of inclusions:

$$\begin{array}{ccc} Y \cap N & \xrightarrow{i'} & Y \\ {\scriptstyle i_y}\uparrow & & \uparrow{\scriptstyle i_S} \\ \{y\} & \xrightarrow[k_y]{} & S \,. \end{array}$$

Then we get by the isomorphism (6.44):

$$k_y^* i_S^* \mathcal{G}[-2s] \simeq i_y^* i'^! \mathcal{G} \quad \text{and} \quad k_y^! i_S^! \mathcal{G}[2s] \simeq i_y^! i'^* \mathcal{G} . \tag{6.62}$$

Consider the cartesian diagram

$$\begin{array}{ccccc} f^{-1}(U') & \xrightarrow{k} & f^{-1}(Y \cap N) & \xrightarrow{i} & X \\ \downarrow & & \downarrow & & \downarrow f \\ U' & \longrightarrow & Y \cap N & \xrightarrow[i']{} & Y , \end{array}$$

with $U' := Y \cap B$ the intersection of Y with a small open ball B around y in N (in some local coordinates). Note that the inclusion i is locally given by $s = dim(S)$ holomorphic equations, since this is the case for the inclusion $Y \cap N \to Y$. We can also assume, that U' is so small that U' is contained in the open neighborhood U of y given by the assumption (a.) of the theorem. Let

$$g : f^{-1}(U) \to \mathbb{R}^{\geq 0}, \quad \text{with} \quad (f, g) : V \to U \times \mathbb{R}^{\geq 0} \quad \textbf{proper},$$

be the corresponding **q-convex $\mathbf{C}^\infty$-function**. Suppose the open ball B is given by $B := \{r < r_0\}$, with r the usual distance function

$$r(z) := \sum_{i=1}^{d} z_i \bar{z}_i \quad \text{in some local coordinates } (N, y) \simeq (\mathbb{C}^d, 0).$$

Then also $f^{-1}(U')$ is **q-convex**, more precisely:

$$g' := g + \frac{1}{r_0 - r} \circ f : f^{-1}(U') \to \mathbb{R}^{\geq 0}$$

is a **proper q-convex function**. Here $1/(r_0 - r)$ is 0-convex (compare [SV, prop.2.2]) so that the Levi form of $1/(r_0 - r) \circ f$ is positive semi-definite. This implies the q-convexity of g' (compare with the proof of [SV, prop.2.4]).

Moreover, by assumption (b.) of the theorem, g' has also a **subanalytic extension** (!) to the compact subanalytic space

$$\overline{V} \cap \bar{f}^{-1}(Y \cap \{r \leq r_0\}) \quad (0 < r_0 << 1) .$$

"**$\leq$**": Assume $\mathcal{F} \in {}^m D^{\leq n}(X)$, with $Rf_* \mathcal{F}$ complex analytically constructible. Then we get

$$i_y^* i'^! (Rf_* \mathcal{F}) \simeq R\Gamma(U', i'^! Rf_* \mathcal{F}) \simeq R\Gamma(f^{-1}(U'), k^* i^! \mathcal{F}) , \tag{6.63}$$

with

$$k^* i^! \mathcal{F} \in {}^m D^{\leq n+s}(f^{-1}(U'))$$

by proposition 6.0.2(1). So by theorem 6.0.1 and (6.62) we get

$$k_y^* i_S^* (Rf_* \mathcal{F}) \in D^{\leq n+s+q'}[2s] = D^{\leq n-s+q'} .$$

"≥": Assume $\mathcal{F} \in {}^m D^{\geq n}(X)$, with $Rf_!\mathcal{F}$ complex analytically constructible. Then we get

$$i_y^! i'^*(Rf_!\mathcal{F}) \simeq R\Gamma_c(U', i'^* Rf_!\mathcal{F}) \simeq R\Gamma_c(f^{-1}(U'), k^* i^* \mathcal{F}) , \tag{6.64}$$

with

$$k^* i^* \mathcal{F} \in {}^m D^{\geq n-s}(f^{-1}(U'))$$

by proposition 6.0.2(2). So by theorem 6.0.1 and (6.62) we get

$$k_y^! i_S^!(Rf_!\mathcal{F}) \in D^{\geq n-s-q'}[-2s] = D^{\geq n+s-q'} .$$

This finishes the proof by remark 6.0.2(1). □

The same proof applies also to corollary 6.0.8. Here we do not need any compactification, since we can use corollary 6.1.2 and corollary 6.1.3.

Similarly, theorem 6.0.3 implies directly theorem 6.0.4, since an **affine algebraic** map is by definition also an "analytic affine" map in the sense of (6.36). And for such a factorization (6.36) one can use the **0-convex function**

$$g := \Big(\frac{1}{r_0 - r} \times r'\Big) \circ i' : f^{-1}(U) \to \mathbb{R}^{\geq 0} ,$$

with $U := \{r < r_0\}$ an open ball as before, and r' the corresponding distance function on $\mathbb{C}^N$ (compare [SV, prop. 2.2, prop. 2.3]). Of course, this g has a suitable **subanalytic compactification** so that theorem 6.0.4 also implies corollary 6.0.9.

Last but not least, let us explain the example 6.0.16. Factorize the corresponding map γ as

$$\begin{array}{ccccc}
E \times_X E^* & \xrightarrow{\ i'\ } & E \times_X (E^* \times \mathbb{C}) & & \\
\gamma \downarrow & & \downarrow j' & & \\
E^* \times \mathbb{C} & \xleftarrow{\ \pi\ } & \mathbb{P}(E \times \mathbb{C}) \times_X (E^* \times \mathbb{C}) & = & \mathbb{P}\big((E \times \mathbb{C}) \times_X (E^* \times \mathbb{C}) \big) ,
\end{array}$$

with $\mathbb{P}(E \times \mathbb{C})$ the **projective completion** of the vector bundle E over X (i.e. the projective bundle of the vector bundle $E \times \mathbb{C}$). Here j' is the canonical open embedding and π the canonical projection. Finally, the closed embedding i' is defined as $i'(x, y) := (x, y, y(x))$. Then γ is of course an "analytic affine" map in the sense of (6.36) (use a local trivialization of the vector bundles).

Let C be a closed **conic** analytic subset of the vector bundle $E \times_X E^* \to E^*$, i.e. an analytic subset invariant under the natural $\mathbb{C}^*$-action on the fibers of E. Then $(j' \circ i')(S)$ is a **locally closed constructible** subset of $\mathbb{P}(E \times \mathbb{C}) \times_X (E^* \times \mathbb{C})$, in fact

$$(j' \circ i')(C) = \mathbb{P}(C \times \mathbb{C}^2) \cap \{([x, \lambda], y, \lambda') \mid y(x) - \lambda \cdot \lambda' = 0,\ \lambda \neq 0\} ,$$

with

$$\mathbb{P}(C \times \mathbb{C}^2) \subset \mathbb{P}(\,(E \times \mathbb{C}) \times_X (E^* \times \mathbb{C})\,)$$

the **projective completion** of the complex analytic **cone** $C \times \mathbb{C}$.

Fix now a complex analytically constructible **monodromic** complex $\mathcal{F} \in D_{mon}(E)$. Then $\mathcal{F}$ is constructible with respect to a complex analytic Whitney b-regular stratification $E_\bullet$ of E. By Thom's isotopy lemma, this is a **cs-stratification** in the sense of [HS, def. 1.1]. By [HS, cor. 3.4], there is a **unique coarsest cs-stratification**

$$X_0 \subset \cdots \subset X_n = E$$

such that $\mathcal{F}$ is constructible with respect to this filtration. Each connected component of $X_i \backslash X_{i-1}$ is then a union of strata of $E_\bullet$ ([HS, def. 3.1]). Especially, the X_i are **closed complex analytic** subsets. Since $\mathcal{F}$ is monodromic, this implies by the definition of the unique coarsest cs-stratification, that all X_i are **invariant** under the $\mathbb{C}^*$-action.

Let $p_1 : E \times_X E^* \to E$ be the projection. Then $p_1^*\mathcal{F}$ is constructible with respect to the induced filtration

$$X_0 \times_X E^* \subset \cdots \subset X_n \times_X E^*$$

by closed **conic** analytic subsets of the vector bundle $E \times_X E^* \to E^*$. So their images under $j' \circ i'$ are locally closed constructible subsets of the larger projective bundle

$$\mathbb{P}(\,(E \times \mathbb{C}) \times_X (E^* \times \mathbb{C})\,)\,.$$

Then there is a complex analytic Whitney b-regular stratification of this projective bundle, whose pullback to $E \times_X E^*$ refines this induced filtration. But this implies that

$$(j' \circ i')_! p_1^*\mathcal{F} \quad \text{and} \quad R(j' \circ i')_* p_1^*\mathcal{F}$$

are complex analytically constructible.

Bibliography

[ABV] Adams, J., Barbasch, D., Vogan Jr., D.A.: *The Langlands classification and irreducible characters for real reductive groups.* Progress in Mathematics 104 Birkhäuser, Boston Basel Berlin (1992)

[AVG] Arnold, V.I., Varchenko, A.N., Gusein-Zade, S.M.: *Singularities of differentiable mappings.* Vol.I,II Birkhäuser, Basel (1986,88)

[AtBo] Atiyah, M.F., Bott, R.: The moment map and equivariant cohomology. Topology 23, 1–28 (1984)

[AtSeg] Atiyah, M.F., Segal, G.B.: The index of elliptic operators: II. Ann. Math. 87, 531–545 (1968)

[BaLer] Bates, L., Lerman, E.: Proper group actions and symplectic stratified spaces. Pacific J. Math. 181, 201–229 (1997)

[BB] Beilinson, A.A., Bernstein, J.: A proof of Jantzen Conjectures. Adv. in Soviet Math. 16, Part 1, 1–50 (1993)

[BBD] Beilinson, A.A., Bernstein, J., Deligne, P.: *Faisceaux pervers.* Astérisque 100 (1982)

[Bekka1] Bekka, K.: (c)-régularité et trivialité topologique. In: *Singularity Theory and its Applications.* Warwick, part I, Lect. Notes Math. 1462, 42–62. Springer, Berlin Heidelberg (1991)

[Bekka2] Bekka, K.: Regular stratification of subanalytic sets. Bull. London Math. Soc. 25, 7–16 (1993)

[Bekka3] Bekka, K.: Regular quasi-homogeneous stratifications. In: *Stratifications, Singularities and Differential Equations II.* ed. Trotman,D.Wilson,L., 1–14. Hermann, Paris (1997)

[Bekka4] Bekka, K.: Continuous vector fields and Thom's isotopy theorem. preprint.

[BeKoi] Bekka, K., Koike, S.: The Kuo condition, an inequality of Thom's type and (C)-regularity. Topology 37, 45–62 (1998)

[Ben] Benedetti, R.: Density of Morse functions on a complex space. Math. Ann. 229, 135–139 (1977)

[BL] Bernstein, J., Lunts, V.: *Equivariant Sheaves and Functors.* Lect. Notes Math. 1578. Springer, Berlin Heidelberg (1994)

[BCR] Bochnak, J., Coste, M., Roy, M.F.: *Real algebraic geometry.* Erg. der Mathematik und ihrer Grenzgebiete, Vol. 36 Springer, Berlin Heidelberg (1998)

[Bo] Borel, A. et al.: A*lgebraic D-Modules.* Perspectives in Math. 2, Acad. Press (1987)

[BH] Borel, A., Haefliger, A.: La classe d'homologie fondamentale d'un espace analytique. Bull. Soc. Math. France 89, 461–513 (1961)

[BM] Borel, A., Moore, J.: Homology theory for locally compact spaces. Michigan Math. J. 7, 137–159 (1960)

[BoM] Borho, W., MacPherson, R.: Partial resolutions of nilpotent varieties. Astérisque 101, 23–74 (1982)

[Bott] Bott, R.: The work of Robert D. MacPherson. Notices of the AMS 40, 325–326 (1993)

[BG] Bradon, T., Grinberg, M.: Perverse sheaves on rank stratifications. Duke Math. J. 96, 317–362 (1999)

[BSchwa] Brasselet, J.P., Schwartz, M.H.: Sur les classes de Chern d'un ensemble analytique complexe. In [CEP], Exposé 6.

[Bre] Bredon, G.E.: *Introduction to compact transformation groups.* Academic Press, New York, London (1972)

[BMM] Briançon, J., Maisonobe, P. Merle, M.: Localisation de systèmes différentiels, stratificatons de Whitney et condition de Thom. Invent. Math. 117, 531–550 (1994)

[BrBe] Bröcker, L., Bernig, A.: Lipschitz-Killing Invariants. Math. Nachr. 245, 5–25 (2002)

[BrKu] Bröcker, L., Kuppe, M.A.: Integral Geometry of Tame Sets. Geometriae Dedicata 82, 285–323 (2000)

[Br] Brønsted, A.: *An introduction to convex polytopes.* Grad. texts in Math. 90, Springer, Berlin Heidelberg (1983)

[Bry] Brylinski, J.L.: *Transformations canonique, Dualité projective, Théorie de Lefschetz, Transformation de Fourier et sommes trigonométriques.* Astérisque 140–141, 3–134 (1986)

[BDK] Brylinski, J.L., Dubson, A., Kashiwara, M.: Formule d'indice pour les modules holonomes et obstruction d'Euler locale. C. R. Acad. Sci. Paris 293, 573–576 (1981)

[BMV1] Brylinski, J.L., Malgrange, B., Verdier, J.L.: Transformation de Fourier géométrique I. C. R. Acad. Sci. Paris 297, 55–58 (1983)

[BMV2] Brylinski, J.L., Malgrange, B., Verdier, J.L.: Transformation de Fourier géométrique II. C. R. Acad. Sci. Paris 303, 193–198 (1986)

[BV] Burghelea, D., Verona, V.: Local homological properties of analytic sets. Manus. Math. 7, 55–66 (1972)

[CEP] *Caractéristique d'Euler-Poincaré.* Séminaire E.N.S. 1978-1979. Astérisque 82–83 (1981)

[Ch] Cheniot, D.: Sur les sections transversales d'un ensemble stratifié. C.R. Acad. Sci. Paris ser. A275, A915–A916 (1972)

[ChGi] Chris, N., Ginzburg, V.: *Representation Theory and Compex Geometry.* Birkhäuser, Basel (1997)

[Cornea] Cornea, O.: Cone-decompositions and degenerate critical points. Proc. London Math. Soc. 77, 437–461 (1998)

[D'AMZ] D'Agnolo, A., Marastoni, C., Zamperi, G.: Non smooth Lagrangian sets and estimations of micro-support. J. Math. Soc. Japan 48, 255–258 (1996)

[De] Deligne, P.: Le formalisme des cycles évanescents. In [SGA7, Part II], exposé XIII.

[De2] Deligne, P.: Cohomology à support propre. In [SGA4, Part III], exposé XVII.

[De3] Deligne, P.: Théorème de finitude en cohomologie l-adique, with an appendix by L.Illusie. In [SGA4$\frac{1}{2}$], 233–261

[Dem] Demazure, M.: Classification des germs à points critiques isolés. *Séminare Bourbaki 1973-74.* In Lect. Notes Math. 431, 124–142. Springer, Berlin Heidelberg (1975)

[DenLoe] Denef, J., Loeser, F.: Motivic exponential integrals and a motivic Thom-Sebastiani theorem. Duke Math. J. 99, 285–309 (1999)

[Dimca] Dimca, A.: *Sheaves in Topology.* (to appear)

[DiNe] Dimca, A., Némethi, A.: Thom Sebastiani construction and monodromy of polynomials. In: *Monodromy in Problems of Algebraic Geometry and Differential Equations.* Proc. of the Steklov Inst. of Math. 238, 97–114 (2002)

[Du1] Dubson, A.: Classes caractéristiques des variétés singulières. C. R. Acad. Sci. Paris 287, 237–240 (1978)

[Du2] Dubson, A.: Formule pour l'indice des complexes constructibles et D-modules holonomes. C. R. Acad. Sci. Paris 298, 113–116 (1984)

[DS] Durfee, A.H., Saito, M.: Mixed Hodge structures on the intersection cohomology of links. Compositio Math. 76, 49–67 (1990)

[EvMi] Evens, S., Mirković, I.: Characteristic cycles for the loop Grassmannian and nilpotent orbits. Duke Math. J. 97, 109–126 (1999)

[EvMi2] Evens, S., Mirković, I.: Fourier transform and the Iwahori-Matsumoto involution. Duke Math. J. 86, 435–464 (1997)

[FK1] Fieseler, K.H., Kaup, L.: Intersection Cohomology of q-complete spaces. In *Proc. Conf. Algebraic Geometry (1985)*, 83–104. Teubner Leibzig (1986)

[FK2] Fieseler, K.H., Kaup, L.: Theorems of Lefschetz type in intersection cohomology. Rev. Roumaine Math. Pures Appl. 33, 175–195 (1988)

[FK3] Fieseler,K.H.,Kaup,L.: Vanishing theorems for the Intersection Homology of Stein spaces. Math. Z. 197, 153–176 (1988)

[Fu] Fu, J.G.H.: Curvature measures of subanalytic sets. Amer. J. Math. 116, 819–880 (1994)

[Fu2] Fu, J.G.H.: On Verdier's specialization formula for Chern classes. Math. Ann. 291, 247–251 (1991)

[FuMC] Fu, J., McCrory, C.: Stiefel-Whitney classes and the conormal cycle of a singular variety. Trans. Amer. Math. Soc. 349, 809–835 (1997)

[FM] Fulton, W., MacPherson, R.: *Categorical framework for the study of singular spaces.* Memoirs of the AMS 243 (1981)

[Ga] Gabrielov, A.M.: Intersection matrices for certain singularities. Funct. Anal. Appl. 7, 182–193 (1973)

[GelMa] Gelfand, S., Manin, Y.: *Methods of homological algebra.* Springer, Berlin Heidelberg (1996)

[Gerg] Gergondey, R.: Le functor image direct a support propre. In [SHS], exposé 3

[GWPL] Gibson, C.G., Wirthmüller, K., du Plessis, A.A., Looijenga, E.J.N.: *Topological Stability of Smooth Mappings.* Lect. Notes Math. 552. Springer, Berlin Heidelberg (1976)

[Gi1] Ginsburg, V.: Characteristic cycles and vanishing cycles. Inv. Math. 84, 327–402 (1986)

[Gi2] Ginsburg, V.: g-Modules, Springer's representations and bivariant Chern classes. Adv. in Math. 61, 1–48 (1986)

[GKM] Goresky, M., Kottwitz, R., MacPherson, R.: Equivariant cohomology, Koszul duality, and the localisation theorem. Inv. Math. 131, 25–83 (1998)

[GM] Goresky, M., MacPherson, R.: *Stratified Morse theory.* Springer, Berlin Heidelberg (1988)

[GM2] Goresky, M., MacPherson, R.: Local contribution to the Lefschetz fixed point formula. Inv. Math. 111, 1–33 (1993)

[GM3] Goresky, M., MacPherson, R.: Morse theory and intersection homology. Astérisque 101, 135–192 (1982)

[GM4] Goresky, M., MacPherson, R.: Intersection Homology II. Inv. Math. 72, 77–129 (1983)

[GM5] Goresky,M.,MacPherson,R.: Intersection homology theory. Topology 149, 155–162 (1980)

[GS] Goresky, M., Siegel. P.: Linking pairings on singular spaces. Comment. Math. Helv. 58, 96–110 (1983)

[GrM] Grinberg, M., MacPherson, R.: Euler characteristics and Lagrangian intersections. In: *Symplectic geometry and topology.* IAS/Park city Math. Ser. 7, 265–293. Amer. Math. Soc., Providence (1999)

[Groth] Grothendieck, A.: Esquisse d'un Programme. In: *Geometric Galois Actions I.* London Math. Soc. Lecture Notes Series 242, 5–48 (1997)

[Groth2] Grothendieck, A.: Sur quelques points d'algèbre homologique. Tohoku Math. J. 9, 119–221 (1957)

[GMT] Guaraldo, F., Marcì, P., Tancredi, F.: *Topics on Real analytic Spaces.* Advanced Lectures in Mathematics, Vieweg (1986)

[HS] Habegger, N., Saper, L.: Intersection cohomology of cs-spaces and Zeeman's filtration. Inv. Math. 105, 247–272 (1991)

[Hajto] Hajto, Z.: On the equivalence of Whitney (b)-regularity and (b_s)-regularity. Rocznik Nauk.-Dydakt. Prace Math. 10, 77–86 (1982)

[Ha] Hamm, H.A.: Affine varieties and Lefschetz theorems. In: *Singularity Theory. (Trieste 1991)*, 248–262. World Scientific Publishing (1995)

[HL] Hamm, H.A., Lê, D.T.: Rectified homotopical depth and Grothendieck conjectures. In: Cartier,P. et al. (eds.) *Grothendieck Festschrift II*, 311–351 Birkhäuser, Boston (1991)

[HL2] Hamm, H.A., Lê, D.T.: Vanishing theorems for constructible sheaves I. J. Reine angew. Math. 471, 115–138 (1996)

[HL3] Hamm, H.A., Lê, D.T.: Vanishing theorems for constructible sheaves II. Kodai Math. J. 21, 208–247 (1998)

[HL4] Hamm, H.A., Lê, D.T.: Théorème d'annulation pour les faisceaux algébriquement constructibles. C. R. Acad. Sci. Paris 327, 759–762 (1998)

[HL5] Hamm, H.A., Lê, D.T.: Relative homotopical depth and a conjecture of Deligne. Math. Ann. 296, 87–101 (1993)

[HNP] Hilgert, J., Neeb, K.H., Plank, W.: Symplectic Convexity Theorems and Coadjoint Orbits. Compositio Math. 94, 129–180 (1994)

[Hiro] Hironaka, H.: Stratification and flatness. In: *Real and complex singularities. Nordic Summer School (Oslo 1976)*, 199–265. Sijthoff-Noordhoff, Groningen (1977)

[Hiro2] Hironaka, H.: Normal cones in analytic Whitney stratifications. Publ. Math. IHES 36, 127–138 (1969)

[Hoch] Hochshild, G.: *The structure of Lie groups.* Holden-Day inc., San Francisco, London, Amsterdam (1965)

[Hou] Houston, K.: Global Topology of Images of finite complex analytic maps. Math. Proc. Cambridge Philos. Soc. 122, 491–501 (1997)

[IC] Borel, A. et al: *Intersection cohomology.* Progress in Math. 50, Birkhäuser, Boston (1984)

[Illma] Illmann, S.: Every proper smooth action of a Lie group is equivalent to a real analytic action. Ann. of Math. Stud. 138, 189–220 (1995)

[Ill] Illusie, L.: Autour du théorème de monodromie locale. Astérisque 223, Exp. I, 9–57 (1994)

[Ill2] Illusie, L.: Catégories dérivées et dualité, traveaux de J.-L.Verdier. L'Enseignement Math. 36, 369–391 (1990)

[Iv] Iversen, B.: *Cohomology of sheaves.* Springer, Berlin Heidelberg (1987)

[J] Joshua, R.: Vanishing of odd-dimensional intersection homology. Math. Z. 195, 239–253 (1987)

[J2] Joshua, R.: The trace formula for equivariant D-modules and perverse sheaves. Math. Z. 206, 293–302 (1991)

[J3] Joshua, R.: Riemann-Roch for weakly-equivariant D-modules II. Math. Z. 214, 297–324 (1993)

[KamTr] Kambouchner, A., Trotman, D.: Whitney (a)-faults which are hard to detect. Ann. Sci. Ec. Norm. Sup. 12, 465–471 (1979)

[Kan] Kankaanrinta, M.: Proper real analytic actions of Lie groups on manifolds. Ann. Acd. Sci. Fenn. Diss. 83 (1991)

[KarMac] Karpinski, M., Macintyre, A.: A generalization of Wilkie's theorem of the complement, and an application to Pfaffian closure. Sel. math. New. Ser. 5, 507–516 (1999)

[Ka1] Kashiwara, M.: Index theorem for maximally overdetermined systems of linear differential equations. Proc. Japan Acad. 49, 803–804 (1973)

[Ka2] Kashiwara, M.: *Systems of microdifferential equations.* Progress in Mathematics 34, Birkhäuser, Boston Basel Berlin (1983)

[Ka3] Kashiwara, M.: Index theorem for constructible sheaves. Astérisque 130, 193–209 (1985)

[KS] Kashiwara, M., Schapira, P.: *Sheaves on Manifolds.* Springer, Berlin Heidelberg (1990)

[KS2] Kashiwara, M., Schapira, P.: *Microlocal study of Sheaves.* Astérisque 128 (1985)

[KM] Kato, M., Matsumoto, Y.: On the connectivity of the Milnor fibre of a holomorphic function at a critical point. In *Proc. of 1973 Tokyo manifolds conf.*, 131–161. Univ. Tokyo Press, Tokyo (1973)

[KL] Katz, N.M., Laumon, G.: Transformation de Fourier et majoration des sommes exponentielles. Publ. Math. IHES 62, 145–202 (1985)

[Kaup] Kaup, L.: Weak Lefschetz theorems and Topological q-Completeness. In: *Singularities* (ed. Brasselet, J.P.), 175–210. London Math. Soc. Lecture Notes Series 201 (1994)

[KV] Keller, B., Vossieck, D.: Sous les catégories dérivées. C. R. Acad. Sci. Paris 305, 225–228 (1987)

[Ken] Kennedy, G.: MacPherson's Chern classes of singular varieties. Com. Algebra. 9, 2821–2839 (1990)

[Ken2] Kennedy, G.: Specialization of MacPherson's Chern classes. Math. Scand. 66, 12–16 (1990)

[King] King, H.: Topology of isolated critical points of functions on singular spaces. In: *Stratifications, Singularities and Differential Equations.* ed. Trotman, D., Wilson, L., 63–72. Hermann, Paris (1997)

[King2] King, H.: Sums of isolated critical points. preprint.

[Klain] Klain, D.A.: An Euler Relation for Valuations on Polytopes. Adv. in Math. 147, 1–34 (1999)

[Klei] Kleiman, S.L.: The development of Intersection Homology Theory. In: *A Century of Mathematics in America, Part II.* AMS, History of Math. Vol 2, 543–585 (1989)

[Ko] Kobayashi, S.: Fixed points of isometries. Nagoya Math. J. 13, 63–68 (1958)

[La] Laumon, G.: Comparaison de characteristiques d'Euler-Poincaré en cohomologie l-adique. C. R. Acad. Sci. Paris 292, 209–212 (1981)

[LawYau] Lawson, H.B., Yau, S.T.: Holomorphic symmetries. Ann. Sci. Ec. Norm. Sup. 20, 557–577 (1987)

[Le] Lê, D.T.: Complex analytic functions with isolated singularities. J. Algebraic Geometry 1, 83–100 (1992)

[Le2] Lê, D.T.: Some remarks on relative monodromy. In: *Real and complex singularities. Nordic Summer School (Oslo 1976)*, 397–403. Sijthoff-Noordhoff, Groningen (1977)

[LeTe] Lê, D.T.,Teissier,B.: Limites d'espaces tangents en géométrie analytique. Com. Math. Helv. 63, 540–578 (1988)

[LiRo] Lion, J.M., Rolin, J.P.: Volumes, Feuilles de Rolle de Feuilletages analytiques et Théorème de Wilkie. Annales de Toulouse 7, 93–112 (1998)

[LdM] López de Medrano, S.: A splitting lemma for C^r functions, $r \geq 2$. In: *Singularity Theory. (Trieste 1991)*, 444–450. World Scientific Publishing (1995)

[MP] MacPherson, R.: Chern classes for singular varietes. Ann. of Math. 100, 423–432 (1974)

[MP2] MacPherson, R.: *Intersection Homology and Perverse Sheaves.* preprint (1990)

[Ma] Massey, D.B.: *Lê Cycles and Hypersurface Singularities.* Lect. Notes Math. 1615. Springer, Berlin Heidelberg (1995)

[Ma2] Massey, D.B.: The Sebastiani-Thom Isomorphism in the derived category. Compositio Math. 125, 353–362 (2001)

[Ma3] Massey, D.B.: *Numerical Control over Complex Analytic Singularities.* Memoirs of the AMS 163, Nr. 778 (2003)

[Mather] Mather, J.: *Notes on Topological Stability.* Mimeographed notes, Harvard University (1970)

[McPa] McCrory, C., Parusiński, A.: Complex monodromy and the topology of real algebraic sets. Compositio Math. 106, 211–233 (1997)

[McPa2] McCrory, C.Parusiński, A.: Algebraically Constructible Functions. Ann. Sci. Éc. Norm. Sup. 30, 527–552 (1997)

[Mi] Milnor, J.: *Singular Points of Complex Hypersurfaces.* Annals of Math. Studies. 61, P.U.P (1968)

[MoePr] Moerdijk, I., Pronk, D.A.: Orbifolds, Sheaves and Groupoids. K-Theory 12, 3–21 (1997)

[MF] Monteiro Fernandes, T.: Microlocalisation of D-modules along a submanifold. Bull. Soc. Math. France 123, 293–327 (1995)

[Mor] Morelli, R.: A theory of Polyhedra. Adv. in Math. 97, 1–73 (1993)

[Na] Narasimhan, R.: The Levi problem for complex spaces. Math. Ann. 142, 355–365 (1961)

[NaTr] Navarro Aznar V., Trotman D.: Whitney regularity and generic wings. Ann. Inst. Fourier, Grenoble 31, 87–111 (1981)

[Nee] Neeman, A.: *Triangulated Categories.* Annals of Math. Studies 148, Princeton University Press (2001)

[Nem1] Némethi, A.: The Milnor fiber and zeta function of the singularities of type f=P(h,g). Compositio Math. 79, 63–97 (1991)

[Nem2] Némethi, A.: Generalised local and global Sebastiani-Thom type theorems. Compositio Math. 80, 1–14 (1991)

[Nem3] Némethi,A.: The zeta function of singularities. J. Algebraic Geometry 2, 1–23 (1993)

[NemSt] Némethi, A., Steenbrink, J.: Spectral pairs, mixed Hodge modules and series of plane curve singularities. New York Journal of Math. (1995)

[OY] Ohmoto, T., Yokura, S.: Product Formulas for the Milnor Class. Bul. Pol. Acad. Sci. Math. 48, 387–401 (2000)

[Oka] Oka, M.: On the homotopy types of hypersurfaces defined by weighted homogeneous polynomials. Topology 12, 19–32 (1973)

[O1] Orro, P.: Tangents limites, cone de Whitney et regularity par intersection. Ann. Inst. Fourier, Grenoble 40, 739–756 (1990)

[O2] Orro, P.: Espaces conormaux et densité des fonctions de Morse. C. R. Acad. Sci. Paris 305, 269–272 (1987)

[OP] Orro, P., Pelletier, F.: Indice local pour des ensembles symétriques par arcs. Math. Z. 233, 817–828 (2000)

[OTr] Orro, P., Trotman, D.: On the regular stratifications and conormal structure of subanalytic sets. Bull. London Math. Soc. 18, 185–191 (1986)

[OTr2] Orro, P., Trotman, D.: Cône normal et régularités de Kuo-Verdier. Bull. Soc. math. France 130, 71–85 (2002)

[Perkal] Perkal, N.: On proving the geometric versions of Whitney regularity. J. London Math. Soc. 29, 343–351 (1984)

[PeSpSt] Petrezil, Y., Speissegger, P., Starchenko, S.: Adding Multiplication to an o-minimal Expansion of the Additive Group of Real Numbers. In: *Logic Colloquium '98*, 357–362. Lect. Notes in Logic 13 (2000)

[PeRa] Petrie, T., Randall, D.: Finite-order algebraic automorphisms of affine varieties. Com. Math. Helv. 61, 203–221 (1986)

[Pflaum] Pflaum, M.J.: *Analytic and Geometric Study of Stratified Spaces.* Lect. Notes Math. 1768. Springer, Berlin Heidelberg (2001)

[Pi] Pillay, A.: On groups and fields definable in o-minimal structures. J. Pure Appl. Algebra 53, 239–255 (1988)

[PK] Pukhlikov, A.V., Khovanskii, A.G.: Finitely additive measures on virtual polytopes. St. Petersburg Math. J. 4, 337–356 (1993)

[Ri] Rickard, J.: Derived categories and stable equivalences. J. Pure and Appl. Algebra 61, 303–317 (1989)

[Rie] Riedtmann, C.: Lie algebras generated by indecomposables. Journal of Algebra 170, 526–546 (1994)

[Ro] Rohlfs, J.: Lefschetz numbers for arithmetic groups. In: *Cohomology of arithmetic groups and automorphic forms.* Lect. Notes Math. 1447, 303–313. Springer, Berlin Heidelberg (1990)

[RoSpWi] Rolin, J.P., Speissegger, P., Wilkie, A.J.: Quasianalytic Denjoy-Carleman classes and o-minimality. to appear in Journal of the AMS

[Sab] Sabbah, C.: Quelques remarques sur la géométrie des espaces conormaux. Astérisque 130, 161–192 (1985)

[Sai] Saito, M.: Mixed Hodge Modules. Publ. RIMS 26, 221–333 (1990)

[Sai2] Saito, M.: Introduction to mixed Hodge Modules. Astérisque 179–180, 145–162 (1989)

[Saito] Saito,M.: A localization theorem for D-modules. Tôhoku Math J. 43, 213–234 (1991)

[Sa] Sakamoto, K.: The Seifert matrices of Milnor fiberings defined by holomorphic functions. J. Math. Soc. Japan 26, 714–721 (1974)

[Scha] Schapira, P.: Operations on constructible functions. J. Pure Appl. Algebra 72, 83–93 (1991)

[ScSc] Schapira, P.,Schneiders,J.P.: Index theorem for elliptic pairs. Astérisque 224 (1994)

[Schei] Scheiderer, C.: *Real and Étale Cohomology.* Lect. Notes Math. 1588. Springer, Berlin Heidelberg (1994)

[SchSt] Scherk, J., Steenbrink, J.: On the Mixed Hodge Structure on the Cohomology of the Milnor fiber. Math. Ann. 271, 641–665 (1985)

[SchVi] Schmid, W., Vilonen, K.: Characteristic cycles for constructible sheaves. Inv. Math. 124, 451–502 (1996)

[Schr] Schröter, M.: Künnethformeln für die Garben der verschwindenden Zykeln in der étalen Kohomologie. Abh. Math. Sem. Univ. Hamburg 65, 1–27 (1995)

[Sch] Schürmann, J.: Some remarks on characteristic classes of Lagrangian cycles (preliminary version). preprint.

[Sch2] Schürmann, J.: A general intersection formula for Lagrangian cycles. preprint. mathAG/0201300, to appear in Composito Math.

[Sch3] Schürmann, J.: Endlichkeits- und Verschwindungssätze für (schwach) konstruierbare Garbenkomplexe auf komplexen Räumen. J. Reine angew. Math. 466, 27–43 (1995)

[Sch4] Schürmann, J.: Lectures on Characteristic Classes of Constructible functions. preprint written up by Pragacz,P. and Weber,A. (2002), available at http://www.impan.gov.pl/ pragacz/notes.htm

[Schwa] Schwartz, M.H.: Classes caractéristiques définies par une stratification d'une variété analytique complexe. C. R. Acad. Sci. Paris 260, 3262–3264 and 3535–3537 (1965)

[Schwa2] Schwartz, M.H.: *Champs radiaux sur une stratification analytique complexe.* Travaux en cours 39, Hermann, Paris (1997)

[Schwa3] Schwartz, M.H.: *Classes de Chern des ensembles analytiques.* Actualités Mathématiques. Hermann, Paris (2000)

[Schw] Schwarz, M.: *Morse Homology.* Progress in Math. 111, Birkhäuser, Boston (1991)

[ST] Sebastiani, M., Thom, R.: Un résultat sur la monodromie. Inv. Math. 13, 90–96 (1971)

[Seg] Segal, G.: Equivariant K-theory. Publ. Math. IHES 34, 129–151 (1968)

[SGA2] Grothendieck, A.: *Cohomologie local des faisceaux cohérents et théorème de Lefschetz loceaux et globaux.* North-Holland (1968)

[SGA4] *Sém. géométrie algébrique (1967–69). Théorie des topos et cohomologie étale.* Part III Lect. Notes Math. 305. Springer, Berlin Heidelberg (1973)

[SGA4$\frac{1}{2}$] *Sém. géométrie algébrique. Cohomologie étale.* Lect. Notes Math. 569. Springer, Berlin Heidelberg (1977)

[SGA5] *Sém. géométrie algébrique (1965–68). Cohomologie l-adique et Fonction L.* Lect. Notes Math. 589. Springer, Berlin Heidelberg (1973)

[SGA7] *Sém. géométrie algébrique (1967–69). Groupes de monodromie en géométrie algébrique.* Part II Lect. Notes Math. 340. Springer, Berlin Heidelberg (1973)

[SHS] *Sém. Heidelberg-Strasbourg (1966–67). Dualité de Poincaré.* Publ. I.R.M.A. 3, Strasbourg (1969)

[Shiota] Shiota, M.: *Geometry of subanalytic and semialgebraic sets.* Progress in Math. 150, Birkhäuser, Basel (1997)

[SjLer] Sjamaar, R., Lerman, E.: Stratified symplectic spaces and reduction. Ann. of Math 134, 375–422 (1991)

[SV] Sorani, G., Villani, V.: q-complete spaces and Cohomology. Trans. Amer. Math. Soc. 125, 432–448 (1966)

[Sp] Spaltenstein, N.: Resolutions of unbounded complexes. Compositio Math. 65, 121–154 (1988)

[Speiss] Speissegger, P.: The Pfaffian closure of an o-minimal structure. J. reine angew. Math. 508, 189–211 (1999)

[Su] Sullivan, D.: Combinatorial invariants of analytic spaces. In: *Procedings of Liverpool Singularities-Symposium I.* Lect. Notes Math. 192, 165–168. Springer, Berlin Heidelberg (1970)

[Ta1] Ta, L.L.: Verdier and strict Thom stratifications in o-minimal structures. Ill. Jour. of Math. 42, 347–356 (1998)

[Ta2] Ta, L.L.: Thom stratification for functions definable in o-minimal structures on (R,+,·). C.R. Acad. Sci. Paris Série I 324, 1391–1394 (1997)

[TaZah] Ta, L.L., Zaharia, A.: Bifurcation sets of functions definable in o-minimal structures. Ill. Jour. of Math. 42, 449–457 (1998)

[Te] Teissier, B.: Varietes polaires II. Multiplicités polaires, sections planes, et conditions de Whitney. Lect. Notes Math. 961, 314–491, Springer, Berlin Heidelberg (1982)

[Te2] Teissier, B.: Tame and Stratified Objects. In: *Geometric Galois Actions I.* London Math. Soc. Lecture Notes Series 242, 231–242 (1997)

[Te3] Teissier, B.: A bouquet of bouquets for a birthday. In: *Topological methods in modern Mathematics.* eds.: Goldberg,L. et. al., 93–122. Publish Perish, Inc. (1993)

[Tibăr] Tibăr, M.: Topology at infinity of polonomial mappings and Thom regularity condition. Compositio Math. 111, 89–109 (1998)

[tDieck] Tom Dieck, T.: *Transformation Groups.* de Gruyter Studies in Math. 8, Walter de Gruyter (1987)

[Tr1] Trotman, D.: Comparing regularity conditions on stratifications. In: *Proc. Sympos. Pure Math. 40, Singularities - Aracta 1981, Part 2*, 575–586. Amer. Math. Soc., Providence Rhode Island (1983)

[Tr2] Trotman, D.: Espaces stratifiés reéls. In: *Stratifications,Singularities and Differential Equations II.* ed. Trotman,D.,Wilson,L., 63–72. Hermann, Paris (1997)

[Tr3] Trotman, D.: Geometric versions of Whitney regularity for smooth stratifications. Ann. Sci. Ec. Norm. Sup. 12, 461–471 (1979)

[Tr4] Trotman, D.: Une version microlocale de la condition (w) de Verdier. Ann. Inst. Fourier Grenoble 39, 825–829 (1989)

[TrWi] Trotman, D., Wilson, L.: Stratifications and finite determinancy. Proc. London Math. Soc. 78, 334–368 (1999)

[vDr] van den Dries, L.: *Tame Topology and o-minimal structures.* London Math. Soc. Lect. Notes Ser. 248, Cambridge Univ. Press (1998)

[vDr2] van den Dries, L.: O-minimal Structures and Real Analytic Geometry. In: *Current Developments in Mathematics*, 105–152 (1998)

[vDrMi] van den Dries, L., Miller, C.: Geometric categories and o-minimal structures. Duke Math. J. 84, 497–540 (1996)

[vDrSp] van den Dries, L., Speissegger, P.: The real field with convergent generalized power series. Trans. AMS 350 (11), 4377–4421 (1998)

[vDrSp2] van den Dries, L., Speissegger, P.: The field of reals with multisummable series and exponential function. Proc. London Math. Soc. 81, 513–565 (2000)

[Ver] Verdier, J.L.: Stratification de Whitney et théorème de Bertini-Sard. Inv. Math. 36, 295–312 (1976)

[Ver2] Verdier, J.L.: Caractéristique d'Euler-Poincaré. Bull. Soc. math. France 101, 441–445 (1973)

[Ver3] Verdier, J.L.: Classe d'homologie associée à un cycle. Astérisque 36-37, 101–151 (1976)

[Ver4] Verdier, J.L.: *DES CATÉGORIES DÉRIVÉES DES CATÉGORIES ABÉLIENNES.* Astérisque 239 (1996)

[Verona] Verona, A.: *Stratified Mappings - Structure and Triangulability.* Lect. Notes Math. 1102. Springer, Berlin Heidelberg (1984)

[Viro] Viro, O.: Some integral calculus based on the Euler characteristic. Lect. Notes Math. 1346, 127–138 Springer, Berlin Heidelberg (1989)

[Wilkie] Wilkie, A.J.: A theorem of the complement and some new o-minimal structures. Sel. math. New. Ser. 5, 397–421 (1999)

Index

Monografie Matematyczne

[1] S. Banach, *Théorie des opérations linéaires*, 1932
[2] S. Saks, *Théorie de l'integrale*, 1933
[3] C. Kuratowski, *Topologie* I, 1933
[4] W. Sierpiński, *Hypothèse de continu*, 1934
[5] A. Zygmund, *Trigonometrical Series*, 1935
[6] S. Kaczmarz, H. Steinhaus, *Theorie der Orthogonalreihen*, 1935
[7] S. Saks, *Theory of the integral*, 1937
[8] S. Banach, *Mechanika*, T. I, 1947
[9] S. Banach, *Mechanika*, T. II, 1947
[10] S. Saks, A. Zygmund, *Funkcje analityczne*, 1948
[11] W. Sierpiński, *Zasady algebry wyższej*, 1946
[12] K. Borsuk, *Geometria analityczna w n wymiarach*, 1950
[13] W. Sierpiński, *Działania nieskończone*, 1948
[14] W. Sierpiński, *Rachunek różniczkowy poprzedzony badaniem funkcji elementarnych*, 1947
[15] K. Kuratowski, *Wykłady rachunku różniczkowego i całkowego*, T. I, 1948
[16] E. Otto, *Geometria wykreślna*, 1950
[17] S. Banach, *Wstęp do teorii funkcji rzeczywistych*, 1951
[18] A. Mostowski, *Logika matematyczna*, 1948
[19] W. Sierpiński, *Teoria liczb*, 1950
[20] C. Kuratowski, *Topologie* I, 1948
[21] C. Kuratowski, *Topologie* II, 1950
[22] W. Rubinowicz, *Wektory i tensory*, 1950
[23] W. Sierpiński, *Algèbre des ensembles*, 1951
[24] S. Banach, *Mechanics*, 1951
[25] W. Nikliborc, *Równania różniczkowe*, Cz. I, 1951
[26] M. Stark, *Geometria analityczna*, 1951
[27] K. Kuratowski, A. Mostowski, *Teoria mnogości*, 1952
[28] S. Saks, A. Zygmund, *Analytic functions*, 1952
[29] F. Leja, *Funkcje analityczne i harmoniczne*, Cz. I, 1952
[30] J. Mikusiński, *Rachunek operatorów*, 1953
∗ [31] W. Ślebodziński, *Formes extérieures et leurs applications*, 1954
[32] S. Mazurkiewicz, *Podstawy rachunku prawdopodobieństwa*, 1956
[33] A. Walfisz, *Gitterpunkte in mehrdimensionalen Kugeln*, 1957
[34] W. Sierpiński, *Cardinal and ordinal numbers*, 1965
[35] R. Sikorski, *Funkcje rzeczywiste*, 1958
[36] K. Maurin, *Metody przestrzeni Hilberta*, 1959
[37] R. Sikorski, *Funkcje rzeczywiste*, T. II, 1959
[38] W. Sierpiński, *Teoria liczb* II, 1959

* [39] J. Aczél, S. Gołąb, *Funktionalgleichungen der Theorie der geometrischen Objekte*, 1960
[40] W. Ślebodziński, *Formes extérieures et leurs applications*, II, 1963
[41] H. Rasiowa, R. Sikorski, *The mathematics of metamathematics*, 1963
[42] W. Sierpiński, *Elementary theory of numbers*, 1964
* [43] J. Szarski, *Differential inequalities*, 1965
[44] K. Borsuk, *Theory of retracts*, 1967
[45] K. Maurin, *Methods of Hilbert spaces*, 1967
[46] M. Kuczma, *Functional equations in a single variable*, 1967
[47] D. Przeworska-Rolewicz, S. Rolewicz, *Equations in linear spaces*, 1968
[48] K. Maurin, *General eigenfunction expansions and unitary representations of topological groups*, 1968
[49] A. Alexiewicz, *Analiza funkcjonalna*, 1969
* [50] K. Borsuk, *Multidimensional analytic geometry*, 1969
* [51] R. Sikorski, *Advanced calculus. Functions of several variables*, 1969
[52] W. Ślebodziński, *Exterior forms and their applications*, 1971
[53] M. Krzyżański, *Partial differential equations of second order*, vol. I, 1971
[54] M. Krzyżański, *Partial differential equations of second order*, vol. II, 1971
[55] Z. Semadeni, *Banach spaces of continuous functions*, 1971
[56] S. Rolewicz, *Metric linear spaces*, 1972
[57] W. Narkiewicz, *Elementary and analytic theory of algebraic numbers*, 1974
[58] Cz. Bessaga, A. Pełczyński, *Selected topics in infinite dimensional topology*, 1975
* [59] K. Borsuk, *Theory of shape*, 1975
[60] R. Engelking, *General topology*, 1977
[61] J. Dugundji, A. Granas, *Fixed point theory*, 1982
* [62] W. Narkiewicz, *Classical problems in number theory*, 1986

The volumes marked with * are available at the exchange department of the library of the Institute of Mathematics, Polish Academy of Sciences.